5th International Conference on Advanced Sciences (ICAS5)

AA007000

IOP Conference Series: Materials Science and Engineering
Volume 1046

Hurghada, Egypt
10 - 12 November 2019

ISBN: 978-1-7138-3334-5
ISSN: 1757-8981

Printed from e-media with permission by:

Curran Associates, Inc.
57 Morehouse Lane
Red Hook, NY 12571

Some format issues inherent in the e-media version may also appear in this print version.

No changes have been made to the content of these proceedings. There may be changes to pagination and minor adjustments for aesthetics.

Printed with permission by Curran Associates, Inc. (2021)

For permission requests, please contact the Institute of Physics at the address below.

Institute of Physics
Dirac House, Temple Back
Bristol BS1 6BE UK

Phone: 44 1 17 929 7481
Fax: 44 1 17 920 0979

techtracking@iop.org

Additional copies of this publication are available from:

Curran Associates, Inc.
57 Morehouse Lane
Red Hook, NY 12571 USA
Phone: 845-758-0400
Fax: 845-758-2633
Email: curran@proceedings.com
Web: www.proceedings.com

TABLE OF CONTENTS

IOP Conf. Series: Materials Science and Engineering 1046 (2021) 011001 doi:10.1088/1757-899X/1046/1/011001

In year2019, we celebrated the ranking of our university "Beni-Suef University" as the 1st Egyptian University, 601-800 world wide and 301-400 in the physical sciences, 101-150 in the young universities as Times Higher Education published. It is also a great pleasure to announce that for the 1st time we are included at Shanghai ranking (901-1000). We are willing to compete next year on advanced positions.

Faculty of Postgraduate studied for advanced sciences (PSAS) is a high-class faculty, a rich research-based education platform that transfers you from multidisciplinary, interdisciplinary to transdisciplinary research in FOUR main departments working cooperatively to achieve the best prototypical education on both regional and international level.

Faculty of Postgraduate Studies for Advanced Sciences (PSAS)

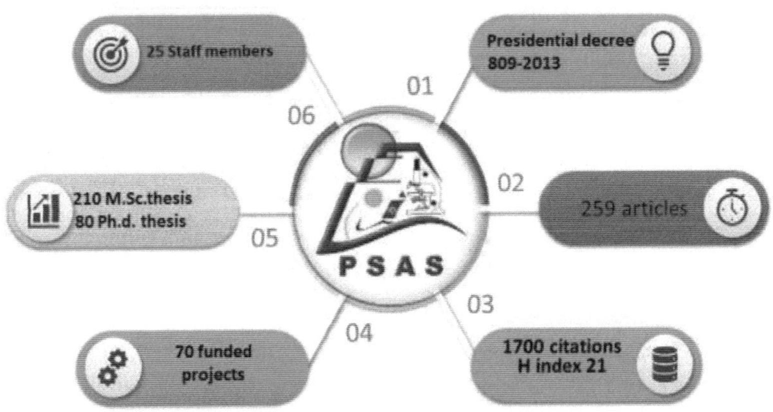

PSAS is distinguished by expert qualified team working together with large number of research facilities in the specified laboratories.

IOP Conf. Series: Materials Science and Engineering 1046 (2021) 011001 doi:10.1088/1757-899X/1046/1/011001

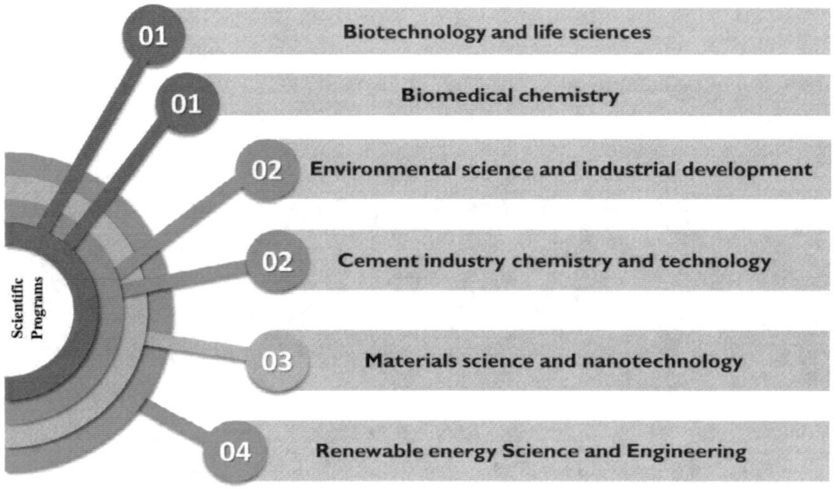

The vision of the faculty is to become a unique scientific school in carrying out researches and advanced studies among universities, local, regional and global institutions. Additionally, to contribute in solving the industrial problems and achieving sustainable development goals tackling EGYPT vision 2030.

It is gratifying to note that the agenda of the conference covers a wide range of very interesting items related to the nanotechnology, biotechnology, environmental sciences and all energy related subjects and especially those directly related to aspects of all multidisciplinary sciences.

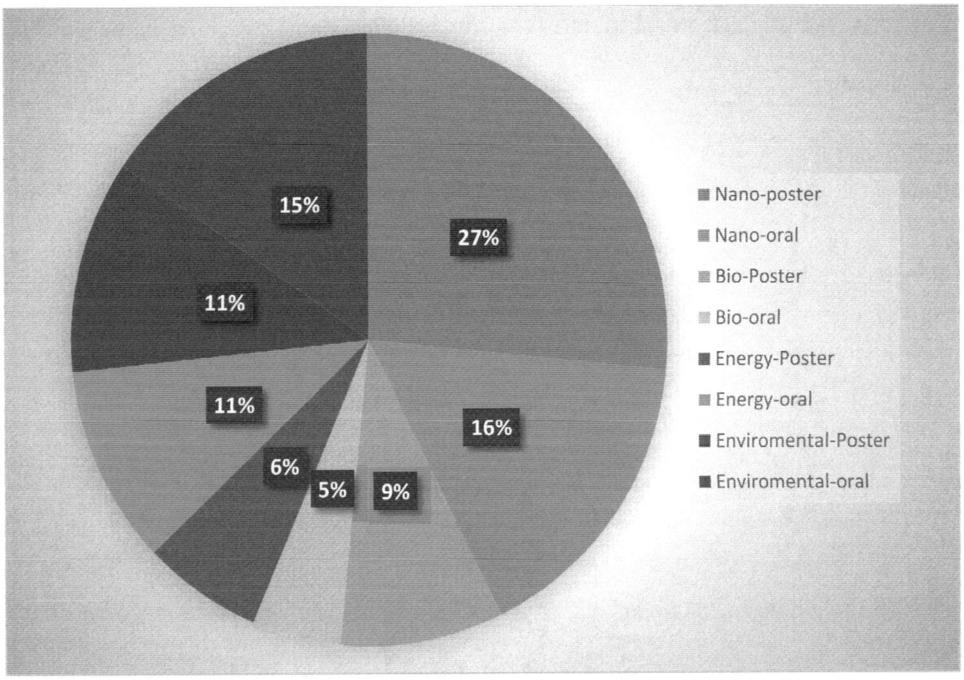

We are aware of the tremendous effort made, by all faculty members as organizers and members of the scientific committee from day 1 up till now.

No matter how much we can do by ourselves on the national level, whether it be research or development, it is never enough. In a spirit of true cooperation, we were proud to announce for the second time reviewing and publishing international proceedings in IOP conference series: materials science and engineering. The editorial board of this issue

Prof. M.H.Khedr -Chemistry of nanomaterials

Prof. A.A.Farghali- Nanocatalysis and mutifunctional materials

Ass. Prof. S.I.El-Dek Materials Physics and magnetism

Ass. Prof. Waleed Elrouby Nanomaterials in Electrochemical applications

Ass. Prof. A.H.Zaki Water treatment using nanostructures

IOP Conf. Series: Materials Science and Engineering 1046 (2021) 011001 doi:10.1088/1757-899X/1046/1/011001

Dr. Ahmed Gomaa Biomedical use of nanoparticles

Dr. Mohamed Taha Modelling and simulation

Dr. Abdalla Abdelwahab Energy applications of nanomaterials

IOP Conf. Series: Materials Science and Engineering 1046 (2021) 011001 doi:10.1088/1757-899X/1046/1/011001

We must track PSAS as a multi-dimensional concept, encompassing the economic, social, institutional and physical elements of the excellent research educational platform, in a wider sense, it would be relatively easy to affect the necessary adjustments for a truly effective cooperation on the regional and international basis. This is fully consonant with our official position taken and the full support of the concept of internationalization

This special issue on "Materials sciences and engineering" resulted from the 2019, 5th International conference on advanced sciences ICAS5, held in Hurghada, Egypt on 10-12th November, 2019. This international conference is the 5th one organized by our group at the Faculty of postgraduate studies for advanced sciences brought together academics, scientists, engineers and postgraduate students to exchange and share their knowledge and research results on all aspects of multidisciplinary sciences and to discuss practical challenges and solutions. The topics covered include:

- Materials Science and Nanotechnology
- Biotechnology and life sciences
- Renewable energy science and engineering
- Environmental science and industrial development.

New programs will be opened soon such as Stem cell biology, water and energy for sustainable environment, petrochemicals science

The event was coorganized with Paris Sud University-Paris and Misr International University (MIU) and sponsored by many Egyptian industrial communities and Scientific companies. Following the submission of abstracts to the international conference, 29 extended full papers have been reviewed from 35 categorized in Materials science and nanotechnology session and chosen for inclusion in this special issue. Collectively, these papers have addressed several key issues. I am therefore delighted to present these papers for all readers.

IOP Conf. Series: Materials Science and Engineering 1046 (2021) 011002 doi:10.1088/1757-899X/1046/1/011002

Peer review declaration

All papers published in this volume of *IOP Conference Series: Materials Science and Engineering* have been peer reviewed through processes administered by the Editors. Reviews were conducted by expert referees to the professional and scientific standards expected of a proceedings journal published by IOP Publishing.

- **Type of peer review: Single-blind / Double-blind / Triple-blind / Open / Other**

Single-anonymous: authors' identities are known to the reviewers, reviewers' identities are hidden from authors

Double-anonymous: author and reviewer identities are hidden to each other

Triple-blind: author and reviewer identities are hidden to each other, and from the Editor(s)

Open: author and reviewer identities are known to each other

- **Describe criteria used by Reviewers when accepting/declining papers. Was there the opportunity to resubmit articles after revisions?**

- **Conference submission management system:**

- **Number of submissions received: 37**

- **Number of submissions sent for review: 37**

- **Number of submissions accepted: 29**

- **Acceptance Rate (Number of Submissions Accepted / Number of Submissions Received X 100): 78**

- **Average number of reviews per paper: 23**

- **Total number of reviewers involved: 63**

- **Any additional info on review process (ie plagiarism check system):** All articles were checked against plagiarism using ithenticate at the academy of the scientific research (ASRT-Egypt). The overall percentage was less than 20%

- **Contact person for queries:** samaa@psas.bsu.edu.eg;mtaha@psas.bsu.edu.eg

IOP Conf. Series: Materials Science and Engineering 1046 (2021) 012001 doi:10.1088/1757-899X/1046/1/012001

Ultracompact gas-sensor based on a 2D photonic crystal waveguide incorporating with tapered microcavity

A Harhouz[1], A Hocini[1,*] and H Tayoub[1,2]

[1]Laboratoire d'Analyse des Signaux et Systèmes, Department of Electronics, University of M'Sila BP.166, Route Ichebilia, M'Sila, 28000, Algeria.
[2]Research Center in Industrial Technologies CRTI, P.O.BOX :64, Cheraga 16014, Algiers, Algeria

*E-mail: abdesselam.hocini@univ-msila.dz

Abstract. In this study, a new ultra compact gas-sensor, based on a 2D photonic crystal waveguide incorporating with tapered microcavity, is designed to detect small refractive index changes. The refractive index (RI) sensor is formed by a point-defect resonant cavity in the sandwiched waveguide on Si slab with triangular lattice. The properties of the sensor are simulated by using the plane wave expansion (PWE) method and the finite-difference time-domain (FDTD) algorithm. The transmission spectra of the sensor with different ambient refractive indices ranging from n = 1.0 to n = 1.01 are calculated. The calculation results show that a change in ambient refractive index of $\Delta n=1\times10^{-4}$is apparent. The proposed sensor achieves a sensitivity ($\Delta\lambda/\Delta n$) of 523.2 nm/RIU. It was found that the resonance wavelength is a linear function of the refractive index in under study range. The sensor is appropriate for detecting homogeneous media.

1. Introduction

Photonic crystal (PC) sensors attracted attention of many researches due to the locally confined optical state into tiny volume [1, 2]. One among the types of optical sensors are refractive index-based sensors whose application has become widely covering many functions, for instance, measurement of parameters such as humidity [3], temperature [4], gas [5, 6], chemical composition|, and biosensing. The refractive index (RI) biosensors have received substantial attention in the development of label-free biomedical and biochemical measure. Accordingly, the target analytes can be sensed in their natural forms with no need of any adjustments [7-9]. Besides, RI change is directly associated with the concentration of liquids or gases, or their existence rather than the entire sample volume, therefore, sensitive free-label detection with small sample volumes can be realized by the RI sensor.

Photonic crystal is a periodic dielectric structure with the potential of directing and manipulating light at the optical wavelength scale [10]. Specifically, the periodic air hole microstructure of hole-type PC is a natural candidate for housing target liquids or gases. PCs are appealing sensing platforms since they provide strong light confinement. Unlike many platforms that use the interaction between the small evanescent tail of of the electromagnetic field and the analyte, PCs can be designed to localize the electric field in the low RI region which makes the biosensors extremely sensitive to a small RI change produced by infiltrated analytes on the air holes [11].

The quality factor Q of wavelength-sized microcavities in two dimensional (2D) photonic crystal slab structures [12, 13] was formerly considered as being limited by vertical radiation loss. Yet, a pletora of studies have indicated that these structures can achieve a high Q value [14, 15] in case of using a proper design intended to decrease the perpendicular radiation. The PCs heterostructure concept is used, a line-defect structure whose width is locally adjusted, to achieve high and ultrahigh quality-factor cavities [16].

This study puts forward a novel ultra-compact gas-sensor based on new design for the microcavities in a PCs;namely, a line-defect structure whose width is locally adjusted. The mode gap, existing in the waveguide, results in the confinement found in the direction of a line defect. The mode gap's location can be changed by changing the geometrical parameters of line-defect waveguides [17-19], this enables the creation of a local confinement potential within theoretically lossless waveguides. The properties of the sensor are simulated using the plane wave expansion (PWE) method and the finite-difference time-domain (FDTD) algorithm (RSoft Photonic Suite).

2. Structure design

First, a precise geometry of 2D photonic crystal Si slab infiltration is proposed, specifically, targeting the waveguide where most of the mode energy exists as an optical detector.

We consider a 2D PC slab with a triangular array of air holes (n_{air}= 1) in a silicon slab (n_{si}= 3.42) with h=230 nm of thickness, as illustrated in figure 1, making use of the triangular PCs is practically important due to their large transverse electric band gap, which is expected to serve as a good platform for photonic integrated circuits and ultra-compact optical sensors [20].The structure has holes of radius=140 nm with lattice constant a=430 nm.

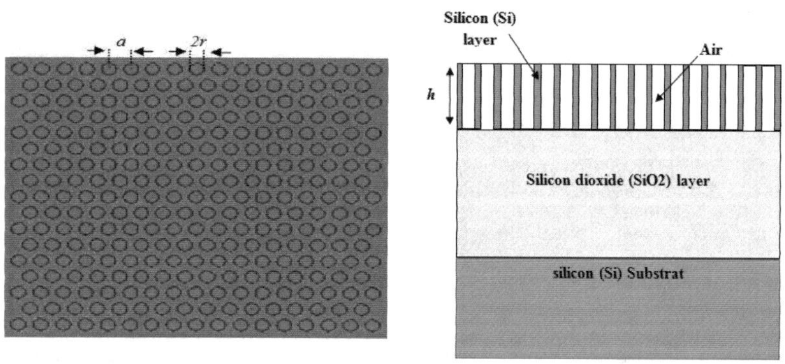

Figure 1. The host structure of PC, 2D triangular lattices, consisting of air holes with a lattice constant of a=430 nm, air hole radius r, is equal to 140 nm and size of photonic crystal is 23×23.

To support the Si Slab, Silicon dioxide (SiO_2) layer with a thickness of 1500 nm is used as a base. The low index of SiO2 layer located underneath the high index of Si slab enables to confine light within the cavity core, hence, avoiding optical losses into the lower substrate. Consequently, the light confinement in the vertical direction is ensured by total internal reflection.

In 3D-FDTD calculation, CPU time and a large computer memory are requisite. Meantime, 2D-FDTD calculation using effective RI approximation method notably cuts down both the computer memory and the simulation times, leading to a reasonable estimation. In 2D-FDTD method, the vertical confinement achieved in z direction is totally ignored; however, the magnitude of the loss in z direction is very low. To reduce the computational efforts required for the full 3D calculations, the PC is replaced by a 2D system with the background dielectric medium having the effective refractive index of 2.838 that corresponds with the efficient index of the fundamental guided TM mode at wavelength of 1550 nm.

The following procedure was the calculation of a photonic bandgap for TM polarization, which ranges of1241.69nm to 1659.93 nm (figure2). It was calculated along the Γ-K-M-Γ edge for the Brillouin zone using a 2D plane wave expansion (PWE) method of the RSoft (BandSOLVE) software and the FDTD algorithm of the RSoft (full wave).

A PC slab waveguide can also be formed by removing one row of air holes in the ΓK direction. Light, which propagates in the waveguide with a frequency within the bandgap of the crystal, is confined too, and can be directed along the waveguide (figure 3).

During the simulation process, the light source was positioned in the input line defect waveguide head, as for the monitor, it was placed at the end of the output line waveguide. The normalized central frequency of Gauss pulse source is centered around 0.26(2πc/a). Since there must be a careful consideration of the boundary conditions at the spatial edges of the computational domain, one-spatial unit thick perfectly matched layer (PML) encircling the simulated area is set to absorb the fields enabling the simulated region to implement reflections.

The transmission spectrum of the PC waveguide, obtained by using of a 2D FDTD method, has shown in figure 4. For PC structures, the nature of waveguide modes depends on its location in the bandgap. One can observe that light with a frequency within the bandgap can propagate in the waveguide.

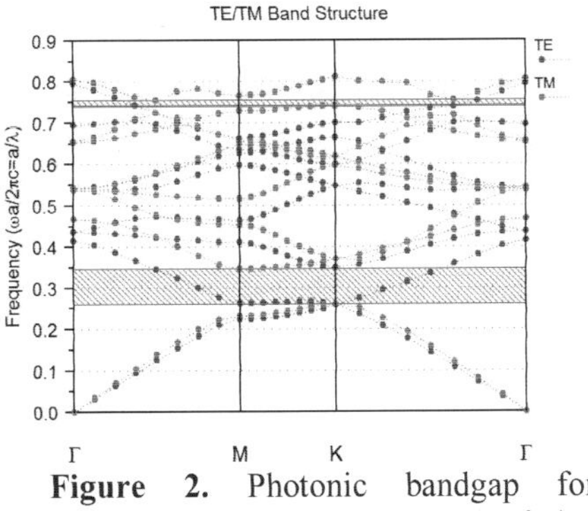

Figure 2. Photonic bandgap for TM polarization, using PWE method of the RSoft (BandSOLVE) software.

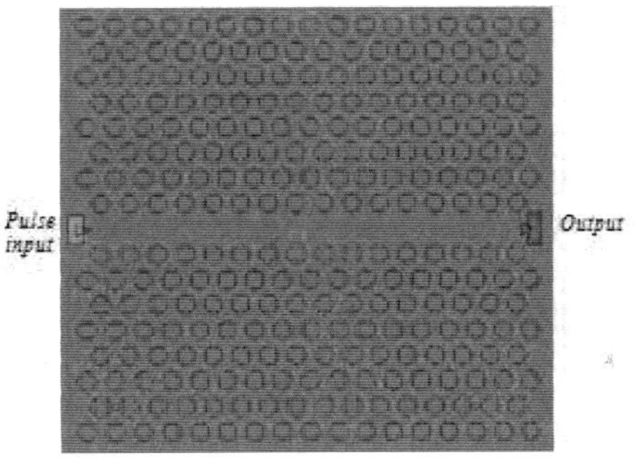

Figure 3. Schematic view of the proposed waveguide in Si slab.

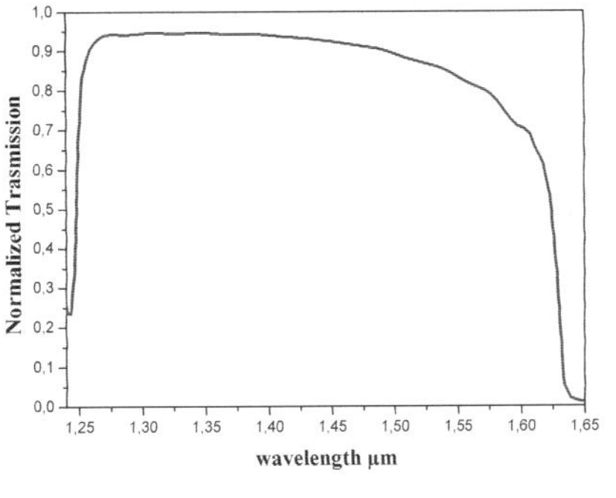

Figure 4. Transmission spectrum for the PC waveguide; using FDTD algorithm of the RSoft (full wave).

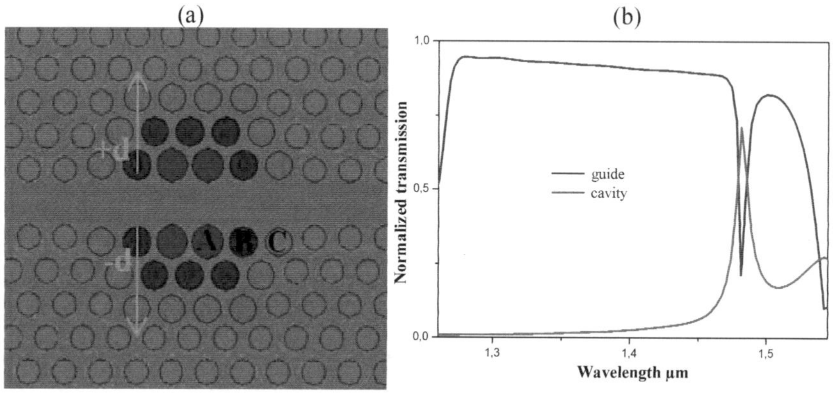

Figure 5. (a)The reference design of the proposed RI sensor.The hole's radiuses are: r_A=0.442a nm, r_B= 0.395a nmandr_C=0.372a.(b)The transmission spectra of the designed structure.

Sensors based on PC waveguides incorporating with microcavities hold numerous merits in compactness, high quality (Q) factor and sensitivity, easy extension to sensor arrays and various choices of materials, and in parallel measurement capacity [21-26]. Furthermore, a tapered-shift structure along the line defect increases the quality factor of the cavity [27]. Thus, the proposed RI sensor is created by means of two waveguide couplers and microcavity in the PC with a triangular lattice of air holes (see figure5).

The microcavity is designed by changing of the size of 30 air holes in the center of the line defect. The holes radiuses areas r_A=0.442anm, r_B=0.395anm an dr_C=0.372 anm. Another defect is realized by increasing the waveguide width at the center of the microcavity by moving the surrounding holes A, B, C away from the waveguide by a distance of -d, +d.

The FDTD method has been applied by the considered grid-size (Δx, Δz) of 43 nm when the lattice constant is 430 nm, so the resolution is set to 10 (that is, with a grid size of a/10, where a represents the lattice constant). A Gaussian exciter has been used with a central wavelength of 1.550μm. Pulsed light is launched from the left (arrow).The output is measured at the right (time monitor), and the time step of 0.035 a/c is employed. All the simulations are carried out with the same mesh size and time step for future equivalent results. The FDTD algorithm depends on structure meshing, accordingly, the prominent impact to be noticed is the dependence of the results on mesh size. Reducing the mesh size should bring the results to a given limit, however, reduction of mesh size can also reduce time step which can lead to deviations in the comparison of the results. In the X and Z directions, when the mesh is reduced from 43 to 21.5 nm (the half), the results remain almost constant indicating that the method has converged.

3. Simulation results and discussion

The sensing principle is based on the shift of resonance wavelength λ_0, which occurs due to the change in RI of the sensor when the PC's air holes are full of homogenous gas. The effective RI of the slab and the RI contrast between the holes and the slab region are changed, by varying the local RI. Since the transmitted wavelengths depend directly on the effective index of the slab [23], the device can be used as a sensor by detecting the output transmission spectrum. With the presence of gas, the effective RI changes and hence causes a shift in the cut-off wavelength or resonance wavelength λ_0 (waveguide or waveguide incorporating with micro-cavities) of the output spectrum. This phenomenon can be analyzed by using the first-order electromagnetic perturbation theory. This latter is due to a small perturbation $\Delta\varepsilon \approx 2n\Delta n$ [28].

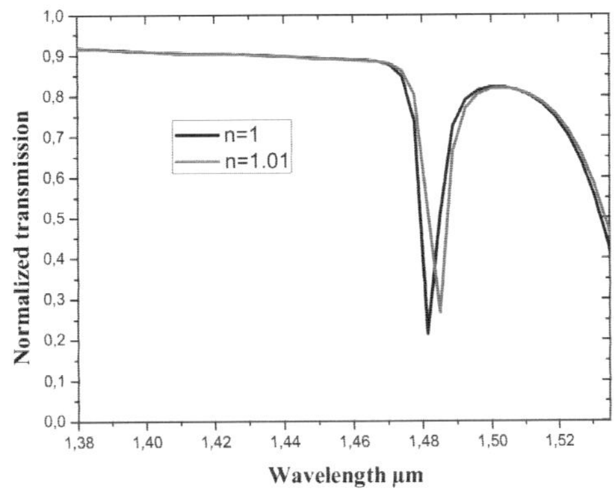

Figure 6. Calculated transmission spectra for TM polarization of the reference RI sensor.

The microcavity gives a shift of 3.667nm, equivalent to the sensitivity of 366.7 nm/RIU for a change in RI value of 0.01. (figure6). concerningRI measurements, the sensitivity (S) is calculated as the ratio of the resonance wavelength λ_0 shift and the change in the RI and its unit is given by nm/RIU. Thus, equation can be written as:

$S=\Delta \lambda_0/\Delta n$ (2)

where$\Delta n=n-n_0$ and $n_0=1$.

To improve the sensitivity of the sensor, we present a new design (figure7 (a)). It is necessary to highlight here that ring-shaped holes configuration does not only offer more flexibility, indesigning new structures compared to circular shaped ones, but also enhance the optical field confinement and, hence, the light-fluid (gas) interactions within the ringed low-dielectric material area[9,29,30].

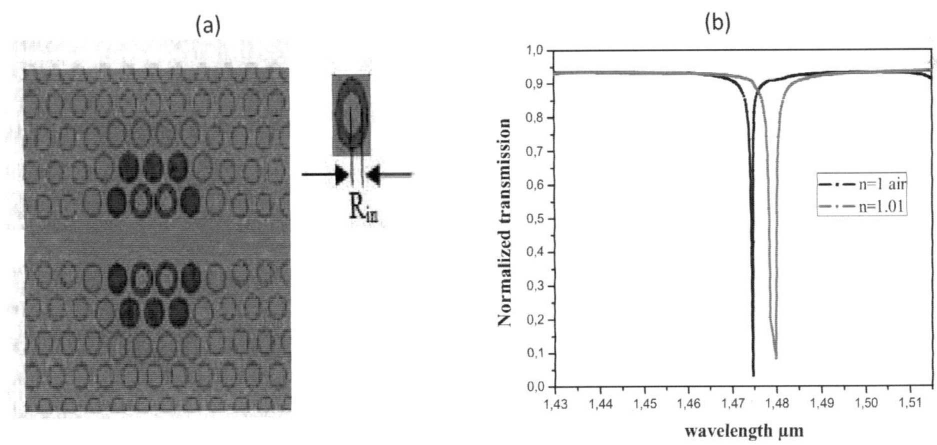

Figure 7. (a)The design of the proposed RI sensor. (b) The transmission spectra of the designed device for R_{in}=0.325a.

The optical free-label detection properties of the designed structure has been quantitatively estimated by the sensitivity parameter S ($\Delta\lambda/\Delta n$), which is defined as the ratio of shift in the wavelength ($\Delta\lambda_0$) to the change in the RI due to gas infiltration (Δn). According to the obtained results the output resonant wavelength is red-shifted due to the increase in the ambient RI of the sensing area (figure 7 (b)), which confirms the gas identification. A change in the RI of $\Delta n=0.01$ results in a spectral red-shift of 5.225 nm which corresponds to a RI sensitivity of 522.5 nm/RIU.

(a) (b)

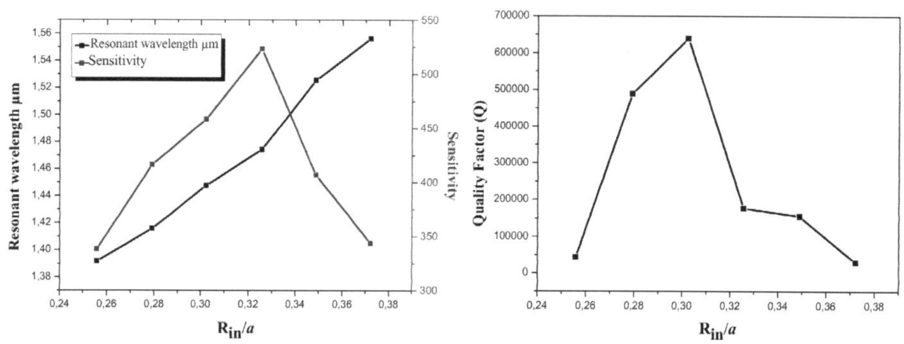

Figure 8.(a)The sensitivity and the resonant wavelength variations according to the change of the inner radius (R_{in}) from 0.255a to 0372a. (b) The quality factor Q as afunction of the inner radius change (R_{in}).

Table 1. Resonant wavelength (λ_0), sensitivity (S) as function of Refractive index (RI).

RI	λ_0 (µm)	*S (nm/RIU)*
1	1.474625	-----
1.000139	1.474700	539.56
1.00026	1.474750	480.76
1.00047	1.474870	521.27

The sensor, additionally, needs to have high transmission efficiency and high Q factor, which can increase the detection feasibility and make it easier. The transmission efficiency can be denoted asA_0/A_i, where A_0 is the output amplitude of magnetic-field at resonance peak, A_i is the input amplitude of magnetic-field at the same frequency. The Q factor is defined as ω_0/ω_x, where ω_x is the full width at half-maximum (FWHM) of the resonator's Lorentz a response, and ω_0, is the resonance frequency [38].

Table 2. Comparison of the proposed gas sensor with various PC designs.

References	Sensing structure	Sensitivity (nm/RIU)	Q factor	Detection limit (RIU)
[34]	Photonic crystal waveguide utilizing width modulated microcavity	80	40 000	10^{-4}
[35]	Photonic crystal microcavity	433	3000	10^{-4}
[36]	Air-slot Photonic crystal nanocavity	510	2.6×10^4	10^{-5}
[37]	RI biosensor formed by two Waveguide and one microcavity (Double-Hole Defect)	500	$1.7\ 890\ 10^4$	0.0001
This work	Photonic crystal waveguide utilizing width modulated microcavity	539	$1.7735 10^5$	10^{-5}

Figure 8a shows the sensitivity and the resonant wavelength variations according to the change of the inner radius (R_{in}) from 0.255a to 0372a and the Figure 8b shows the quality factor as a function of R_{in},which enlarges as the inner ring radius (R_{in}) from 0.256a to 0.372a and reaches its maximum of

$6.388x10^5$ at the resonant mode located at 1.47425nm for R_{in}=0.302a,then followed by a sharp decrease. The increase in the inner radius value is goes along with the decrease of the annular air area. When the R_{in}reaches a value that is more than 0.302athe rings area becomes seriously thin. This results in a reduction in the light confinement within the cavity area and a significant performance of the optical leakages, which have a negative effect on the photon lifetime within the cavity area that results in a sharp decrease in the Q factor value.

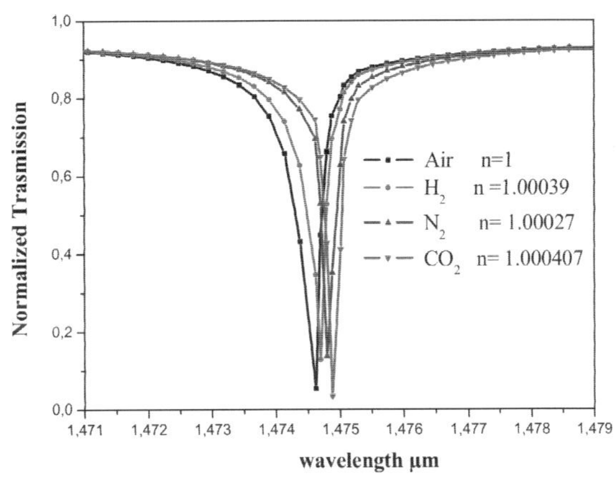

Figure 9. Resonant wavelength changes of the designed sensor as function as a function of changes in the refractive index corresponding to the three infiltration states of gas.

To assess the sensing potential of the sensor, we expose the sample to the gases of different refractive indices, such as, He (1.000139), N_2 (1.000270), and CO_2 (1.000407) [31-33]. The refractive index values were obtained in an atmospheric pressure and were recalculated for λ=1570 nm in room temperature T=20° C using the ideal gas model [35]. The FDTD simulation result show that the resonance wavelength shifts upped by the change of the gas refractive index (figure9).

4. Conclusion

In summary, we have simulated an ultra-compact RI sensor based on tapered-shift defect microcavity introduced to the center of waveguides in the 2D triangular lattice of air holes. A 5.225 nm resonance wavelength shift was observed corresponding to a sensitivity of522.5 nm/RIU for a change in RI value of 0.01.The high-enough sensitivity makes this design useful for realizing photonic crystal-based refractometric and biochemical sensors. And because of the minimized size of the proposed RI sensor, it can be used for measurement in some harsh environments and exceptional conditions.

Acknowledgments

This work is part of the PRFU project No. A25N01UN280120180001 fully funded by the Algerian Ministry of Higher Education and Scientific Research and La Direction Générale de la Recherche Scientifique et du Développement Technologique (DGRSDT).

References

[1] Fan X, White I M, Shopova S I, Zhu H, Suter J D and Sun Y 2008 *Anal. Chim. Acta* **620** 8
[2] Säynätjoki A, Mulot M, Vynck K, Cassagne D, Ahopelto J and Lipsanen H 2008 *Photonics Nanostruct. -Fundam Appl.* **6** 42
[3] Lopez-Torres D, Elosua C, Villatoro J, Zubia J, Rothhardt M, Schuster K and Arregui F J 2017 *Sensor. Actuat. B-Chem.* **251** 1059

[4] Hocini A and Harhouz A 2016 *J. Nanophotonics* **10** 016007
[5] Xu H, Wu P, Zhu C, Elbaz A and Gu Z Z 2013 *J. Mater. Chem. C* **1** 6087
[6] Islam M I, Ahmed K, Sen S, Chowdhury S, Paul B K, Islam M S, Miah M B and Asaduzzaman S 2017 *Photonic. Sens.* **7** 234
[7] Gouveia C, Jorge P A, Baptista J M and Frazao O 2011 *IEEE Sens. J.* **12**1 7
[8] Harhouz A and Hocini A 2015 *J. Electromagn. Waves. Appl.* **29** 659
[9] Arafa S, Bouchemat M, Bouchemat T, Benmerkhi A and Hocini A 2017 *Opt. Commun.***384** 93
[10] Zamani M and Hocini A2016 *Opt. Mater.* **58** 306
[11] Boufenar R, Bouamar M and Hocini A 2017 *Photonics. Nanostruct.* **24** 47
[12] Johnson SG and Joannopoulos JD 2001*Photonic Crystals: the Road from Theory to Practice* (Kluwer Academic, Norwel, MA, USA: Springer Science & Business Media)
[13] Hocini A, Moukhtari R, Khedrouche D, Kahlouche A and Zamani M 2017 *Opt. Commun.* **384** 111
[14] Akahane Y, Asano T, Song B S and Noda S 2003 *Nature.* **425** 944
[15] Song B S, Noda S, Asano T and Akahane Y2005 *Nat. Mater.* **4** 207
[16] Di Falco A, O'faolain L and Krauss T F 2009 *Appl. phys. Lett.* **94** 063503
[17] Notomi M, Yamada K, Shinya A, Takahashi J, Takahashi C and Yokohama I 2001 *Phys. Rev. Lett.* **87** 253902
[18] Notomi M, Shinya A, Yamada K, Takahashi J I, Takahashi C and Yokohama I 2002 *IEEE. J. Quantum. Electron.* **38** 736
[19] Maache M, Hocini A and Khedrouche D 2017 *Chin.J.Phys.* **55** 2318
[20] Kahlouche A, Hocini A and Khedrouche D 2014 *J. Comput. Electron.* **13** 490
[21] Wang X, Xu Z, Lu N, Zhu J and Jin G 2008 *Opt. Commun.* **281** 1725
[22] Joannopoulos J D, Johnson S G, Winn J N and Meade R D 2008 *Molding the Flow of Light*. (Princeton, NJ, USA: Princeton Univ. Press).
[23] Mandal S and Erickson D 2008 *Opt. Express.* **16** 1623
[24] Huang L, Tian H, Yang D, Zhou J, Liu Q, Zhang P and Ji Y 2014 *Opti. Commun.* **332** 42
[25] Huang L, Tian H, Zhou J, Liu Q, Zhang P and Ji Y *Opt. Commun.* **335** 73
[26] Zhou J, Tian H, Yang D, Liu Q and Ji Y 2014 *Opt. Commun.* **330** 175
[27] Kuramochi E, Notomi M, Mitsugi S, Shinya A, Tanabe T and Watanabe T 2006 *Appl. Phys. Lett.* **88** 041112
[28] Joannoupoulos J D, Meade R D and Winn J N 1995 *Photonic Crystals* (Princeton, NJ, USA: Princeton Univ. Press).
[29] Dutta H S, Goyal A K and Pal S2014*J. Nanophotonics.* **8** 083088
[30] Zouache T, Hocini A, Harhouz A and Mokhtari R 2017 *Acta Phys. Pol. A* **131** 68
[31] Le Thomas N, Houdré R, Kotlyar M V, O'Brien D and Krauss T F 2007 *J. Opt. Soc. Am. B* **24** 2964
[32] Canadian Liquid Air Ltd Division scientifique 1976 *Encyclopédie des Gaz* (Elsevier).
[33] Simmons A C 1978 *Opt. Commun.* **25** 211
[34] Sünner T, Stichel T, Kwon S H, Schlereth T W, Höfling S, Kamp M and Forchel A 2008 *Appl. Phys. Lett.* **92** 261112
[35] Wang X, Lu N, Zhu J and Jin G 2008 *Proc. SPIE.* **6831** 68310D
[36] Jágerská J, Zhang H, Diao Z, Le Thomas N and Houdré R 2010 *Opt. let.* **35** 2523
[37] Shiramin L A, Kheradmand R and Abbasi A 2012 *IEEE. Sens. J.* **13** 1483
[38] Villeneuve P R, Fan S and Joannopoulos J D1996 *Phys. Review B* **54** 7837

Influence of nanosilica, metakaolin and Polypropylene Fiber on dynamic and physicomechanical properties of Portland cement mortar

Ibrahim M El-Kattan[1*], **T A Tawfik**[2], **K A Metwally**[2], **A S** Faried[3] and W **Zakid**[4]

[1] *Environmental Science and Industrial Development Department, Faculty of Postgraduate Studies for Advanced Sciences, Beni-Suef University, Beni-Suef, Egypt
[2] Department of Construction and Building, Higher Institute of Engineering on [6]thOctober, Giza, Egypt
[3] Department of Civil Engineering, Faculty of Engineering, Fayoum University, Fayoum, Egypt
[4] Department of Civil Engineering, Faculty of engineering, Beni-Suef University, Beni-Suef, Egypt

*E-mail: ibrahim.elkattan@psas.bsu.edu.eg

Abstract. This study presents an experimental research conducted to inspect the impact of partial substitution through the usage of nano silica (NS) and metakaolin (MK) on cement mortars (CMs).Besides to, the effect of the additive usage such as polypropylene (PP) on the dynamic and physicomechanical properties of CMs as single and combined effect. CM was formed through the usage of different substitution ratios of nano silica varying from 2 to 6% and metakaolin varying from 10 to 30 wt %, by cement weight, where the considered content for PP fiber was at 0.2%. Scanning electron microscopy (SEM), ultrasonic pulse velocity (UPV), flexural strength, and Compressive strength analysis are demonstrated in this paper. Results illustrated that, physicomechanical and dynamic properties of CMs positively enhanced as a result of using a mixture of PP, nano silica, and metakaolin. The CMs that included 0.2% PP, 4% NS, and 20% MK, showed the highest impact on physicomechanical properties compared to other mortar specimens.

1. Introduction

The occurrence of phenomena and structural elements in nano and micro scale affects the concrete mechanical behavior. So nanotechnology can be used in order to enhance characteristics of material's bulk through modifying the concrete molecular structure. Also, nanotechnology can enhance the concrete sustainability, durability, volume stability, and mechanical performance. Recently, concrete materials unprecedented uses have been emerged as the cement and concrete long-lasting products, high-performance and cost-effective have been developed due to the impact of nanotechnology usage. The nanomaterials is characterized by the ability to provide mechanical reinforcement for structural materials based on cement which is considered from the most desired nanomaterials characteristics in construction sector. Three main advantages for nanomaterials usage are recognized. Firstly, the reduction in the duration of construction is observed due to the production of high-strength concrete accompanied by less curing time [1]. Secondly, the reduction in the cement quantity which is required in concrete to reduce the construction materials cost and the environmental effect, in addition, obtaining similar strengths. Thirdly, manufacturing of high-strength concrete for a certain application is considered as another advantage. The nano-sized solid particles mainly impact cementetious based materials performance such as the porosity of nano-sized within the interfacial transition zone between aggregate particles and cement, or calcium–silicate–hydrates particles (C–S–H). Typical characteristics influenced by voids or particles of nano-sized are steel-bond, shrinkage, durability, and strength [2]. The spaces between C– S–H gel particles can be filled by a nano-filler which is SiO2 Nano-particles (NS). Moreover, a high matrix densification was formed due to the increase in of C–S–H amount during the reaction of pozzolanic accompanied by calcium hydroxide that causes improvement in the material durability and strength. The nano-particles addition changes hardened and fresh and state properties, even in comparison with additions of conventional mineral according to previous research [3-9]. C3S hydration process was affected by amorphous silica colloidal particles [4]. Nano-silica reduced the mortar setting time in comparison with silica fume (SF) [5, 6] besides, the reduction in segregation and bleeding water during enhancing the admixtures cohesiveness within the fresh state [7]. The combination accompanied by ultra-fine fly ash ensures the performance improvement compared to that accomplished by the usage of silica fume lonely [8]. Also, the concrete or mortar compressive strength accompanied by silica fume was enhanced compared to the formula in the absence of the addition [9].

Recently, the introduction of highly effective and active pozzolan such as metakaolin has been recognized as a partial substitution for cement within concrete which is considered as ultrafine material that manufactured through kaolin precursor dehydroxylation by heating within temperature range of 700–800 ◦C [10,11]. At ambient temperature, CSH gel was formed due to the reaction of metakaolin which is considered as a silica-based product and Ca (OH)2. In addition, alumina which was presented in metakaolin formed a reaction with CH which generated additional alumina-containing phases whereC3AH6, C2ASH8, and C4AH13were included [12,13]. A significant enhancement in the early strength was observed due to the metakaolin incorporation within concrete [14]. Metakaolin enhances concrete resistance against alkali-silica reaction [15], and its impact on sulfate resistance grows with the increment of the cement substitution portion by metakaolin [16]. As high-reactivity metakaolin is introduced to the admixture, high-performance steel fiber-reinforced concrete (HPSFRC) toughness or energy absorption grows .Thus, the usage of high-reactivity metakaolin concrete is considered as added value for applications that needs high toughness and improved durability[17].Nevertheless, the water demand increases as the substitution amount of metakaolin increases according to other research. The need of water could be overcome through the addition of water reducer to preserve the flow characteristics and workability [18]. In this article, polypropylene (PP) fiber was utilized as another supplementary cementetious materials (SCMs). Generally, in concrete construction, PP and steel fibers were considered as the most common utilized material [19,20]. At low volume portions (e.g., smaller than 0.3%), PP fibers are utilized within various applications. For instance, controlling reinforcement of shrinkage control reinforcement for concrete members is considered as one of the applications [21]. The fiber-reinforced concrete (FRC) mechanical characteristics was investigated by a lot of researchers especially concerning the hybrid usage of PP together accompanied by other fibers, for example the carbon/glass fibers, glass/PP fibers, or carbon/PP fibers mixtures. Nevertheless, combination of steel/PP FRC was studied in the most research [22].

The additives performance within conjunction accompanied by PP fiber was not fully reported even though various cement-based materials aspects that include PP fibers and by-product materials was investigated. Particularly, the impact of PP fiber as materials that can enhance the toughness on the workability as well as MK and NS impact needed a detailed study. Besides, mortar durability properties and pore structure is considered in need of further detailed study.

This aim of the research to inspect the combination usage of NS/MK cements substitution materials and PP fibers (as addition) as well as considering their role on the resulted CMs dynamic and physicomechanical characteristics. The optimum percentages for fibers and mixtures within mortars were figured out through the hardened concrete experiments such as ultrasonic pulse velocity test, flexural strength, and compressive strength. A better indication for the impact of studied materials could be obtained through the usage of mortar instead of concrete within this study due to the material behavior of mortar which is considered as straightforward. For example, the ability to control experimental tests on mortar is more compared to those on concrete [23]. Most of the observations that was concluded from CM within this research can be applied to concrete that owns coarse aggregates as a supplementary stage.

2. Experimental program

2.1. Materials

2.1.1. Cement. Misr Beni-Suef Company (Beni-Suef city, Egypt) is considered as a producer for the Ordinary Portland cement (OPC) type I (42.5N), which is recognized as the used cement in this paper. A number of tests are carried out on a cement specimen to guarantee testing quality of final setting time, initial, soundness of cement, fineness, and specific weight. According to these tests result, the cement specimen physical characteristic and chemical composition are shown in tables1 and 2.

Table 1:- Physical properties of cement.

Property	Specific gravity	Fineness cm^2/g	Initial setting time minutes	Final setting time minutes	Soundness mm	Compressive strength Kg/cm^2		
						3 days	7 days	28days
Result	3.15	2700	85	180	1	202	308	438

Table 2. Chemical composition of the cement.

Oxides	SiO_2	Al_2O_3	Fe_2O_3	MgO	CaO	SO_3	K_2O	Na_2O	F.L	LOI
Wt,(%)	21.16	5.50	3.21	0.69	63.40	2.40	0.50	0.10	2.70	2.30

* LoI: Loss on ignition / F.L: Free lime

2.1.2. Fine aggregates. In this study, sand was utilized and undergone the tested Following ASTM C128 where the sand physical characteristics is given in table 3.

Table 3:- Standard sand physical properties.

Property	Specific gravity	Bulk density, kg/m^3	Percent of Clay and other Fine Materials
Result	2.65	1650	0%

2.1.3. Metakaolin. The Physical properties of metakaolin is shown in table 4 where metakaolin is characterized by 3 μm mean particle size and 99.9% particles <16 μm. The metakaolin chemical analysis is shown in table 5.

Table 4:- Physical properties of metakaolin.

Property	Specific gravity	Bulk density, kg/m^3	Physical form	GE Brightness	Color
Value	2.57	0.4	Powder	80	Off-white

Table 5:- Chemical analysis of metakaolin

Oxides	SiO_2	TiO_2	Al_2O_3	Fe_2O_3	MgO	CaO	SO_3	K_2O	Na_2O	LOI
Wt,(%)	51.52	51.52	40.18	1.23	0.12	2.0	0.0	0.53	0.08	12.01

2.1.4. Nano- silica. In this study, the used nano silica was produced at the laboratories of Beni-Suef central. Nano-silica (NS) physical and chemical characteristics are illustrated in tables 6 and 7, respectively. Figure 1 shows the TEM and SEM micrographs of nanomaterials particle of NS.

Table 6:- Nano-silica physical properties.

Property	Particle size (μm)	Bulk density, kg/m^3	Specific gravity	Specific surface area (m^2/kg)	Color
result	7.00	345	2.15	17.8x10^3	Light gray

Table 7:- Chemical properties of nano-silica.

Oxides	SiO_2	TiO_2	Al_2O_3	Fe_2O_3	MnO	MgO	CaO	P_2O_5	K_2O	Na_2O	LOI
Wt,(%)	99.65	0.02	0.01	0.012	<0.01	<0.01	<0.01	<0.01	<0.01	<0.01	0.25

a) TEM b) SEM

Figure 1. TEM and SEM images of NS

2.1.5. Superplasticizer. In this research, the usage of superplasticizer which is chemical mixture that is characterized by a modified polycarboxylates high-range water-reducing is studied following (types F) ASTM C494. The total performance is controlled by chemical mixture which permits delayed absorption of cement particles and spread them. A concrete with a high quality is produced through the usage of superplasticizer of high-range water reducer through the strength incremental. Table 8 illustrates superplasticizer properties [19].

Table 8:- Technical data properties of the superplasticizer.

Form	Appearance	Density (kg/liter)	pH value
Aqueous solution of modified polycarboxylates	Brown liquid	1.185	4.5-4.9

2.1.6. Polypropylene fibers (pp). Polypropylene fibers with the diameter and length of 0.019 and 12 mm, respectively, were used within the study. The fibers properties and the geometry are given in figure 2 and table 9, respectively.

Figure 2:- Shape and dimension of polypropylene fiber

Table 9:- Properties of polypropylene fiber.

Property	Shape of fiber	Length l (mm)	Diameter d (mm)	Aspect ratio L/d	Density (g/cm^3)	Elastic modulus (GPa)	Tensile strength (MPa)
value	Straight	6	0.019	631	0.91	3.5	350

2.2. Mixes composition of CMs

Fresh CMs were formed through the usage of0.35 water/binder (W/B) portion and 3:1sand/binder weight portion (S/B) as well as the same W/B portion was used in cement paste preparation [33]. Paste and mortars were formed through the usage of 30%, 20% , 10% and 0% metakaolin (first group) while the second group composed of a constant metakaolin percentage (20wt%) accompanied by various percentage of nanosilica (NS)as a replacement for cement 6%, 4%, 2% and0% in weight. Third group composed of the usage of PPF amount in 0 and 0.2% of the binder, fourth group composed of a constant metakaolin percentage (0.2) accompanied by various metakaolin percentage as a replacement for cement 10%, 20%, and 30% MK in weight, and the fifth group composed of a constant metakaolin percentage (20wt%) and 0.2 PP, accompanied by various nanosilica percentage as a replacement for cement 0%, 2%, 4% and 6% NS in weight. In all groups, the SP quantity was 1.5 wt% of the binder cement. These mixtures are illustrated within tables 10 and11.

Table 10. Mix composition of CMs.

Mix component	Cement	Sand	Water	SP
Weight (g)	250	500	87.5	3.75

Table11. Mix composition of the CMs with or without SF, MK and PP.*

Group	Mix No.	Mix Notation	O.P.C	MK	NS	PP
A	1	M0	100	0	0	0
	2	Mk_{10}	90	10	0	0
B	3	MK_{20}	80	20	0	0
	4	MK_{30}	70	30	0	0
	5	$MK_{20}NS_2$	78	20	2	0
C	6	$MK_{20}NS_4$	76	20	4	0
	7	$MK_{20}NS_6$	72	20	6	0
D	8	PP 0.2	100	0	0	0.2
	9	$PP_{0.2}MK_{10}$	90	10	0	0.2
E	10	$PP_{0.2}MK_{20}$	80	20	0	0.2
	11	$PP_{0.2}MK_{30}$	70	30	0	0.2

*M0: control cement mortar; PP: polypropylene; MK: Metakaolin; NS: Nano-silica.

2.3. Production of specimens

The large surface area of nanoparticles affects their uniform distribution within the mix [20, 21] which directly impacts the mortars mechanical characteristics. The production procedure of samples utilized during this research was conducted following the ASTMC305 standard [22]. Nevertheless, the necessity of conducting changes within the procedure of mixing was observed because of nanoparticles presence. During the mixing, direct contact was avoided through the usage of gloves and Masks. Firstly for 1 min, the MK and cement were dry blended within the mixer at 80 rpm (moderate) velocity. After that these ingredients were blended for 90 s at 120 rpm (high) velocity accompanied by the nanoparticles, the specified fibers quantity and 90% of the water. Then while the mixer was running at 80 rpm (moderate) velocity, the addition of the sand was gradually conducted over a period of 30 s. ultimately, the addition of the remaining water and superplasticizer were carried out and stirred at 120 rpm (high) velocity for 30 s. Then the mix was left to rest for 90 s and after that the blending was continued for one minute at 120 rpm (high) velocity. This procedure of mixing was pursued to aid the fibers and the nanoparticles distribution within the mortar.

(50 x 50 x 50) mm cubes of casted fresh mortar were used for ultrasonic and compressive strength tests and 40×40×160mmprismatic samples for test of flexural. The samples were condensed through the usage of tamping rod to eliminate the air bubbles from the mortar. After that the samples were cured and casted within water at 23 ± 3 ◦C then demoulded 24 hours till the specimens were tested.

2.4. Testing procedures

The test of Compressive strength was carried out following the ASTM-C109 standard [23] through the usage of a hydraulic testing machine at 1350 N/s loading rate. The same specimens were dried for 3 hours at the laboratory environment then CHT-225A device was used to measure ultrasonic pulse velocity (UPV) before carrying out the test of compressive. The test of three-point for loading flexural (i.e. center-point) was conducted accompanied by180 mm span at 44 N/s loading rate. The results of compressive were detected at 3, 7, 28 and 90 curing days however results of flexural strength were detected at 28 and 90 days. The three samples average value for the test was recorded.

3. Results and discussion

3.1. Compressive strength

The samples compressive strength in the absence and presence of polypropylene fiber at 3, 7, 28 and 90 days, respectively is shown in figures 3, 4 and 5. For both curing times, the compressive strength incremented when raising the MK substitution from 10 to 20% after that the compressive strength starts to de drop at a substitution portion higher than 20%. Nevertheless, in spite of the decrement, the mortar compressive strength including MK alone is still greater compared to the control samples. The possible reason that MK is considered as a pozzolanic material that besides cement hydration acceleration also interacts with (C—H) calcium hydroxide produced from the Portland cement hydration. Subsequently, the concrete microstructure and hydration products chemistry are changed by pozzolanic reactions through the consumption of discharged calcium hydroxide (C—H) besides the production of extra calcium silicate hydrate (C—S—H) which produce in a decremented porosity and incremented strength. Thus, stronger bond between the fine aggregates and the cement paste is arisen from the raise within compressive strength within MK20 and MK10. The cement matrix homogeneity and density is enhanced due to the cause that MK is finer compared to OPC through the restriction of growth of C—H crystal and the reservation of better C—S—H distribution that causes the existent pores blockage and thus, alter the structure of pore. MK is characterized by a fine divided material which is considered as a thermal activated aluminosilicate material that forms a reaction within the moisture existence and with slaked lime at ambient temperature to form a strong slow-hardening cement which is formed through clay calcinations to obtain kaolin at temperature ranged from 700 to 850 °C without production of CO_2 [24, 34].

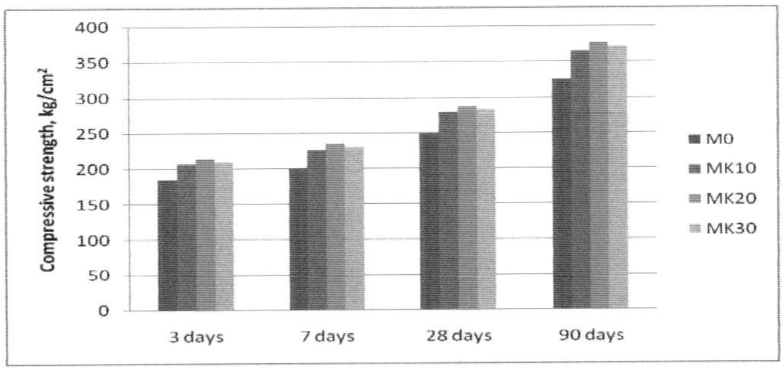

Figure 3:- Compressive strength of specimens with replacement different % MK at 3, 7, 28 and90 days.

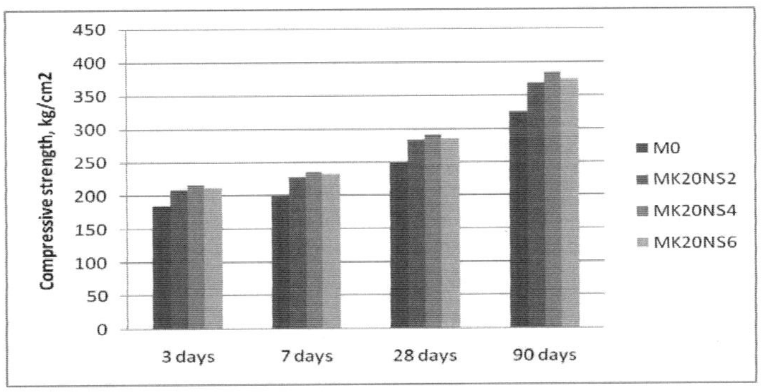

Figure 4:-Compressive strength of specimens with constant 20% MK and with different % of NS at 3, 7, 28 and90 days.

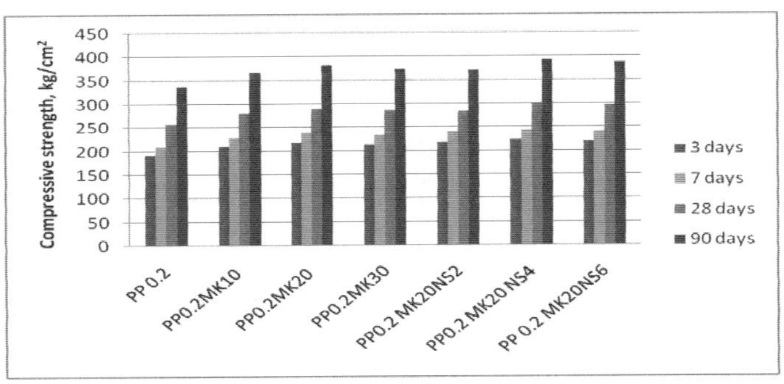

Figure5:-Compressive strength of specimens with polypropylene at 3, 7, 28 and 90 days.

The compressive strength reduction from MK20 to MK30 could be possibly explained by twofold. Firstly, the development of an inadequate hydration is observed due to the inadequate water quantity because of the extreme need of MK for water. Secondly, The non existence of sufficient C—H to form a reaction at a high content rate with the MK, for instance, within the usage of MK 30, some of the MK behaved as a filler only and did not participate in the reaction as shown within the photograph of SEM in figures 6(a). The high C—H concentration which is accessible for pozzolanic reaction is formed due to the pozzolanic reaction high amount within cement pastes accompanied by a low substitution amount with [25]. Table 5 is considered as illustration for chemical composition where the Al and Si emitted from the MK are attributed to the huge purities of aluminosilicate of this mineral.

Besides, the creation of convenient conditions for the composition of certain C—A—S—H quantities such as binding gels phases, in spite of the low MK proportions as well as fostering aluminosilicate portion akin to the filler impact due to the presence of the MK particles. According to other studies, the strength was enhanced and reached the highest level at 20% of MK substitution, however, the reduction in compressive strength was observed at MK >30% [26, 27].

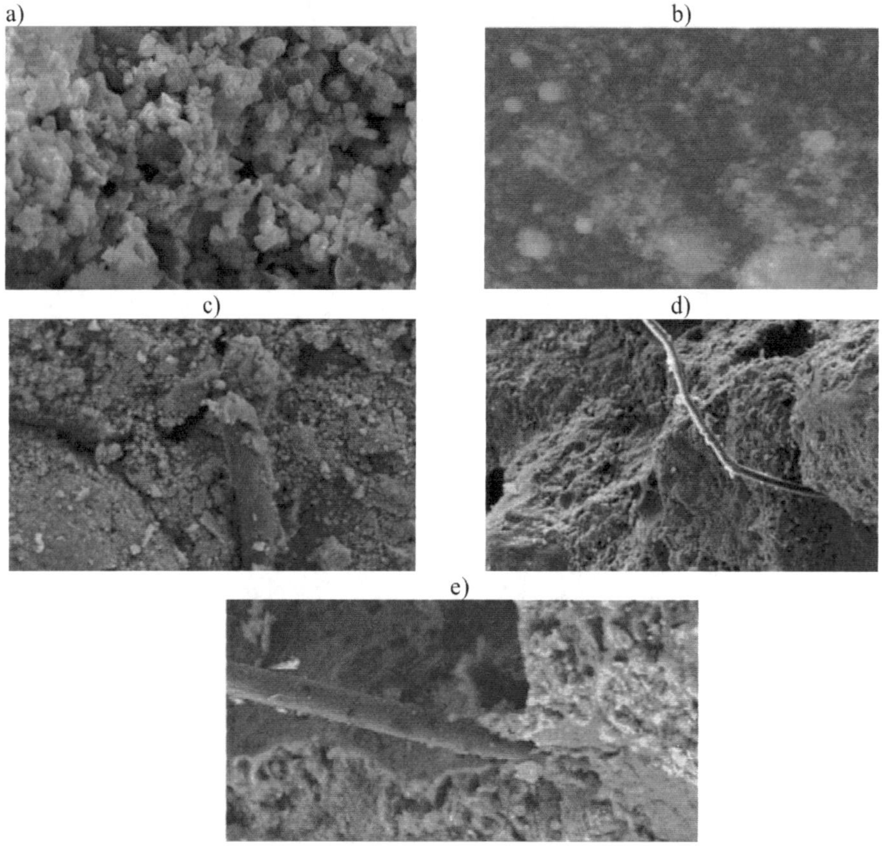

Figure 6:- SEM micrograph of cement mortars containing 0.2% PP, 20% MK and 4% SF singly and combined. a) SEM micrograph of mortar with 20% MK ;b) SEM micrograph of mortar with 20% MK +4% NS; c) SEM micrograph of mortar with PP 0.2 d) SEM micrograph of mortar with PP0.2MK20; e) SEM micrograph of mortar with PP0.2MK20 NS4.

The compressive strength was enhanced the usage of NS content till 4% accompanied by a constant value of MK 20% as shown in figure 3; nevertheless, the strength decreased as NS content increased. This could be explained as illustrated in Fig.5-b, for admixture includes6% NS. The reaction between the nanoparticles and calcium hydroxide within lime solution is considered as the reason for the increment of compressive strength due to the production of additional C–S–H gel [28]. Ghafari et al. [29] reported that in order to accomplish the highest compressive strength, the optimal cement substitution quantity by nanosilica was 4wt% that coincides with the current study. However, Massana et al. [30] reported the nanosilica addition till 7.5wt% is recognized to increment the compressive strength.

The compressive strength raised consistently by adding polypropylene fiber (PP) accompanied by similar trend for 90, 28, 7 and 3 days as shown in figure 4. The highest strength was figured out within

PP0.2 MK20 NS4accompanied by a 40% raise compared to the control admixture. The negative impact of the addition of cement containing high NS could be compensated though adding PP fiber. The reason for the compressive strength increment as the following: the ability of fibers to postpone the cracks growth rate, alter the cracks direction, minimize stress concentration extent at cracks tip, and control the cracks extension [31]. This coincides with Hsie et al. outcomes [19], which stated that the usage of polypropylene hybrid fiber enhanced the samples compressive strength. Mortar specimen SEM micrograph accompanied by 0.2% PP and 20% MK is illustrated in figure 5. A compact hydration products formation and raised C-S-H crystals content are illustrated in 0.2PPMK20NS4 micrograph. PPF are recognized to be able for holding the cracks propagation and bridging across the cracks and voids in the concrete.

3.2. Flexural strength
The mortar samples flexural strength in the absence and presence of polypropylene fiber (PP) at 90 and 28 days was illustrated in figures 7, 8 and 9. In accordance with the results, the flexural strength values improved slightly within admixtures including MK compared to the control mortar. As the addition of NS and substitution content of MK raised, the admixture flexural strength reduced and achieved a minimum within MK20NS6 that reduced by around18.7% compared to the control samples at both 90 and 28 days. Generally, the flexural strength was improved through the substitution with constant 20% MK accompanied by various NS portion weather at 90 or 28 days. The MK impact as partial substitution for cement within concrete was investigated by Murali and Sruthee [32]. The flexural strength was enhanced within concrete by 9.3% through the usage of MK at the optimum ratio which approached 7.5%. The flexural strength was increased significantly that was illustrated in Fig. 9 which demonstrated the PP fibers importance within improving the mortars flexural strength. A growth of 24% in the flexural strength was recorded which was considered the highest compared to the control specimens at 90 and 28 days. The increment within flexural strength was caused by the fibers bridging impact due to the appropriate mechanical bonding between fibers and paste of the cement. Thus, within the fiber-reinforced mortars fracture process, the failure planes and the cracks of fibers bridging within the matrix could afford certain impedance for the crack propagation prior being stressed to or pulled out the rupture point.

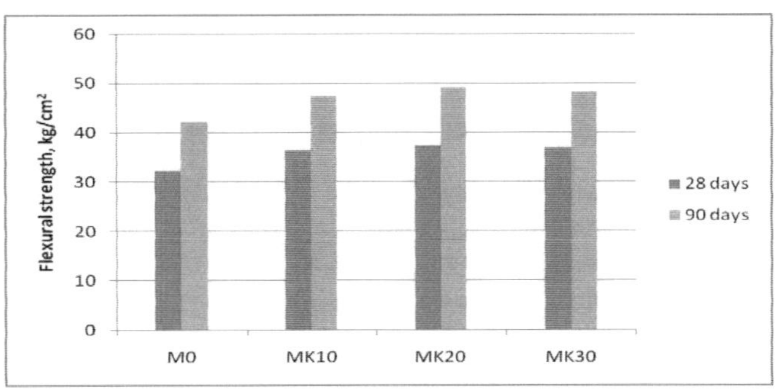

Figure7:- Flexural strength of specimens with replacement different % MK at 3, 7, 28 and 90 days

3.3. Ultrasonic pulse velocity
Two main functions of ultrasonic Pulse Velocity (UPV) were determined; Firstly establishing a relationship within nondestructive way between mortar internal structure and water absorption, Secondly, providing information about the samples internal structure. To carry out the UPV test, 50 × 50 × 50mm samples were utilized within this research. This technique depends on the assessment of the pulse propagation velocity for vibration energy that crossed a medium of concrete. According to the results, combining 20, 10% MK, without PP or NS, enhanced the UPV by22% and 13.5 %,

respectively, compared to the control sample, while the 15% MK combination reduced the pulse velocity by 2% compared to MK10 (figure 10) which coincidence with Khatib et al. results[25]. The highest level of UPV achieved at 28 curing days for all mortar admixtures at 20% MK while the UPV declined for values of MK higher than 20%. The hydrated phases amount such as calcium silicate hydrates were considered from the factors that impact the enhancement of cement mechanical properties significantly. The pozzolanic materials usage such as MK could chip within the hydrated phase composition. The occurrence of pozzolanic reactions was recognized within mortars that included main products of MK which own C—A—S—H, C—S—H, and a less extent, C—A—H. Combining 4% of NS content could enhance the UPV particularly. According to the results of all the samples, the UPV was reduced for NS content higher than 4. Thus, 4% is considered an optimum ratio for NS substitution. The enhancement in the properties of the specimens combined with NS could be due to the NS physical and chemical effects. For the physical effect that could be recognized as filler impact, the solid materials packing was encouraged by NS particles through stuffing the gaps between the grains of cement. The chemical effect depends on reactions of pozzolanic between the calcium hydroxide (C—H) formed by the hydration of cement to produce calcium-silicate-hydrates (C—S—H) and NS amorphous silica. Besides, a huge number of nucleation sites were generated from the addition of small particles for the hydration products precipitation [35]. Thus, the reaction would be accelerated which lead to the formation of smaller C—H crystals.

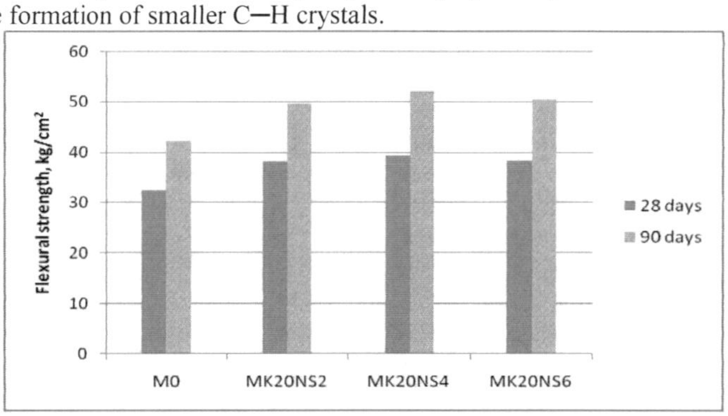

Figure 8:- Flexural strength of specimens with constant 20% MK and with different % of NS at 3, 7, 28, and 90 days.

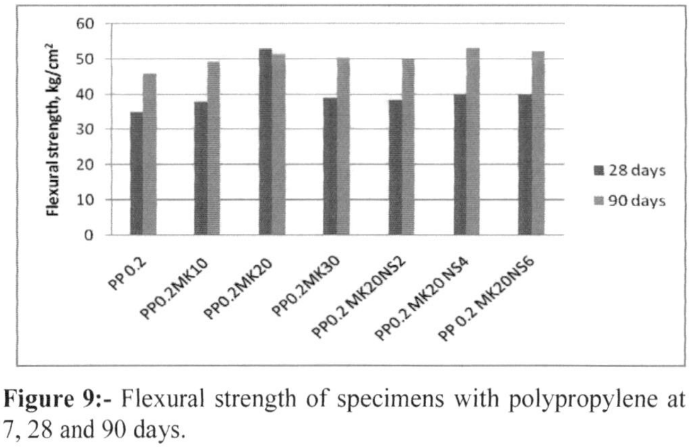

Figure 9:- Flexural strength of specimens with polypropylene at 3, 7, 28 and 90 days.

The Transformation probability for the continuous pores into discontinuous ones and the reduction in the large pores numbers from the causes of NS usage. Therefore, all these mechanisms produce the paste microstructure denser and homogeneous. A micrograph of the admixture including substitution

(NS and MK) was illustrated in figure 5 (SEM) that includes reduced C—H crystals content and compact hydration product composition that causes a reduction for the total porosity amount till the minimum compared to the control sample mortar. Nevertheless, the UPV was reduced through the addition of 6 % NS.

Generally, even though UPV decreased with raising NS by a value higher than 4% but it was still greater than the control sample UPV. Due to the low value of NS specific gravity compared to cement, raising the content of NS provides an increase to a lower density of mortar and thus a lower UPV.

As illustrated in figure 10 for polypropylene (PP), mixtures accompanied by4% NS illustrated the largest value of UPV compared to the other mortar samples. The total velocity reduced as the content of NS raises in the PP contained samples that illustrate a decline within the propagation pulse velocity of vibration energy that crossed the mortar. Nevertheless, UPV was specified by a high value in all mixtures compared to the control admixture.

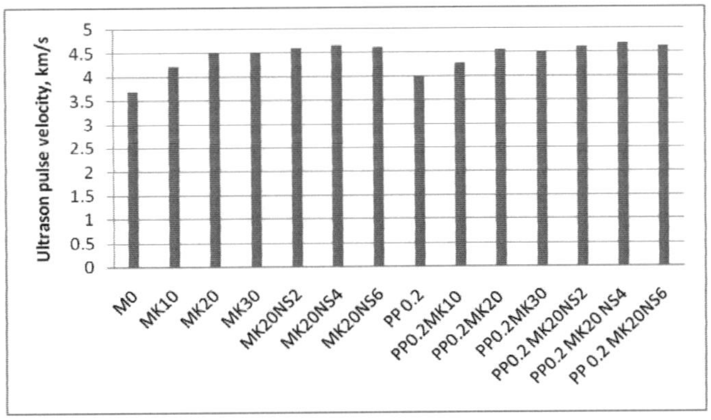

Figure 10. Ultrasonic pulse velocity of the specimens with and without polypropylene at 28 days.

4. Conclusions

An improvement in the compressive strength has been observed at 3, 7, 28 and 90 days due to the usage of a mixture of 20% MK and 4% NS. The compressive strength has been enhanced slightly when 0.2% PP fibers has been added. For instance, PP0.2 MK20 NS4sample enhanced the compressive strength by 19:35% compared to the control sample. Generally, a non considerable impact effect on the flexural strength for partial substitution through the usage of MK and addition NS was observed. The highest flexural strength was acquired at MK20 within without PP samples. A substantial enhancement within the flexural strength of specimens was detected due to the PP integration into mortars as an additive. For instance, PP0.2 MK20 NS4enhanced the flexural strength by 24 %, compared to the control sample mortar. Particularly, similar trend within flexural strength could be detected at both 90 and 28 days. Regards to UPV, the replacement of NS up to 4 % and 20%MK enhanced the velocity of ultrasonic pulse. Also PP addition is recognized to have a similar trend. Besides, an increment velocity was noticed due to the addition of PP fibers where a significant rate was observed for 20% NS and MK specimens. The 4% is considered as an optimal content for both 20 % NS and MK in samples with or without PP.

References

[1] Morsy M S, Alsayed S H and Aqel M 2011 *Construct. Build. Mater.* **25** 145
[2] Collepardi S, Borsoi A, Olagot J O, Troli R, Collepardi M and Curzio A Q *Proceedings of the International Conference held at the University of Dundee*, Scotland, UK on 7 July 2005, Thomas Telford Publishing, pp. 55-66.

[3] Marchon D, Kawashima S, Bessaies-Bey H, Mantellato S and Ng S 2018 *Cem.Conc. Res.* **112** 96
[4] Gaitero J J, Campillo I and Guerrero A 2008 *Cem.Conc. Res.* **38** 1112
[5] Qing Y, Zenan Z, Deyu K and Rongshen C 2007 *Construct. Build. Mater.* **21** 539
[6] Jalal M, Pouladkhan A, Harandi O F and Jafari D 2015. *Construct. Build Mater.* **94** 90
[7] Li H, Zhang M H and Ou J 2006 *Wear* **260**1 262
[8] Jo B W, Kim C H, Tae G H and Park J B 2007 *Construct. Build. Mater.* **21** 1351
[9] Li G. 2004.*Cem.Conc. Res.* **34** 1043
[10] Klimesch D S and Ray A 1998 *Adv. Cem. Bas. Mat.* **7** 109
[11] Abdel-Gawwad H A, El-Enein S A, Heikal M, Abd El-Aleem S, Amer A A and El-Kattan I M 2018 *Construct. Build Mater.* **176** 676
[12] Siddique R and Klaus J 2009 *App.Cl.Sci.* **43** 392
[13] Khatib J M and Hibbert J J 2005 *Construct. Build Mater.* **19** 460
[14] Kim H , Lee S and Moon H 2007 *Construct. Build Mater.* **21** 1229
[15] Lindgård J, Thomas M D, Sellevold E J, Pedersen B, Andiç-Çakır Ö, Justnes H and Rønning T F 2013 *Cem. Conc. Res.* **53** 68
[16] Abdel-Gawwad HA, Mohammed MS and Alomayri T 2019 *Constr. Build.Mater.* **228** 116827
[17] Akcaya B and Tasdemir M A 2018 *Construct. Build Mater.* **185** 436
[18] Madandoust R and Mousavi S Y 2012 *Constr. Build. Mater.* **35** 752
[19] Hsie M, Tu C H and Song P S 2008 *Mater.Sci. Eng. A* **494** 153
[20] Tawfik T A, Abd EL-Aziz M A, Abd El-Aleem S and Serag Faried A, 2018 *J. Chi.Adv.Mats.* Soc. **6** 409
[21] Sobolev K, 2006 *NSF Workshop on Nanomodification of Cementitious Materials, Portland Cement Concrete and Asphalt Concrete.*
[22] ASTM C305 2011 (*Annual Book of ASTM Standards, ASTM)* Philadelphia *PA*
[23] ASTM 2007 *(ASTM C109-99)* West Conshohocken PA
[24] Zheng Z and Feldman D 1995 *Pro. Poly. Sci.* **20** 185
[25] Rashad A M 2015 *Mater.* Des. **52** 143
[26] Khatib J M, Negim E M and Gjonbalaj E 2012 *Wor. J. Chem.* **7** **7**
[27] Vejmelkova E, Pavlikova M, Keppert M, Keršner Z, Rovnaníková P, Ondráček M and Černý R 2010 *Construct. Build Mater.* **24** 1404
[28] Weng T L, Lin W T and Cheng A 2013 *Sci. World. J.* **2013** 606524
[29] Ghafari E, Costa H, Júlio E, Portugal A and Durães L 2014 *Mater. Des.* **59** 1
[30] Massana J, Reyes E, Bernal J, León N and Sánchez-Espinosa E 2018 *Construct. Build Mater.* **165** 93
[31] Mohseni E, Yazdi M A, Miyandehi B M, Zadshir M and Ranjbar M M, 2017 *J. Mater. Civ. Eng.* **29** 04017025
[32] Shivram B and Nagesh P 2007 *Cem. Compos.* **21** 38
[33] Abo-EL-ENein S A, Abdel-Gawwad H A, Hodhod OA and El-Kattan I M 2016 *J. Am. Sci.* **12** 51
[34] Abdel-Gawwad H A, Hassan H S, Vásquez-García, S R, Israde-Alcántara, I, Ding Y C, Martinez-Cinco M A, Abd El-Aleem S, Khater H M, Tawfik T A and El-Kattan I M 2020 *J. Clean. Prod.* **252** 119875.
[35] El-Kattan I M, Abdelzaher M A and Farghali A A 2020 *J. Mater. Res. Tech.* https://doi.org/10.1016/j.jmrt.2020.05.087

Design and analysis of near infrared high sensitive metal-insulator-metal plasmonic bio-sensor

H Ben salah [1, 2], A Hocini[1,*], N Melouki[1] and D Khedrouche[1]

[1]Laboratoire d'Analyse des Signaux et Systèmes, Department of Electronics, University of M'Sila BP.166, Route Ichebilia, M'Sila, 28000, Algeria.

[2]Université Yahia fares Médéa, Algeria.

*E-mail: abdesselam.hocini@univ-msila.dz

Abstract. In this paper, a high sensitivity metal-insulator-metal (MIM) waveguide based plasmonic sensor, coupled by a hexagonal ring resonator is proposed. The sensing characteristics of the device are analyzed by the finite-difference time-domain (FDTD) method embedded in the commercial simulator R-Soft. From results, there is a linear relation between the material's refractive index and its wavelength resonances. Moreover, the maximum linear sensitivity is S = 1743 nm/RIU for the second mode and it is S = 836 nm/RIU for the first mode, its corresponding sensing resolution is 5.73×10^{-6} RIU for mode 2 and 1.19×10^{-5} RIU for mode 1. The proposed sensor can be implemented in high performance nano-sensors and bio-sensing devices. The positions of transmission peaks can be easily manipulated by adjusting the inner side lengths of the hexagonal ring resonator, making this structure a dynamically controllable band pass filter. In addition, introducing another small hexagonal ring within the base resonator decreased the full width of half maximum (FWHM) of the resonance peak.

1. Introduction

The Surface Plasmon Polaritons SPPs are electromagnetic fields which travel along the metal-insulator interface. The capability to modulate nano-scale light and to break the diffraction limit takes the interest of many researchers over several decades [1]. Furthermore, as its volume is so tiny several applications will be realizable such as, optical processing and manipulation in nanoscale optical communication [2,3]. SPP-based waveguide designs, for instance, filters [4], diodes [5], optical switches and sensors [6,7] have been thoroughly tested for optical instrumentation. Sensors based on Surface Plasmon Resonance (SPR) have been developed for a variety of applications such as proteins [8], DNA [9] and drugs [10] in the biomedical and clinical research sector. SPR instruments are commonly utilized to analyze and assess the index of refraction for a variety of substances because of their high sensitivity, easy production and nanoscale size [11,12].Among different plasmonic structures, metal-insulator–metal (MIM) based structure is one of the most promising choices for realization of the mentioned requirements [13]. The confinement of light in sub-wavelength insulator regions is highly achievable, so a prominent transmission efficiency and simplicity in fabrication procedure, made the MIM-based devices a good candidate in nanoscale photonic devices [14]. As one of the most innovating devices in optical circuits, MIM waveguide based plasmonic sensors and filters for wavelength selection has been investigated numerically and experimentally, and different types of the plasmonic sensors and filters such as nanodisk resonators [15], ring resonators [16,17,18,19] and tooth-shaped waveguide [7-20].

In integrated circuits, plasmonic ring resonators are key elements. Owing to their compact size, excellent efficiency, and ease of production, they have gained interest in latest years [21]. Besides, wavelengths of resonance may be finely tuned by changing the ring geometric parameters [22, 23]

Sensitivity of the sensor is among the most critical merit for evaluating the sensor functionality for various substances according to the variation in the index of refraction. The aim is to enhance the sensitivity aspect by using various designs in a limited space and to evaluate the efficient sensitivity and transmission features.

In this work, a MIM waveguide with structures based on stub and hexagonal resonators are used as a refractive index based plasmonic nano-sensor. In order to determine the impact on the sensor functionality and refractive index sensibility, we have varied the inner side length of the hexagonal resonator with the distance between the hexagonal ring resonator and the waveguide. The boundary condition of absorbing perfectly matched layers (PML) is imposed and computed using the FDTD algorithm which is implemented in the commercial software R-SOFT.

2. Structural and theoretical analysis

The proposed MIM based plasmonic structure is depicted in figure 1. It consists of a slit (waveguide) with w =50 nm as a width and fixed to that value so the only excited mode is (TM_0) in the MIM waveguides [24], and a hexagonal ring resonator (HRR), with L_1=200nm and L_2 =125nm as an outer and inner side lengths respectively, g =10 nm, is the distance between the HRR and the waveguide, The gray area represents the silver layer, and the white one represents the dielectric material. The frequency dependent dielectric constant of silver is calculated and expressed by Lorentz–Drude model [25]:

$$\varepsilon_m(\omega) = \varepsilon_\infty - \frac{\omega_p^2}{\omega(\omega + i\gamma)} \tag{1}$$

where $\varepsilon_\infty$is the dielectric constant at the infinite angular frequency with a value of 3.7, ω_p is the bulk plasma frequency (ω=1.38 x 10^{16}*Hz*), and γ is the electron collision frequency (γ=2.73 x 10^{13}*Hz*) and $\omega$is the angular frequency of the incident wave in vacuum [26].

The incident light for excitation in our simulation process of the SPP Mode is TM-polarized; this led to the magnetic field being parallel to y axis [27].

The resonance wavelength λ_m of the single plasmonic hexagonal cavity can be obtained theoretically by [28]:

$$\lambda_m = \frac{(6L_{eff})x\ \mathrm{Re}(n_{eff})}{m - \varphi 2\pi} \tag{2}$$

Where L_{eff} is side lengths of hexagonal cavity, Re (n_{eff}) is the real part of the effective refraction index of the hexagonal shaped cavity. The positive integer m represents the order of the standing SPP wave in the cavity; φ also presents the total phase shift in the corners of the hexagonal cavity.

3. Simulation results and discussion

The commercial simulator R-Soft, is used in all carried out simulations, it's based on the FDTD method [29]. The Perfectly Matched Layers (PML) is imposed and applied on the structure to simulate its transmission characteristics, as the absorbing boundary condition for all sides of the computational window to absorb outgoing waves.

The grid size in both the x and z directions was chosen as $\Delta x = \Delta z = 5$ nm [7], with a continuous TM polarized Gaussian modulated wave based input type. The transmitted light was collected at the right side of the waveguide which is defined as $T = P_{out}/P_{in}$. The incident power P_{in} and the transmitted power P_{out} are observed at positions A and B respectively. Initially, both the insulator in the dielectric core and the cavities have a refractive index n of 1. Figure 2(a) Shows the transmission spectra of both the hexagonal based cavity (side length L_1=200nm) and the proposed hexagonal ring resonator (L_1=200nm and L_2=125nm).

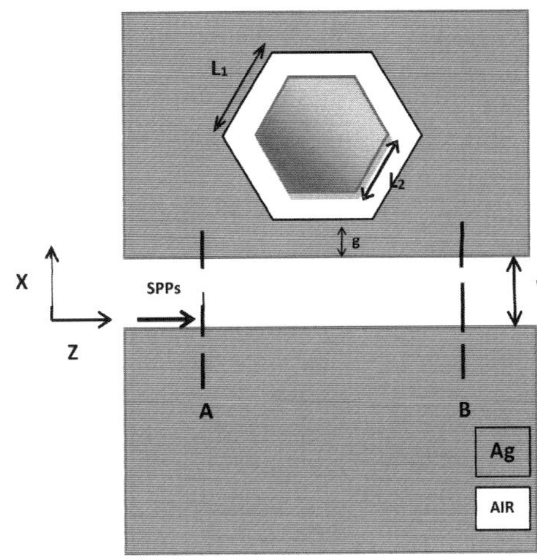

Figure 1. Schematic diagram of the proposed MIM Based plasmonic structure.

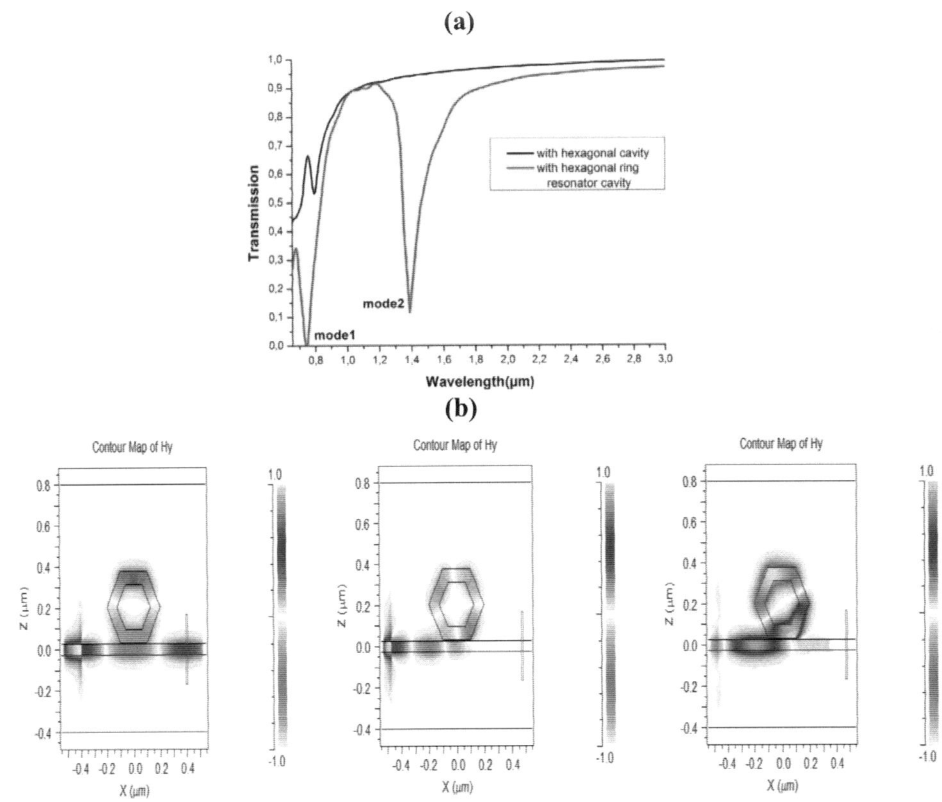

Figure 2. (a) The transmission spectra of MIM plasmonic structure with n = 1, L_1=200nm, L_2=125nm and g=10nm. (b) Magnetic-filed patterns of the MIM plasmonic structure at resonance wavelengths of 742.6 nm (center), 1338.4 nm (right) and at wavelength 1200nm (left).

The structure with hexagonal shaped cavity exhibits a low transmission peak at the resonance wavelengths of 742.6 nm, whereas the case of the MIM waveguide is coupled with the HRR, the two transmission peaks is observed at the resonance wavelengths of 742.6 nm and 1388.4 nm, corresponding to the first and second resonance modes of the HRR. Figure 2(b) depict the field distributions of $|H_y|$ for the on and off resonance wavelengths of the proposed structure. It's clearly noticed that the incident light within the wavelengths of both 742.6 nm and 1388.4 nm, couples to the HRRs resonance wavelengths, so it's prohibited to pass through the waveguide, contrary to the rest of the wavelengths which are transmitted to the output.

To investigate the influence of the coupling distance g on the transmission spectra of the proposed MIM based plasmonic structure, the distance gis set as a state variable from 10 nm to 20 nm with a step of 5 nm, while keeping the other structural parameters fixed (L_1=200nm, L_2 =125nm, and n = 1.0). From figure 3, it's noted that when the coupling distance g is increased, the wavelengths of the peaks shift slowly towards short wavelengths and the bandwidth become narrower. In the other hand, the maximum transmittances are reduced accordingly, due to the fact that the MIM coupler losses are increased also with increasing of the distance g [30].

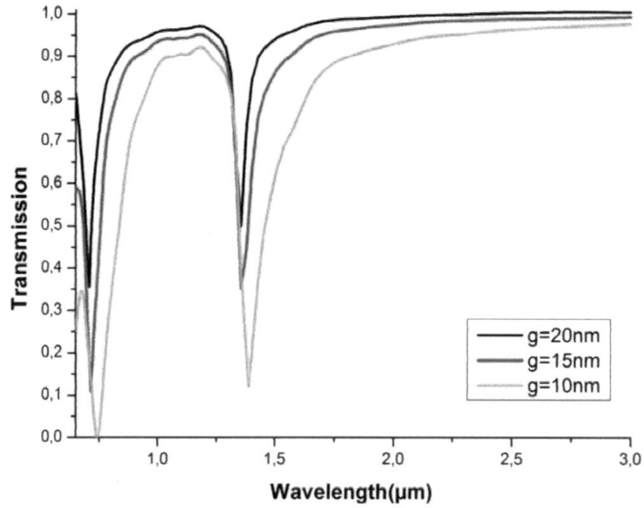

Figure 3. The transmission spectrum of the structure versus wavelength with different coupling distance g at n = 1, L_1=200nm, and L_2=125nm.

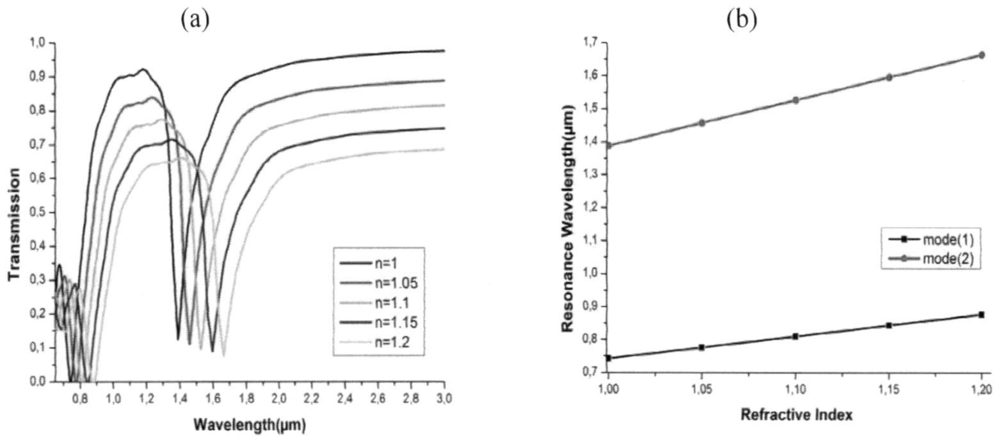

Figure 4. (a) The transmission spectrum of the structure for different refractive index n, at g = 10nm, L_1=200nm, and L_2=125nm. (b) The resonance wavelength versus the refractive index n of the material under sensing.

Figure 4(a) illustrates the transmission spectra of the proposed structure for different refractive indices varying from 1 to 1.2 in steps of 0.05. Increasing the refractive index n, results in a red-shift of resonance wavelengths of the transmitted spectra, with a larger shift in mode 2 in comparison to mode

1. Figure 4(b) represents the maintained linearity relation between the resonance wavelengths and the refractive indices for both the first and the second modes.

Sensitivity (S), is an important aspect in the designing and analyzing of the SPR based sensors, it can be calculated as $S = \Delta\lambda/\Delta n$ (nanometer per refractive index (nm/RIU))[31], where $\Delta\lambda$ denotes the shifting rate of resonant peak wavelength of transmittance, and Δn represents the changing rate in the refraction index in the plasmonic sensor structure. Keeping the other existing parameters fixed, and increasing the refraction index between n = 1 and n = 1.2, the sensitivity for the first mode is around 670 nm/RIU and for the second mode is around 1380 nm/RIU.

In Figure 5, the increase in the value of L_2 from 125 to 145 nm with a step of 5 nm is associated with an increase in the resonance wavelength. Meaning, the value of L_2 causes a shift in resonance wavelength, and the shift for mode 2 is larger than that for mode 1 as shown in figure 5(a). According to figure 5(b) the linearity of wavelength resonances versus the refractive indices is maintained in spite of the increase in the value of L_2. Moreover, an enhancement of the sensor's sensitivity is achieved by increasing the value of L_2, for mode 1 the sensitivity is 836 nm/RIU whereas in the case of mode 2, the sensitivity was 1743 nm/RIU, i.e. sensing resolution of 1.19×10^{-5} RIU for mode1 and 5.73×10^{-6} RIU for mode2, is achieved corresponding to value of L_2 =145 nm, in comparison to sensitivity of 670 nm/RIU for mode1 and 1380 nm/RIU for mode2 where the value of L_2 was 125 nm. In figure 5(c) it is clearly shown that the greater the value of L_2 is, the better the sensor's performance.

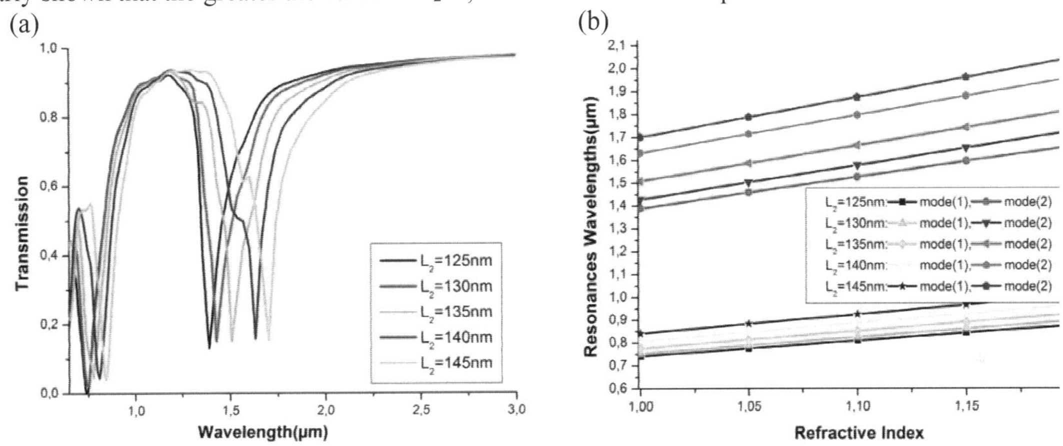

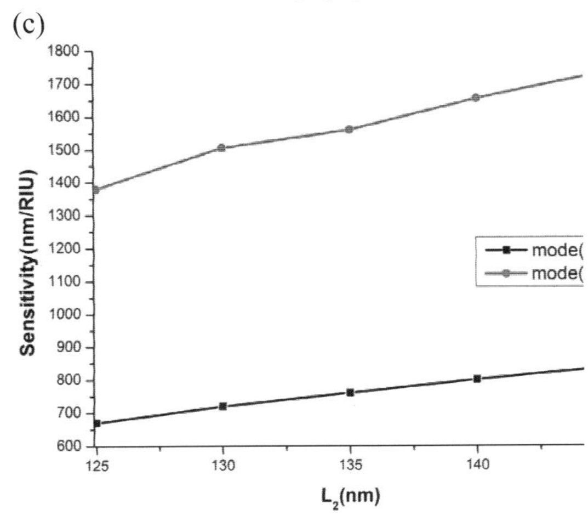

Figure 5. Sensing properties as functions of L_2. (a) Transmission spectra of index 1 for L_2 varying from 125 to 145 nm. (b) The resonance wavelengths versus the refractive index for different L_2. (c) Sensitivities of the plasmonic sensors for L_2 varying from 125 to 145 nm.

For a better characterization of a plasmonic sensor's performances, both the sensitivity (S) and figure of merit ($FOM=S/\Delta\lambda_{FWHM}$) are considered to be the most important parameters, and preferably being as high as possible for a better performing sensor [32].

By using a secondary hexagonal ring inside the main one, as illustrated in figure 6, the proposed design with double-ring resonator where L_1 =200 nm, L_2 =145 nm, L_3=100nm, L_4=70nm and g=10nm are the main parameters of this structure, narrowed down the bandwidth, hence the higher FOM as illustrated in figure 6(b) at the resonance wavelength of 1701 nm, where the FWHM when using the proposed design decreased from 112 nm to 55nm, in addition to the results from both the sensitivity and the FOM value for the first proposed design (single hexagonal ring resonator (S= 1743 nm/RIU, FOM=15.6 (mode2)), and the second proposed MIM based plasmonic sensor with double hexagonal ring resonator (S= 1743 nm/RIU, FOM=31.7(mode2)), shows a promising results for enhancing the FOM, and consequently the sensor's performance.

Figure 6(c) depict the field distributions of $|H_y|$ for the on and off resonance wavelengths states of the proposed structure. It's clearly observed that the incident light within the wavelength of 841.5nm and 1701 nm, couples the hexagonal shaped double-ring resonator and it's prohibited to pass through the waveguide, while the rest of the wavelengths are transmitted to the output freely. Table 1 compares the sensitivity (S) for different reported MIM plasmonic sensors in literatures with our present work.

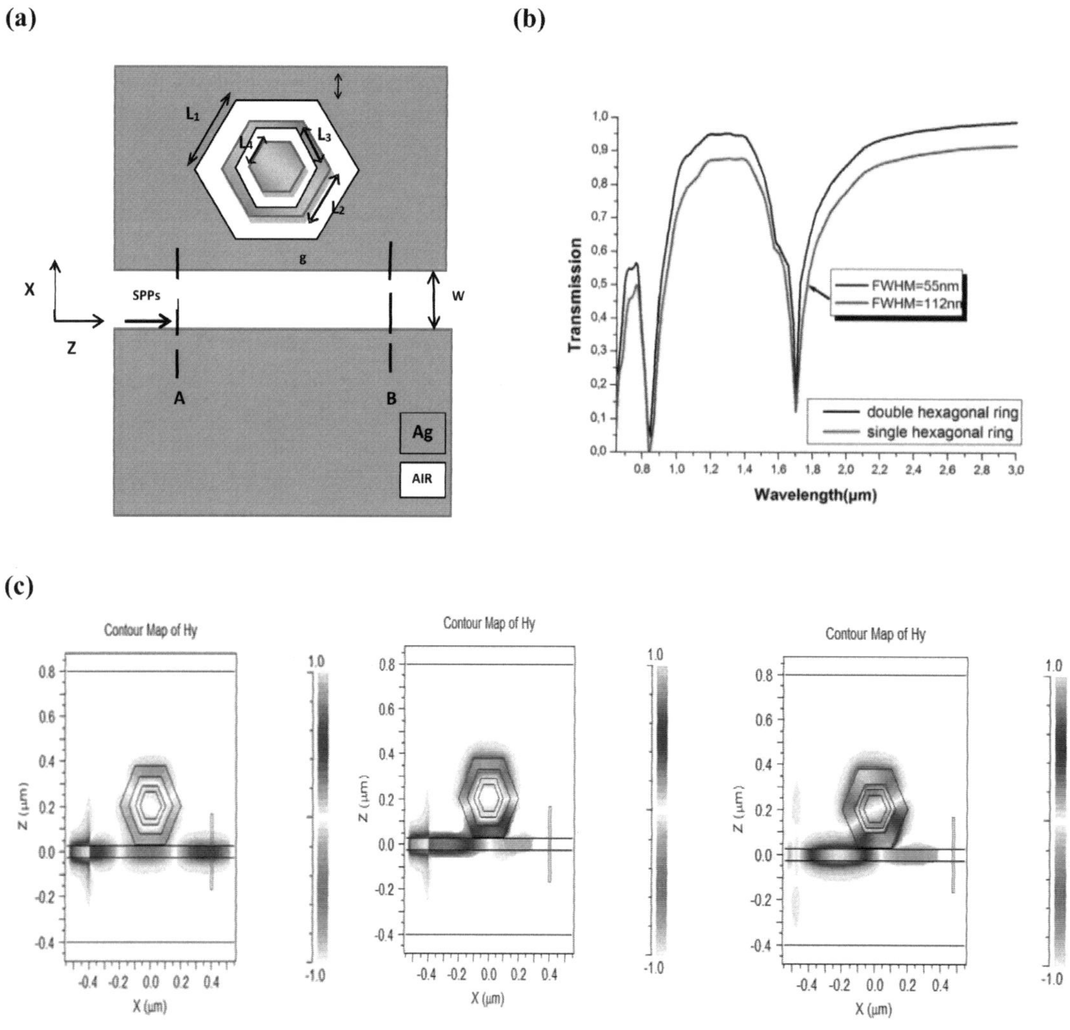

Figure 6. (a) schematic of the proposed structure with double-hexagonal ring resonators, (b) transmission of the structure proposed with double-hexagonal rings with corresponding FWHM, (c) Magnetic-filed patterns of the structure proposed at resonance wavelengths of 841.5 nm (center), 1701 nm (right) and at wavelength 1200nm (left).

Table 1. Sensitivity comparison of different sensor structures.

Reference	Sensitivity	*Year*
[33]	1125	2017
[22]	1540 for mode 1 and 1010 for mode 2	2018
[18]	1160	2018
[7]	2602.5	2019
This work	1743 for mode 2 and 836 for mode 1	2019

4. Conclusion

An MIM based waveguide slab coupled by a hexagonal shaped ring resonator for a near-infrared refractive index sensor is proposed. The finite-difference time-domain (FDTD) method embedded in the commercial simulator R-Soft is used to characterize all design variations. The positions of transmission peaks are of a linear nature in respect to the refractive index of the material under sensing, and can be easily manipulated by adjusting the inner side length of HRR. In this proposed structure, the maximum linear sensitivity is S = 1743 nm/RIU and S = 836 nm/RIU for both mode 2 and mode 1 respectively, after optimizing the structural parameters. In addition, when introducing another hexagonal shaped ring inside the main one, the figure of merit (FOM) is enhanced, which makes our proposed design an attractive candidate for a good performing bio-sensor.

Acknowledgments
This work was supported by the Algerian Ministry of Higher Education and Scientific Research and Scientific Research and La Direction Générale de la Recherche Scientifique et du Développement Technologique (DGRSDT) via funding through the PRFU project No. A25N01UN280120180001.

References

[1] Barnes W L, Dereux A and Ebbesen T W 2003 *Nature* **424** 824
[2] Yin Y, Qiu T, Li J and Chu P K 2012 *Nano Energy* **1** 25
[3] Ozbay E 2006 *Science* **311** 189
[4] Gao Y, Ren G, Zhu B, Huang L, Li H, Yin B and Jian S. 2016 *Plasmonics* **11** 291
[5] Fan C, Shi F, Wu H and Chen Y 2015 *Opt. Lett.* **40** 2449
[6] Hocini A, Ben salah H, Khedrouche D and Melouki N 2020 *Opt. Quant. Electron.* **52** 336
[7] Ben salah H, Hocini A, Temmar M N and Khedrouche D 2019 *Chin. J. Phys.* **61** 86
[8] Etezadi D, Warner I V J B, Ruggeri F S, Dietler G, Lashuel H A and Altug H 2017 *Light. Sci. Appl.* **6** e17029
[9] Hameed M F, Saadeldin A S, Elkaramany E M and Obayya S S 2017 *J. Lightwave Technol.* **35** 4851
[10] Kong W, Wu D, Hu N, Li N, Dai C, Chen X, Suo Y, Li G and Wu Y 2018 *Biosens. Bioelectron.* **99** 653
[11] Jin X P, Huang X G, Tao J, Lin X S and Zhang Q 2010 *IEEE Trans. Nanotechnol.* **9** 134
[12] Zou S, Wang F, Liang R, Xiao L and Hu M 2014 *IEEE Sen. J.* **15** 646
[13] Wang Y, Wang T, Han X, Zhu Y and Wang B 2016 *Opt. Eng.* **55** 027108
[14] Lei J, Ji B and Lin J. 2017 *Plasmonics* **12** 151
[15] Wang G, Lu H, Liu X, Gong Y and Wang L 2011 *Appl. Opt.* **50** 5287
[16] Zheng G, Chen Y, Xu L, Lai M and Liu Y 2013 *Opt. Commun.* **305** 164
[17] Nezhad VF, Abaslou S and Abrishamian M S 2013 *J. Opt.* **15** 055007
[18] Zhang X, Qi Y, Zhou P, Gong H, Hu B and Yan C 2018 *Photonic. Sens.* **8** 367
[19] Rakhshani M R and Mansouri-Birjandi M A 2016 *IEEE. Sens. J.* **16** 3041
[20] Xiang D and Li W2014 *J. Mod. Opt.* **61** 222
[21] Rafiee E, Emami F and Nozhat N 2014 *Opt. Eng.* **53** 123108

[22] Ghorbani S, Dashti MA and Jabbari M 2018 *Laser. Phys.* **28** 066208
[23] Wu T, Liu Y, Yu Z, Peng Y, Shu C and Ye H 2014 *Opt. Express.* **22** 7669
[24] Gai H, Wang J and Tian Q 2007 *Appl. Opt.* **46** 2229
[25] Taflove A and Hagness S C 2005 *Computational Electrodynamics: The Finite-Difference Time-Domain Method* (Boston, MA, USA: Artech house)
[26] Zafar R and Salim M 2015 *IEEE. Sens. J* **15** 6313
[27] Wei PK, Huang Y C, Chieng C C, Tseng F G and Fann W 2005 *Opt. Express.* **13** 10784
[28] Xie YY, Huang Y X, Zhao W L, Xu W H and He C 2015 *IEEE Photonics J.* **7** 1
[29] Zhang Q, Huang X G, Lin X S, Tao J and Jin X P 2009 *Opt. Express.* **17** 7549
[30] Zhao H, Guang X G and Huang J 2008 *Physica E: Low-dimensional Systems and Nanostructures* **40** 3025
[31] Wu T, Liu Y, Yu Z, Peng Y, Shu C and He H 2014 *Opt. Commun.* **323** 44
[32] Kwon S H 2013 *IEEE Photonics J.* **5** 4800107
[33] Tang Y, Zhang Z, Wang R, Hai Z, Xue C, Zhang W and Yan S 2017 *Sensors* **17** 784

Impact of using nano-hydroxyapatite on postoperative hypersensitivity of two bleaching techniques – randomized controlled clinical trial

H Mohamed[1], A Fayed[1,*], D Elkassas[1] and O Hassanien[2]

[1]Faculty of Oral and Dental Medicine, Misr International University, Cairo, Egypt
[2]Faculty of Dentistry, Cairo University, Cairo, Egypt

*E-mail: dr.fayed@live.com

Abstract. To evaluate the impact of using Nano-hydroxyapatite on post-bleaching hypersensitivity after using two different bleaching techniques. Twenty-eight patients were enrolled in this study according to inclusion and exclusion criteria. The patients were randomly assigned into four groups (n=7). G1; bleaching was performed using photo-catalyzed bleaching technique (Philips, zoom), followed by application of Amorphous Calcium phosphate based desensitizing agent (ACP, Relief ACP). G2; bleaching was performed using chemo-catalyzed bleaching technique (Philips, Dash), followed by application of ACP. G3: bleaching was performed using zoom, followed by application of nano-hydroxyapatite based desensitizing agent (Nano p), G4: bleaching was performed using Dash, followed by application of Nano-hydroxyapatite. Post bleaching hypersensitivity was assessed using "Visual analogue scale". Hypersensitivity was evaluated after: bleaching immediately, application of the desensitizing agent immediately, 24 hours, 48 hours and after one week. Cases with severe hypersensitivity were relieved immediately after Nano p application, meanwhile in case of ACP, cases with severe pain were relieved after 24 hours after its application. Statistically significant difference was evident with the photo catalyzed group immediately after Nano p application than before its application. Reduction in hypersensitivity by time was evident in both the chemo and photo catalyzed groups. In addition, it was found that after one week, all pain scores disappeared with all tested groups. Nano hydroxyapatite particles have an immediate (just post application) relief effect of severe pain with both used bleaching systems. However, the post bleaching hypersensitivity diminished completely after 1 week, irrespective of the desensitizing agent used. The application of Nano hydroxyapatite paste after bleaching is a valid method for diminishing severe post bleaching hypersensitivity when applied with either a photo or chemo catalyzed bleaching agents.

1. Introduction

The dream for a more esthetic smile has grown amazingly during the last few decades. Consequently, tooth color is currently believed to be one of the biggest today's patient's interest [1,2]. Tooth bleaching has been recognized as simple, effective and well-accepted method for solving the problem of teeth discoloration. There are three main ways to bleach vital teeth. In office bleaching, which is done under clinician supervision, "night guard" bleaching which is administrated by the clinician for patient use at his home, and commercial bleaching products which are applied by the patient without clinician supervision [3]. In comparison between the three techniques, in –office bleaching has shown many advantages over the other techniques such as the professional control throughout the whole procedure, rapid initial results, reduced treatment time in addition to avoiding any accidental material ingestion and discomfort in wearing bleaching trays. This made the in-office bleaching a pioneer in bleaching treatment [4].

In office whitening technique which was named power bleaching by many authors [5–8], is achieved via the use of high concentrations of hydrogen peroxides. This technique can be classified according to the mode of activation of the bleaching gel into photo catalyzed and chemo catalyzed in-office bleaching systems [2,9]. Post bleaching hypersensitivity is the most prevalent drawback of in-office bleaching technique. Although being the most prevalent drawback, the etiology of post bleaching hypersensitivity is not yet fully understood [10,11].

Currently, in the dental market, many desensitizing agents are available [12]. Recently, Nano technology has paved the way for introduction of materials at Nano scale having the same chemical composition of either organic or inorganic analogue of the hard tooth structure making them biomimetic [13–15]. Consequently appeared the revolutionary role of using the Nano hydroxyapatite. Gopinath et al [14]. and Jena et al. [10], reported the desensitizing effect of Nano hydroxyapatite paste, however their studies were carried out on virgin dentin. They attributed this to its role to act as filler to fill and repair the small holes and pores in enamel. However, the impact of using Nano hydroxyapatite paste on post-bleaching hypersensitivity was not tackled enough in the literature.

The objectives of this study were to compare the post-bleaching hypersensitivity induced by two different bleaching techniques and to evaluate the impact of using Nano-hydroxyapatite on post-bleaching hypersensitivity using two different bleaching techniques.

2. Materials and Methods

2.1. Materials

Material's composition, manufacturer and Lot numbers are presented in table 1. Two types of inoffice bleaching agents were used in this study, Photo catalyzed bleaching agent (Philips ZOOM! Chairside tooth whitening system, Discus Dental, LLC Ontario, CA 91761 USA) containing 25% hydrogen peroxide gel and Chemo catalyzed bleaching agent (Philips Dash, Discus Dental, LLC Los Angeles, CA 90094 USA) containing 30% hydrogen peroxide gel. Two types of desensitizing agents were used in this study, amorphous calcium phosphate-based desensitizing agent. (Relief ACP, Discus Dental) and Nano-hydroxyapatite-based desensitizing agent (Desensitizable Nano P, FGM, Portugal, Brazil).

2.2. Methods

2.2.1. Study design. The study is a unicentered randomized controlled clinical trial. Twenty-eight participants with brown or yellow stains in their six anterior sound maxillary teeth were included in this study. Participants were divided into two equal groups (14 in each group) according to the type of bleaching agent used. A group received a photo catalytic bleaching agent and the other group received a chemo catalytic bleaching agent. Each group was divided into two equal sub-groups, (7 in each sub-group) according to the desensitizing agent used after bleaching. The first sub-group received amorphous calcium phosphate-based (ACP) desensitizing agent after bleaching (control group), while the second subgroup received Nano-hydroxyapatite-based (Nano p) desensitizing agent after bleaching (test group). The trial was approved by the institutional review board (IRB) of Misr International University (MIU- IRB-1516-0011). Reporting of this trial follows the Consolidated Standards of Reporting Trials (CONSORT) guidelines to ensure transparent and complete reporting.

Table 1. The material's composition, manufacturer and Lot number.

Character	Materials	Composition	Manufacturer	Lot no.
In-office photo catalyzed bleaching agent.	Philips ZOOM	25% hydrogen peroxide, water, poloxomer, Glycerin, Glycol, Potassium nitrate, potassium hydroxide, Mentha pipe Rita, Eugenol, and Ferrous gluconate.	Philips, Discus Dental.	2513
In-office chemo catalyzed bleaching agent.	Philips Dash	30% hydrogen peroxide, water, glycerin, hydroxyethyl acrylate, sodium acryloyldimethyltaurate copolymer, etidronic acid, potassium stannate, ammonium hydroxide.	Philips Discus Dental	15267001
Amorphous calcium phosphate-based Desensitizing agent.	Relief ACP	Potassium Nitrate, Sodium Fluoride, water, Poloxamer, 338, Natural Mentha pipe Rita, Calcium Nitrate, Sodium Phosphate, Sodium saccharin.	Philips, Discus Dental	113241
Nanohydroxyapatite based Desensitizing agent.	Nano P paste	Nanometer-sized (20 nm) calcium phosphates (in the form of hydroxyapatite), calcium fluoride, potassium nitrate, distilled water, surfactant, thickener, flavors, preservatives.	FGM	141015

2.2.2. Setting, participants and recruitment. The study took place at the outpatient clinic of Misr International University. Participants were recruited according to the inclusion and exclusion criteria summarised in table 2. All participants signed computerized informed consents after being completely aware of the aim, settings, procedures, benefits and potential side effects of the study. The study information and consent forms were written in Arabic language to be well understood by all the participants. Patients were enrolled into the study only after signing the informed consent.

2.2.3 Initial study visit. After screening for eligibility and prior to the study, eligible patients were identified by the primary investigator (DK). Teeth were re-assessed for confirmation. From 36 identified patients, 28 patients were included. Eligible patients were given a serial number (ID) at this initial study visit, from 1 to 28, according to the order for enrolment into the study. A preoperative base line scoring of hypersensitivity was recorded. All participating patients then passed through scaling and polishing procedures. After scaling and polishing procedures, for ACP desensitizing agent application after bleaching (in the control group), preliminary impression of the maxillary arch was taken using alginate impression material. Impression was poured and soft vacuum trays were obtained. In the photo catalyzed bleaching system (Zoom), the kit was supplied with an "Iso prep retractor" for cheek and lip retraction. Meanwhile, in the chemo catalyzed bleaching agent (Dash), the kit was supplied with a "Cheek retractor" for cheek retraction.

Table 2. Inclusion and exclusion criteria.

Inclusion criteria	Exclusion criteria
Patient related criteria	
➢ Adult patients (age: 18 to 45 years) of both genders. ➢ Good oral and systemic health. ➢ Able to tolerate bleaching procedures. ➢ Willing to sign the informed consent. ➢ Accepts the follow-up period.	➢ Bruxer patients. ➢ Smokers. ➢ Alcoholism. ➢ Patient taking any anti-inflammatory or antioxidant drugs. ➢ Patient who had any previous whitening procedure. ➢ Pregnant or breast feeding women.
Teeth related criteria	
➢ Presence of six anterior sound maxillary teeth. ➢ Teeth having brown or yellow stains. ➢ Pre-operative shade of central incisor should be A2 or darker according to the value – oriented shade guide (Vita easy shade).	➢ Existing anterior crowns or large restorations. ➢ Teeth with exposed roots. ➢ Severe internal tooth discoloration. ➢ Patient with history of teeth hypersensitivity. ➢ Teeth with congenital anomalies or surface defects.

2.2.4 Treatment visit. For both bleaching techniques, isolation was done following the concept of "No pink" according to the manufacturer instructions. The Photo catalyzed bleaching agent (ZOOM) was dispensed out of its syringe and mixed with a brush to be activated. The whitening gel was applied on the upper and lower six anterior teeth. The photo catalyzed bleaching agent (ZOOM) whitening LED accelerator device, was used for 15 minutes (Each patient received three sessions, fifteen minutes each according to the manufacturer instructions). The gel was applied according to the manufacturer's instructions. After the end of the 3 sessions, the Lamp was moved away and the whitening gel was carefully removed with a high volume suction tip and damp gauze. All means of isolation were removed. For the chemo catalytic bleaching system (Dash group), same isolation steps were done as the photo catalyzed bleaching system (Zoom). A whitening accelerator liquid (from the kit contents) was applied on the teeth using a cotton swab before the application of the whitening gel which accelerated the whitening chemical reaction according to the manufacturer recommended protocol. The whitening gel was then injected on the teeth surfaces using a flocked tip. The gel was applied for 3 sessions each is of 15 minutes according to the manufacturer instructions. After the 3 sessions were done, all isolation means were removed. For the control group, who received the amorphous calcium phosphate-based desensitizing agent (ACP), the ACP was placed into a vacuum tray and set on patient's teeth for 30 minutes according to the manufacturer instructions immediately after bleaching. For the test group, who received the Nano-hydroxyapatite- based desensitizing agent (Nano p), it was placed also according to the manufacturer instructions immediately after bleaching. A pea-sized application was applied and rubbed on enamel for 10 seconds using a rubber cup mounted on a low speed hand piece. Afterwards, the material was left undisturbed for five minutes then it was removed by a slight dry cotton roll. (It's worth mentioning that the lower arch was submitted to dental bleaching but was not used for data assessment). Hypersensitivity was assessed immediately after bleaching procedure [3,16,17].

2.2.5 Data collection and follow-up examination. Patients were recalled for follow-up after bleaching sessions by 24 hours, 48 hours and 1 week to measure hypersensitivity using the visual analogue scale which was a qualitative verbal scale (no pain, mild, moderate, severe). The degree of sensitivity was evaluated using light air jet over the labial surface of the teeth near the root area for one second at distance from two to five mm [4,6]. Assessment was performed by one blinded, calibrated outcome assessor for each tooth who had not been involved in the treatment at all visits.

2.2.6 Sequence generation, allocation and blinding. In this clinical trial, quasi randomization was done by tossing a coin to choose the amorphous calcium phosphate-based desensitizing agent (ACP) and Nano-hydroxyapatite- based desensitizing agent (Nano p).

2.2.7. Statistical analysis. The obtained data of this clinical study was collected and tabulated for the statistical analysis. Qualitative data for hypersensitivity were presented as frequencies and percentages. Fisher's Exact test was used to compare between the techniques. Friedman's test was used to study the effect of time. Dunn's test was used for pair-wise comparisons when Friedman's test is significant. The significance level was set at $P \leq 0.05$. Statistical analysis was performed with IBM® SPSS® Statistics Version 20 for Windows.

3. Results

From table 3 and figure 1, the following data were analyzed, in the photo catalyzed bleaching group (Zoom) who received Nano p, before application of Nano p, cases with severe pain score presented 57.1%, while cases with moderate pain score presented 42.9%, with 0% for both scores of mild and no pain. Immediately after application, cases with severe pain score presented 0%, while cases with mild and moderate pain scores equally presented 42.9%. Meanwhile, cases with no pain score presented 14.3%. At day one, cases with moderate and severe pain scores presented 0%, while cases with mild pain score presented 14.3%, meanwhile cases with no pain score presented 85.7%. At day two and seven, all pain scores presented 0% except no pain score which presented 100% of cases indicating disappearance of pain. Before Nano p application, there was a statistically significant difference in the severity of pain, but immediately after Nano particles application there was no statistically significant difference till day seven indicating the gradual disappearance of pain. In the photo catalyzed bleaching group (Zoom) who received ACP, before application of ACP, cases with severe pain score presented 57%, while cases with moderate pain score presented 42%, with 0% for both scores of mild and no pain. Immediately after application, cases with severe pain score presented 57%, while cases with moderate pain score presented 28.6%, cases with mild pain score presented 14.3%.

Meanwhile, cases with no pain score presented 0%. At day one, cases with severe pain score presented 0%, while cases with moderate and mild pain scores equally presented 28.6%, meanwhile cases with no pain score presented 42.9%. At day two, cases with severe and moderate pain scores equally presented 0%, while cases with mild pain score presented 14.3%, meanwhile, cases with no pain score presented 85.7%. At day seven, all pain scores presented 0% except no pain score which presented 100% indicating disappearance of pain. Before and immediately after ACP application, there was no statistically significant difference in degree of pain severity, no statistically significant difference between day two and day seven but they were statistically significant from before and immediately after application of ACP, considering day one as a midway in significance between immediately after application and day two. In the chemo catalyzed bleaching group (Dash) who received Nano p, before Nano p application, cases with moderate and severe pain scores equally presented 42%, while cases with mild pain scores presented 0%, meanwhile cases with no pain scores presented 14%. Immediately after Nano p application, cases with severe pain score presented 0%, while cases with mild and moderate pain scores equally presented 42.9%, meanwhile cases with no pain score presented 14.3%. At day one, cases with severe and moderate pain scores equally presented 0%, while cases with mild pain score presented 14.3%, meanwhile, cases with no pain score presented 85.7%.

Table 3. The frequencies, percentages and results of Friedman's test for comparison between prevalence of hypersensitivity at different follow up periods.

Group	Hypersensitivity	Before NP		Immediately after		D1		D2		D7		*P*-value
		n	%	n	%	n	%	n	%	n	%	
	No pain	0	0.0	1	14.3	6	85.7	7	100.0	7	100.0	<0.001*
	Mild	0	0.0	3	42.9	1	14.3	0	0.0	0	0.0	
Zoom + NP	Moderate	3	42.9 a	3	42.9 b	0	0.0 b	0	0.0 b	0	0.0 b	
	Severe	4	57.1	0	0.0	0	0.0	0	0.0	0	0.0	
Zoom	No pain	0	0.0 a	0	0.0 a	3	42.9 ab	6	85.7 b	7	100.0 b	<0.001*
+ ACP	Mild	0	0.0	1	14.3	2	28.6	1	14.3	0	0.0	
	Moderate	3	42	2	28.6	2	28.6	0	0.0	0	0.0	
	Severe	4	57	4	57.1	0	0.0	0	0.0	0	0.0	
	No pain	1	14.3	1	14.3	6	85.7	7	100.0	7	100.0	<0.001*
	Mild	0	0.0	3	42.9	1	14.3	0	0.0	0	0.0	
Dash + NP	Moderate	3	42.9 a	3	42.9 a	0	0.0 ab	0	0.0 b	0	0.0 b	
	Severe	3	42.9	0	0.0	0	0.0	0	0.0	0	0.0	
	No pain	1	14	0	0.0	4	57.1	6	85.7	7	100.0	<0.001*
	Mild	0	0.0	2	28.6	2	28.6	1	14.3	0	0.0	
Dash + ACP	Moderate	3	42 a	2	28.6 a	1	14.3 ab	0	0.0 b	0	0.0 bc	
	Severe	3	42	3	42.9	0	0.0	0	0.0	0	0.0	

Superscripts with different letters are statistically significant. P≤0.05.

At day two and seven, all pain scores presented 0% except no pain score which presented 100% of cases indicating disappearance of pain. No statistically significant difference was found at all time intervals till day one, day two and seven showed a statistically significant difference from other time interval indicating gradual disappearance of pain. In the chemo catalyzed bleaching group (Dash) who received ACP, before application of ACP, cases with moderate and severe pain scores equally presented 42%, while cases with mild pain scores presented 0%, meanwhile cases with no pain scores presented 14%. Immediately after ACP application, cases with severe pain score presented 42.9%, while cases with mild and moderate pain scores equally presented 28.6%, meanwhile, cases with no pain score presented 0%. At day one, cases with severe pain score presented 0%, while cases with moderate pain score presented 14.3%, cases with mild pain presented 28.6%, meanwhile, cases with no pain score presented 57.1%. At day two, cases with moderate and severe pain scores equally presented 0%, while cases with mild pain score presented 14.3%, meanwhile, cases with no pain score presented 85.7%. At day seven, all pain scores presented 0% except no pain score which presented

100% indicating the gradual disappearance of pain. No statistically significant change in pain severity between before ACP application and immediately after application, day one is considered a mid-way between the degree of significance which occurred in the time interval between immediately after application and at day two. Day two was statistically significant from all previous time intervals.

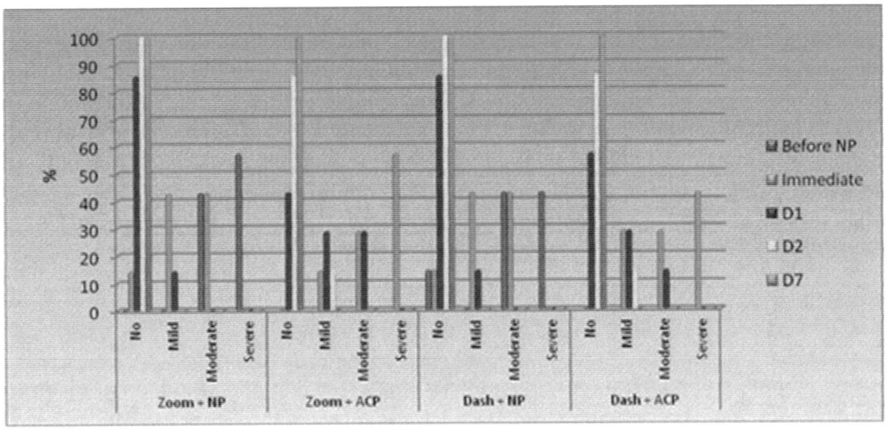

Figure 1. Bar chart representing prevalence of hypersensitivity at different follow up periods.

4. Discussion

Brighter smile became the latest trend in dental fashion. Thus, bleaching treatment is appealing as this type of treatment is in parallel with the physiology of conservative and minimal invasive dentistry [4]. Dental bleaching can be categorized into three approaches, in-office bleaching with high peroxide concentrations but for short time, at home bleaching with low peroxide concentration but for a long time and over the counter whitening. In the present study, two in-office bleaching products were evaluated in terms of post-bleaching hypersensitivity. A photo catalyzed system (Philips ZOOM), with blue LED light activated technology, is considered a cost effective alternative to laser and halogens as it requires less energy to generate light with less heat production [3]. The other was a chemo catalyzed alternative (Philips Dash), which is claimed by the manufacturer to be formulated for stability and efficacy, eliminating the need for syringe to syringe mixing or refrigeration. These two systems are the products of the same company. The selection of light or chemical activated products might help to reach a proper decision in different clinical situations. Yet remains the choice difficult, as it may be related to the patient's choice, financial status and the cost effectiveness of both tested techniques.

Post bleaching hypersensitivity is a major drawback following the whitening process which is time dependent as it is an instant problem that disappears by time, however it might force the operator to terminate the whole whitening process. In an attempt to overcome this problem, most in-office bleaching kits now are supplied with desensitizing agents having dual effects of remineralization and desensitization. In the present study, both bleaching kits were supplied with Relief ACP desensitizing agent, which is based on the amorphous calcium phosphate technology. However, the Nano-p is gaining wide acceptance as being biomimetic with nano-sized particles similar to enamel building blocks and paved the way for manufacturers to use it as a remineralizing and desensitizing agent [15,18]. Some studies highlighted its desensitizing effect, but unfortunately, they were done on exposed virgin dentin [14,16] which is not the condition in case of dental bleaching. Additional numerous trials have proven its remineralizing effect [15,19]. Thus, this study was carried out to evaluate clinically post bleaching hypersensitivity of two different bleaching techniques with their corresponding desensitizing agent (ACP) and Nanohydroxyapatite paste.

The selection of patients in this study was done according to inclusion and exclusion criteria [20,21]. In addition, the selection was limited to those having obvious stains A2 or darker [22]. The chief complain of the selected patients was teeth discolorations which hinders their esthetic appearance, self-confidence and esteem [3,11,12]. In this study, "quasi randomization" was done for patient selection using the coin Tossing method (heads and tails) to decide which patients will be in the tested groups [22].

All patients went through primary steps of scaling and polishing before the actual bleaching treatment to ensure a clean, debris free surface that may interfere with pre-operative shade of the teeth. In both bleaching techniques, all patients received three whitening sessions each is of 15 minutes with a total time equals 45 minutes according to the manufacturer instructions. Hypersensitivity was assessed immediately after whitening procedure to monitor the post bleaching sensitivity of each system before applying the desensitizing agent then immediately after application of the desensitizing agent, after 24 hours, after 48 hours and after one week. It was assessed with a light air jet over the labial surface near the root area for one second at a distance from two to five mm to monitor the effect of different desensitizing agents [6]. Being an experience, pain is best assessed by subjective evaluation. This evaluation was done using the visual analogue scale by a qualitative verbal scale (absent, mild, moderate, severe) [3,16,23]. Results of the present study revealed that both bleaching techniques induced post bleaching hypersensitivity. No statistically significant difference was found between both groups. This observation is in line with Mondelli et al. [24], who found no difference in sensitivity between light activated and chemical activated systems after 24 hours. Kossatz et al. [25] also compared the post bleaching sensitivity of 15% hydrogen peroxide with halogen activated 25% hydrogen peroxide and found no difference in sensitivity.

Although the etiology of post bleaching hypersensitivity is not fully understood [3,6,21], hypersensitivity as a result of tooth whitening occur despite the lack of dentin exposure. Different theories are proposed, in accordance with De paula et al. [21], post bleaching hypersensitivity could be attributed to direct activation of the intra dental nerves via chemo sensitive channels which are sensitive to a variety of oxidizing products, including hydrogen peroxide. This assumption was further clarified by Thiensen et al. [26], who claimed that sensitivity after whitening occurs as a result of the functional properties of a chemo-sensitive ion channel called "TRPA1" (transient receptor potential ankyrin 1). This ion channel is associated with the pain caused by oxidants since the afferent fibers of the pulp contain this ion channel so it may be directly involved in the pain. Furthermore, this could explain why the pain after bleaching differs from the pain of normal tooth sensitivity as reported by patients who described the pain (which is of very short duration and variable frequencies) as sensations that resembled "needles " or "Shock like" and that were not triggered by thermal stimuli. Many studies reported the presence of acute inflammatory reaction in the dental pulp of human bleached teeth [27]. They suggested that some released mediators of inflammatory reactions such as bradykinin or substance P maybe involved in the etiology of post bleaching hypersensitivity. Basting et al. [28], assumed that the increase in enamel and dentin permeability have paved the way for easy passage of peroxide molecules, triggering the inflammatory reaction. According to Markwotiz et al. [27], hypersensitivity during bleaching is related to the hydrodynamic theory. Free radicals from hydrogen peroxide diffuse through the enamel and dentin, promoting the movement of fluid present in the dentinal tubules, stimulating nerves, and generating sensitivity. Huynh et al. [29] suggested that deep peroxide penetration into the pulp cells may cause pulp inflammation, which in turn leads to the release of cell derived factors such as adenosine triphosphate (ATP), neuropeptides and prostaglandins which excite pulpal nociceptors causing the post bleaching hypersensitivity.

The use of Nano p paste caused immediate significant reduction of post bleaching hypersensitivity with the photo catalyzed bleaching system and diminished the prevalence of severe pain with the chemo catalytic system. This finding is in line with Browning et al. [16] and Vano et al. [30], this could be assumed to three different approaches. First, this product contains potassium nitrate which is according to the neural theory, potassium nitrate is claimed to diminish the excitability of nerve fibers in the dental pulp, and in addition it's assumed to be responsible for the sustained depolarization of sensory nerves leading an "axonal accommodation" phenomenon [31-33]. Potassium nitrate is claimed to have a calming effect on the nerves by preventing it from repolarizing after it was depolarized in a pain cycle. It raises the extracellular potassium ion concentration to prevent any action potential by axonal accommodation. Meanwhile, potassium ion concentration above 8 mM to 16 mM (0.08-0.16% as KNO_3) around the axons is needed to sustain nerve depolarization [34,35]. Second, these products which contain Nano hydroxyapatite particles have a high surface area and strong affinity to enamel by strong adsorption [33,36,37]. In case of the photo catalyzed system with minimal surface alterations, the particle size of the Nano hydroxyapatite paste (20-40 nm) could be relatively larger than the surface micro and macro gaps, so it might have only covered the surface, resulting in instant fixation of enamel surface density and covering the surface gaps thus blocking any pathway for any stimuli that might induce post bleaching hypersensitivity according to the hydrodynamic theory. Third, this product contains calcium fluoride crystals which might have induced granular precipitates to block the dentinal tubules and fill the micro porosities induced as a result of enamel alternatives.

Although that the constitutional composition of ACP is in the same line with that of the Nano p, but it took the ACP 24 hours to act on relieving severe hypersensitivity pain cases. It contains potassium nitrate which acts on nerve depolarization. It also contains calcium, phosphate and fluoride compounds which enable them to be in active form once contacted saliva. [38,39] ACP might also induce enamel remineralization and block the dentinal tubules in an attempt to diminish the effect of any stimuli according to the hydrodynamic theory. However, it could be assumed that the form of supply and application might have played an important role as the Nano p is applied in a paste form which is to be scrubbed and left undisturbed on the tooth surface for five minutes according to the manufacturer instructions. This scrubbing action for the paste increased its" thixotropic phenomenon", which might have enhanced deeper penetration of various items released among which is the potassium nitrate and calcium fluoride thus acting faster in relieving severe hypersensitive pain efficiently. It's worth mentioning that post bleaching hypersensitivity diminished by time, irrespective of the desensitizing agent used. This might denote that pain alleviation is a time- dependent process.

5. Conclusion

Under the parameters of the given study, it was concluded that the used Nano hydroxyapatite particles showed immediate relief of severe pain in 100 % of the patients with both used bleaching systems, however, the post bleaching hypersensitivity diminished completely in 100 % of the patients after 1 week, irrespective of the desensitizing agent used.

Clinical relevance

The application of Nano hydroxyapatite paste after bleaching is a valid method for diminishing severe post bleaching hypersensitivity when applied with either a photo or chemo catalyzed bleaching agents.

References

[1] da Costa Soares M U S, Araujo N C, Borges B C D, Sales W D S and Sobral A P V 2013 *J. Acta Odontol. Scand.* **71** 343
[2] Machado L S, de Oliveira F G, Rocha E P, dos Santos P H, Briso A L F and Sundefeld M L 2013 *Int. J. Periodont. Rest. Dent.* **33** 209
[3] Pale M, Mayoral J R, Lopis J, Valles M, Basilio J and Roig M 2014 *J. Odontl.* **102** 203
[4] Sabbagh J, El Masri L, El Hajj M and Khafaja S 2016 *Int. J. Oral Health* Dent. **8** 153
[5] Alqahtani M Q 2014 *Saudi Dent. J.* **26** 33
[6] Klaric E, Par M, Profeta I, Marović D and Tarle Z 2011 *J. Acta. Stomatol. Croat.* **45** 100
[7] Camargo S E A, Cardoso P E, Valera M C, de Araujo M A M and Kojima A N 2009 *Eur. J. Esthet. Dent.* **4** 82
[8] Calatayud J O, Calatayud C O, Zaccagnini A O and Box M J C 2010 *Eur. J. Esthet. Dent.* **5** 216
[9] Kugel G, Ferreira S, Sharma S, Barker M L and Gerlach R W 2009 *J. Esthet. Restor. Dent.* **21** 336

[10] Jena A and Shashirekha G 2015 *J. Conserv. Dent.* **18** 389
[11] Majeed A, Farooq I, Grobler S R and Rossouw R J 2015 *J. Coll. Physicians Surg. Pak.* **25** 891
[12] Goldberg M, Grootveld M and Lynch E 2010 *J. Clin. Oral Invest.* **14** 1
[13] Jin J, Xu X, Lai G and Kunzelmann K 2013 *Eur. J. Oral Sci.* **121** 382
[14] Gopinath N M, John J, Nagappan N, Prabhu S and Kumar E S 2015 *J. Int. oral Heath* **7** 118
[15] Li L, Pan H, Tao J, Xu X, Mao C and Tang R 2008 *J. Mater. Chem.* **40** 79
[16] Browning W D, Cho S D and Deschepper E J 2012 *J. Esthet. Restor. Dent.* **24** 268
[17] Vano M, Derchi G, Barone A and Covani U 2014 *J. Quintessence Int.* **45** 703
[18] Ajami S, Pakshir H R and Babanouri N 2016 *J. Prog. Orthod.* **17** 1
[19] Vyavhare S, Sharma D S and Kulkarni V K 2015 *J. Clin. Pediatr. Dent.* **39** 149
[20] Martos J and Kinalski M A 2014 *J. Restor. Dent.* **2** 149
[21] De Paula E A, Nava J A, Rosso C, Benazzi C M, Fernandes K T and Kossatz S 2015 *J. Dent.* **43** 424
[22] Zhao K, Zong L, Zhang Q and Att W 2013 *J. Quintessence Int.* **44** 601
[23] Cartagena A F, Parreiras S O, Loguercio A D, Reis A and Campanha N H 2015 *Braz. J. Oral Sci.* **29** 1
[24] Mondelli R F L, Azevedo J F D, Francisconi P A S and Ishikiriama S K 2009 *Eur. J. Esthet. Dent.* **4** 396
[25] Kossatz S, Dalanhol A P, Cunha T, Loguercio A and Reis A 2011 *J. Oper. Dent.* **36** 251
[26] Thiesen C H, Rodrigues F R, Prates L H M and Sartori N 2013 *Braz. J. Oral Sci.* **27** 517
[27] Markowitz K 2010 *J. Med. Hypoth.* **74** 835
[28] Basting R T, Amaral F L B, Franca F M G and Florio F M 2012 *J. Oper. Dent.* **37** 464
[29] Huynh M P and Yagiela J A 2003 *J Calif. Dent. Assoc.* **31** 419
[30] Vano M, Derchi G, Barone A, Genovesi A and Covani U 2015 *Int. J. Dent. Hyg.* **13** 301
[31] Frechoso S C, Menendez M, Guisasola C, Arregui I, Tejerina J M and Sicilia A 2003 *J. Clin. Periodontol.* **30** 315
[32] Nanjundasetty J K and Ashrafulla M 2016 *J. Conserv. Dent.* **19** 207
[33] Coceska E, Gjorgievska E, Coleman N J, Gabric D, Slipper I J and Stevanovic M 2016 *J. Microsc.* **262** 232
[34] Seong J, Macdonald E, Newcombe R G, Davies M, Jones S B and Johnson S 2013 *J. Clin. Oral Invest.* **17** 195
[35] Tawfik.S, Khairy. M, El Baz. M and El Korashy. M 2019 *F1000 Research* **8** 1762
[36] El Baz M and Abo El Enien K 2018 *Future Dent. J. (FDJ)* **4** 262
[37] Severa E., Budimirb Z., Cerovacb M., Stambukb M., Parc M., Vranicd D. and Tarlea Z 2018 *J. Acta Odontol. Scand.* **76** 30
[38] Clark D and Levin L 2018 *J. Quintessence Int.* **49** 147
[39] Mehta D, Jyothi S, Moogi P, Finger W and Sasaki K 2018 *J. Esthet. Restor. Dent.* **30** 254

Synthesis, identification, antibacterial, and dyeing applications of complexes of hexadentate (N_4O_2 donor) Schiff base ligands derived from curcumin with some transition and non–transition metal cations

A M Ali* and T H Al-Noor*

Department of Chemistry.Ibn -Al-Haithem College of Education for Pure Science, Baghdad University, Iraq

*E-mail: mahmoodamzali8@gmail.com (AMA); drtaghreed2@gmail.com (THA)

Abstract. Schiff base ligand (H_2CANPT) was prepared by two steps: first, by the condensation of curcumin with 4-amino antipyrin produces4,4'-(((1E,3Z,5Z,6E)-1,7-bis(4-hydroxy-3-methoxyphenyl)hepta-1,6-diene-3,5-diylidene)bis(azanylylidene))bis(1,5-dimethyl-2-phenyl-1,2-dihydro-3H-pyrazol-3-one) (CANP). Second, by the condensation of (CANP) with L-tyrosine produces2,2'-(((3Z,3'Z)-(((1E,3Z,5Z,6E)-1,7-bis(4-hydroxy-3-methoxyphenyl)hepta 1,6-diene-3,5-diylidene)bis(azanylylidene))bis(1,5-dimethyl-2-phenyl-1,2-dihydro-3-H-pyrazole-4-yl-3-ylidene))bis(azanylylidene))bis(3-(4-hydroxyphenyl)propanoic acid) (H_2CANPT). The resulted Schiff comported as hexadentate coordinated with (N_4O_2) atoms, then it was treated with some transition and non-transaction metal ions, such as (Mn(II), Fe(III),Co(II), Ni(II) ,Cu(II), Cd(II), Hg(II) chloride and Ag(I) as nitrate to synthesize complexes. All the prepared compounds were identified using FT-IR, UV-visible, elemental analysis (C.H.N), ^{1}H and^{13}C-NMR, molar conductivity and magnetic susceptibility. Schiff base and its complexes were tested against two types of bacteria (*Staphylococcus aureus* and *Escherichia Coli*). Besides, they were tested as dyes on cotton fibres.

1. Introduction

Schiff bases are organic compounds carry azomethine (C=N) group, known for the first time by Hugo Schiff in1864 [1,2]. Schiff bases were prepared by several methods, such as Reflux, Fusion, Microwave and Grading [3,4]. They have many applications in organic chemistry field [5]. Azomethine compounds and copper complexes are Schiff bases involved in the manufacturing of ink and dyes [6,7]. Schiff base ligands form colored complexes with many metal ions; thus, they are selective and sensitive reagents compounds for estimating metals [8,9]. Many derivatives of 4-amino antipyrine estimated in medical fields such as analgesic, anti-inflammatory, antimicrobial and anticancer activities [10,11]. Schiff bases complex of amino acids and metal ions play an important role in industries and in pharmaceutical and biological fields [12]. This study aimed to synthesize and identify the behaviour of Schiff bases that have four azomethine groups from curcumin, 4-amino antipyrine and L-Tyrosine amino acid, to test the efficient biological activity of the ligand and its complexes against two types of bacteria and finally, to investigate the usage of these Schiff bases and their complexes in dyeing process.

2. Experimental

2.1. Chemicals
All chemical compounds used with no need to purification and they were purchased from (BDH) company.

2.2. Material and Methods
The melting point was determined by using (MPA160 Digi Melt) melting point apparatus. FT-IR spectra measurements were determined by Fourier transform infrared spectra (4000-400 cm^{-1}) as KBr pellets on (Shimadzu FT-IR 8400S) spectrophotometer. Ultraviolet-visible (UV) spectra were recorded by using a solvent (DMSO) at 1x 10^{-3}M for the Schiff base (ligand) and its complexes by device of type (U.V 160A Shimadzu apparatus). Elemental analyses (C, H, N) were performed by (Eager300 for EA1112 Thermo Finnegan C.H.N.S 2400) elemental analyzer. The Λ_m measurements were recorded for complexes with at 10^{-3}mol.L^{-1} in dimethyl sulfoxide by using digital conductivity Series (Ino.Lab.720). Magnetic susceptibility of all the complexes was measured by (Balance Johnson Mattey). ^{1}H and ^{13}C-NMR spectra were recorded using dimethyl sulfoxide (DMSO) -d^6 Solution (Inova 500).

2.3 . Synthesis of Schiff base ligand (H_2CANPT)

*2.3.1. Step I (CANP)***.** A weight of 0.368 gm (1mmol) of curcumin was dissolved in 20 mL ethanol, 4 drops of glacial acetic acid were added and 406 gm (2 mmol) of 4-amino antipyrine dissolved in 20 mL of ethanol was gradually added. This mixture was refluxed with stirring at 70 °C , for 5 hours, The reaction was followed up by TLC technique. A yellow precipitate was separated and recrystallized from hot ethanol, as shown in scheme 1.

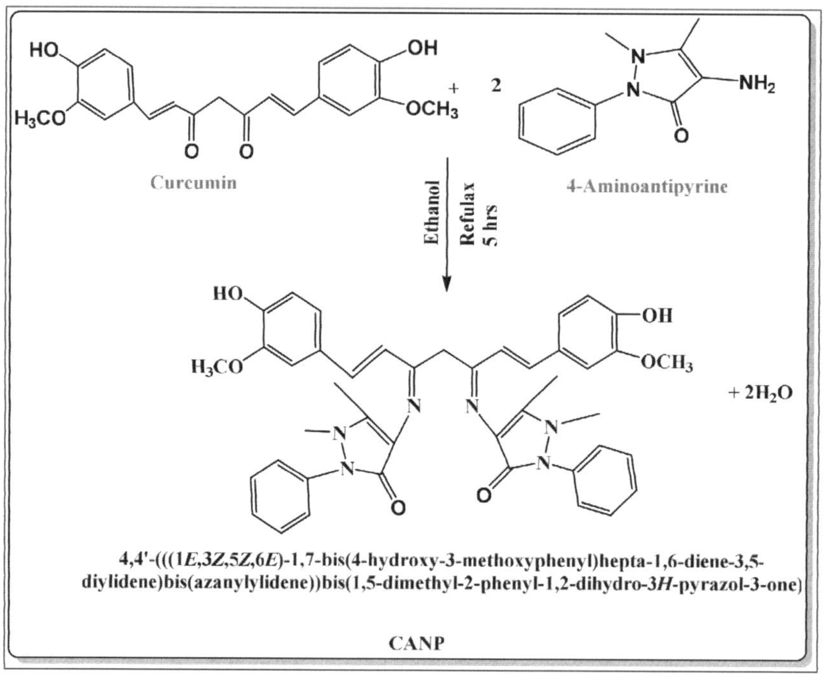

Scheme 1. A diagram of synthesis of Schiff base (CANP).

2.3.2. Step II ligand (H_2CANPT). A weight of 0.730 gm (1 mmol) of (CANP) was dissolved in 20 mL ethanol and mixed with 0.362 gm (2 mmol) of L-Tyrosine (dissolved in 20 mL ethanol in a base medium). This mixed solution was refluxed with stirring at 70°C for 12 hours and the reaction followed up by TLC technique until a brown-yellow precipitate formed, then separated, washed several times with hot ethanol and finally recrystallizing from hot ethanol under room temperature. The yield: 78%. Elemental Analysis: Found (Calc.): C, 68.78(67.92); H, 5.68 (5.92); N, 10.52 (10.60); O, 15.02 , as shown in scheme 2.

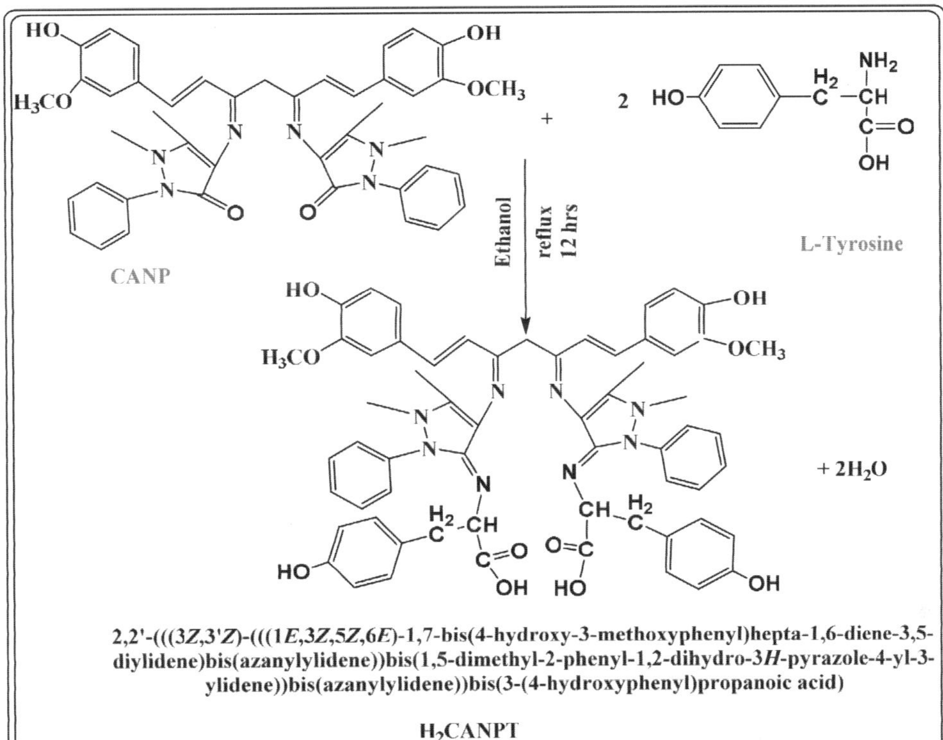

Scheme 2. A schematic diagram for the synthesis of Schiff base (H_2CANPT).

2.4. Preparation of Schiff base ligand (H_2CANPT) with metal complexes
A weight of 1.05 gm (1mmol) of the Schiff base (ligand)) was dissolved in 20 mL ethanol and then 0.112 gm (2 mmol) KOH was added. The mixture was gradually added to (1 mmol) of $MnCl_2.4H_2O$, $FeCl_3$, $CoCl_2.6H_2O$,$NiCl_2.6H_2O$, $CuCl_2.2H_2O$, $AgNO_3$, $CdCl_2.2H_2O$ and $HgCl_2$ (dissolved in 10 mL ethanol) with constant stirring for 30 minutes. The resulted precipitates were separated, washed several times with a mixture of solvent (distilled water: ethanol 50%) and then recrystallized from hot ethanol. Scheme 3 showed the suggested geometery of the recrystalled structures.

3. Results and Discussion

Generally, all complexes were prepared by reacting to the respective metal salts with the Schiff base (H_2CANPT) using 1:1 mole ratio (metal salts: H_2CANPT). All complexes were intensively coloured solid compounds. The analytical data in table 1 agreed with the suggested formulas of 1:1 metal : (H_2CANPT) molar ratio. The compounds were insoluble in water, but fairly soluble in some common solvents, such as ethanol, methanol, acetone, hydrochloric acid, dimethylformamide, chloroform and dimethyl sulfoxide. Most of the prepared compounds have a relatively high melting point and high stability at room temperature. Table 1 shows some the physical and chemical properties. The molar conductance of 1x10^{-3} M of the complexes (dissolved in DMSO) in a range 5.4 to 16 Ω^{-1} cm^2 mol^{-1} indicated that these complexes were nonelectrolytes; except Fe(III) and Ag(I) complexes that are electrolytes at 32.2 and 34.2 Ω^{-1} cm^2 mol^{-1}, respectively [13].

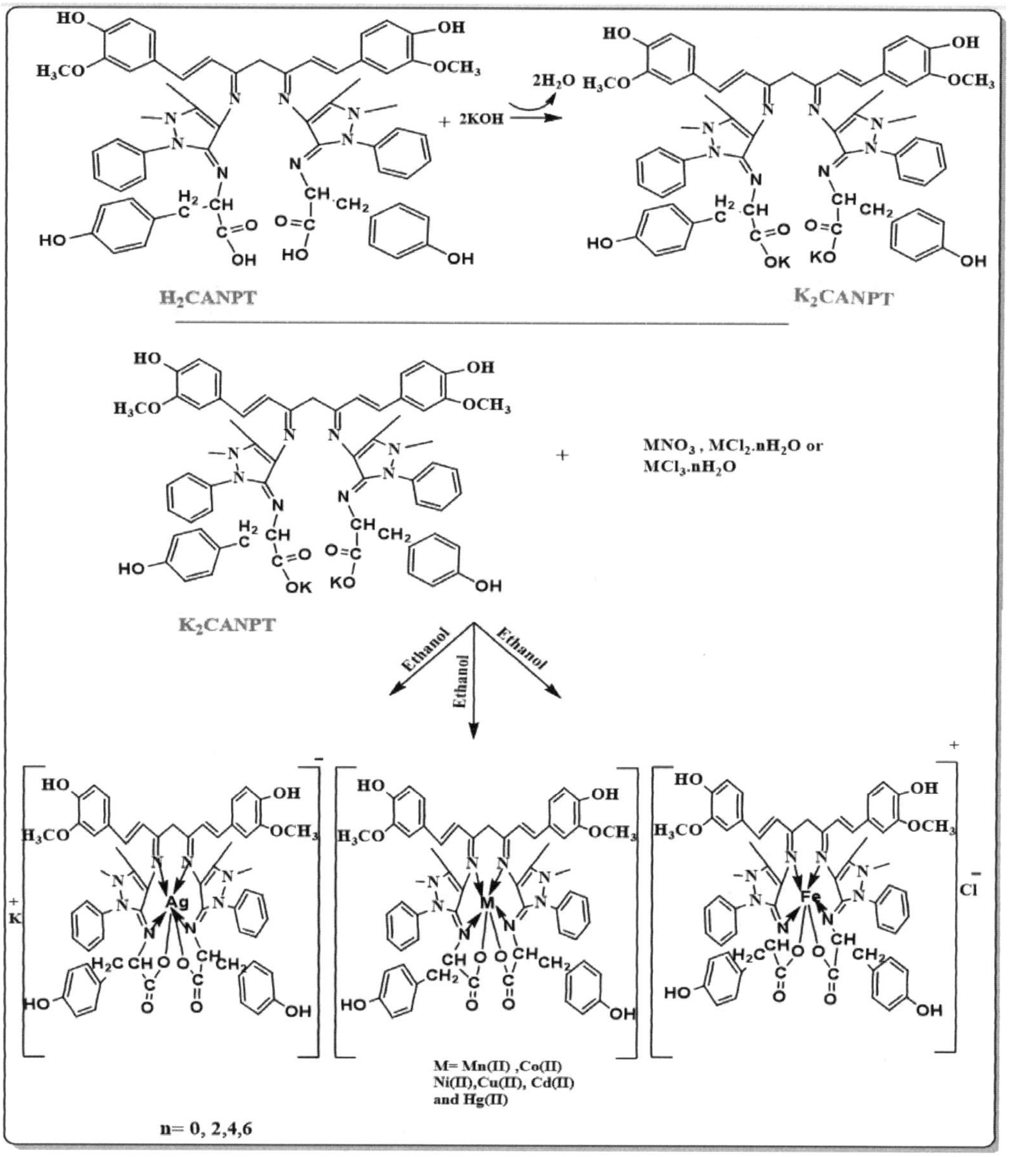

Scheme 3. The proposed geometry structures of the Schiff base(H_2CANPT) complexes.

Table 1. The physical and chemical properties of Schiff base and its complexes.

Compounds molecular formula	M.Wt	M.P °C	Color	Yield%	Λ m $\Omega^{-1}.cm^2.mole^{-1}$	Metal% (exp)
H_2CANPT (Schiff base) $C_{61}H_{60}N_8O_{10}$	1065.20	118-120	Brown-yellow	78	------	----
[Mn (CANPT)] $C_{61}H_{58}N8O_{10}Mn$	1118.13	196-198	Orange	78	16	4.91(4.99)
[Fe(CANPT)] Cl $C_{61}H_{58}N8O_{10}FeCl$	1154.54	162-163	black	90	32.2	4.83(4.87)
[Co(CANPT)] $C_{61}H_{58}N8O_{10}Co$	1122.13	190-192	Pale yellow	70	9.5	5.25(4.88)
[Ni(CANPT)] $C_{61}H_{58}N8O_{10}Ni$	1122.13	151-152	Pale yellow	75	6.3	5.23(4.73)
[Cu (CANPT)] $C_{61}H_{58}N8O_{10}Cu$	1126.74	170-172	Brown	84	10.4	5.63(5.63)
K [Ag(CANPT)] K $C_{61}H_{58}N8O_{10}Ag$	1210.15	190-192	Deep brown	93	34.2	8.91(8.30)
[Cd (CANPT)] $C_{61}H_{58}N8O_{10}Cd$	1175.41	216-217	Brown	89	9.1	9.56(9.73)
[Hg(CANPT)] $C_{61}H_{58}N8O_{10}Hg$	1263.79	129-130	Reddish brown	88	5.4	15.87(15.60)

3.1. FT-IR spectra

The synthesized ligand (Schiff base) and its complexes showed distinctive functional groups and bands, such as -OH (hydroxyl group), C=N(azomethine groups),COO-(carboxylic group) and metal bands (M-N, M-O) , where the frequencies of absorbing changes were identfied using FT-IR spectroscopy. Two strong bands at (3420 cm^{-1}) and (1610-1631 cm^{-1}) belong to –OH and C=N, respectively were observed [11,14]. The C=N band was detected in all complexes to be shifted to lower and higher frequencies according to the proportions, such as involvement in the process of coordination with metal ions through the nitrogen atom of (azomethine) band[15]. The asymmetric stretching of $v_{as}COO^-$ was assigned at 1589 cm^{-1} while the symmetric stretching v_s COO^- diagnosed at 1417cm^{-1} [16,17]. The difference between frequencies Δ v= (v $_{asymmetric}$ (COO^-) - v $_{symmetric}$ (COO^-)) of free ligand was at 172 cm^{-1}. When the ligand coordinated with a metal ion at its carboxyl group, the Δ v observed between 213-259 cm^{-1}, which supports the monodentate coordination nature of (COO^-) carboxylate anion [16]. As well as, a band of v (O-CH_3) appeared at 1246 cm^{-1} [18], but the band (1456 cm^{-1}) shifted back to (C=C) aromatic [18]. New bands assigned at 574-586 cm^{-1} and 416-445cm^{-1} for stretching frequencies of M-O and M-N, respectively[12,14]. Table 2 includes the most important FT-IR data in this study.

3.2 . UV-vis spectra

The electronic spectra of the ligand and its complexes were collected at $1x10^{-3}$ mol/L in DMSO at room temperature (table 3). The electronic spectrum of the Schiff base exhibited three bands. The first absorption band in UV-region appeared at 275 nm (36363 cm^{-1}) which could be appointed to π - π^* of the benzene ring and double bonds in the ligand structure [12,18]. The second absorption band appeared at 344nm (29069 cm^{-1}) could be appointed to the n– π^* transitions in azomethine (C=N) groups[12,14]. The third band at 422 nm (23696 cm^{-1}) could be allotted to the n– π^* transitions ,which possessed non-shared electronic pairs of groups (hydroxyl, amine, and methoxy) [15,18]. Herein, shiffting of the absorption bands to higher wavelength indicate a successful preparation of the complexes and the appearance of new bands could refer to (d-d) orbital transitions.

Table 2. The main FT-IR spectra (cm^{-1}) of compounds under stufy.

Compound	ν O-H	ν C=N	ν as COO	ν s COO	Δν as-s	ν M-O	ν M-N
H_2CANPT	3420	1631 1610	1589	1417	172	-------	--------
[Mn(CANPT)]	3408	1612 1585	1570	1325	245	580	422
[Fe(CANPT)] Cl	3444	1612 1593	1565	1328	237	574	430
[Co(CANPT)]	3441	1629 1587	1568	1355	213	586	418
[Ni(CANPT)]	3402	1643 1633	1544	1321	223	586	416
[Cu(CANPT)]	3400	1643 1633	1554	1305	249	586	418
k[Ag(CANPT)]	3398	1643 1633	1564	1305	259	586	445
[Cd(CANPT)]	3442	1612 1587	1570	1328	242	574	428
[Hg(CANPT)]	3400	1625 1587	1548	1300	248	586	418

Four bands of manganese ion complex [Mn(CANPT)] were observed at 276 nm (36231 cm^{-1}), 345nm (28985 cm^{-1}), 397nm (25188 cm^{-1}) and 817nm (12239cm^{-1}), which could be addressed as ligand field (L.F), charge transfer (C.T) and $^6A_1g \rightarrow ^4T_1g(G)$, respectively [19]. The magnetic moment (μeff) of the Mn (II) d^5 (Term ^6S) the complex was (5.19 B.M) corresponding to (five unpaired) electrons, see table 3.

The spectrum of [Fe(CANPT)]Cl exhibited four transitions; the first two high intense bands at 280 nm (35714 cm^{-1}) and 431 nm (23201 m^{-1}) assigned to the ligand field (L.F) and intra ligand charge transfer (INCT) ,respectively. The other two bands at 822 nm (12165 cm^{-1}) and 951 nm (10515cm^{-1}) referred to $^6A_1g^{(S)} \rightarrow ^4T_1g$ (υ3) and $^6A_1g^{(S)} \rightarrow ^4T_2g$ (υ2) transitions respectively, that corresponded to their octahedral(O.h) geometry [20,21]. The spectrum of [Co(CANPT)] complex revealed four bands at 275nm (36363cm^{-1}),345 (28985cm^{-1}) and 420nm (23809cm^{-1}) assigned to (L.F) and (C.T) transitions, while d-d transition at 744 nm(13440 cm^{-1}) assigned to the ^3T_1g (F) → ^3A_2g(F) v_3 transition [22].

The solution spectrum of [Ni(CANPT)] complexes illustrated two intensive bands at 281 nm (35587cm^{-1}) and 348nm (28735 cm^{-1}), which assigned to the ligand field and charge transfer ,respectively. Another band observed at 424nm (23584 cm^{-1}) assigned to ^3A_2g (F) → ^3T_1g(F) transition. The magnetic moment of the Ni(II) d^8 (Term ^3F) complex (2.98 B.M) indicated a octahedral configuration of this complex [23,24].

The UV-visible spectrum of [Cu (CANPT)] was exhibited three bands. The first band was located in the ultraviolet region at 277 nm and related to the ligand filed and other three bands were appeared in the visible region at 346nm, 362nm related to Charge transfer.In addition,the value of ($\mu_{eff.}$) that had been measured for [Cu (CANPT)] complex (1.85 B.M) indicated a mononuclear octahedral geometry[25] .

The UV-vis spectrum of k[Ag(CANPT)] showed two absorption bands; a band at 277 nm (36101 cm^{-1}) assigned to the exist of intra ligand, but a second band at 424 nm (23584 cm^{-1}) assigned to (C.T) transition type (M→L) [26,27].The absorption has included electronic transitions for all other

IOP Conf. Series: Materials Science and Engineering 1046 (2021) 012005 doi:10.1088/1757-899X/1046/1/012005

complexes in (table 3) that supported by magnetic sensitivity values to reach the proposed geometric shapes.

Table 3. UV- visible outputs and magnetic moments of the H_2CANPT and its complexes.

Compound	λ max nm	υ' cm^{-1}	€ max $Mol^{-1}\cdot L\cdot cm^{-1}$	Assignments	μeff B.M	Geometric shape
H_2CANPT	275	36363	1851	$\pi\rightarrow\pi^*$		___
	344	29069	2295	$^*n-\pi$	----	
	422	23696	1511	$^*n-\pi$		
[Mn (CANPT)]	276	36231	2058	Ligand field	5.19	O.h
	345	28985	1981	Charge transfer		
	397	25188	1209	Charge transfer		
	817	12239	5	$^6A_1g\rightarrow{}^4T_2g(G)$		
[Fe(CANPT)] Cl	280	35714	760	Ligand field	5.72	O.h
	431	23201	170	Charge transfer		
	822	12165	5	$^6A_1g^{(S)}\rightarrow{}^4T_1g$ (υ3)		
	951	10515	2	$^6A_1g^{(S)}\rightarrow{}^4T_2g$(υ2)		
[Co(CANPT)]	275	36363	1782	Ligand field	4.46	O.h
	345	28985	2262	Charge transfer		
	420	23809	1672	Charge transfer		
	744	13440	21	$^4T_1g(F)\rightarrow{}^4A_2g(F)$		
[Ni(CANPT)]	281	35587	550	Ligand field	2.98	O.h
	348	28735	627	Charge transfer		
	424	23584	507	$^3A_2g\ (F)\rightarrow{}^3T_2g(F)$		
[Cu(CANPT)]	277	36101	1860	Ligand field	1.85	O.h
	346	28901	1475	Charge transfer		
	362	27624	934	Charge transfer		
k[Ag(CANPT)]	277	36101	1231	$\pi\rightarrow\pi^*$	Dia.	O.h
	424	23584	1556	intra-ligand $^*n-\pi$		
[Cd(CANPT)]	276	36231	1755	$\pi\rightarrow\pi^*$	Dia.	O.h
	345	28985	1969	$^*n-\pi$		
	416	24038	1240	C.T		
[Hg(CANPT)]	277	36101	1484	$\pi\rightarrow\pi^*$	Dia.	O.h
	385	25974	968	$^*n-\pi$		
	819	12210	1	C.T		

Dia = Diamagnetic O.h = octahedral

3.3 . The ^1H and ^{13}C -NMR spectra of the ligand

The Schiff base ligand was dissolved in (DMSO) -d^6 solution to get spectra of ^1H and ^{13}C-NMR . Figure 1 showed the ^1H-NMR spectrum, of ligand (H_2CANPT), a sharp peak at (12.32) ppm δ was assigned to (-COOH) carboxyl groups , the (C-OH) phenol protons were observed at δ (9.46 and,9.06) ppm[16,17]. Furthermore, multiple signals at range of 5.93-7.51 ppm were owned to the δ (C-H) aromatic protons [28]. The δ ($-OCH_3$), δ H (-CH) and δ H ($-CH_2$) protons have appeared at (3.83, 3.62 and 2.7, 2.95) ppm, respectively [22,29].

Figure 2 shows the structure of the Schiff base (ligand) by ^{13}C–NMR. Through the ^{13}C –NMR spectrum of H_2CANPT was elucidated a peak at δ (159.73) ppm belong to the carbon atom (CH=N) azomethine groups [30,31]. A peak at 172.21 ppm indicated a carbon atom of δ (-COOH) carboxyl groups [32], then peaks at range (111.83-130.33) ppm were attained to an aromatic carbon, as well as, peaks at (135.6, 55.70, 21.3 and 9.92) ppm referred to δ (C-OH) phenol, δ ($-OCH_3$), δ H ($-CH_2$) and δ H (-CH), respectively [31,32]. The chemical shifts of ^{1}H and ^{13}C –NMR are listed and assigned in (table 4).

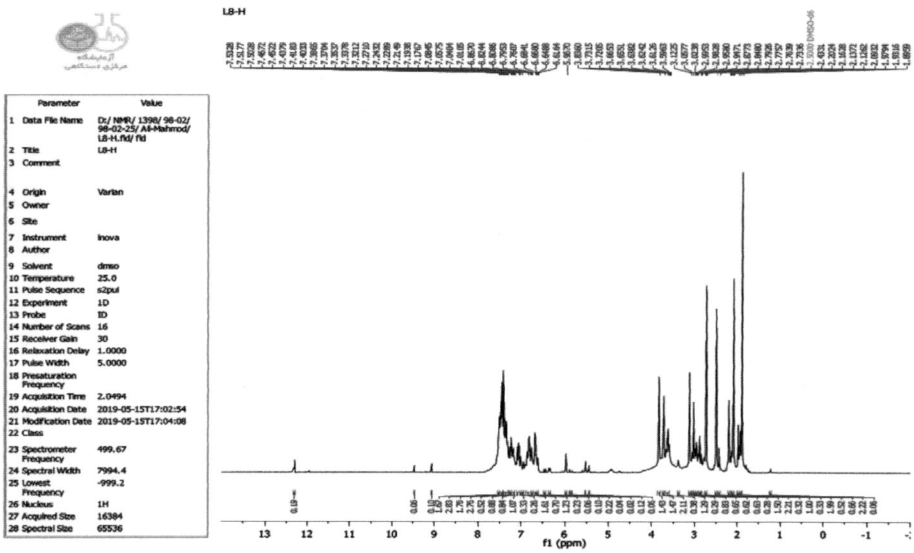

Figure 1. ^{1}H-NMR spectrum of H_2CANPT.

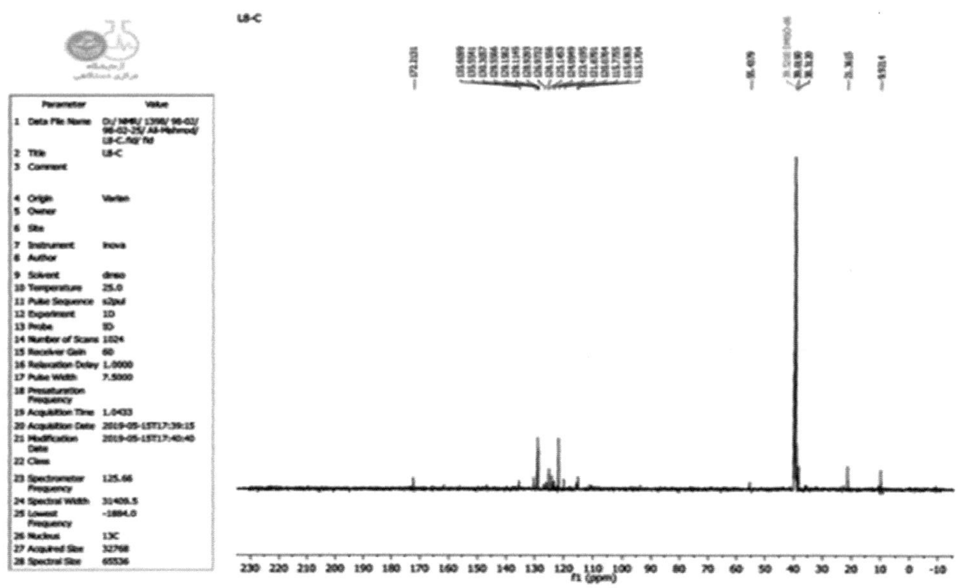

Figure 2. ^{13}C-NMR spectrum of H_2CANPT.

Table 4. Analysis of the ^{1}H and ^{13}C -NMR spectrum of H_2CANPT.

Functional groups	^{1}H-NMR δ (ppm)	^{13}C –NMR δ (ppm)
COOH) carboxyl(- δ	12.32	172.21
C-OH) phenol(- δ	9.46 ,9.06	135.6
For δ (-C=N)	------	159.73
δ (C-H) aromatic	5.93-7.51	111.83-130.33
For δ (-OCH_3)	3.83	55.70
for δ (-CH)	3.62	21.36
for δ (-CH_2)	2.7, 2.95	9.92
δ DMSO	1.90-2.50	39.58

3.4. Biological studies

Antibacterial activities of the synthesized compounds were carried out by two bacterial strains of Gram-negative *Escherichia coli* and Gram-positive *Staphylococcus aureus* using the agar diffusion method[30]. The antibacterial activity of each compound was measured according to the diameter of the inhibition zone (IZ) in mm and compared with DMSO as a control or as a solvent when for a concentration $1x10^{-3}$ M [25]. *Escherichia coli*, and *Staphylococcus aureus* were selected here due to their known effects on a number of diseases; they have different resistance nature to the antibiotics and therapeutic chemicals [33,34]. The ligand and its complexes showed a very good zone of inhibition, except the nickel complex, which showed no efficacy for both types of bacteria. The order of the inhibition potential can be arranged as follows for both types of bacteria: *Escherichia coli* : H_2CANPT > Co > Hg > Ag > Cd > Cu = Mn = Fe = Ni; *Staphylococcus aureus* : H_2CANPT = Fe > Ag > Hg> Cd > Cu = Co > Mn >Ni. The antibacterial activities of all the prepared compounds are included in figure 3 and image 1.

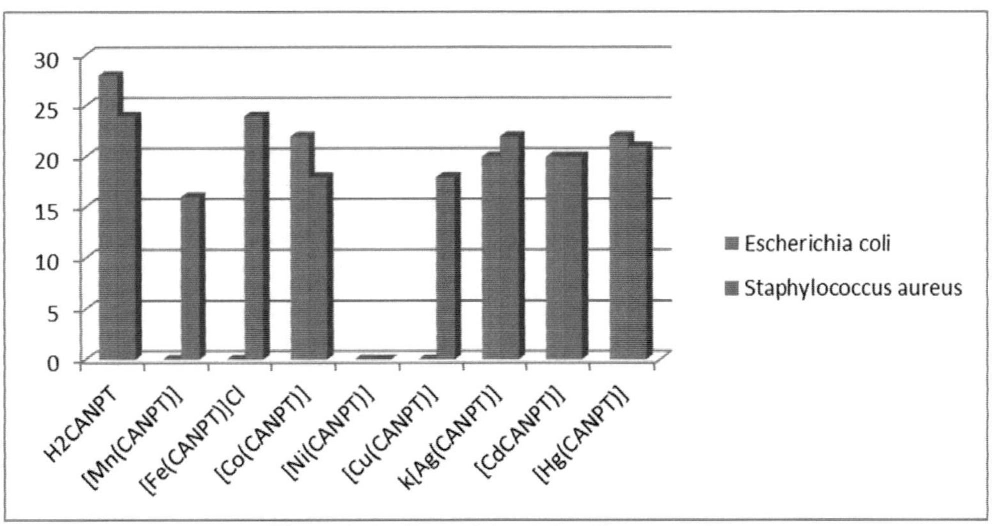

Figure 3. Antibacterial activity of compounds.

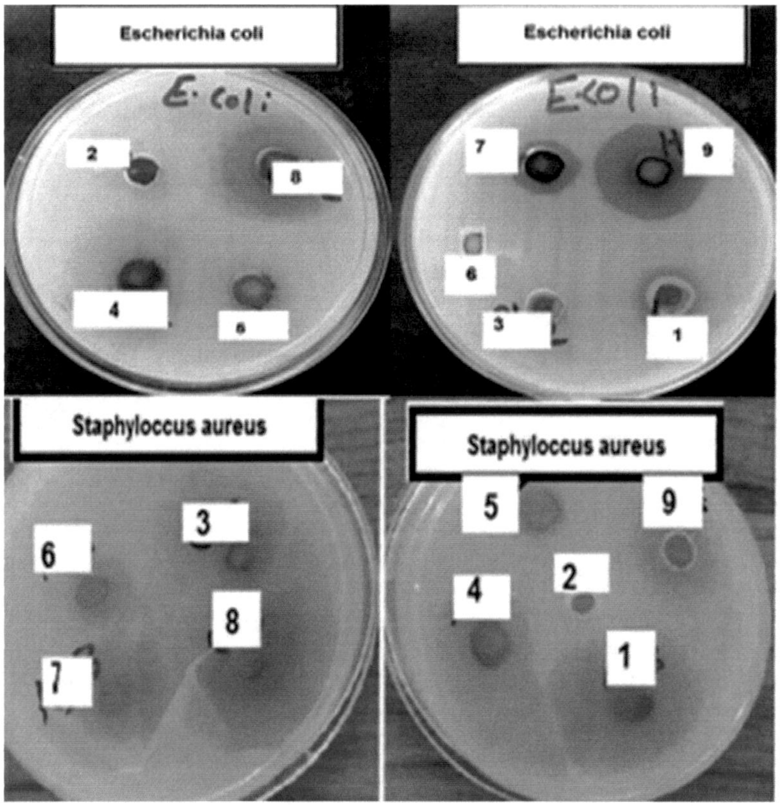

Image 1. Photograph of antibacterial activity of componnds:1- H_2CANPT, 2-[Mn(CANPT)], 3-[Fe(CANPT)]Cl,4-[Co(CANPT)], 5- [Ni(CANPT] 6-[Cu (CANPT)] , 7-k[Ag (CANPT)] ,8-[Cd(CANPT)] ,9-Hg(CANPT)].

3.5. In dyes

Schiff base and their complexes can participate in the dyeing process [35]. The presence of some ions of exist transitional elements promotes positive properties in the dyeing technique [36]. Some azomethine complexes such as cobalt and chromium used in the dyeing process because of their obvious pigments that give constant colors for food packages, wool and leather [37].This species of dyes based on many methods depending on the chemical structure, the chromophore system and other systems [38].

3.5.1. Dyeing procedure. The dyeing process was conducted according to the optimized provision which was tested approving to ISO 105- C10 method [39]: (1) Preparation of dyeing solution: The ligand and its complexes were dissolved in ethanoic aqueous solution and soaked in a dyeing bath with mordant (alum or sodium carbonate), then some drops of (NaOH) solution was added. (2) Dyeing after treatment: The cotton fibers was delivered in the mixture which was prepared in the step 1, then heated to 50 °Cfor 20 minutes with continuous stirring until getting a homogenous dye. It was extracted from the mixture, then washed with cold water and left to dry. (3) The cotton was separated from the mixture, washed with water and left to dry. (4) The washing fastness was examined with 5 g/L (soap liquid) at 40 °C for 30 minutes. After washing tests for dyeing cotton fibers, some samples showed good results as explained in image 2 and table 5 for manganese, nickel, copper, silver, cadmium, and mercury complexes, while the results of ligand complexes and other ions were weak.

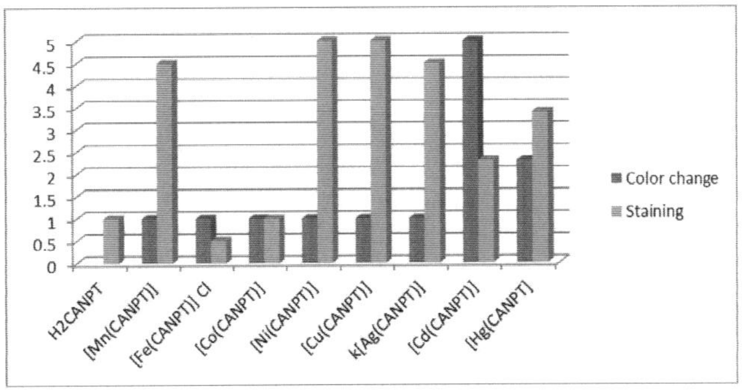

Figure 4. The zone of inhibition in mm of (H_2CANPT) Schiff base and its complexes.

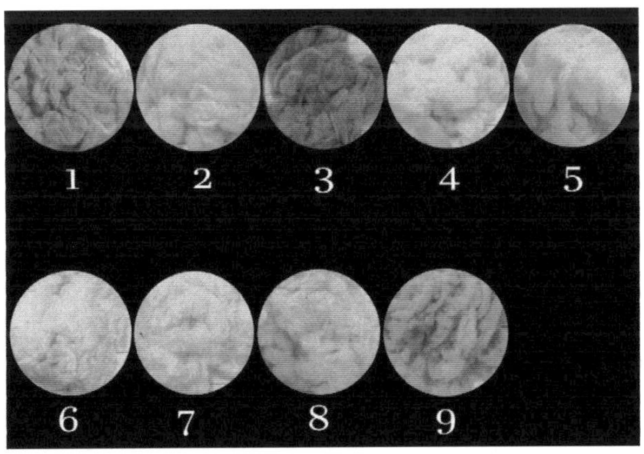

Iimage 2. Shapes of cotton fibers dyeing to Schiff base and complexes; 1-H_2CANPT, 2- [Mn(CANPT)], 3-[Fe(CANPT)]Cl,4-[Co(CANPT)], 5-[Ni(CANPT], 6-[Cu (CANPT)], 7-k[Ag (CANPT)], 8-[Cd(CANPT)] ,9-[Hg(CANPT)].

Table 5. The results of the colorfastness rating of the cotton fabric.

Compounds	Color change	Staining
H_2CANPT	1	1
[Mn (CANPT)]	1	4.5
[Fe(CANPT)] Cl	1	0.5
[Co(CANPT)]	1	1
[Ni(CANPT)]	1	5
[Cu (CANPT)]	1	5
K[Ag(CANPT)]	5	4.5
[Cd (CANPT)]	2.3	2.3
[Hg(CANPT)]	2	3.4

* Wash fastness; 5 = very good, 4-4.5 = good, 1-2 and 2-2.3 = poor

4. Conclusion

Synthesis of Schiff base was done by two steps, the reacting of curcumin 1mmol with 4-aminantipyrine 2 mmol to produce compound (CANP) and reacting of (CANP) 1mmol with L-tyrosine (2 mmole) to produce Schiff base ligand (H_2CANPT). It was performed as hexadentate coordinates by the atoms (N_4O_2). All complexes prepared to assume octahedral geometric shape based on the available data from the such as FT-IR, UV-Visible, micro elemental (C.H.N), ^{13}C,^{1}H-NMR, molar conductivity, and magnetic susceptibility. Some of Schiff bases and their complexes were observed to have color stability when used in the dyeing process on fiber cotton. The prepared compounds had an acceptable inhibition in varying proportions when assessing their biological efficacy against two types of bacteria (*Staphylococcus aureus* and *Escherichia Coli*), except for nickel complex that was not recorded inhibition of two types of bacterial species, in addition to (Iron, manganese and copper) complexes toward *Escherichia Coli.*

References

[1] Kajal A, Bala S, Kamboj S, Sharma N and Saini V 2013 *J. Catal.* **2013** 893512
[2] Hussain Z, Yousif E, Ahmed A and Altaie A 2014 *Org. Med. Chem. Lett.* **4** 1
[3] Kailas H K, Sheetal P J, Anita P and Apoorva H 2016 *W. J. Pharma. Pharmace. Sci.* **5** 1055
[4] Yang, Z and Sun P 2006 *Molbank* **2006** 514
[5] Abu-Khadra A S, Farag R S and Abdel-Hady A E D M 2016 *Am. J. Anal.Chem.* **7** 233
[6] Kumar S, Dhar D N and Saxena P N 2009 *J. Sci. Ind. Res.* **68** 181
[7] Grechukha N M, Gorshkova K O, Panov M S, Tumkin I, Kirillova E O, Lukianov V and Kochemirovsky V A 2017 *Appl. Sci.* **7** 991
[8] Nagata T, Koyanagib M, Tsukamoto H, Mutt E Schertler G F, Deupi X and Terakita A 2019 *Commun. Bio.* **2** 1
[9] Raman N, Sakthivel A and Rajasekaran, K 2007 *Mycobiology* **35** 150
[10] Selvi E T and Mahalakshmi S 2017 *Int. J. Advan.Res. Devel.* **2** 2
[11] R Jayalakshmi, D Priya, V Jayakkumar and R Rajavel 2017 *J. Eng. Res. Tech.* **6** 1
[12] Singh H L 2016 *Rasayan. J. Chem.* **9** 614
[13] Geary W J 1971 *Coord. Chem. Rev.* **7** 81
[14] K Savithri, B C V Kumar, H K Vivek and H D Revanasiddappa 2018 *Int. J. Spectrosc.* **2018** 8759372
[15] Kavitha T, Sarojini P and Navaneethakumar A 2017 *Int. J. Adv. Res.* **6** 202
[16] Singh H L and Singh J 2014 *Bioinorg. Chem. Appl.* **2014** 716578
[17] Anacona J R, Rodriguez J L and Camus J 2014 *Spectroc. Acta Pt. A-Molec. Biomolec. Spectr.* **129** 96
[18] Călinescu M, Fiastru M, Bala D, Mihailciuc C, Negreanu-Pîrjol T and Jurcă B 2019 *J. Saudi Chem. Soc.* **23** 817
[19] Mishra A and Sharma R 2018 *J. Phys.Conf. Ser.* **1003** 012016
[20] Abdel-rahman L H, El-khatib R M, Nassr L A E and Abu-dief A M 2013 *J. Mol. Struct.* **1040** 9
[21] Syamal A, Singh M and Kumar D 1999 *React. Funct. Polym.* **39** 27
[22] Kareem A, Arshad M, Nami S A, and Nishat N 2016 *J. Photochem. Photobio. B* **160** 163
[23] Thamaraikannan T, Rubiga M, Jayalakshmi R and Rajavel R 2017 *IOSR-J. Appl. Chem.* **10** 1
[24] Chaudhary N K and Mishra P 2017 *Bioinorg. Chem. Appl.* **2017** 6927675
[25] Zare N and Zabardasti A 2019 *Appl. Organomet. Chem.* **33** 4687
[26] Singh K, Thakur R and Kumar V 2016 *Beni-Suef Univ. J. Basic. Appl. Sci.* **5** 21
[27] Lever ABP 1968 *Inorganic Electronic Spectroscopy* (Netherlands-Elsevier Publishing) p 246
[28] Kavitha N Anantha P V and Lakshmi 2017 *J. Saudi Chem. Soc.* **21** 457
[29] Nassar M Y, Ahmed I S, Dessouki H A and Ali S 2018 *J. Basic Environ. Sci.* **5** 60
[30] Jayaseelan P, Akila E, Rani M U and Rajavel R 2016 *J. Saudi Chem. Soc.* **20** 625
[31] Chaudhary N K and Mishra P 2018 *J. Saudi Chem. Soc.* **22** 601

[32] Balić T, Marković B and Medvidović-Kosanović M 2015 *J. Mol. Struct.* **1084** 82
[33] Morgan C R, Opio C and Migabo S 2020 *S. Afr. J. Bot.* **129** 243
[34] Roth N, Hofacre C, Zitz U, Mathis G F, Moder K, Doupovec B and Domig K. J 2019 *Poul. Sci.* ***98*** 2598
[35] Abauamer K M, Maihub A and Elarbi F M 2018 *Asian J. Advan. Basic Sci.* **6** 77
[36] Ren Y, Fu R, Fang K, Chen W, Hao L , Xie R and Shi Z 2019 *Ind . Crops Prod.* **142** 111863
[37] Prakash A and Adhikari D 2011 *Int. J. Chem. Tech. Res.* **3** 1891
[38] Abuamer K M , Maihub A , El-Ajaily M , Etorki A M , Abou-Krisha M and Almagani, M A 2014 *Int. J. Org. Chem.* **4** 9
[39] Ren Y, Gong J, Fu R, Li Z, Li Q, Zhang J and Cheng X 2017 *Dyes Pigm.* **138** 147

Synthesis, spectroscopic study, biological activity and dyeing application of curcumin - Schiff base with various metal ions complexes

A M Ali* and T H Al-Noor*

Department of Chemistry.Ibn -Al-Haithem College of Education for Pure Science, Baghdad University, Iraq

*E-mail: Cult.rel.unt@ihcoedu.uobaghdad.edu.iq (AMA); drtaghreed2@gmail.com (THA)

Abstract. A tetradentate (N_2O_2) Schiff base (H_2Ldfm) was successfully synthesized via condensation of curcumin / diferuloylmethane (dfm) and L-leucine amino acid (HL). There were three different methods that used for synthesizing H_2Ldfm; (refluxing, grading, and fusion). Ten different metal complexes were also successfully synthesized by combination of the Schiff base (H_2Ldfm) and 1,10-phenanthroline (phen) ligand to form a hexadentate (N_4O_2) mixed ligands (Ldfm , phen) with ten different metal salts (M) where{ M= Al(III), Mn(II), Fe(III), Co(II), Ni(II), Cu(II), Ag(I), Cd(II), Hg(II), and Pb(II)}. The molar ratio of reactants was (1:1:1) (M: H_2Ldfm : phen). The new Schiff base and its new complexes were characterized by different physicochemical techniques such as FT-IR, UV-Visible, elemental analysis (C.H.N), ^{13}C-NMR, ^{1}H-NMR, molar conductivity, mass spectroscopy, magnetic susceptibility, and thermal analysis. All complexes exhibited octahedral geometry. The solution conductivity of (Al (III), Fe (III), and Ag(I)) complexes in DMSO shows an electrolytic behavior, while the solution conductivity of (Mn(II), Co(II), Ni(II), Cu(II), Cd(II), Hg(II) and Pb(II)) complexes shows non-electrolytic attitude. The Schiff base (H_2Ldfm) and its metal complexes were used for industrial and biological application such as dyeing cotton fibers testing and two types of bacteria testing like (*Staphylococcus aureus* and *Escherichia Coli*).

1. Introduction

Mixed ligand and Schiff bases (azometine compounds) have played important role in the development of coordination chemistry and have been implicated as a significant point in the development of inorganic biochemistry, some Schiff base compounds containing N and O are used in pharmaceutical, industrial, agricultural, biological and medical applications [1]. Curcumin ($C_{21}H_{20}O_6$) = ((1E,6E)-1,7-bis (4-hydroxy-3-methoxyphenyl)-1,6- heptadiene-3,5-dione)obtained from the rhizome of turmeric (*curcuma longa*),and it has a yellow coloring natural pigment [2] . In addition , it has many functional groups such as a polyphenol and ortho–methoxy. Compounds have widely used of diverse fields like antibacterial, antiviral, anticancer, antifungal, antioxidant activities and anti-inflammatory [3][4]. Many studies dealing with curcumin complexes have been published within the past 20 years have been used as a traditional medicine, food additive dye , and indicator [2]. Curcumin has a conjugated β-diketone moiety can be a powerful natural chelating agent with its safe evaluation even administered at high doses by human [3]. Recently, complexation of curcumin with metal ions have attracted the attention of researchers around the world as one of the useful requirements for the treatment of Alzheimer's disease and *in vitro* antioxidant activity [3][5]. An amino acid is a compound that generally includes two different functional groups such as (a carboxyl and an amino), both grafted to the same carbon atom in α position [6], they used in the synthesis of the Schiff base compounds [7], and amino acid based Schiff base complexes have biological importance[8]. In addition, they are used with metal ions the synthesis of complexes [9]. Moreover, Schiff base compounds are formed from the condensation any primary amine with an aldehyde or a ketone [10]. They used the synthesize complexes with different metal ions [11]. They are well known due to their applications in different fields such as analytical chemistry, coordination chemistry and agricultural applications [12]. 1,10-Phenanthroline widely used as coordinating agent for preparing of stable compounds [13].

The present work are focusing on synthesis and study Schiff base from curcumin with L-leucine, then preparing of (transitional and non-transitional) metal complexes with mixed ligands of a new Schiff base (H_2Ldfm) and 1,10- phenanthroline (phen). The biological efficacy with two types of bacteria and tested as dyes on cotton fibers were studied

2. Experimental

2.1. Materials and methods

All chemical materials were used without further purification. The metallic ions $MCl_n.XH_2O$, (M=Al(III), Mn(II), Fe(III), Co (II), Ni(II), Cu(II), Cd(II), Hg(II) and $M(NO_3)_n$ M= Ag(I), Pb(II)), curcumin, L-leucine, DMSO and C_2H_5OH were Analar grade (BDH).

2.2. Physical measurements

Elemental analysis (C, H , N) of ligand were performed by Eager300 for EA1112 Thermal Finnegan C.H.N.S 2400 elemental analyzer. The (AAS) analysis in complexes were recorded by using a fair agreement method by used the device from type Shimadzu (A.A 620) atomic absorption spectrophotometer. The Λ_m measurements of the complexes with 10^{-3}mol/L in dimethyl sulfoxide (DMSO) were reached by using a device digital conductivity series Ino.Lab.720. Magnetic measurements of the complexes were measured by using Balance Johnson Mattey. The mass spectrum of ligand was determined by using Ion source: Electron Impact (EI) 70eV mass spectra register on manufacturer company: Agilent Technology 1. MS Model: 5973pectrometer. Ultraviolet-visible spectra were obtained by using solvent (DMSO) (10^{-3}M) for the ligand and its metal complexes by double-beam UV-visible spectrophotometry type U.V 160A (Shimadzu), between (200-1100) nm with a 10^{-1}cm length of quartz cell. The Fourier-transform infrared spectra were recorded using KBr pellets on Shimadzu FT-IR 8400S spectrophotometry has a range (400-4000 cm^{-1}). ^{1}H and ^{13}C-NMR spectra were obtained from (DMSO) -d^{6}Solution using an Inova 500 MHz, and thermal analysis (TGA-DSC) of the complexes were performed by SDT Q600 V20.9 Build 20 DSC/TG thermal system under the N_2 flow with a heating rate of 10 C/min from 25 to 700°C.

2.3. Synthesis of Schiff base (N_2O_2) tetradentate

It was synthesized by reflux, grinding, and fusion methods.

2.3.1. Reflux. The Schiff base ligand (H_2Ldfm) was synthesized with improvement to the common method in the literature [14] . It was prepared by condensation of HL (0.262 gm, 2 mmol) dissolved in hot ethanoic aqueous solution (25 mL) with dfm (0.368 gm. 1mmole) dissolved in hot ethanoic solution (25 mL) in a basic medium of KOH. The mixture refluxed for about eight hrs until a yellow-brown polycrystalline precipitate was obtained, the reaction monitored by TLC technique. The precipitate was filtered and washed with ethanoic aqueous solution, then recrystallized from hot ethanol. Elemental Analysis: call (found) C, 66.65(66.40); H, 7.12 (6.99); N, 4.71 (4.60); O, 21.52 (---), yield (80%). The synthesis pathway is shown in Scheme 1.

2.3.2. Grinding. The H_2Ldfm was also prepared by the mixing of L-leucine (0.262 gm, 2 mmol) and curcumin (0.368 gm, 1mmol) in a manual mortar [15]. Four drops of glacial acetic acid was added, then the mixture was grinded for about (0.5-1) hour. The reaction was observed with the TLC technique. The yellow Schiff base was purified by repeated washing with ethanoic aqueous solution and recrystallized from hot ethanol afforded yield (82%).

2.3.3. Fusion. A molar ratio of (1:2) (dfm : HL) was mixed in a round bottom flask, then 4 drops of glacial acetic acid was added [16]. The mixture was heated on oil bath for about 10 minutes until fused. The resulted yellow-brown Schiff base was dissolved in ethanol, filtered and dried, then recrystallized from hot ethanol, afforded yield (87%).

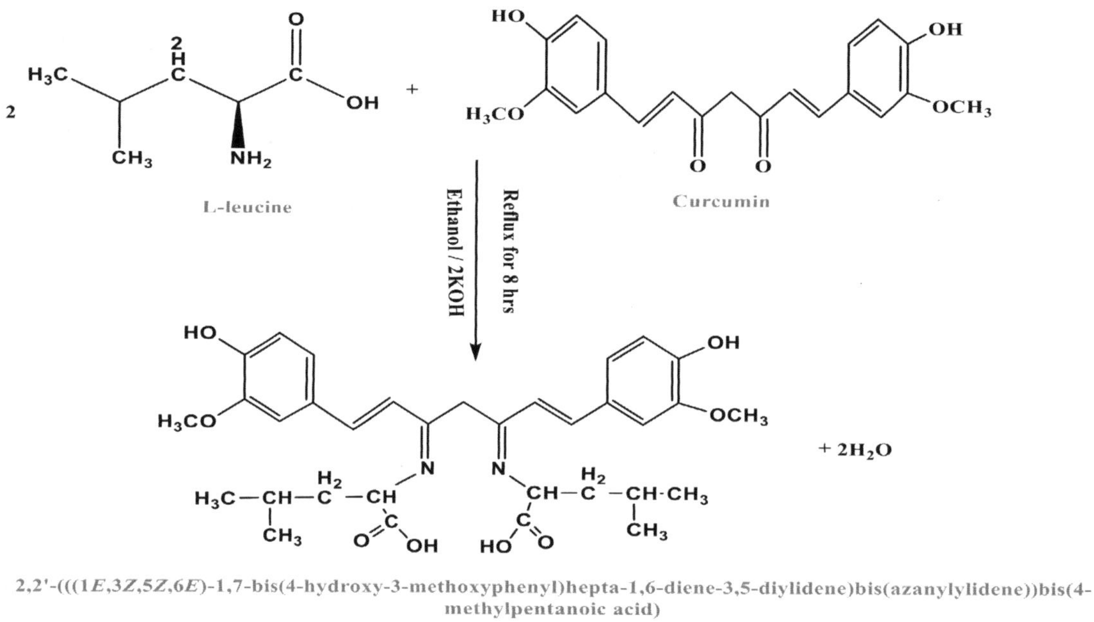

Scheme (1): Shows the preparation process of (H_2Ldfm) Schiff base ligand.

2.4. Preparation of metal complexes

As shown in Scheme 2, the metal complexes were synthesized by mixing of the ligands H_2Ldfm phen and some metal ions M in the ratio of 1:1:1 (H_2Ldfm: phen: M). A (0.594 gm, 1mmol) of H_2Ldfm was dissolved in 15 mL of absolute ethanol, then (0.112 gm of potassium hydroxide) was added. A (0.180 gm, 1mmol) of phen was dissolved in 15 mL of absolute ethanol. The metal salts were dissolved the in 10 mL of absolute ethanol. 0.133g of aluminum(III) chloride, 0.197gm of manganese(II) chloride, 0.270 gm of iron(III) chloride, 0.237gm of cobalt(II) chloride, 0.237gm of nickel(II) chloride, 0.170 gm of copper(II) chloride, 0.169 gm of silver(I) nitrate, 0.183 gm of cadmium(II) chloride, 0.271 gm of mercury(II) chloride and 0.331 gm of lead(II) nitrate were used . The solutions of the ligands were gradually added to the above-mentioned salt solutions with constant stirring about half an hour to obtain precipitates of the complexes. The precipitates were separated and washed several times with an ethanoic aqueous solution, then re-crystallization from the hot ethanol.

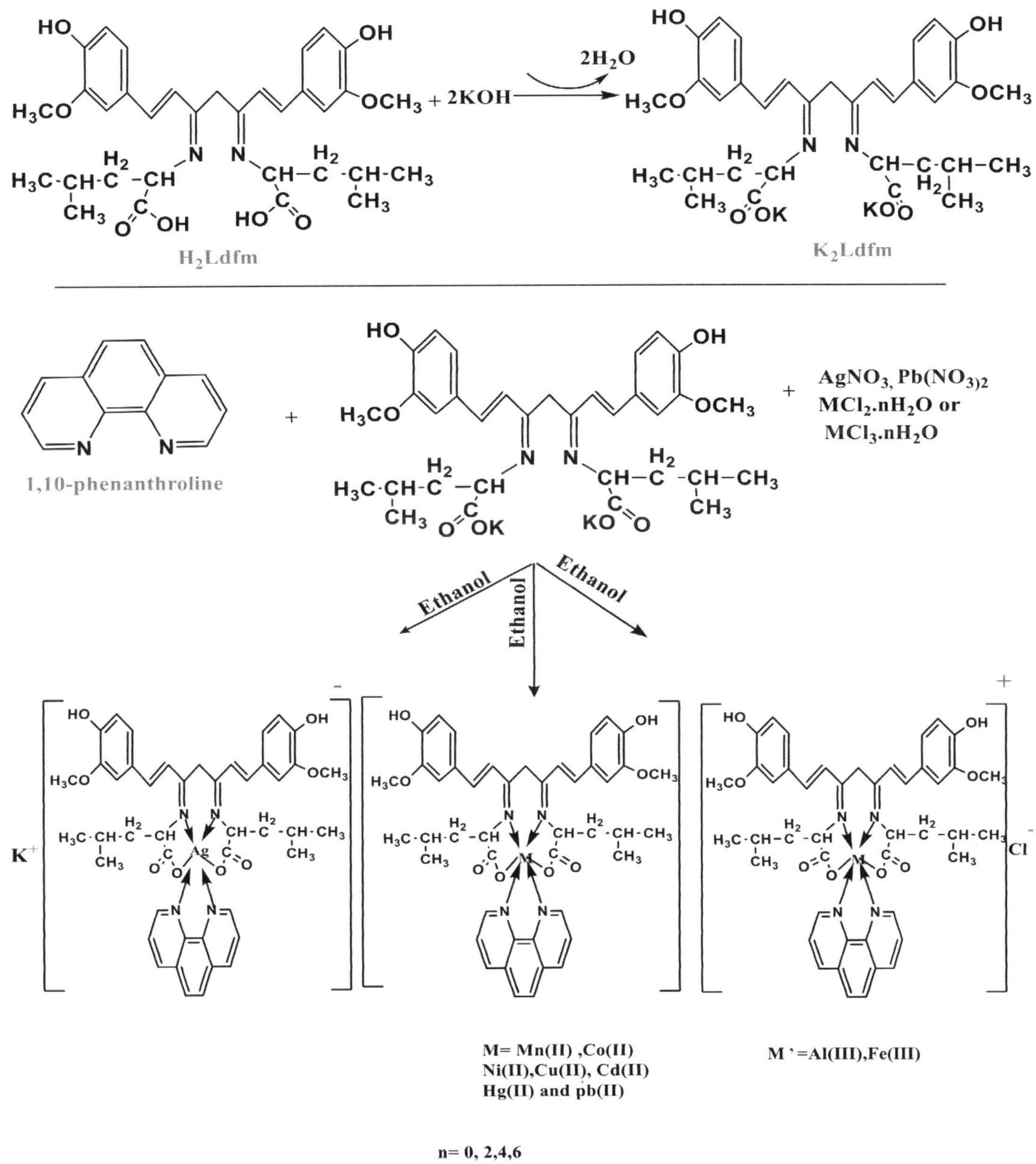

Scheme 2. Preparation of the metal complexes

3. Results and Discussion

Generally, the Schiff base ligand (H_2Ldfm) was attained by condensation of one equivalent of dfm with two equivalent of HL, while its complexes were prepared by reacting the metal salts with the Schiff base (H_2Ldfm) and phen ligands by using (1:1:1) molar ratio. The H_2Ldfm (primary ligand) behaves as a tetradentate ligand, while the phen (secondary ligand) behaves as bidentate, therefore, the resulted mixed ligand exhibited a hexadentate coordination sphere around the metal ions in the new prepared complexes. The researchers notice most of the prepared compounds have distinguished by deep color because they had a partially filled electronic arrangement (Cotton F. A,1987). Moreover, the molar conductance (Λ_m) values which were recorded in DMSO (10^{-3}M) in the range of (6.5-19.9)

$\Omega^{-1}.cm^2.mole^{-1}$ gave an evidence that these complexes are non-electrolytes and exception of type Fe(III), Al(III) and Ag(I) complexes are electrolytes the range of(36.2-40.3) $\Omega^{-1}.cm^2.mole^{-1}$ [17]. All complexes possess high melting points ranging between (199- >250) and they are all stable at room temperatures over time. Some physicochemical characteristics of complexes are shown in (table 1).

3.1. Solubility test

Solubility test was performed in different types of solvents such as water, dimethyl sulfoxide, chloroform, ethanol, methanol, dimethylformamide, hexane, and acetone. All compounds were not dissolved in water, but they were easily dissolved in methanol, ethanol, chloroform, dimethylformamide, dimethyl sulfoxide and hydrochloric acid. In addition, the Al(III), Co(II), Ni(II), Cd(II), and Hg(II) complexes were poorly soluble in acetone and hexane solvents.

Table 1. Physicochemical characteristics data of the compounds.

Compounds	M.Wt	Color	M.P	Yield%	Λ mΩ^{-1}·cm^2·mol^{-1}	Metal % (exp)
$C_{33}H_{42}N_2O_8$)(H_2Ldfm	594.71	Yellow brown	170-172	80	---	----
[Al(L dfm) (Phen)]Cl	835.5	Pale yellow	>260	56	36.2	3.25(3.05)
[Mn(Ldfm)(Phen)]	827.84	Dark brown	223-224	62	10.1	6.63(5.22)
[Fe(Ldfm)(Phen)] Cl	864.24	Dark brown	>260	77	40.3	6.46(6.30)
[Co(Ldfm)(Phen)]	831.83	Pale yellow	>260	61	15.5	7.08(6.96)
[Ni(Ldfm)(Phen)]	831.59	Pale yellow	>260	59	8.6	7.05(6.88)
[Cu(Ldfm)(Phen)]	836.45	Dark yellow	228-230	88	6.5	7.59(7.05)
K [Ag(Ldfm)(Phen)]	919.77	Pale brown	199 Dec	90	39.2	11.72(10.01)
[Cd(Ldfm)(Phen)]	885.31	Dark brown	>260	78	18.9	12.69(10.35)
[Hg(Ldfm)(Phen)]	973.49	Pale yellow	238-240	70	19.9	20.05(20.12)
[Pb(Ldfm)(Phen)]	980.10	Brown	228-229	79	12.2	21.14(20.11)

Phen=1,10-phenanthroline = ($C_{12}H_8N_2$). (H_2Ldfm) = $C_{33}H_{40}N_2O_8$
Dec = decomposition

3.2 . UV-vis spectra

The UV–vis spectra of the compounds were performed in DMSO (10^{-3} M) solution as shown in (table 2). There was a shifting to higher wavelength that could observed with the electronic spectra of all complexes , which supports the formation of these complexes [18]. The UV-Visible spectrum of H_2Ldfm has showed two bands, the first band at 270 nm which can be assigned to $\pi \rightarrow \pi^*$ transition in the aromatic ring and double bonds in curcumin structure, and the second band at 434 nm which can be assigned to $n \rightarrow \pi^*$ transitions in azomethine. whilst the UV-Visible spectrum of phen (second ligand) showed only one band at 271nm which is attributed to $\pi \rightarrow \pi^*$transition [19].

The UV-visible spectrum of [Al(Ldfm)(phen)]Cl complex displayed three bands, the first band with a high intense at 268 nm due to the intra ligand charge transfer (INCT), and other two bands at 430 nm and 942 nm due to the 3p transitions in the Al→ that were observed by monitoring 4s→3p emission from Al(4s) atoms produced by pre dissociation of the excited complex [19]. The UV-visible spectrum of [Mn(Ldfm)(Phen)] showed four bands, the first band was found in the ultraviolet region at 270 nm which assigned to ligand field while the other three bands at 426nm, 450nm and 805 nm were assigned to L→M charge transfer, L→M charge transfer , and $^6A_1g\rightarrow^4T_2g(G)$, respectively. The Mn(II) complex displayed the octahedral geometry depending on the magnetic susceptibility value (5.34 B.M) [20].

Table 2. Electronic spectral main data and magnetic measurements of compounds in (DMSO).

Compounds	λ_{max} nm	υ' cm^{-1}	€ max $mol^{-1}\cdot L\cdot cm^{-1}$	Assignments	(μeff) B.M	Geometry shape
H_2Ldfm	270	37037	839	$\pi\rightarrow\pi^*$	----	---
	434	23041	2182	$n\rightarrow\pi^*$		
Phen ($C_{12}H_8N_2$)	271	36900	801	$\pi\rightarrow\pi^*$	----	----
[Al(Ldfm)(Phen)] Cl	268	37313	369	C.T	Dia.	O.h
	430	23255	81	The 3d ← 3p transitions in the Al-N		
	942	10615	7			
[Mn(Ldfm)(Phen)]	270	37037	1946	Ligand field	5.34	O.h
	426	23474	1835	Charge transfer		
	450	22222	1553	Charge transfer		
	805	11764	7	$^6A_1g\rightarrow^4T_2g(G)$		
[Fe(Ldfm)(Phen)]Cl	446	22421	1850	LMCT	6.02	O.h
	961	10405	1335	$^6A_1g\ ^{(s)}\rightarrow^4T_2g$		
[Co(Ldfm)(Phen]	267	37453	523	Ligand field	4.38	O.h
	441	22675	249	$^4T_1g(F)\rightarrow^4T_2g(F)$		
	788	12690	5	$^4T_2g(f)\rightarrow^4A_2g(F)$		
	976	10245	5	$^4T_1g(F)\rightarrow^4T_1g(P)$		
[Ni(Ldfm)(Phen)]	271	36900	404	Ligand field	3.79	O.h
	989	10111	6	$^3A_2g\ (F)\rightarrow\ ^3T_2g(f)$		
[Cu(Ldfm)(Phen)]	272	36764	1604	Ligand field	1.71	O.h
	437	22883	1868	LMCT		
	458	21834	1764	LMCT		
	825	12121	6	$^2B_1g\rightarrow^2A_1g$		
k[Ag(Ldfm)(Phen)]	268	37313	1594	$\pi\rightarrow\pi^*$	Dia.	O.h
	428	23368	1072	$^*n-\pi$		
	666	15015	216	C.T		
[Cd(Ldfm)(Phen]	270	37037	1744	$\pi\rightarrow\pi^*$	Dia.	O.h
	428	23364	634	$^*n-\pi$		
	858	11655	3	C. T		
[Hg(Ldfm)(Phen)]	270	37037	1819	$\pi\rightarrow\pi^*$	Dia.	O.h
	426	23474	190	$n\rightarrow\pi^*$		
	813	12300	25	C.T		
[Pb(Ldfm(Phen)]	270	37313	1384	$\pi\rightarrow\pi^*$	Dia.	O.h
	427	23417	1218	$^*n-\pi$		
	756	13227	17	C.T		

Dia. = Diamagnetic O.h = Octahedral

The UV-Visible spectrum that outputted from [Fe(Ldfm)(phen)]Cl complex exhibited two bands, one at 446 nm and the other at 961 nm which were due to intra ligand metal charge transfer (LMCT) and $^6A_1g^{(s)} \rightarrow ^4T_2g$ transition, respectively. The (μeff) value of [Fe(Ldfm)(phen)]Cl [d^5] complex was 6.02B.M, that a tentative interpretation expects to be taken the structure of Fe(III) to be an octahedral geometry which matches the data reported by several research workers [20,21].
The UV-Visible spectrum of [Co(Ldfm)(phen)] complex was displayed four bands. The first band at 267 nm with high energy was related to the charge transfer transitions, and the other three bands were related to (d-d transitions) which appeared at 441 nm, 788 nm and 976 nm and were considered as $^4T_1g \rightarrow T_1g(p)$ υ_3, $^4T_1g \rightarrow ^4A_2g(f)$ υ_2 and $^4T_2g(f) \rightarrow ^4A_2g(F)$ $\upsilon_1$they also supported by magnetic sensitivity (4.38 B.M) which was a good evidence that this complex displayed an octahedral geometry [20-22].

The UV-visible spectrum of [Ni(Ldfm)(phen)] was revealed only two bands, the first band was wide and appeared at 271 due to the ligand field, and the other band was weak and related to (d-d transitions) at 989 nm attributed to ^3A_2g (F) $\rightarrow ^3T_2g(f)$ transition, and via measuring magnetic sensitivity (3.79 B.M) i.e., it had an electronic arrangement of three unpaired electrons that bolsters the octahedral geometry [22,23].

The UV-visible spectrum of [Cu(Ldfm)(phen)] was exhibited four bands. The first band was located in the ultraviolet region at 272 nm and related to the ligand filed and other three bands were appeared in the visible region at 437nm, 458nm and 825nm related to (LMCT, LMCT and $^2B_1g \rightarrow ^2A_1g$), respectively. In addition, it had magnetic sensitivity (1.17 B.M) that supported the octahedral geometry [21,24].

The [Pb(Ldfm)(phen)] complex exhibited three bands in the UV-Visible spectrum, the first band was at 270 nm and the second was at 427 nm assigned to $\pi \rightarrow \pi^*$ and n– π^* transitions, respectively, due to conjugated π system, while the third band was observed at 756 nm which ascribed to L-M charge transfer (C.T) transitions of the heterocyclic and COO- groups [20,25]. The spectra of Ag(I), Cd(II) and Hg(II) complexes were exhibited bands back to $\pi \rightarrow \pi^*$, n$\rightarrow \pi^*$ transitions. All the ions had an electronic arrangement with shill (d^{10}); therefore, do not give the type (d-d) transition [18,23].

3.3 . IR Spectra

The FT-IR spectra data of H_2Ldfm (as shown in figure 1) and its complexes with their respective assignments as shown in (table 3).The spectrum of H_2Ldfm ligand showed a wide stretching vibration recognized at 3508 cm^{-1} which could be corresponded to (O-H) carboxylic groups, while the broad band at 3415cm^{-1} indicated (O-H) phenolic groups [26,8,22].

The azomethine (HC=N) groups of H_2Ldfm appeared at 1622 cm^{-1} [20,23], where it was recorded closer to the frequency of this ligand in the three different methods of synthesis .It was shifted to a lower value at (1602-1618cm^{-1}) in all complexes, which gives an evidence that it coordinated with different metals' ions via the nitrogen atom of (HC=N) group [23]. These results indicated that H_2Ldfm is acting as a tetradentate (NNOO) ligand when coordinating with the metal ions. Moreover, the two bands shown at 1589 cm^{-1} and 1405 cm^{-1} belong to (COO^-) asymmetric and (COO^-) symmetric, respectively [25].

The different values of the stretching vibration ($\Delta\upsilon$) of carboxylic group (COO^-) asymmetric and symmetric of all complexes were greater than (200 cm^{-1} which indicated to the monodentate coordination with the metal atoms [8,25,27]. The frequencies of phenolic (O-H) group remained without any change in most complexes indicated that OH groups did not incorporate in the process of consistency with metallic ions, while the disappearance of the frequencies of the carboxylic hydroxyl groups was a good evidence for its involvement in the process of coordination with metals. In addition, the appearance of new bands at all complexes' spectra which range from 534 - 570 and 441-466 could be assigned to the ν (M-O) and ν (M-N),respectively [27,28] .

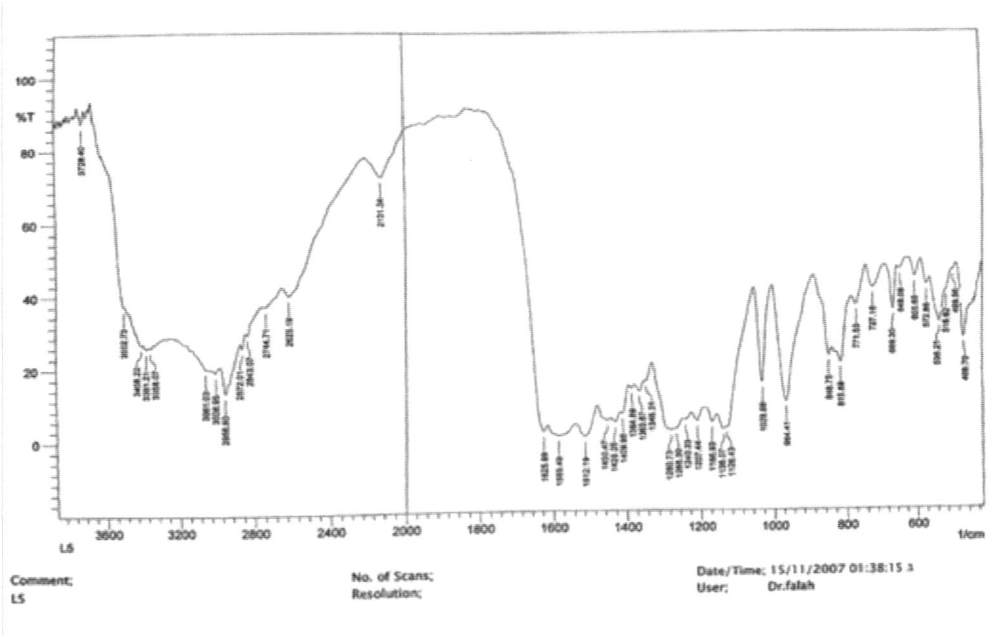

Figure 1. FT-IR spectrum of (H_2Ldfm) Schiff base.

Table 3. Reprehensive results for the test of FT- IR spectra of compounds by (cm^{-1}).

Comp	ν (O-H) carboxylic (O-H)phenolic&	ν C=N	ν as COO	ν s COO	Δν	ν M-O	ν M-N
H_2Ldfm (Reflux)	3508&3415	1622	1589	1405	184	----	----
H_2Ldfm (Grinding)	3502 & 3400	1625	1583	1408	175	----	----
H_2Ldfm (Fusion)	3512&3408	1625	1585	1409	176	----	----
Phen	----	1649	----	---	----	----	---
[Al(Ldfm) (Phen] Cl	3410	1608	1583	1361	222	534	441
[Mn (Ldfm)(Phen]	3387	1618	1585	1361	224	551	447
[Fe(Ldfm) (Phen] Cl	3388	1618	1589	1342	247	555	459
[Co(Ldfm) (Phen]	3417	1608	1585	1359	226	557	443
[Ni(Ldfm) (Phen)]	3427	1602	1582	1382	200	534	443
[Cu(Ldfm) (Phen)]	3406	1612	1589	1338	251	570	443
K[Ag(Ldfm) (Phen)]	3419	1607	1582	1365	218	536	441
[Cd (Ldfm) (Phen)]	3417	1618	1581	1352	229	549	466
[Hg(Ldfm) (Phen)]	3417	1604	1568	1354	214	549	460
[Pb(Ldfm) (Phen)]	3417	1616	1581	1374	207	551	460

3.4. ^{1}H-NMR Spectrum of H_2Ldfm ligand

The ^{1}H-NMR spectrum of the Schiff base (H_2Ldfm) was recorded in the DMSO-d^6 solvent. The ^{1}H NMR spectrum of H_2Ldfm, (figure 2), was exhibited multiple signals at 7.56-6.66 that allotted to the aromatic protons of the benzene ring [29]. The three sharp signals at 3.8, 3.3 and 2.08 ppm were assigned to the OCH_3, CH_2 and CH_3 groups, respectively [30], also the proton's signal of COOH group was absent in the spectrum due to hydrogen bonding [31].

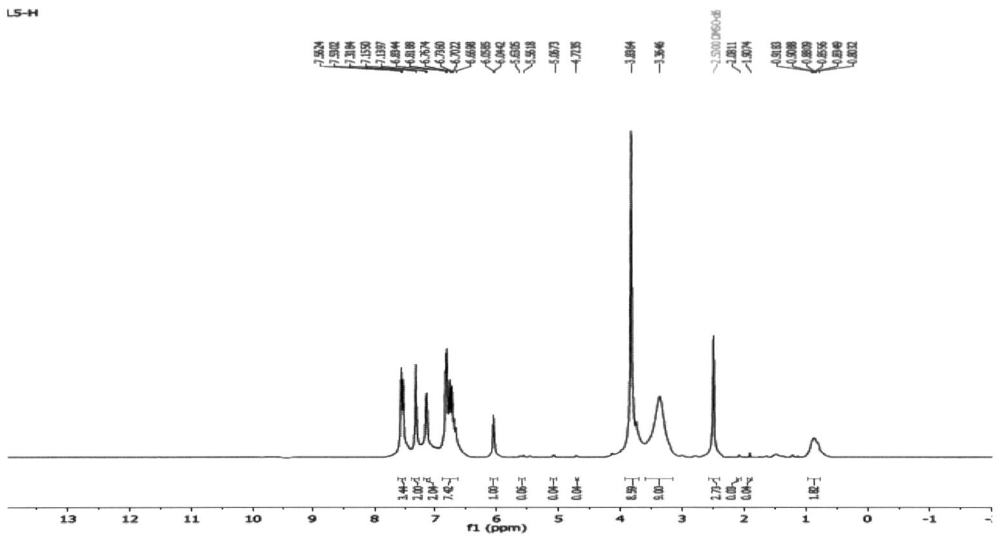

Figure 2. ^{1}H-NMR spectrum of H_2Ldfm ligand.

3.5. ^{13}C- NMR Spectrum of H_2Ldfm

The ^{13}C -NMR spectrum of (H_2Ldfm) showed two signals at (183.20 and 159.73 ppm) ascribed to C=O (carboxylic) and C=N (azomethine) groups, respectively [22, 32]. The other bands that appeared at the region(149 and 140 ppm) correspond to the carbons that attached with =C-N and =C-C groups, respectively [22]. The bands in the range at 100.83-130.33 ppm were attributed to aromatic ring groups [33] .In addition, the band at (55.70 ppm) was assigned to two methyl carbons [34], as shown in (figure 3).

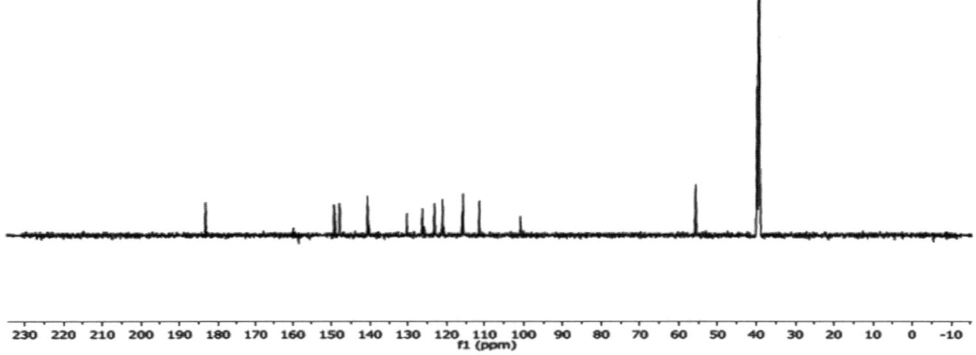

Figure 3. ^{13}C-NMR spectrum of H_2Ldfm ligand.

IOP Conf. Series: Materials Science and Engineering 1046 (2021) 012006 doi:10.1088/1757-899X/1046/1/012006

3.6. Mass Spectrum of H_2Ldfm Ligand

The mass spectrum fragmentations of H_2Ldfm ligand is shown in (figure 4). The peak that appeared at m/e+ = 594 due to the ligand [$C_{33}H_{42}N_2O_8$] molecular weight. The peaks at m/e+ = 579, 564, 524, 491 and 472 corresponded to the fragments $C_{33}H_{42}N_2O_7$, $C_{33}H_{44}N_2O_6$, $C_{32}H_{48}N_2O_4$, $C_{32}H_{46}N_2O_2$ and $C_{31}H_{40}N_2O_2$, respectively. This proves the structure of Cur-leucine ligand (H_2Ldfm).

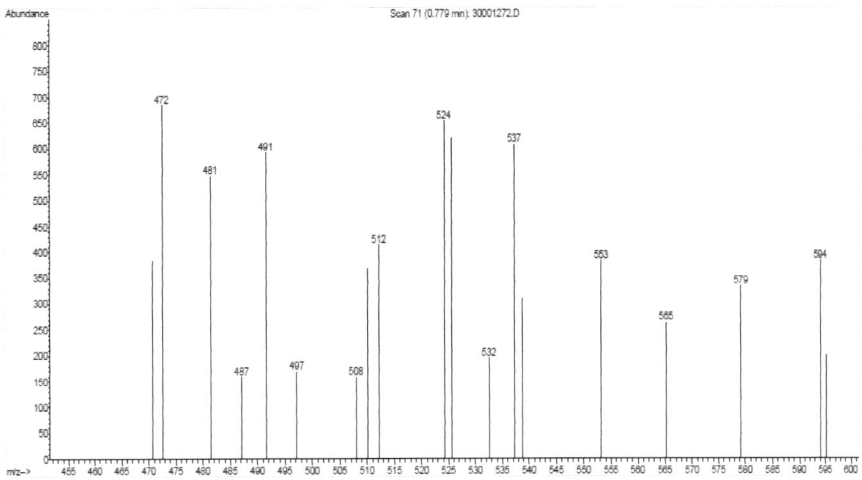

Figure 4. Mass spectrum of H_2Ldfm ligand.

3.7. Thermal Analysis

TGA is used to investigate the thermal degradation of inorganic compounds [26]. The TGA and DSC analysis for all the prepared complexes are studied where the heating rates are suitably controlled at 10 deg/min under N_2 gas flow and weight loss is measured from ambient temperature to 600 °C. Figures (5-8) showed the TGA& DSC curves and the thermal degradation were conducted to (Mn (II), Co(II), Ni(II) and Cu(II)) complexes. TGA curve of the [Mn(Ldfm)(phen)] showed four decomposition peaks, the first peak observed at 42 to 110 °C, with accompanying by a mass loss of 3% (calc.2.99) to elimination of (-OH and H_2) molecules. The second peak detected at 185 to 250 °C, with accompanying by a mass loss 22% (calc19.99) due to the elimination of (CH_3OH,CO_2 and CH_4) molecules. Whilst the third peak observed at 328-468 °C with elimination of 21% (calc.20.95) corresponding to the decomposition of ($C_{12}H_8N_2$) molecule, and the forth peak located at 516-600 °C, with accompanying by a mass loss of 6% (calc 5.96) due to the decomposition of (C_2H_2 and H_2) molecules. The DSC curve showed a peak at 378.64 °C, which represented its melting point. Moreover, the TGA curve of [Co(Ldfm)(phen)] complex displayed four decomposition peaks, the first peak located at 216 to 274 °C, with accompanying by a mass loss of 16% (calc.16) related to loss of (CO and CH_4) molecules. The second peak observed at 302 to 348 °C, with accompanying by a mass loss of 10% (calc.10) corresponded to elimination of (CO_2, CH_3OH and H_2) molecules. Whilst the third peak was detected at 399-423 °C, and matched a mass loss of 8.5% (calc.8.4) that compatible with loss of (CH_3COOCH_3) molecule and the forth peak was located at 524-540 °C, with accompanying of a mass loss of 4.5% (calc. 4.4) that compatible with the loss of (CO and CH_4) molecules.

The TGA curve of the [Ni(Ldfm)(phen)] showed two peaks. The first peak detected at 68 to 127 °C, with accompanying of a mass loss of 6.8 % (calc. 6.8) which related to mass loss of (OH and H_2) molecules. The second peak detected at 185 to 250 °C, with accompanying of a mass loss of 53% (calc. 52.99) which related to decomposition of (phen, CO, CO_2, CH_3COOCH_3, C_2H_2 and CH_4) molecules. While the DSC curve exhibited a peak in 306.70°C which represented its melting point. The TGA curve of the [Cu(Ldfm)(phen)] showed only one peak at 221 to 240 °C, with accompanying of a mass loss of

47% (calc. 47) to an approximate mass loss of (phen,CH_4 and C_2H_2) molecules. The DSC curve displayed a peak in 221.68°C which represented melting point.

Table 4. Thermal analysis (TGA) abridgment data of Mn(II) , Co(II), Ni(II) and Cu(II)) complexes.

Compounds	Stage	T_{max} (°C)	(%)Estimated (calculated)		Assignment
			Mass loss	Total mass loss	Loss
[Mn(Ldfm)(Phen)]	I	62.90	3(2.99)	49.51	OH,$3H_2$
	II	212.79	22(19.99)	(49.89)	$2CH_3OH,2CO_2,4CH_4$
	II	378.64	21(20.95)		Leaving ($C_{12}H_8N_2$)
	IV	530.46	6(5.96)		$2C_2H_2$, H_2
[Co(Ldfm)(Phen)]	I	238.71	16(16)	37.79	$3CO,3CH_4$
	II	315.69	10(10)	(38.89)	CO_2,CH_3OH, $3H_2$
	III	411.49	8.5(8.4)		CH_3COOCH_3
	IV	529.13	4.5(4.4)		CO,CH_4
[Ni(Ldfm)(Phen)]	I	90.21	6.8(6.8)	59.60	$3OH,4H_2$
	II	301.67	53(52.99)	(59.79)	Phen,CO,CO_2, $CH_3COOCH_3,2C_2H_2,CH_4$
[Cu(Ldfm)(Phen)]	I	221.69	47 (47)	47.92	Phen,$5CH_4,5C_2H_2$

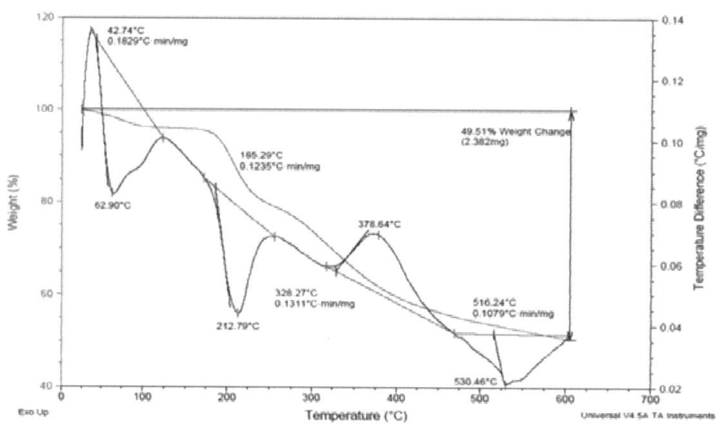

Figure 5. TGA -DSC curves of [Mn(Ldfm)(Phen].

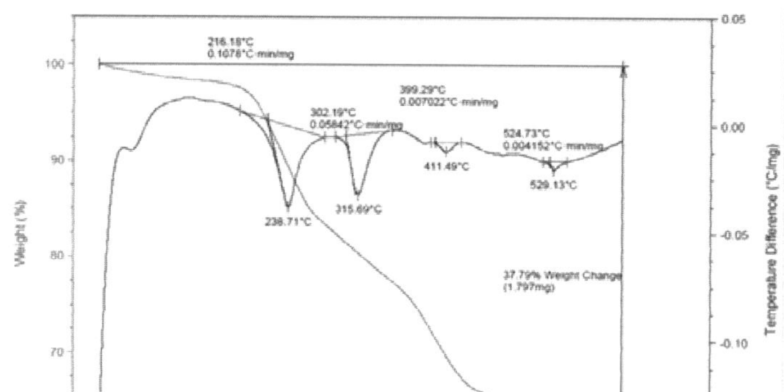

Figure 6. TGA -DSC curves of [Co(Ldfm)(Phen)].

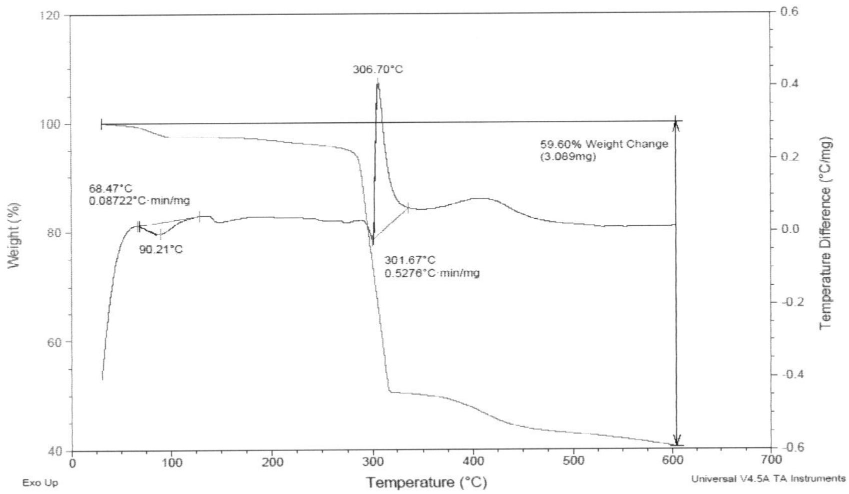

Figure 7. TGA- DSC curves of [Ni(Ldfm)(Phen)].

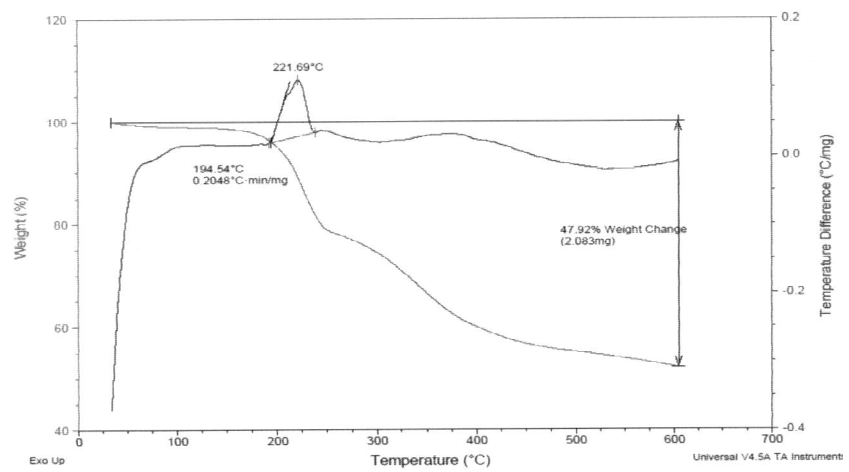

Figure 8. TGA -DSC curves of [Cu(Ldfm)(Phen)].

3.8. Antibacterial activities

The prepared compounds were screened against the growth of two types of bacteria *Staphylococcus aureus* and *Escherichia Coli.* The sensitivity of the prepared compounds was measured by using the propagation method and incubated at a temperature of 37 °C for 24 hours. It was noted that most compounds have the ability to inhibit the two type of bacteria by using the disk diffusion method [35].

It is known, that anion ligands were reducing the polarity of the central atom that carries a positive charge because they were participating by partial charges with metal in the coordination process. This characteristic can be taken as advantage by entering the compounds of the fatty layers of the membrane cell and affecting on the minerals that present in the structure of enzymes of microorganisms [36,37]. Furthermore, it can increase the biological activity of the compounds like solubility, conductivity, and bond distance between the central atom and the ligand. It was expected that the ligand and its complexes are involving in the formation of a H- bonding through (azomethine CH=N) groups with active centers of the cell and so do hydroxyl groups that have almost the same effect [37]. The results which gained from the experimental section showed that the Cu(II), Cd (II) and Hg(II) complexes have higher inhibition (as shown in figure 9) for both types of bacteria than the other complexes. The (Fe(III), Ag(I),

Cd(II), and Hg(II)) complexes that had a higher degree of toxicity towards bacteria cells. Therefore, they had high inhibition the direction of bacteria. The antibacterial activity of all compounds is depicted on the figure 10.

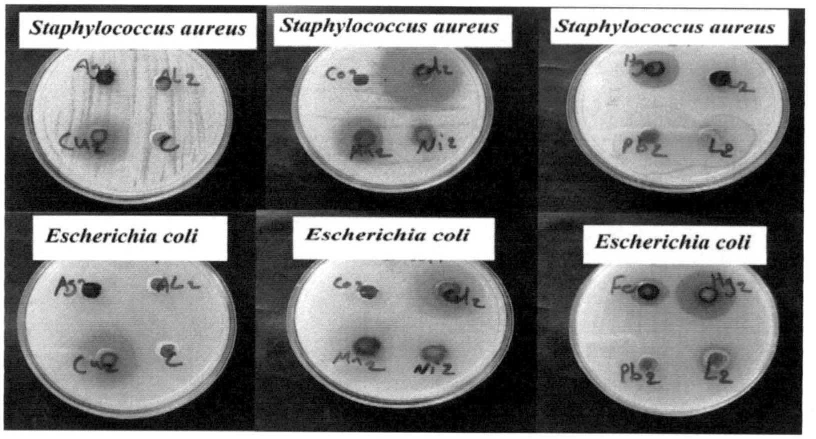

Figure 9. Photograph of zone of inhibition in mm for antibacterial activity of the prepared compounds.

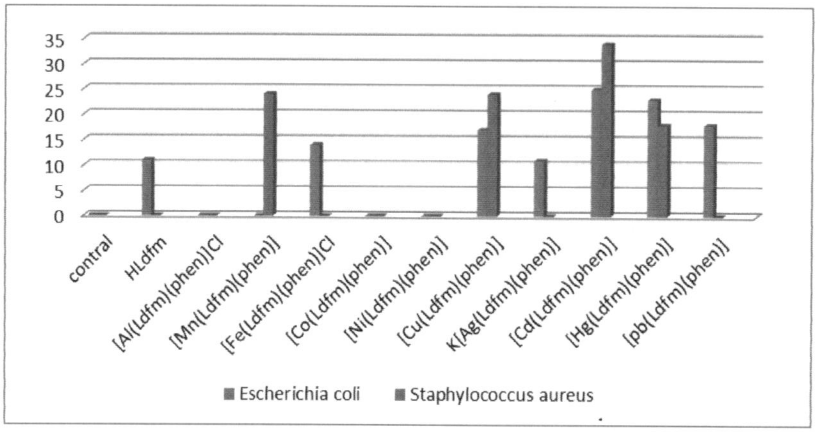

Figure 10. Histogram of antibacterial activities of compounds.

3.9. Dyeing realization

Several studies have demonstrated the use of Schiff bases and complexes in the manufacture of dyes and inks [12,38]. In this study, the prepared compounds were used in the process of dyeing cotton fibers. Wash stabilization dyes of the ligand and its complexes were tested according to ISO 105-CO3 method via three steps: (1) Production of dyeing solution – the ligand and its complexes were dissolved in ethanoic aqueous solution and soaked in a dyeing bath with mordant (alum or sodium carbonate), then some drops of (NaOH) solution was added. (2) Dyeing process – the cotton fibers was delivered in the mixture which was prepared in the step 1, then heated to 40 °C for 20 minutes with continuous stirring until getting a homogenous dye. It was extracted from the mixture, then washed with cold water and left to dry. (3) Dye stability was tested by using a solution containing (5 g/L (soap liquid) through 38°C for 30 min. After dyeing of the cotton fibers, some compounds such as ligand and (Co(II), Ag((I), Hg(II) and Mn(II)) complexes gave good results as clarified in figure (11,12) and table 5.

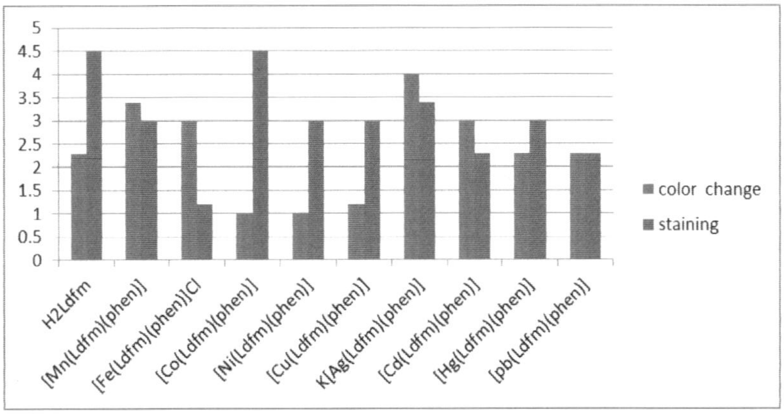

Figure 11. Fastness properties of cotton fibers for compounds.

Table 5. Fastness properties of cotton fibers dyeing for compounds.

No.	Comp.	Color change	Staining
1	H_2 Ldfm	2.3	4.5
2	[Mn(Ldfm)(Phen)]	3.4	3
3	[Fe(Ldfm)(Phen)]Cl	3	0.5
4	[Co(Ldfm)(Phen)]	1	4.5
5	[Ni(Ldfm)(Phen)]	1	3
6	[Cu(Ldfm)(Phen)]	0.5	3
7	K[Ag(Ldfm)(Phen)]	4	3.4
8	[Cd(Ldfm)(Phen]	3	2.3
9	[Hg(Ldfm)(Phen)]	2.3	3
10	[pb(Ldfm)(Phen)]	2.3	2.3

Wash fastness 5- very good , 4-4.5 good , 0.5-1 , 2-2.3 poor

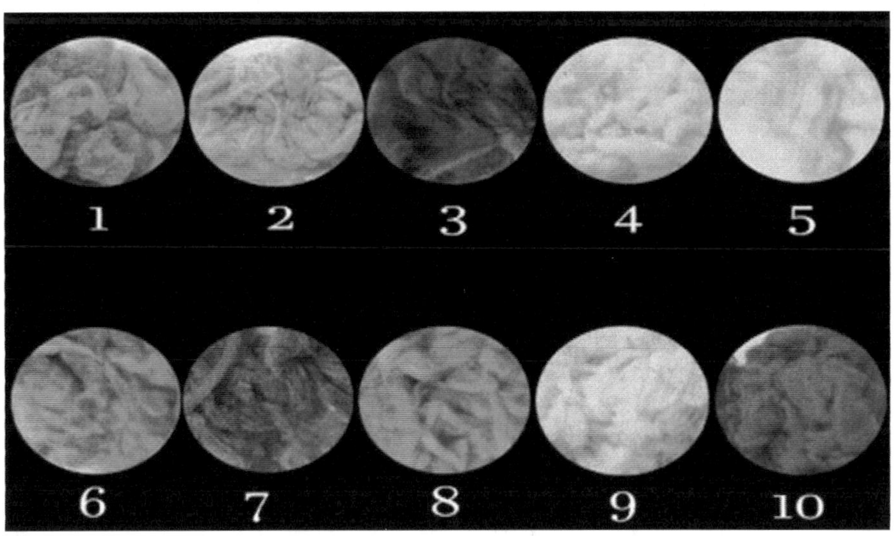

Figure 12. Shapes of cotton fibers dyeing to Schiff base and complexes. 1-H $_2$Ldfm, 2-[Mn(LdfmL)(Phen], 3-[Co(Ldfm)(Phen)], 4-[Fe(Ldfm)(Phen)]Cl, 5- [Ni(Ldfm)(Phen)]6-[Cu(Ldfm)(Phen)],7-K[Ag(Ldfm)(Phen)],8- [Cd(Ldfm)(Phen)], 9- [Hg(Ldfm) (Phen)], 10-[Pb(Ldfm)(Phen)].

IOP Conf. Series: Materials Science and Engineering 1046 (2021) 012006 doi:10.1088/1757-899X/1046/1/012006

4. Conclusions

A Schiff base ligand was prepared by using three different ways (refluxing, fusion, and grading) and its metal complexes with another ligand were prepared and identified. The ligands coordinated with metals ions through the nitrogen atom of azomethine group and oxygen of carboxylic atoms. All complexes possessed an octahedral geometric shapes. Applications of this Schiff base and its complexes have been carried out, including biological activity against two types of bacteria (*Staphylococcus aureus and Escherichia*) which showed greater effectiveness of Fe (III), Cd (II) and Hg(II) complexes. All compounds were tested as dyes on the cotton fibers and gave distinct results, especially the Schiff base and (Ag(I), Fe(III), and Cd(II)) complexes.

References

[1] Kavitha N and Lakshmi P A 2017 *J. Saudi Chem. Soc.* **21** 457
[2] Noureddin S A, El-Shishtawy R M and Al-Footy K O 2019 *Eur. J. Med. Chem.* **182** 111631
[3] Rodrigues F C, Anilkumar, N V and Thakur G 2019 *Eur. J. Med. Chem.***1 77**
[4] Zorofchian Moghadamtousi S, Abdul Kadir H, Hassan darvish P, Tajik H, Abubakar S and Zandi K 2014 *Bio. Med. Res. Int.* **2014**
[5] Zhao X Z, JiangT, Wang L, YangH, Zhang S and Zhou P 2010 *J. Mol. Struct.* **984** 316
[6] Hamdi M Z and Mustafa I A 2013 *Sci. J. Anal .Chem.* **1** 21
[7] O' Donnel and M J 2019 *Tetrahedron* **75** 3667
[8] Zhao P, Zhai S, Dong J, Gao L, Liu X, Wang L and Li L 2018 *Bioinorg. Chem. Appl.* **2018**
[9] Özdemir Ö, Gürkan P, Özçelik B and Oyard Ö 2016 *J. Mol. Struct.* **1106** 181
[10] Pervaiz M, Ahmad I, Yousaf M, Kirn S, Munawar A, Saeed Z and Rashid A 2019 *Spectrochim. Acta Part A: Mol. Biomol. Spectrosc.* **206** 642
[11] Singh K, Thakur R and Kumar V 2016 *J. Basic App. Sci.* **5** 21
[12] Nakagame R, Tsaturyan A, Haraguchi T, Pimonova Y, Lastovina T, Akitsu T and Shcherbakov I 2019 *Inorg. Chim. Acta* **486** 221
[13] Fayad N K, Al-Noor T H, Mahmood A and Malih I K 2013 *Chem. Mater. Res.* **3** 66
[14] Berrones-Reyes J, Muñoz-Flores B M, Gómez-Treviño A, Treto-Suárez M A, Páez- Hernández, D, Schott E and Jiménez-Pérez V M 2019 *Mater. Chem. Phys.* **233** 89
[15] Kailas H K, Sheetal P J, Anita P and Apoorva H 2016 *W. J. Pharma. Pharmace. Sci.***5** 1055
[16] Abood R G 2014 *J. Basrah. Res. (Sci.)* **40** 95
[17] Geary W J 1971 *Coord. Chem. Rev.* **7** 81
[18] Lever A P 1984 *Inorganic Electronic Spectroscopy Chemistry* (American Chemical Society, Washington) p 1081
[19] Cotton C A, Albert F, Wilkinson G M 1999 *Advanced Inorganic Chemistry* (New York: Wiley Interscience: Bochmann) p 555
[20] Hamdani H E L, Amane M E L, Atmani Z and Haddad M 2016 *J. Mater. Environ. Sci.* **7** 3100
[21] Syamal A, Singh M and Kumar D 1999 *Rea. Func. Polym.* **39** 27
[22] Kareem A, Arshad M, Nami S A and Nishat N 2016 *J. Photochem. Photobio. B: Bio* **160** 163
[23] Chaudhary N K and Mishra P 2017 *Bioinorg. Chem. App.* **2017**.
[24] Singh H L 2016 *Rasayan J. Chem.* **9** 614
[25] Bingöl M and Turan N 2020 *J. Mol. Struct.* **1205** 127542.
[26] Housecroft C E 2012 *Inorganic Chemistry (Barcelona, Spain :Arte sobre papel)*p 92
[27] Anacona J R, Calvo J and Almanza O A 2013 *Int. J. Inorg. Chem.* **2013**.
[28] Shakila D and Geetha 2016 *IOSR J. App. Chem.* **9** 81
[29] Basha M T, Alghanmi R M, Shehata, M R and Abdel-Rahman L H 2019 *J. Mol. Struct.* **1183** 298
[30] Özdemir Ö 2019 *J. Photochem. Photobio. A: Chem.* **380** 111868.
[31] Ossowicz P, Janus E Schroeder G and Rozwadowski Z 2013 *Molecules* **18** 4986
[32] Joseyphus R S, Shiju C, Joseph J, Dhanaraj C J and Bright K C 2015 *Der. Pharm. Chem.* **7** 265

IOP Conf. Series: Materials Science and Engineering 1046 (2021) 012006 doi:10.1088/1757-899X/1046/1/012006

[33] Singh H L and Singh J B 2012 *Int. J. Inorg. Chem.* **2012** 568797
[34] Shah A D, Raval D J, Jatakiya V P, Sen D J and Badmanaban R 2012 *Int. J. Drug Dev. Res.* **4** 205
[35] Akila E, Usharani M, Ramachandran S, Jayaseelan P, Velraj G and Rajavel R 2017 *Arab. J. Chem.* **10** 2950
[36] Abou-Hussein A and Linert W 2014 *Mol. Biomol. Spectrosc.* **117** 763
[37] Abu-Khadra A S, Farag R S and Abdel-Hady A E D M 2016 *Am. J. Anal. Chem.* **7** 233
[38] Dhar D N and Taploo C L 1982 *J. Sci. Ind. Res.* **41** 501

IOP Conf. Series: Materials Science and Engineering 1046 (2021) 012007 doi:10.1088/1757-899X/1046/1/012007

Green data collection strategy in the science domain for environmental sustainability

Davidson E Egirani[1,*] and Nabila Shehata[2]

[1]Head of EAAWRE, Niger Delta University, Wilberforce Island, Nigeria
[2]Environmental Science and Industrial Development Department, Faculty of Postgraduate Studies for Advanced Science (PSAS), Beni-Suef University, P.O. Box 62511, Beni-Suef, Egypt.

*E-mail: eenonidavidson@yahoo.com

Abstract. There is a recognition by modern communities world-wide, that the steadily increasing level of activities in the science domain and population growth, lead to an increase in environmental pressure on critical areas. There is a need, therefore, for stakeholders to critically appraise principles, policies, and practices that could lead to the best practices required for sustainable science activities and environmental management that is driven by green data collection, green principles, and green practices. The models involved in green data collection in science include finance-environment-community-science domain model (FECS), environment-community-science domain-finance model (ECSF), and the science domain, finance, environment-community model (SFEC). SFEC is a novel framework and is being applied in the management of active science domain, abandoned science domain, thus, providing reasonable economic and environmental benefits to the science players and non-science communities. The green data collection strategy is for enhancing social, economic, and environmental resource performance. This strategy has been tested in the Philippines. The formulation of the green data collection strategic framework involved a desktop review of previous and existing local, regional, and national data collection strategies. Consequently, visits are made to active science domain and discussions held with stakeholders. In this paper, the role of information, education, and communication have been revisited and fits adequately into the novel "green data collection model" being advanced for all players in the science domain. The absence of this framework poses the most challenging aspect of prosperity offered by the quality, quantity, diversity, and the sustainability of the science domain. Therefore, this paper has discussed the content of the green data collection strategic plan, anchored on a sustainable-controlled planning cycle. This approach could drive and sustain the harmony between man, science, and the environment.

1. Introduction

The primary concern of stakeholders in the science domain was to ensure the sustenance of financial gains from the scientific activities to the general economy. Some of these stakeholders include the departments of education, ministries of education and their agencies, universities, and research institutes. The strategy of money first, environment and sustainability as secondary drove the image of the science domain underground with so many adverse consequences [1].

This engagement was aimed at crafting a strategic science-sectoral plan anchored on best practices at the science domain [2-4]. In crafting this strategic plan, there was a need for a balance shift of emphasis on the sustainability of the science sector that is derived from responsible scientific activities. The sustainability of the scientific environment and the development of the scientific communities form parts of best practices required for productive activities at science centers. Therefore, the collection of scientific data from the science sites must be seen to be environmentally friendly by the science community, the regulator, and the regulated [4-8].

The models involved in information, education, and communication strategy include a finance-environment-community-science-domain model (FECS); environment-community-science domain-finance model (ECSF); and the science domain, finance, environment-community model (SFEC). The perception for or against large scale scientific activities exist world-wide. IEC form, therefore, the focal point for the development of the science domain. IEC aimed at addressing existing obstacles and challenges take many forms, and it is dependent on the model to be adopted [9-12]. In all these models, the most important component of interest comes first, descending the line in the order presented. A FECS, ECSF, and SFEC models, therefore, consider finance, environment, and science domain as most important respectively (tables 1 and 2).

Table 1. Information, education and communication strategies for science stakeholders

Source of obstacle	Nature of obstacle	Form of IEC
Community/General public	Scepticism of stakeholders commitment	1. Community-based documentary video on good practice during festivals 2. Newsletter on the benefits of scientific activities 3. Radio program on meet the miner 4. Community consultations 5. Billboards on good practice 6. Community-based interactive theatre show
Science sectors	Conflict at science centres despite common effects of activities	1. Reduce single-mindedness, intolerance and competitiveness through inter-sectoral forum 2. Do inter-sectoral seminar workshop and IEC to create common ground for co-existence
Regulator-Partners (education, environment, community)	Lack of political will and moral values to process and implement science-based policies	1. Reduce bureaucracy through regular inter-agency and inter-sectorial forum
Regulated Partners such as scientist and science processors	Illegal activities and non-compliance to basic scientific standards and laws	1. How community-based documentary video on the implications of illegal scientific activities 2. Non-compliance to other stakeholders 3. Appraise procedures for legalizing operations within a given time frame

Before the crafting and implementation of the science domain-sectorial plan, there was imbalance and misplacement of priority in the kind of model to be adopted for data collection strategy in the science domain. The science centers lacked basic amenities for safe data collection.

The waste generated at science centers lacked effective treatment, management, and utilization. Herein, there was evidence of the lack of effective information, education, and communication strategy by the stakeholders.

In revisiting the IEC strategy and subsequent implementation of the SFEC model, professionalism and improved scientific data collection practices became the targets of interest [13-17]. Thus, the use of the SFEC model has resulted in greater performance, productivity, and sustainability of the science sector. The author recognizes that sound implementation of scientific principles and policies give birth to sound scientific practices at science centers. A need to, therefore, revisit existing science-related legislation existed. To restore the image of the science domain, protect the environment, and balance sustainability, this paper revisited the activities in the science domain.

Table 2. Some science centres in the Philippines

Operating centers	Location	Activity
Calos P. Lademora	Brgy.3, Karaos, San Francisco	Mineral science processing
Nepthali Salcedo	San Francisco	
Jessie Aguirre	San Francisco	
Emerita Salcedo	Brgy. 5, San Francisco	
Darna Daytoc	Brgy. 2, San Francisco	
Carlo P. Lademora	Brgy. 5, San Francisco	
Generoso Camarin Sr.	San Andres, Bunawan	
PHILSAGA Corporation (PSMC)	Bayugan 3, Rosario	

2. Methodology of Study

The study methods involved a desktop review of previous and existing local, regional, national, and international data collection strategy in the science sector. These activities were supported with visits to science-centers, and discussions were held with stakeholders. These steps were aligned to a novel scientific- data collection model (SFEC) that would promote sustainability of the science sector (Eq. 1):

$$\partial S_s / \partial t = k(f)^{(e)^{(c)}} \tag{1}$$

Herein, S denotes the science domain at center s and time t (d), k denotes a constant that is dependent on the number of stakeholders; f denotes the finance required to sustain the scientific environment *e* and scientific community *c*.

SFEC was used to revisit scientific activities in the Philippines, aimed at strengthening existing scientific regulations and mitigate the environmental effect of small-scale scientific activities. Also, the revisit was to encourage the implementation of small-scale scientific activities as a livelihood source

Government Units are increasing becoming incapacitated to carry out scientific activities due to lack of administrative and technical capability to monitor the science-centers. The department of education, therefore, needed to strengthen government Units' monitoring and law enforcement capabilities. In the use of the SFEC, capability building may, therefore, include training of Government Unit personnel and acquisition of any equipment required in this regard. The principles of SFEC were used to harmonize small-scale science contract covered by small-scale science permits. The former aims to grant contracts to science domain cooperatives while the latter aims to grant a permit to individual, entity, or organization [18-20]. Therefore, SFEC as an IEC model is more than providing information to those who are the sources of obstacles. It is all about starting a participatory process, ensuring that stakeholders support existing scientific policies, identify and involve all stakeholders in decision making, and implementation of these policies. This model ensures the involvement of non-governmental organizations that have local credibility in the campaign, identification, and involvement of individuals in the scientific communities with specific talents and skills. Also, capacitating scientific community to take the lead in the Information, education, and communication strategy, was carefully considered since using this model.

An effective scientific IEC mechanism to appraise some of the common activities in the science and art formed the basis of the inter-sectorial discussions aimed at achieving best-practices. Some of the common activities which cause environmental degradation and social problems to scientific communities were revisited.

Interviews were conducted with stakeholders (ie. based on SFEC model) to ascertain the implementation of existing scientific rules and regulations, the condition at science centers and efforts being made by stakeholders to improve on their performance and sustainability. Some of the stakeholders interviewed included the Mayors of the Municipalities, small scale scientific operators, large scale scientific companies, department of education, tribal associations, environment management bureau, and national commission on indigenous people.

3. Findings

About 2.5% of the inhabitants of the Philippines are engaged in science-related activities. There is a limited large-scale scientific operation currently being done. Regular visits made to these scientific centers and application of SFEC model assessment tool revealed a limited scientific approach to the collection of data and the protection of scientific participants. Also, the tool revealed the level of scientific outcomes and the necessity for secondary review of outcomes.

SFEC tool revealed more risk in granting individuals a permit compared with granting the same to a cooperative, individuals and entities federated to capacitate their technical deficiency. The option of repealing scientific-related laws, therefore, does not arise. Instead, existing laws should be given equal weighting to ease administration of the scientific policies. This means cooperatives and individual permittees should co-exist and political concerns (pressures) should be balanced with political will by the stakeholders to enforce them. The author found that there is enough legislation to ensure the sustainability of the science sector and provide sound practices at the scientific centers.

For effective implementation of existing legislation at science centers, however, there is a need for a science domain code drawn up through a science community-based participation. Therefore, the author adopts the science domain, finance, environment-community model (SFEC) to process a green data collection strategy. In this model, what happens at the scientific centers is the most important. This model is guided by the funding of the science sector by the stakeholders. Also, the sustainability of the science sector is anchored at some point on effective community-based participation engineered by the funding of a sound environmental rehabilitation strategy.

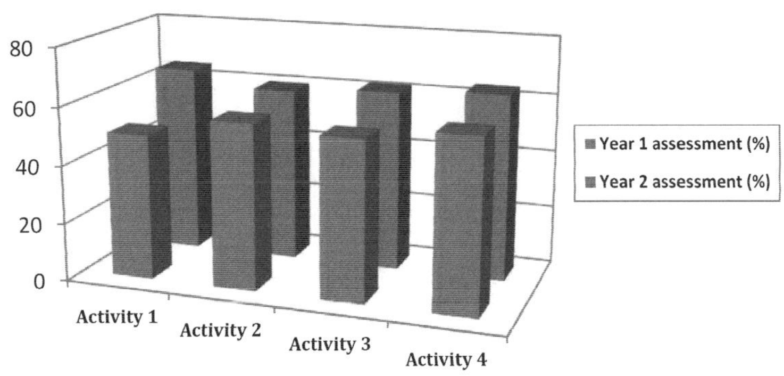

Figure 1. Activities at science centres in the Philippines using the SFEC model where: activity 1: community consultations, activity 2: inter-sectorial seminar-workshop, activity 3: appraise procedures for legalizing operations and activity 4: re-enacting waste utilization strategy.

The use of the SFEC model improved community participation at mine-sites from 50%-65% in the second year of intervention (figure 1). Also, SFEC model was used to promote inter-sectorial seminars and workshops, thus leading to an improvement in the conditions of scientific centers from 57%-60% in the second year of intervention. A revised procedure for securing the scientific permit based on the SFEC model provided a 7% increase in the legalization of scientific operations. A strategy on re-enacting waste utilization gave 5% increase in the recovery of gold ore from tailings

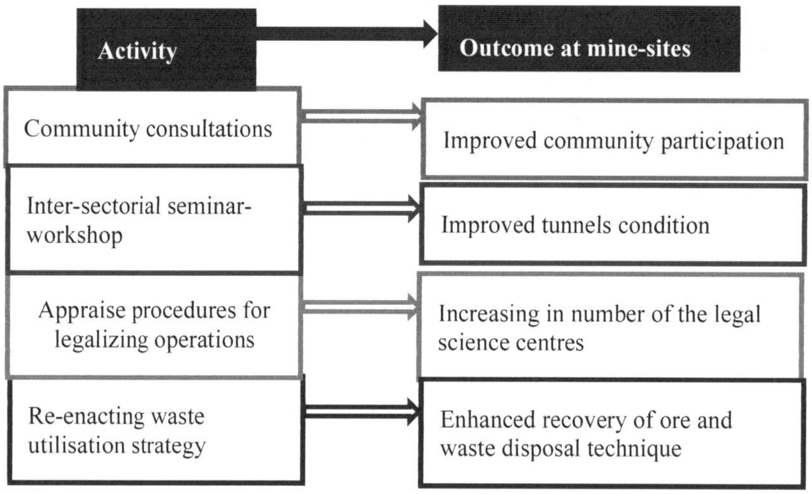

Figure 2. Activities and outcome at science centres in the Philippines using the SFEC model.

4. Discussion

The outcome of the SFEC model (figure 2) is predicated on the fact that scientific communities became incorporated as stakeholders in the science sector. SFEC activities are centered on the benefits science communities and scientists derive from safe data collection practices. A revisit of scientific activities based on the SFEC model ensured the need to secure the scientific permit for scientific activities was made less cumbersome. In using the SFEC strategy, wastes generated by scientific activities are made to undergo natural and physical processes of elimination that are environmentally friendly. At scientific centers, two activities of interest are identified. The condition of the center in terms of safety need be addressed.

Also, toxic waste generated, and the treatment, management, and utilization of the waste should be of interest to stakeholders. Ordinarily, most scientific data collection strategies use chemicals. For example, the gold industry of today applies heap or valley fill leaching technique; agitation leaching followed by carbon-in-pulp (CIP), and agitated carbon-in-leach (CIL). Furthermore, the use of cyanide was appropriate for fine-grain gold disseminated in the ore. In the extractive scientific centers, the waste from the process is turned into waste. Series of tanks are used in carbon-in-pulp and carbon-in-leach cyanidation techniques. In this process, finely meshed gold ore is turned into slurries using a leaching solution. Consequently, the gold-cyanide mix is charged with activated carbon. In carbon in pulp process, the leaching and recovery processes are done in two separate tanks. Also, in carbon in leach cyanidation, the operations are done in a single tank operation with a higher rate of gold recovery. Therefore, a lot of waste containing toxic heavy metals is generated at these science processing sites.

Depending on the data collection method, technology, and type of operation, science and processing operators have to employ waste management mechanism in consonance with sustainable development concepts and as embodied in their environmental compliance certificates. Therefore, appropriate measures and approaches in handling waste that will diminish and mitigate adverse consequences to the environment and the surrounding scientific communities should be implemented using the SFEC model.

The potentials for science and mineral science processing centers to generate high acidity and release contaminants is reduced and controlled; if mill waste is treated and managed by impounding waste below the water table to avoid contact with atmospheric oxygen. The use of clay liners to reduce leakage of leachates, the diversion of wastewater to treatment plants, and treating them with bio-sorbents has been established using the SFEC model. Also, the introduction of seaweeds, carbonate and carbonate beds to reduce metal content and increase the pH of wastewater are additional measures being established using the SFEC model.

The control and management of waste require the usage of a large volume of water. Therefore, it is recommended that all scientific processing operators have a centralized borehole generating water for effective dilution of the waste for easy natural degradation [17]. Also, it shall be the responsibility of science processing operators in coordination with the Government Units to dispose or utilize all treated waste in a manner that will comply with existing environmental standards.

The application of the SFEC to process a green data collection strategy in the science domain has worked out an improved framework for best practices in the science domain in general and the mineral science sector in particular.

5. Conclusion

In this study, stakeholders have critically revisited scientific principles, policies, and practices that govern scientific operations world-wide and applied the same in the Philippines. The author adopts the science domain, finance, environment-community model (SFEC) to process a green data collection strategic framework. The application of this model has promoted the best practices required for sustainable scientific collection strategy since it is community-based and driven by all stakeholders.

This novel regulatory framework is put in place by the author after a two-year investigation. This framework is being applied in the management of active and abandoned scientific centers, thus, providing significant economic and environmental benefits to the mineral science sector. The green data collection strategy is for enhancing social, economic and environmental resource performance. In this work, the role of information, education, and communication strategy have been revisited; and this strategy fits adequately into the novel SFEC model for the science domain. The absence of this framework poses the most challenging aspect of prosperity offered by the quality, quantity, diversity, and sustainability of the science domain in general and the mineral science sector in particular. The application of the SFEC model in the administration of the science domain would balance the ecological requirements and the sustainability of the science domain.

Acknowledgement

The author is grateful to the VS Organization of the United Kingdom for providing sponsorship for the research work. Also, the author is grateful to the Provincial Government of Agusan del Sur for providing the platform and logistic support for the research work.

References

[1] Bernaldez I and Soriano V 2017 *In: International Conference on Applied Human Factors and Ergonomics. Springer* 851

[2] Bernaldez I and Soriano V 2019 *In: International Conference on Applied Human Factors and Ergonomics. Springer* 280

[3] Boquet Y 2017 *In: The Philippine Archipelago. Springer* 779

[4] Borde R and Rasch E D 2018 *J Dev Soc.* **34** 195

[5] Castro E T and Hernandez A A 2019. *In IEEE 15th International Colloquium on Signal Processing & Its Applications* 243

[6] Buenafe-Ze M A, Minter T and Telan W G 2016 *J. Entwickl. Polit.* **32** 67

[7] Cadiz A, Camacho V and Quizon R 2016 *Southeast Asian J. Trop. Med. Public. Heal.* **47** 833
[8] Camba A A 2016 *Austrian J. South-East Asian Stud.* **9** 69
[9] Cedeno K, Edraki M, McIntyre N, Huynh T and Callow I 2019 *Environ. Geochem. Health* **41** 563
[10] Køster-Rasmussen R, Westergaard M L, Brasholt M, Gutierrez R, Jørs E and Thomsen J F 2016 *New Solut. A. J. Environ. Occup. Heal. Policy* **25** 567
[11] Merilles M L D and Silva B B 2018 *Environ. Asia* **11** 91
[12] Migo V P, Mendoza M D, Alfafara C G and Pulhin J M 2018 *In: Water Policy in the Philippines. Springer* 87
[13] Opiso E M, Aseneiro J P J, Banda M H T and Tabelin C B 2018 *Waste Manag. Res.* **36** 269
[14] Simbulan R G 2016 *Peace. Rev.* **28(1)** 29
[15] Soriano V and Bernaldez I 2019 *In: International Conference on Applied Human Factors and Ergonomics. Springer* 298
[16] Thomashausen S, Maennling N and Mebratu-Tsegaye T 2018 *Resour Policy* **55** 143
[17] Trinidad Jr J T. *In: Proceedings of the 2019 3rd International Conference on Digital Signal Processing* 124
[18] Verbrugge B and Besmanos B 2016 *Resour. Policy* **47** 134
[19] Verbrugge B, Cuvelier J and Van Bockstael S 2015 *J. Rural. Stud.* **37** 50
[20] Wayland J 2019 *Ann. Am. Assoc. Geogr.* **109** 1006

Dielectric and dynamic mechanical properties of Rochelle salt / PVA composites

E O Taha[1*] and G M Nasr[2]

[1] Egyptian Petroleum Research Institute, Nasr City, Cairo 11727, Egypt
[2] Department of Physics, Faculty of Science, Cairo University, Giza 12613, Egypt

*Email: [1]eman.omar2006@gmail.com; [2]rrrrrgmail@yahoo.com

Abstract. A new type of PVA composites containing Rochelle salt was prepared using the casting method. The electrical, Dynamic mechanical analysis (DMA) and thermogravimetric analysis (TGA) tests were performed on 0, 20, 30, and 50 weight% PVA based Rochelle salt to identify the effect of salt on the electrical, thermal and mechanical properties of composites. The results showed that the dielectric properties of composites were greatly improved by the addition of salt. The dielectric constant by 28.5%, 98.4% and 236.5% for composite contents 20 wt%, 30 wt% and 50 wt% Rochelle salt, respectively. The observations of dynamic mechanical analysis indicated that the adding of salt reduces both the storage modulus and affect the glass transition temperature of the composites. Moreover, the thermal stability and char yield of the PVA composites were highly affected by the Rochelle salt content.

1. Introduction

A massive number of polymers and polymer blends such as polymethyl methacrylate (PMMA), polyvinylpyrrolidone (PVP), polyvinyl alcohol (PVA), PVP-PVA, polyethylenedioxythiophene: polystyrenesulfonate-polyvinyl alcohol and polypyrrolepoly(ethylene glycol) have been used as dielectric layers recently [1-5]. Mixing physically two or more polymers with inorganic materials in order to create polymer composites with modern, specific properties is one of the most effective ways to improve dielectric behaviour. [4 6]. The dielectric, electrical and mechanical properties of the material composites are important for its applications in electronic devices [7]. Numerous articles on PVA composites were reported. For instance, the studies on the dielectric properties of PVA-InTio [8] and PVA-Al_2O_3 [9] have been reported by Sugumaran et al. Also, PVA/ $BaTiO_3$ by Tuncer et al. [10] and PVA/ Al_2O_3 by Nigrawal et al. [11]. The studies on the mechanical properties of Reduced graphene oxide (RGO) doped PVA has been reported by Aslam et al. [12] and PVA/ Cellulose nanofibrils (CNF) nanocomposite has been reported [13].

In this sense, PVA composites are attractive for electrical applications because of their relatively high dielectric strength, high flexibility, lightweight and better chemical stability [14,15]. Morsi et al. [16] examined the structural properties and the dielectric behavior of polyvinyl alcohol (PVA)/ carboxymethyl cellulose (CMC) blend filled different contents of biosynthesized Au nanoparticles to be used for radio/audio frequency operating microelectronic devices. Anlin Golda et al. [17] have prepared a novel $Bi_{1-x}Sm_xFeO_3$/PVA nanocomposite film with enhanced dielectric properties to be used in the development of transient electronics of microwave devices. Ghoshal et al. [14] have investigated dielectric properties of PVA composites filled different concentrations of the alizarin red and have confirmed its potentially used in optical filtering and flexible optoelectronic applications.

This work also aimed to enhance the dielectric behaviour of the PVA matrix by means of incorporating different concentrations of Rochelle salt in the PVA matrix in additional to improve mechanical properties PVA composites by increase its flexibility to be able to use in applications like organic flexible electronic devices.

2. Experimental Methodology

2.1. Materials and sample preparation

Polyvinyl alcohol (PVA) powder [Sigma- Aldrich] was dissolved in double distilled water at 80℃ and stirred for 4 h in order to ensure uniform dispersion. The Rochelle salt (Potassium sodium tartrate tetrahydrate, $KNaC_4H_4O_6$ $4H_2O$) was papered by mixing the two ingredients; potassium bitartrate ($KHC_4H_4O_6$)and Sodium carbonate (Na_2CO_3) in distill water at temperature approximately 55 ^oC. To prepare Rochelle salt/ PVA composites, different weight percentages (0, 20, 30 and 50 wt. %) are added to the above solution of PVA in water and stirred for 1 h. The mixture was then cast in a glass dish and the sample was left to dry for a week at room temperature.

2.2. Characterizations

Dielectric properties were measured with a Hioki 3532-50 LCR Hitester in frequency range from 50 Hz to 5 MHz. Dynamic mechanical analysis (DMA) was conducted on a Triton Instruments, running in the tension mode at frequency of 1 Hz. over the temperature range from -50 to 80 °C at a scanning rate of 5 °C/min. The thermal stability was performed using Thermogravimetric Analysis (TGA) - Mettler Toledo where the samples were heated in N_2 atmosphere from room temperature to 600 ^oC with 20 ^oC/minute heating rate.

3. Results and Discussion

3.1. Dielectric properties

Complex dielectric constant as function of frequency (f) is represent by equation: $\varepsilon^*(f) = \varepsilon'(f) - i\varepsilon''(f)$. The real part ε'(f) is the relative dielectric constant and the imaginary part ε"(f) is the dielectric loss. The relationship between the imaginary and the real part (ε"/ ε') is called "dissipation factor" and denoted by tan δ which referring to the angle between the voltage and the charging current [18].

The change in the dielectric permittivity, ε', and the dielectric loss, ε", versus frequency of Rochelle salt/PVA composites is shown in figure 1 (a, b) for three concentrations of the Rochelle salt (RS) (20 wt%, 30 wt% and 50 wt%)at room temperature. Dielectric permittivity and dielectric loss reach high values in the low frequency region, which decrease with the increasing frequency. This behavior demonstrates a process of relaxation attributable to an interfacial polarization, known as the Maxwall-Wagner-Silars effect, which is a phenomena of accumulation of charges on interfaces in heterogeneous media. [18]. Also, figure 1 (a, b) shows that the ε' of the PVA composite increased from 63 for pure PVA at frequency 50 Hz to 212, 125 and 81 for composite contents 50 wt%, 30 wt% and 20 wt% Rochelle salt (RS), respectively, which represents an increase in dielectric constant by 236.5%, 98.4% and 28.5% for composite contents 50 wt%, 30 wt% and 20 wt% RS, respectively. This increase might be due to the high dielectric constant of RS which approximately is 2000 at room temperature [19] resulting the accumulation of carriers at the internal surface of a PVA matrix producing interfacial polarizations. The dielectric losses are also influenced by the concentrations of Rochelle salt (RS) in the PVA and increased from 104 for pure PVA at frequency 50 Hz to 395, 541 and 455 for composite contents 50 wt%, 30 wt% and 20 wt% Rochelle salt (RS), respectively.

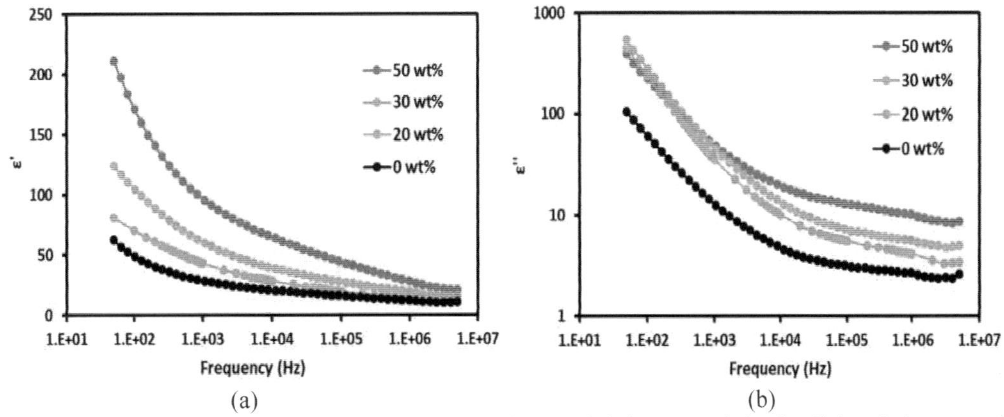

Figure 1. The frequency dependence of (a) dielectric permittivity, ε', and (b) the dielectric loss, ε" of Rochelle salt / PVA composites.

Figure 2 shows AC conductivity for the Rochelle salt / PVA composites as frequency function. The values of conductivity show great dispersion of frequencies. Conductivity corresponds to constant values within the low frequency range and sometimes increases slowly with frequency, whereas it shows an exponential frequency dependence at high frequency. The conductivity tends to be following the universality rule stated in the following Eq.:

$$\sigma_{AC}(f) = \sigma_{DC} + Af^{s} \quad (1)$$

where σ_{DC} is the DC limiting value of conductivity, f the frequency, A fitting parameter and s frequency exponent parameter related to the under test material. The value of 's' could be calculated from the slope of a plot of log σ_{AC} versus log f. The values of 's' were found to be 0.69 for pure and 50 wt% composites and 0.61 for 20 wt % and 30 wt% composites.

AC conductivity sums all processes of dissipation, including dipolars and real ohms due to migration charges for the carrier. The field alternation is sluggish at low frequencies, allowing charges inside the composite to drift over large distances. The insulating existence of the host material imposes heavy restrictions on this movement, thereby contributing to low conductivity values. As frequency rises, field alternation becomes faster, and charges cannot migrate over long distances. Charges can, therefore" jump" between neighboring conductive sites and thus contribute to the overall conductivity reported

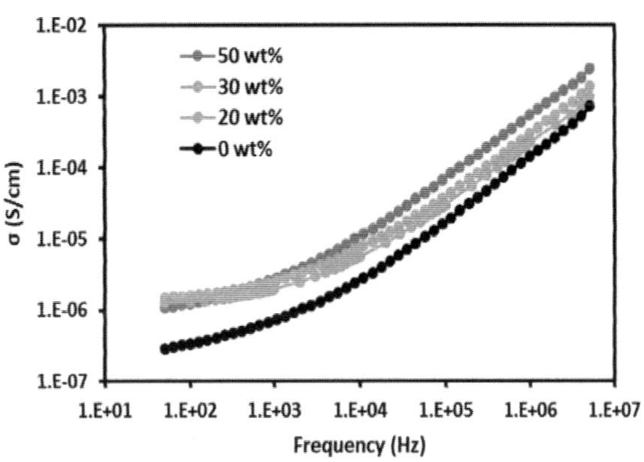

Figure 2. The frequency dependence of conductivity of Rochelle salt / PVA composites.

3.2. Dynamic mechanical properties

The temperature impact of Rochelle salts / PVA composites is illustrated in figure 3(a) on the storage module at 1 Hz frequency. When the temperature rose in the first glassy region, the storage Module decreased in three distinct areas of all PVA composites; due to the reduction in segmental mobility, the Module exhibited the highest module values in the first glassy region. The modulus dramatically reduces the transition temperature of the glass (Tg) for the second transition region. In the third rubbery region, due to the high mobility of polymer chains and their energy dissipation, the modulus has low values. The storage modulus value for purity PVA was higher than all composites (4.9 GPa at 10 ° C), and with the increase in Rochelle salt (RS) the storage modulus decreases.

Figure 3(b) shows a variation of the damping parameter (tan δ) for various Rochelle salt / PVA composites. The temperature of the glass transition (Tg) is a temperature at which maximum parameter of peak damping is attained. The tan δ can shift to a higher temperature by increasing the content of Rochelle salt. The values of Tg are 29.7 °C, 26.4 °C,41.3 °C and 46.9 °C for pure PVA, 20 wt% RS, 30 wt% RS and 50 wt% RS, respectively. For 30 wt% RS and 50 wt% RS composites the peaks appear between -18 °C and -24 °C are the Curie points of RS [20].

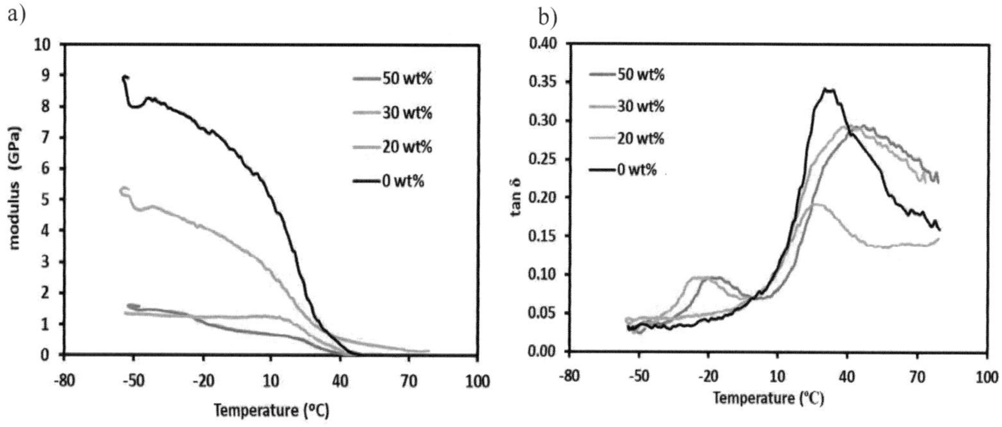

Figure 3. Temperature dependence of (a) storage modulus and (b) tan δ of Rochelle salt / PVA composites.

3.3. Thermal Properties:

Thermogravimetric Analysis (TGA) test was utilized to investigate the thermal stability of Rochelle salt / PVA composites (figure 4). The decomposition temperature of pure PVA is approximately 217 °C and it shifted towards higher temperatures as concentrations of RS increase. The decomposition temperatures of composites content 20 wt% RS, 30 wt% RS and 50 wt% RS are approximately 231 °C, 236°C and 258°C, respectively. Moreover, The Char residue is a significant affected by RS content and increases with RS content and its values are 13 mg, 22 mg and 25 mg for 20 wt%, 30 wt% and 50 wt% RS / PVA composites, respectively.

4. Conclusions

In this paper, 20 wt% RS, 30 wt% RS and 50 wt% RS were well dispersed in the PVA matrix and used to fabricate Rochelle salt / PVA composites using the casting method. The dielectric, dynamic mechanical and thermal properties were examined. The results showed that the dielectric properties of Rochelle salt / PVA composites were improved with increase Rochelle salt content. The dielectric constant increases by 236.5%, 98.4% and 28.5% for composite contents 50 wt%, 30 wt% and 20 wt% RS, respectively. The results also showed that the storage modulus and the glass transition temperature were significantly affected by adding of Rochelle salt where the storage modulus decreases with Rochelle salt and the elasticity of all PVA composites increases. Moreover, the thermal stability of the

Rochelle salt/PVA composites was also highly affected by the Rochelle salt content. The decomposition temperatures are approximately 231°C, 236°C and 258°C for 20 wt% RS, 30 wt% RS and 50 wt% RS PVA composites, respectively.

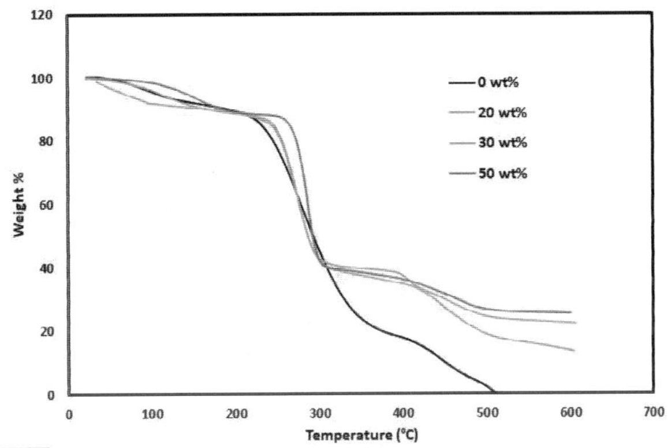

Figure 4. TGA graph of Rochelle salt / PVA composites

References

[1] Chen C H, Kine A and Nelson R D 2015 *Synth. Met.* **206** 106
[2] Deshmukh K, Ahamed M B and Polu A R 2016 *Mater. Sci. Mater. Electron.* **27** 11410
[3] Yee L M, Ekramul Mahmud H N M and Kassim A 2007 *Synth. Met.* **157** 386
[4] Ramamohan K, Achari V B S and Sharma A K 2014 *Ionics.* **21**1333
[5] Van Etten E A, Ximenes E S and Tarasconi LT 2014 *Thin Solid Films.* **568** 111
[6] More S, Dhokne R and Moharil S 2018 *Poly. Bulletin.* **75** 909
[7] Nangia R, Shukla N K and Sharma A 2018 *High Perform. Polym.* **30** 918
[8] Sugumaran S, Bellan C S, Muthu D, Raja S, Bheeman D and Rajamani R 2015 *Polym. Adv. Technol.* **26** 1486
[9] Sugumaran S, Bellan C S and Nadimuthu M 2015 *Iran Polym. J.* **24** 63
[10] Tuncer E, Duckworth R C, Sauers I, James D R and Ellis A R 2007 *Annu. Report Conf. - IEEE.* 225
[11] Nigrawal A and Chand N 2013 *Prog. Nanotechnol Nanomater.* **2** 25
[12] Aslam M, Kalyar M A, and Raza Z A 2017 *Appl. Phys. A.* 123 424
[13] Niazi M B K, Jahana Z, Bergb S S and Gregersen Y W 2017 *Carbohydr. Polym.* **177** 258
[14] Ghoshal D, Bhattacharya D, Mondal D, Das S, Bose N and Basu M 2020 *Appl. Phys. A* **126**
[15] Aslam M, Kalyar M A and Raza Z A 2018 *Polym. Eng. Sci.* 2119
[16] Morsi M A, Oraby A H, Elshahawy A G and Abd El-Hady R M 2019 *J. Mater. Res. Techol.* **8** 5996
[17] Anlin Golda R, Marikani A and Alex E J, 2020 *J. Mater. Sci.* **31** 324
[18] Mittal V 2012 *Characterization Techniques for Polymer Nanocomposites* (Wiley Online Library)
[19] Kandpal B and Upadhyay T C 2016 *J. Appl. Phys.* **8** 53
[20] Lemaire E, Ayela C and Atli A 2018 *Smart Mater. Struct.* **27** 025005

Shear bond strength of bio-active cement versus self-adhesive resin cement with enamel and dentin when bonded to zirconia in wet and dry conditions (in-vitro study)

M A Elattar [1,*], A H Khalil [2] and J F Younis [3]

[1] Department of Fixed Prosthodontics, Faculty of Oral and Dental Medicine, Misr International University, KM 28 Cairo – Ismailia Road Ahmed Orabi District, Cairo, Egypt.

[2] Department of Fixed Prosthodontics, Faculty of Faculty of Oral and Dental Medicine, Cairo University, El Manial, Cairo, Egypt.

[3] Department of Fixed Prosthodontics, Faculty of Oral and Dental Medicine, Misr International University, KM 28 Cairo – Ismailia Road Ahmed Orabi District, Cairo, Egypt.

*E-mail: Mohamed060940@miuegypt.edu.eg; Elattar89@gmail.com

Abstract: ACTIVA bio-active cement is resin-modified glass-ionomer cement that was introduced for re-mineralizing of the disintegrated joint. Due to the lack of data about this bioactive cement, this study evaluated the shear bond strength of bioactive cement and self-adhesive resin cements to tooth substrate (enamel and dentin), in wet and dry conditions when bonded to zirconia. In this study, 40 extracted mandibular molars caries-free were collected. Each tooth was cut by diamond disk (bucco-lingual section) to expose the dentin and enamel. Teeth were randomly divided into two groups enamel and dentin (n=20). Each group was further divided into two subgroups; wet and dry samples (n=10). Each subgroup was divided according to the cement used, into resin cement or bioactive cement (n=5) and bonded to zirconia. Shear bond strength was determined, after thermo-cycling. Statistical analysis of the results showed that there was no statistically significant difference between shear bond strength of the two cement types either with wet enamel or dentin. With both cements, enamel showed statistically significantly higher mean shear bond strength than dentin. It was concluded that, both cements showed comparable results with significant superiority to the self-adhesive resin cement.

1. Introduction

Dissolution of cement is a risk factor that diminishes marginal integrity of the cast restorations and results in bacterial ingress and secondary decay which leads consequently to its failure. Therefore, bio-active cement was introduced to overcome the shortcomings of the contemporary cements to increase the success rate of restorations [1]. The principle of bioactive restorative materials is not recent it existed since the introduction of glass-ionomer into the market that had the ability to release fluoride and resist secondary decay but lacked strength, toughness and instability due to water sorption [2]. Then resin modified glass ionomer (RMGI) that combines the advantages of glass ionomer cement and resin cement, was developed but had questionable biocompatibility and limited bioactivity [3].

ACTIVA bio-active cement was introduced that comprises of RMGI besides modified resin matrix (Blend of diurethane, other methacrylate with modified polyacrylic acid and bioactive nanoparticles), that is capable of penetrating dentin through micro-mechanical interlocking. It forms a combined ionic bond between polyacrylic acid and hydroxyapatite and a micro-mechanical interlocking with dentin tubules and collagen. It does not elicit post-operative sensitivity and can tolerate PH cycles, leach and recharge calcium, phosphate and fluoride according to the manufacturer [4]. In 2015, Pameijer et al. [6] affirmed that Activa bioactive RMGI flexural strength is equivalent to that of flowable composite and greater than that of glass ionomer cement (GIC), concerning biocompatibility Zemener et al. [6] conducted test to assess sealing capability of the new RMGIC and compared it to other four luting agents for complete cast crown cementation and concluded that Activa demonstrated the lowest micro leakage scores. In 2017, Korkut et al. [7] found that ACTIVA cement showed better mechanical and physical properties that was attributed to its content of the bioactive nanoparticles because it fills the matrix gap and the cracks formed under stress are smaller and fewer.

RelyX unicem cement is self-adhesive resin cement with a multifunctional phosphoric acid that required no pretreatment of the tooth or the restoration. This cement is moisture tolerant and capable of releasing fluoride the same as glass ionomer cements. RelyX Unicem cement is recommended for full-coverage restoration, including gold, porcelain-fused-to-metal, or all-ceramic/composite crown and untreated zirconia [8].

Success of adhesive bonding in dentistry is multifactorial; it depends on type of substrate, type of adhesive substance, operator skills to bond the restoration successfully and humidity of the working area. Regarding the different substrates for instance, enamel is highly mineralized, which facilitates bonding procedure. On the other hand, dentin is not as of harmonious structure as enamel, it is composed of inorganic, organic components and water that demands ideal moisture conditions to ensure proper bonding by avoidance of collapse of the organic part of the dentin(collagen fibres) [9].

Zirconia is a polycrystalline material with better mechanical and physical properties that is used in different dental prosthesis such as, single crowns, adhesive bridges and fixed partial dentures. Bonding zirconia-based restorations is a critical issue; it cannot be etched with hydrofluoric acid because of its polycrystalline composition. It can be sandblasted which increases the mechanical bond strength. Another method of enhancing bond strength is, silane coupling agents with tribo-chemical treatment. Self-adhesive resin cements are used with zirconia restorations without a preceding step to prepare the surface [10].

In general, it is difficult to control moisture in the oral cavity due to the presence of diverse contaminants like saliva, gingival fluid, blood and hand piece lubricant especially in sites near gingival margin [11]. This vitro study evaluated and compared the shear bond strength of the bioactive cement (Activa) with self-adhesive resin cement (RelyX) when bonded to enamel and dentin in dry and wet conditions and determined the mode of failure of each sample. The null hypothesis was that there is no difference between shear bond strength of the two cements with tooth structure, in dry and wet conditions and with enamel and dentin.

2. Materials and Methods

The tested cements Activa bioactive cement, Relyx self-adhesive resin cement and zirconia Aidite are shown in table 1.

Table 1. The materials used in this study are listed.

Material	Composition and description	Manufacturer	Batch number
Activa bio-active cement	Blend of diurethane and other methacrylates with modified polyacrylic acid (53.2%) Amorphous silica (3.0%) Sodium fluoride (0.90%)	PULPDENT, USA	161007
RelyX Unicem Dual Cure Self-adhesive Resin cement.	In a capsule form it consists of: Powder: Glass Fillers, Silica, Calcium Hydroxide, Self-Curing Initiators, Pigments, Light-Curing Initiators. Liquid: Methacrylate Phosphoric Esters, Dimethacrylates, Stabilizers, Self-Curing Initiators, Light-Curing Initiators.	3M ESPE, USA	669862
Zirconia based ceramic block	ZrO_2, Y_2O_3, HfO_2, Al_2O_3, Na_2O. Yttria-stabilized tetragonal Zirconia polycrystal blocks Shade A2	Aidite (Qinhuangdao) Technology	S160716ATA2M-7 3D

2.1. Tooth preparation

In this study, 40 caries free, non-identified mandibular molars were collected and stored in distilled water with 0.5% thymol for its antibacterial properties [12].Teeth were sectioned from roots using diamond disc at the cement-enamel junction and were placed in a circular mold (polypropylene plastic tube) of 24 mm diameter and 1.5 cm height filled with epoxy resin (Kemapoxy, JM, Egypt). Epoxy resin was prepared by mixing p/m (3.5/1) by volume according to manufacturer instructions. They were randomly divided into two main groups enamel and dentin (n=20). For Enamel group buccal surfaces of the teeth were prepared by just scratching using a low speed diamond saw (Isomet 4000, Linear Precision Saw). For Dentin group, buccal surfaces of the teeth were prepared using the same diamond saw under copious water to flatten the buccal surface and expose 2 mm of the dentin surface [13]. Each group was divided into two subgroups wet and dry samples (n=10). Each subgroup was farther divided according to the cement used into self-adhesive resin cement as control group and bioactive cement (*n*=5) (figure 1).

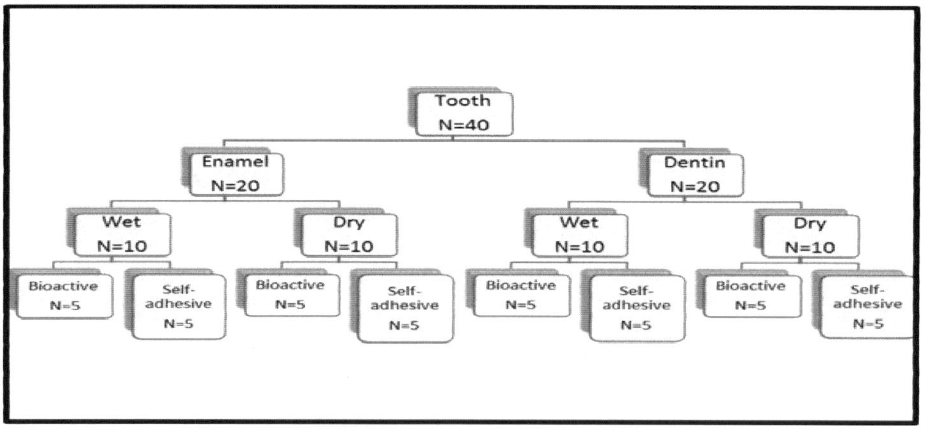

Figure 1. Samples grouping

2.2. Zirconia disc preparation

Zirconia rods were prepared from zirconia blocks (Aidite, Qinhuangdao Technology Co., LTD.) , using computer numerical cutting machine (CNC), then were cut using a water-cooled diamond saw (Isomet 4000, Linear Precision Saw) at low speed. The amount of post-sintering shrinkage was considered, by preparing slightly oversized discs to reach their final dimension (3 mm diameter x 2.5 mm thickness) after sintering at 1450 ^{o}C and cleaned from debris [14]. Zirconia discs were airborne-particle abraded using 50 µm Al_2O_3 at 2.5 bars of pressure (15 seconds) at adjusted distance of 10 mm before bonding.

2.3. *Samples bonding*

For subgroup samples planned to have a dry condition, they were dried with oil-free air for 10seconds. For those to have a wet condition, water was applied once using a micro-brush disposable applicator (Microbrush International, Grafton, USA) for enamel and dentin samples. Excess water was removed with passive application of absorbent paper (Kleenex) only once [10] .Samples of both groups (enamel and dentin) were bonded with bio-active cement or resin cement according to manufacturer instructions where the cement was injected on the previously prepared buccal surfaces of teeth. Then, zirconia discs were placed over the substrate by slight finger pressure, followed by a load of 5 kg applied using a loading device that was held for 10-15 seconds before light curing using (Optilux501, SDS Kerr, Danbury, USA), which was operated in standard mode [10]. During load application, tack curing was applied and then excess cement was removed by a probe. Complete curing was proceeded for 20 seconds around the margins according to the manufacturer instructions.

2.4. *Thermo-cycling and SBS testing*

Samples were thermo-cycled for 3,000 thermal cycles between 5°C and 55°C temperature with 30 seconds dwell time at each temperature in distilled water in thermo-cycling unit which is equivalent to 3 months of clinical life [14] .Shear bonding testing of all groups was measured using universal testing machine data recorded using computer software Bluehill version 3.3 at a cross-head speed of 0.5mm/min and load cell of 5000 Newton in a compression mode using a mono-bevel chisel of 0.5 mm thickness parallel to enamel and dentin surfaces as the shearing element (figure 2) [15]. Measurement of shear bond strength values were calculated by converting the loads into Megapascal (MPa) by dividing the maximum failure load Newton (N) by the bonding area (mm2) [16].

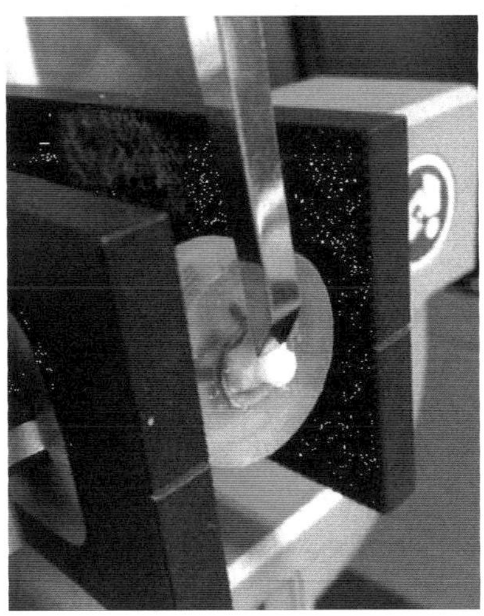

Figure 2. Shear bond testing using mono-bevel chisel.

2.5. *Scanning Electron Microscope Analysis (SEM)*

Eight representative specimens from each sub-group were mounted on aluminum stubs with carbon glue after the critical-point drying process. The specimens were gold sputter-coated and analyzed using field-emission scanning electron microscopy (FE-SEM S-4100; Hitachi, Wokingham, UK) at 10 kV and a working distance of 15 mm and using magnification of 800 x and 1000 x.

2.6. Statistical analysis

Numerical data were explored for normality by checking the distribution of data and using tests of normality (Kolmogorov-Smirnov and Shapiro-Wilk tests). Data showed non-normal (non-parametric) distribution. Data were presented as mean, standard deviation (SD), median and range values. Mann-Whitney U test was used to compare between wet and dry conditions, the two cement types as well as the two substrates. The significance level was set at $P \leq 0.05$. Statistical analysis was performed with IBM® SPSS® Statistics Version 20 for Windows.

3. Results

3.1. Shear bond strength

The mean, standard deviation and median for shear bond strength of the two cement types when bonded to enamel or dentin in two humid conditions were calculated. Comparison between shear bond strength values (MPa) of the two cement types is shown in table 2. Regarding wet condition; either with enamel or dentin, there was no statistically significant difference between shear bond strength of the two cement types. Regarding dry condition; there was no statistically significant difference in shear bond strength of the two cement types with dry enamel. Meanwhile, Rely X cement showed statistically significantly higher mean shear bond strength than Activa cement when bonded to dry dentin.

Table 2. The mean, standard deviation (SD) values and results of Mann-Whitney U test for comparison between shear bond strength values (MPa) of the two cement types.

Humidity	Substrate	Rely X		Activa		*P*-value
		Mean	SD	Mean	SD	
Wet	Enamel	6.15	2.27	3.89	1.44	0.095
	Dentin	1.26	0.66	0.63	0.35	0.421
Dry	Enamel	3.37	0.83	2.53	1.08	0.421
	Dentin	1.64	0.93	0.51	0.06	0.008*

* *Significant at* $P \leq 0.05$

Comparison between shear bond strength values (MPa) in wet and dry conditions is shown in table 3. Using Rely X cement, wet enamel showed statistically significantly higher mean shear bond strength than dry enamel. While there was no statistically significant difference between shear bond strength of wet and dry dentin. Using Activa cement, there was no statistically significant difference between wet and dry enamel or dentin. Either under wet or dry conditions and using both types of cements, enamel showed statistically significantly higher mean shear bond strength than dentin (table 4).

Table 3. The mean, standard deviation (SD) values and results of Mann-Whitney U test for comparison between shear bond strength under different humidity conditions.

Cement type	**Substrate**	**Wet**		**Dry**		**P-value**
		Mean	**SD**	**Mean**	**SD**	
Rely X	Enamel	6.15	2.27	3.37	0.83	0.032*
	Dentin	1.26	0.66	1.64	0.93	0.548
Activa	Enamel	3.89	1.44	2.53	1.08	0.310
	Dentin	0.63	0.35	0.51	0.06	0.151

* *Significant at P ≤ 0.05.*

Table 4. The mean, standard deviation (SD) values and results of Mann-Whitney U test for comparison between shear bond strength of the two substrates.

Humidity	Cement type	Enamel		Dentin		P-value
		Mean	SD	Mean	SD	
Wet	Rely X	6.15	2.27	1.26	0.66	0.008*
	Activa	3.89	1.44	0.63	0.35	0.008*
Dry	Rely X	3.37	0.83	1.64	0.93	0.032*
	Activa	2.53	1.08	0.51	0.06	0.008*

* Significant at P ≤ 0.05.

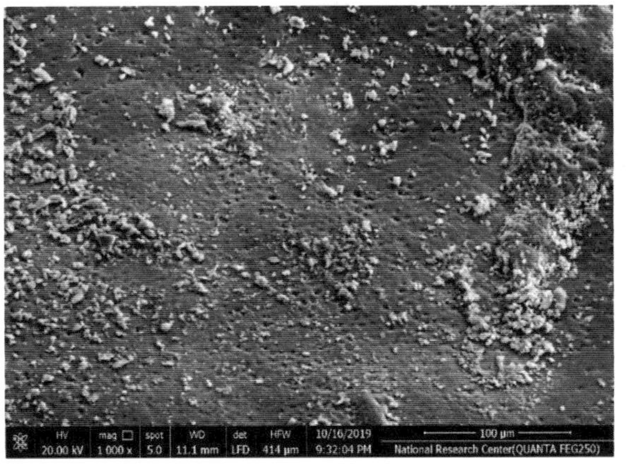

Figure 3. SEM photomicrograph (magnification 1000x): Mixed failure in dry enamel surface with Activa cement

3.2. Scanning electron microscopy

ACTIVA bioactive cement samples demonstrated mixed failure when bonded to enamel. SEM photomicrograph showed clusters of cement debris failed to completely infiltrate through the enamel layer (figure 3). Meanwhile, samples showed adhesive failure when bonded to wet dentin surface with partial obliteration of the dentinal tubules with cement remnants (figure 4). ACTIVA bioactive cement samples showed mixed failure when bonded to dry dentin surface. SEM photomicrograph revealed partial obliteration of the dentinal tubules with cement remnants (figure 5).

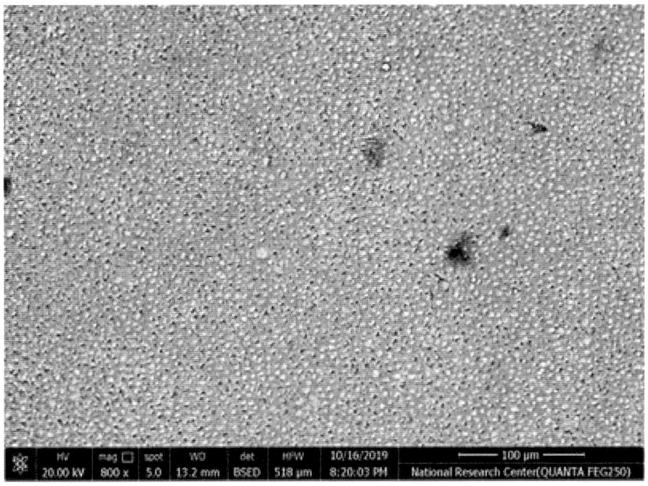

Figure 4. SEM photomicrograph (magnification 800x): Adhesive failure in wet dentin surface with Activa cement.

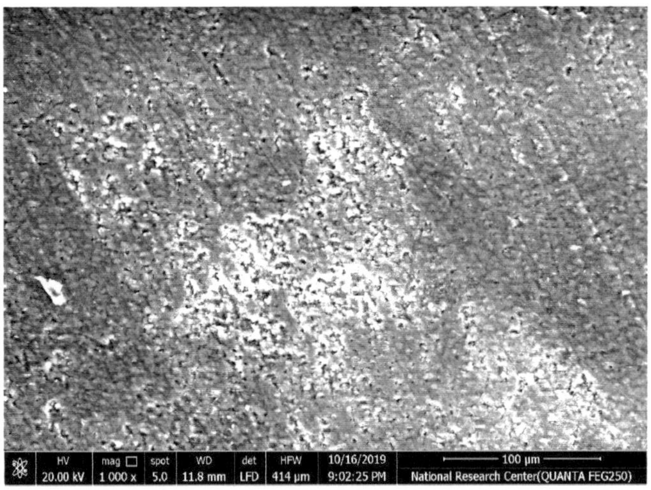

Figure 5. SEM photomicrographs magnification (1000x): Mixed failure in dry dentin surface with Activa cement.

RelyX Unicem demonstrated mixed mode of failure when bonded to dry enamel. SEM photomicrograph revealed clusters of resin remnants on the enamel surface (figure 6). RelyX Unicem demonstrated adhesive failure when bonded to dry dentin. SEM photomicrograph showed partial obliteration of the dentinal tubules with no cement remnants (figure 7).

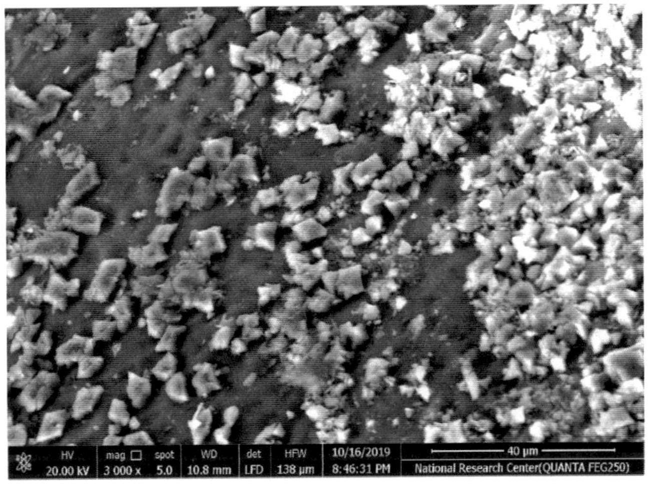

Figure 6. SEM photomicrograph (magnification 1000x): Mixed failure of RelyX Unicem in dry enamel.

Figure 7. SEM photomicrograph (magnification of 800x): Adhesive failure of Relyx Unicem in dry dentin.

4. Discussion

Dental cements are one of the factors that determine the success of indirect restorative restorations as they provide the required retention to dental surfaces and seal tooth restoration margins. Also incorporation of Nano-sized particles could decrease the setting time and enhance compression strength and elastic modulus [17]. Bond strength values obtained in this study were comparable to those found in another study; Hattar et al in 2015, that recorded nearly similar shear bond strength values with RelyX Unicem (6.81 MPa for enamel) [18]. This could be attributed to similar factors such as, bonding tooth substrate and cement type [19].

Regarding the effect of cement type: with dry dentin, RelyX cement showed statistically significant higher mean shear bond strength than Activa cement. Gerth *et al.* [20], in their study showed an increased chemical interaction of RelyX with Calcium from hydroxyapatite, which may explain the higher bond strength of this cement in this study.

Concerning the effect of wet and dry conditions, using Rely X cement with wet enamel showed statistically significantly higher mean shear bond strength than dry enamel which is in accordance with results obtained from Mushashe *et al.* [21]. It was pointed out that self-adhesive cement needs ionizing medium for the chemical reaction to start and enhances the adhesive ability to interact with both substrates enamel and dentin. Similarly, *Fursue et al.* [22] found that all in one adhesive was affected positively by the presence of water and that was rationalized by the fact that all in one adhesive requires ionizing medium that is represented in water to start the reaction. Water could be trapped in the hybrid layer that's the reason why it required cautious air blowing following the adhesive application. While there was no statistical difference between both substrates whether wet or dry using either Activa or RelyX unicem cements.

In the matter of effect of the substrate, it was found that under wet or dry conditions both types of cements (Relyx unicem and Activa) when bonded to enamel, showed statistically higher shear bond strength than with dentin. It was found that Relyx unicem offered high bond strength with enamel due to the chemical adhesion rather than micromechanical bonding which is achieved by the functional monomer (methacrylate phosphoric esters) which chelates calcium ions of hydroxyl-apatite [23] . Conversely, Lee et al. [24] found that enamel showed low bond strength values when bonded to self-adhesive resin cements. De Munk [25] demonstrated that self-adhesive cements characterized with low PH, but it hardly demineralizes tooth surface, because of its high viscosity and poor penetration thus resulting in low bond strength.

In regard to modes of failure using SEM, RelyX and Activa bioactive cements, both, showed mixed and adhesive mode of failure. This indicates that bioactive and self-adhesive cement interacted superficially with minimal penetration in the tooth substrate. Regarding dentin samples, the prevalent adhesive failure could be attributed to the high-water content of dentin that increases water fraction and interferes with complete polarization of the adhesive. Partial obliteration of dentinal tubules with resin plugs revealed in ACTIVA SEM sample (figure 5), may be attributed to the presence of nanoparticles that helped in the flow of cement and formation of hybrid layer. In accordance with the current study, Flurry *et al.* [26], Lin *et al.* [27] and Viotti *et al.* [28] observed mixed and adhesive pattern of failure indicating that no real chemical bond was established.

Despite the modest performance of ACTIVA bioactive cement, it showed comparable results to the well reputed self-adhesive resin cement RelyX. This performance could be attributed to its various setting mechanism; light and chemical activated polymerization and acid-base reaction that involve ionic attraction of two carboxyl groups in the cement to calcium in tooth substrate. The bioactive cement has an ionic resin network and bioactive nanoparticles that fill the gaps in the matrix and minimally penetrate the dentinal tubules that supports ionic bonding and eventually extends the longevity of the restoration [5]. Future clinical long term studies are required to test the material clinical behavior and longevity and to assess the effect of pH fluctuation, load switches and enzymatic challenges on the bonding joint.

5. Conclusion

Within the parameters used and the limitations of this study, the following could be concluded: Both tested cements exhibited comparable shear bond strength to either wet or dry tooth substrate (enamel or dentin), with superiority of self-adhesive resin cement RelyX. Enamel substrate showed improved bond strength than dentin regardless wet or dry conditions and cement type. RelyX and Activa Bioactive cements, showed mixed and adhesive mode of failure

Clinical Relevance

Despite the fact that the resin cement outperformed the bioactive cement, the new bioactive cement containing nano particles showed comparable results and is considered as an advantageous candidate for bonding procedures in the clinic especially when moisture control is unattainable.

References

[1] Piwowarczyc A, Lauer C and Sorensen J A 2005 *Dent. Mater.* **21** 445
[2] Yli-Urpo H, Lassila L V, Närhi T and Vallittu P K 2005 *Dent. Mater.* **21** 201
[3] Choi J Y, Lee H and Kim H 2008 *J. Mater. Sci. Mater. Med.* **19** 3278

[4] PULPDENT 2016 *ACTIVA* (Watertown Oakland) [Brochure].
[5] Pameijer C, Garcia-Godoy F, Morrow B and Jefferies S 2015 *J. Clin. Dent.* **26** 23
[6] Zmener O, Pameijer CH and Hernández S 2014 *Am. J. Dent.* **27** 51
[7] Korkut E, Gezgin O, Tulumbaci F, Özer H and Şener Y 2017 *Araştırma EÜ Dişhek Fak Derg.* **38** 170
[8] Ernst C P, Cohnen U, Stender E and Willershausen B 2005 *J. Prosthet. Dent.* **93** 551
[9] Nanci, A and Ten C A R 2003 *Ten Cate's Oral Histology: Development, Structure, And Function* (St. Louis, Mo: Mosby) p.2
[10] André C, Aguiar T, Ayres P, Ambrosano G and Giannini M. 2013 *BRAZ.ORAL.RES.* **27** 389
[11] Ayyildiz S, Emir F, Tunc E and Sen D 2015 *Appl. Adhes. Sci.* **3** 13
[12] Zheng T L, Huang C, Zhang Z X, Wang S and Zhang G 2005 *Shanghai Kou Qiang Yi Xue.* **14** 147
[13] Huda A A and Walid N A 2012 *TJODS.* **1** 13
[14] Majed F A, Allen S and Fouad S 2001 *J. Clin. Pediatr. Dent.* **25** 221
[15] Lee S, Bae J, Choi J, Jeon Y and Cha V 2015 *MATEG9.* **8** 3306
[16] Rodrigues J and Da Silva J 2015 *Int. J. Esthet. Dent.* **10** 146
[17] De Caluwe T, Vecruysse C, Fraeyman S and Verbeeck R 2014 *Dent. Mater.* **30** 1029
[18] Hattar S, Hatamleh M and Faleh S F 2015 *Saudi. Dent. J.* **27** 70
[19] Sirisha K, Rambabu T, Shankar Y and Ravikumar P 2014 *J. Conserv.Dent.***17** 305
[20] Gerth H U, Dammaschke T, Züchner H and Schäfer E 2006 *Dent. Mater.* **22** 934
[21] Mushashe A, Gonzaga C, DaCunha L, Furuse A Y, Moro A and Correr G M 2016 *Braz. Dent. J.* **27** 537
[22] Furuse AY, Cunha LF, Moresca R and Paganeli G 2011 *Oper. Dent.* **36** 274
[23] Nagakane K, Yoshida Y, Hirata I, Fukuda R, Nakayama Y, Shirai K, Ogawa T, Suzuki K, Van Meerbeek B and Okazaki M 2006; *Dent. Mater. J.* **25** 645
[24] Lee S, Bae J, Choi J, Jeon Y and Virgílio C 2015 *MATEG9.* **6** 3306
[25] De Munck, J, Vargas M, Van Landuyt K, Hikita K, Lambrechts P and Van Meerbeek B 2004 *Dent. Mater.* **20** 963
[26] Flury S, Lussi A, Peutzfeldt A and Zimmerli B 2010 *Dent. Mater.* **26** 855
[27] Lin J, Shinya A, Gomi H and Shinya A 2010 *Int. J. Oral Sci.* **2** 28
[28] Viotti, R G, Kasaz A, Pena C E, Alexandre R S, Arrais C A and Reis A F 2009 *J. Prosthet. Dent.* **102** 306

IOP Conf. Series: Materials Science and Engineering 1046 (2021) 012010 doi:10.1088/1757-899X/1046/1/012010

Study of dark electrical properties and photoelectric performance of organic/inorganic (ZnPc/p-Si) Solar cells

A A El Amin and M M Mahmoud*

Department of physics, Faculty of science, Aswan University, Egypt

*E-mail: mahasen.m.mahmoud@gmail.com

Abstract. A Zinc Phthalocyanine (ZnPc) film was deposited by thermal evaporation on a monocrystalline p-Si substrate to provide a p-ZnPc/p-Si organic/inorganic heterojunction. The electrical characteristics of the fabricated cell were examined by measuring the current density versus voltage (J-V). The fabricated cell was characterized by a contact barrier of 0.56 eV and a rectification ratio of 230 calculated at ± 1 V^{-1}. The dark J-V measurement suggested that the DC current density in this junction contains a thermionic conduction mechanism when the applied voltage is relatively low. At higher voltages another conduction is dominated called the space-charge limited conduction mechanism (SCLC) controlled by a single trapping level. The I-V measurement was performed with an illumination of 100 $mWcm^{-2}$ and the photoelectric conversion characteristics of the junction were also studied, achieving a power conversion efficiency of 2.7%. The other solar cell parameters such as the short-circuit current I_{sc}, the open-circuit voltage V_{oc} and the filling factor FF were evaluated and equaled to 2.5 A, 1.132 V and 0.477 at room temperature. The dependence of these parameters as a function of the temperature has also been studied in the temperature range from 300 to 400 K.

1. Introduction

The photovoltaic effect produced when a semiconductor/metal interface [1] exposed to sunlight or so called the process of solar energy conversation had been studied thoroughly over the years. Semiconductor materials like Si, GaAs and CdS [2] were used because of its stability in hostile environment and its sensitivity towards the spectrum of the sun. But these semiconductor materials are unable to detect visible light region due to their wide band gap, to overcome this problem photoactive dyes were incorporated on top of the semiconductor material surface as a light absorber ;meaning to sensitize Si to the visible light region [3, 4]. Compounds such as phthalocyanines were studied to be light absorbers because of their high absorption coefficient in the solar spectrum, chemical stability, and their various synthesis techniques [5, 6]. Moreover, phthalocyanine compounds could be metal substituted phthalocyanines (MPcs) such as ZnPc, CoPc, and FePc or metal free (H_2Pc). These compounds have been utilized in solar cells, organic photoreceptors, sensors [7], light emitting diodes[8], the display devices in mobile phones[8], photosensitizers in laser printers[8], the active layer in Schottky diodes [9, 10] and in organic thin film transistors[11]. A heterojunction device made of organic material (Pc) and inorganic material (Si) was found to be highly stabilized as compared to the bare Si interface [12]. These heterojunction devices also presented rectification properties in the dark conduction in addition to the photoconversion efficiency ranging from 0.74 to 5.63 %, examples for these heterojunctions are ZnPc/Si [13, 14], H_2Pc/p-Si [15], CuPc/p-Si [16], MgPc/n-Si [17], CoPc/n-GaAs [18], NiPc/p-Si [19], CoPc/p-Si [20], TCVA/p-Si [21], FeTPPCl/p-Si [22], and DBM/p-Si [23].

In the present work, Zinc phthalocyanine (ZnPc) was used as organic material and silicon (Si) as inorganic material to compose the ZnPc/Si heterojunction. The J-V characteristics in the dark and under illumination of the ZnPc/Si heterojunction have been analyzed to determine the photovoltaic parameters and the various conduction mechanisms. Also, the optical energy gap was deduced.

2. Experimental techniques

Zinc phthalocyanine (ZnPc) powder as purchased from Sigma-Aldrich. The p-ZnPc/p-Si heterojunction was composed by using high vacuum coating unit (E306A, Edwards Co. England). The substrate temperature was kept at room temperature during deposition process. The film thickness and rate of deposition were controlled and monitored during deposition. Figure 1 shows a schematic diagram of the device, comprising a top ZnPc thin film and p-Si as a semiconductor absorber with back contact electrode (Al) and the chemical structure of Zinc phthalocyanine [24].

Keithley 2400 SourceMeter was used for the electrical J–V measurements. The p-type Si single crystal wafers with carrier concentration of $10^{22}m^{-3}$. The back side of Si was coated by a thick layer of aluminum electrode.

Measurements of dark current–voltage characteristics within the temperature range 300–400 K were made in air. The cells were exposed to light coming from a light source (white light) to get an intensity of incident power of about 100 mW/cm2. The optical measurements were done using double beam spectrophotometer (JASCO, V-670).

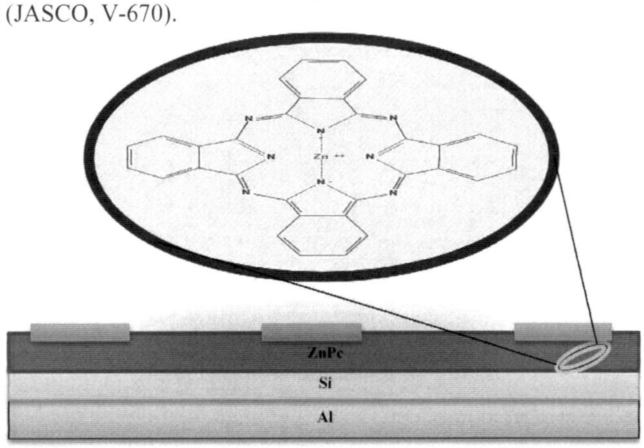

Figure 1. Schematic diagram of the p-ZnPc/p-Si heterojunction and the chemical structure of Zinc phthalocyanine.

3. Results and discussions

3.1. The dark current density–voltage characteristics

Significant information about the properties of ZnPc/p-Si organic/inorganic heterojunction solar cell could be found by analyzing the current density–voltage characteristics (J–V) as a function of temperature. Also, this analysis is important in determining the transport mechanisms operating the conduction in this junction. Figure 2 shows the characteristic (J-V) curves for ZnPc/p-Si organic/inorganic heterojunction cell under the forward and reverse bias measured at different temperatures ranging from 300 to 400 K and in the voltage range −0.6 to +4.25 V. These curves are similar in behavior to the I-V diode characteristic curve. This similarity could be understood in terms of a p–n junction formed between ZnPc and p-Si, the barrier at this interface limits the forward and reverse carrier flow across the junction in which a built-in potential could be developed. These observations suggesting that the device exhibits a good rectifying performance with a rectification ratio of 230 calculated at a bias potential of ± 1 V. The rectification factor RR is the ratio of forward current to reverse current at a certain applied voltage. Figure 2 also shows the current density was increasing as the temperature increased meaning that the diode had a negative resistance temperature coefficient or a positive coefficient of conductivity.

The semilogarithmic scale plots of the forward current density against the applied voltage at different temperatures illustrated in figure 3, is composed of two regions: region (I) at low voltage and region (II) at high voltage. The dark current moving through an ideal junction under forward bias could be represented by a single exponential diode equation as illustrated in[25].

However, for non-ideal diode such as the heterojunction studied here, the J-V curves maybe composed of two (or sometimes more) different regions associated with two varied conduction mechanisms.

That behavior is more complex than to be simply represented by the single exponential expression, for that reason, the summation of these exponential expressions associated with their reverse saturation current density and ideality factors, would be best to accurately model these conduction mechanisms[26]. So, the forward current density for non-ideal junctions in low potential range is given by[27],

$$J = J_{s1}\left[\exp\left(\frac{eV}{n_1 k_B T}\right) - 1\right] + J_{s2}\left[\exp\left(\frac{eV}{n_2 k_B T}\right) - 1\right] + \frac{V - JR_s}{R_{sh}} \quad (1)$$

Where J_s is the reverse-saturation current density, e is the electronic charge, V is the applied voltage, n is the ideality factor of the diode, T is the temperature and k_B is the Boltzmann constant.

Also, the subscripts 1 and 2 indicating the possible presentence of two diode current. And R_s is the series resistance and R_{sh} is the shunt resistance.

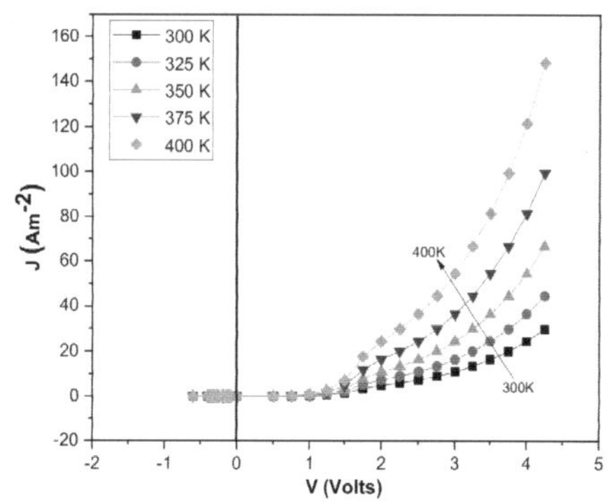

Figure 2. Dark J–V characteristics at different temperature for p-ZnPc/p-Si heterojunction.

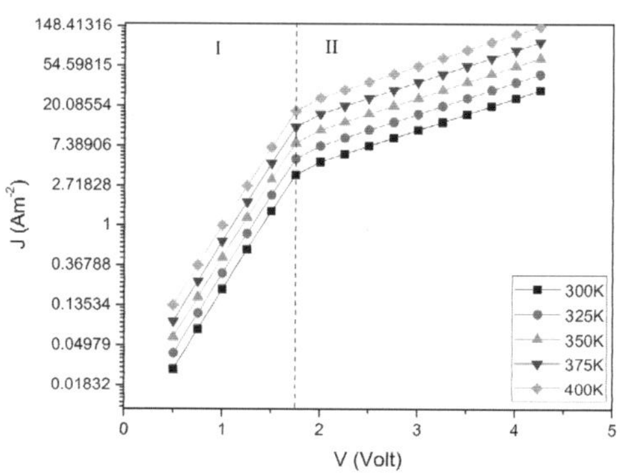

Figure 3. Semilogarithmic scale of forward current density (J) for Al/p-ZnPc/p-Si heterojunction.

The series resistance R_s for ZnPc/p-Si heterojunction was estimated from temperature 300K to 400K, by plotting the diode junction resistance R_j against the applied potential as viewed in figure 4 (a), where $R_j = \partial V/\partial I$[20, 28], that was determined from the current–voltage curves. At sufficiently high forward bias (V ≤ 1.75 V), the junction resistance approaches a constant value R_s that indicate the series resistance. The obtained values of the series resistances R_s at temperatures of 300, 325, 350, 375 and 400 K are determined as1.103 x 10^{-4}, 7.395 x 10^{-5}, 4.957 x 10^{-5}, 3.323 x 10^{-5} and 2.227 x $10^{-5}\Omega$, respectively. It can be observed from figure 4 (a) that as the temperature increases the series resistance decreases. These resistances are one the most playing factors in improving the solar cells performance and design. We used another way to measure R_S. To estimate the series resistance R_S, the I-V curve for example at room temperature would be analyzed as follows: at a certain current (I),the horizontal displacement between the observed curve in figure 4 and the extrapolated linear part, yields the voltage drop, $\Delta V = IR_S$, across the neutral region [18]. Then the resulted straight line after plotting the voltage drop versus the current shown in the inset of figure 4 (b), gives a slope of 32.1 μ Ω which is the value of R_s.

An equation that relate I_s to the reciprocal of temperature and the activation energy is as follows[29],

$$I_s = I_{00} \exp\left(-\frac{E_a}{K_B T}\right) \tag{2}$$

Where I_{00} is a weak function of temperature and E_a is the activation energy of the charge carriers. The low voltage range (V ≤1.75 V) illustrated in figure 5, shows a forward current density behavior that could be represented by a single exponential diode equation as follows,

$$J = J_s\left[\exp\left(\frac{eV}{nK_B T}\right) - 1\right] \tag{3}$$

Both J_s and n would be found from the intercept and slope of figure 5 and written in table 1. The ideality factor listed in table 1 is constant with changing the temperature from 300 to 400, this result is a strong indication that the conduction mechanism is the thermionic emission one. With its saturation current density would be given as [30],

$$J_s = A^{**} T^2 \exp\left(-\frac{e\varphi_b}{k_B T}\right) \tag{4}$$

Where A^{**} is the effective Richardson constant and φ_b is the barrier height.

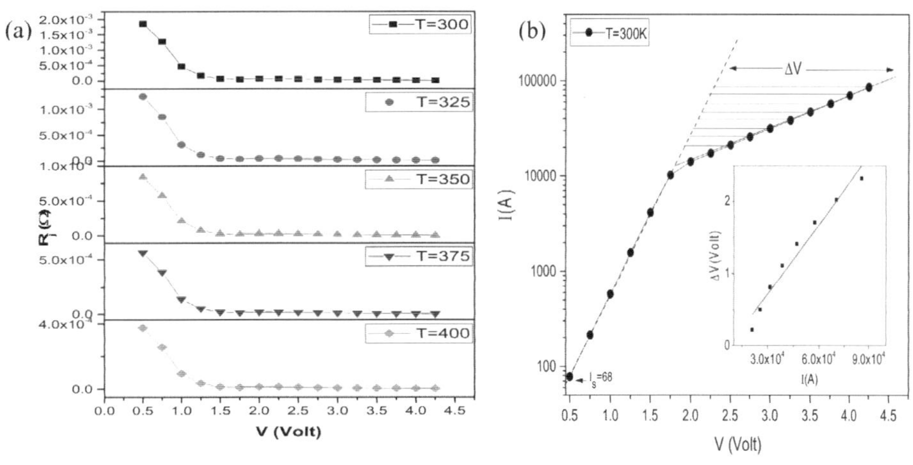

Figure 4. (a) Junction resistance R_j versus V for Al/p-ZnPc/p-Si cell at the temperatures from 300 to 400 K. (b) The junction I-V characteristics of Al/p-ZnPc/p-SI heterojunction at 300 K. Inset is the voltage drop across the series resistance.

Table 1. The ideality factor listed (T, J_s, n) with changing the temperature from 300 to 400 K.

K	J_s	n
300	0.003962025	3.055125928
325	0.00591064	2.820116242
350	0.008817649	2.618679367
375	0.013154387	2.444100743
400	0.019624039	2.291344446

Drawing the logarithm of J_s/T^2 versus the reciprocal of Tas shown in the inset of figure 5, result in straight line, which is a further confirmation of the presence of thermionic mechanism. Also, the intercept and the slope were used to calculate the potential barrier height φ_b, and the effective Richardson constant A^{**}, of values 2.48 J and 0.105 $Am^{-2}K^{-2}$, respectively.

The download curvature at relatively higher voltage (≥1.75), observed in figure 5. indicate the presence of another conduction mechanism. In figure 6 of log I and log$V$at different temperatures, shows that the increase in current form change from an exponential behavior to a behavior of power law relation. In this specific range (≥1.75) which is formed because of the transport properties of organic material, could be interpreted in terms of the space-charge limited conduction mechanism (SCLC) controlled by a single trapping level[31]. Space charge is described as the existing net charge in a material dividing from the condition of charge neutrality.

The J–V relation showing a power law dependence on voltage of $J \sim V^s$which would be explained in terms of the single energy trap level since $s \cong 2$[32], is given by the equation[33, 34],

$$J = \frac{9}{8}\,\varepsilon\varepsilon_o\mu\,\frac{N_v\,V^2}{N_t\,d^3}\exp\left(-\frac{E_t}{K_BT}\right) \tag{5}$$

Where ε and ε_o are the relative permittivity of ZnPc and the permittivity of free space, which were taken to be equal to 3.95 [14] and 8.85 x 10^{-12} Fm^{-1} respectively,μ is the mobility with a value of 7.6 x 10^{-9} m^2V^{-1} s^{-2} for ZnPc [14], d is the film thickness,N_v is the effective density of states in the valance band, and N_t is the total trap concentration at an energy level of E_t above the valence band edge.

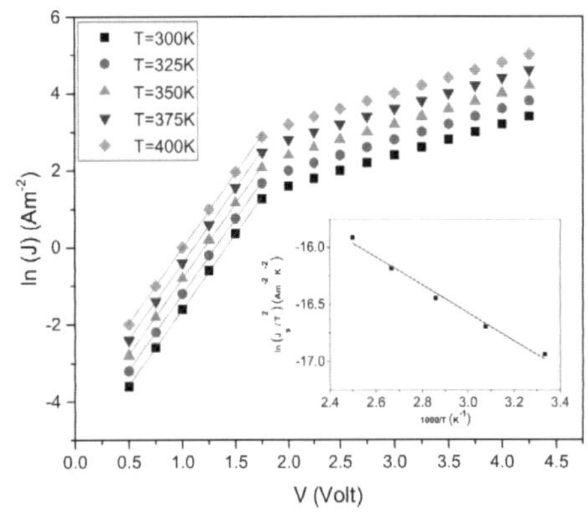

Figure 5. ln(J_s) versus 1/T for Al/p-ZnPc/p-Si/Al heterojunction, the inset is a plot of ln(J_s/T^2) versus 1/T of Al/p-ZnPc/p-Si/Al heterojunction.

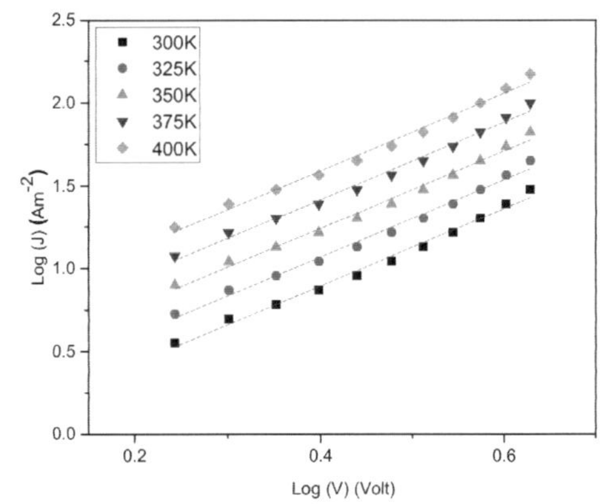

Figure 6. Variation of log (J) with log (V) at higher forward voltage bias for Al/p-ZnPc/p-Si/Al heterojunction

Estimation of further information from the previous equation, were obtained by plotting the logarithm of J against the reciprocal of T at the applied voltages of 1.75, 2, 2.25, 2.5, 2.75, 3, 3.25, 3.5, 3.75, 4 and 4.25 Volts in the SCLC region as viewed in figure 7.

Knowing the N_v for ZnPc to be equals 10^{27} m^{-3}[14, 35-37], leads to the calculation of E_t and N_t as: From the slopes of these straight lines, the trap level $E_t$would be calculated and equals to the value of1.14595 x 10^{-20}.

In the other hand the total trapping concentrations N_t were evaluated from the intercepts and found equal to 3.08524 x 10^{11}, 4.87763 x 10^{11} and 6.12589 x $10^{11}m^{-3}$ at the potential of 2,3 and 4 V respectively.

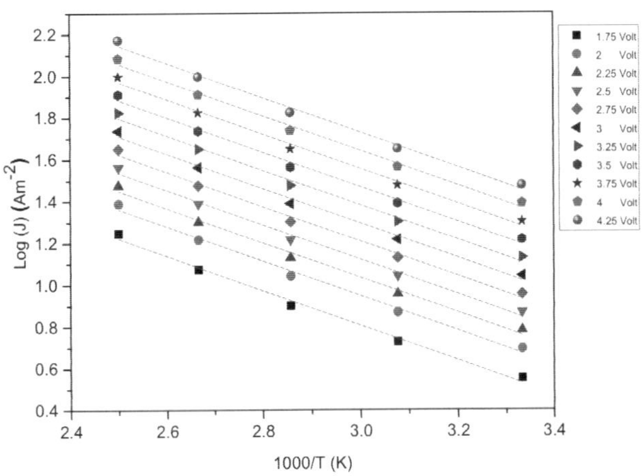

Figure 7. variation of ln (J) with 1000/T in SCLC for Al/p-ZnPc/p-Si/Al heterojunction.

3.2. The illumination current –voltage characteristics

The current -voltage characteristics of Al/p-ZnPc/p-Si/Al heterojunction solar cell observed under the illumination of 100 mW cm^{-2} is viewed in figure 8. In this figure the dark curve was included for comparison. The reverse current increased considerably with illumination as seen in figure 8, this observation is regarded to the production of charge carriers. Also, the short circuit current $I_{sc}$and the open circuit voltageV_{oc}were estimated to be equal to 2.5 mA and 1.132 V respectively.

Further solar cell parameter could be found from this figure such as $I_M = 1.5$mA and $V_M = 0.9$ V, which are the current and the potential at maximum power point, consequently the maximum power could be calculated from

$$P_M = V_M I_M \tag{6}$$

And found equal to 1.35 mAV.

Also, the filling factor evaluated to be equal to 0.477 from

$$FF = \frac{V_M I_M}{V_{oc} I_{sc}} \tag{7}$$

Finally, the experimental efficiency of this organic/inorganic solar cell was given by

$$\eta\% = \frac{P_M}{P_{in}} = \frac{FF\ V_{OC} I_{sc}}{P_{in}}\ X\ 100 \tag{8}$$

And found equal to 2.7 %. Where $P_{in}$is the incident light intensity (100 mW/cm^2).

The parameters of the organic/inorganic p-ZnPc/p-Si heterojunction was compared to the ones of other heterojunctions in previous studies and listed in table 2.

Table 2. The parameters (J_{sc}, V_{oc}, FF, η) of the organic/inorganic p-ZnPc/p-Si heterojunction.

Organic/inorganic cell	J_{sc}(mA/cm^2)	V_{oc}(V)	*FF*	η (%)	reference
n-SnPcCl$_2$/p-Si	14.43	0.500	0.422	3.05	[30]
MgPc/n-Si	3.760	0.35	0.4	1.05	[17]
CoPc/n-GaAs	3.5	0.33	0.34	0.81	[18]
P-DCPF/n-Si	9.13	0.297	0.329	1.27	[38]
NiPc/Si	I_{sc}=186μA	0.32	0.28	1.11	[19]
p-ZnPc/p-Si	I_{sc}=2.5 mA	1.132	0.477	2.7	This study

Table 2 shows values of η lower than that of p-ZnPc/p-Si, in the other hand there are values of η greater than p-ZnPc/p-Si. The lower efficiency of p-ZnPc/p-Si could be attributed to recombination losses in the bulk in addition to the one in the interface region. This result could be due to the series resistance of the junction and dissimilarity between organic thin film and p-Si substrate.

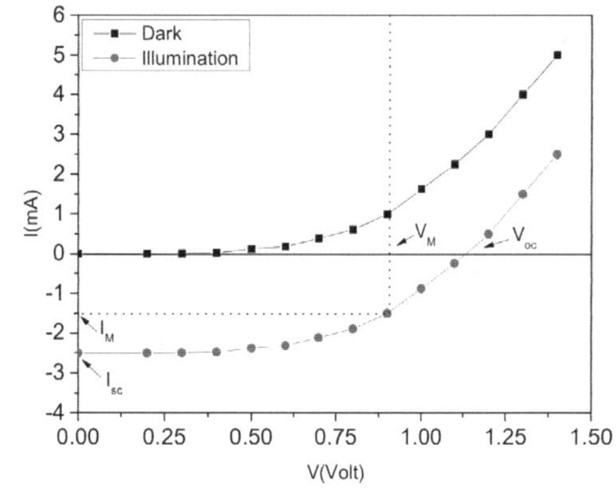

Figure 8. Current–voltage characteristics under illumination for Al/p-ZnPc/p-Si/Al heterojunction.

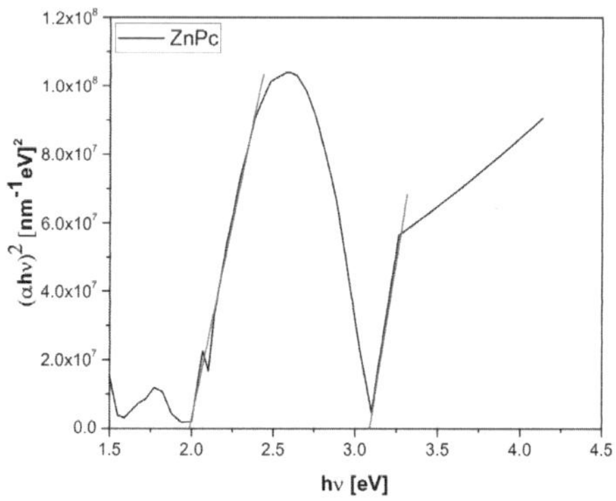

Figure 9. Relation between $(\alpha h\upsilon)^2$ and photon energy (hυ) for ZnPc.

3.3. The optical characterization of p-ZnPc/p-Si

The molecules of the organic phthalocyanine show optical properties because of their structural ring. The zinc phthalocyanine nanoparticles thin films provide two types of electronic transitions; B-band (Soret band) and Q-band.

The absorption coefficient of ZnPc thin film can be used to estimate the type of transition and the value of the optical energy gap E_g according to well-known Tauc's relation [39].

$$(\alpha h\upsilon) = B\left(h\upsilon - E_g\right)^r \tag{9}$$

where B is a constant and r is a constant which depends on the probability of transition; it takes the values r = 1/2, 3/2, 2 and 3 for *direct allowed*, *direct forbidden*, *indirect allowed* and *indirect forbidden* transitions, respectively. The dependence of $(\alpha h\upsilon)^{1/r}$ on photon energy $(h\upsilon)$ was plotted for different values of r. The best fit was obtained for r =1/2 and displayed in figure 9, which pointed out that the transitions are direct allowed transitions [40]. The direct optical band gap values can be determined by the extrapolated linear regression of the curves resulting from plotting of $(\alpha h\upsilon)^2$ versus photon energy $(h\upsilon)$. From the presented graphs, the direct band gap of ZnPc were found to be 1.99 and 3.09 eV.

4. Conclusion

The organic/inorganic p-ZnPc/p-Si heterojunction was investigated after the deposition of the organic ZnPc on the inorganic ⟨100⟩ silicon wafer. The current density versus applied potential data were analyzed to find the junction electrical properties. At low forward applied voltages ($V \leq 1.75$ V), the J-V analysis indicate the presence of thermionic conduction mechanism. However, in the relatively high voltage region the space-charge limited conduction mechanism (SCLC) controlled by a single trapping level manifest itself as a different conduction mechanism. The current- voltage were examined under illumination as compared to the dark I-V data. From which significant solar cell parameters at room temperature were calculated to be equivalent to1.132 V, 2.5 mA, 0.477 and 2.7% for the open circuit voltage, the short circuit current, the fill factor and the power conversion efficiency, respectively. Also studying the optical properties of ZnPc showed it had tow energy gap of 1.99 and 3.09 eV.

References

[1] Chapin D M, Fuller C and Pearson G 1954 *J. Appl. Phys.* **25** 676
[2] Yanagi H, Kataura H and Ueda Y 1994 *J. Appl. Phys.* **75** 568
[3] Matsumura M, Matsudaira S, Tsubomura H, Takata M and Yanagida H 1980 *Ind. Eng. Chem. Prod. Res. Dev.* **19** 415
[4] Memming R 1984 *Prog. Surf. Sci.* **17** 7
[5] Moser F H and Thomas A L 1983 *The Phthalocyanines* (CRC Press)
[6] Simon J, Lehn J M, Andre J J and Rees C W 2012 *Molecular Semiconductors: Photoelectrical Properties and Solar Cells* (Springer Berlin Heidelberg)
[7] Kerp H and Van Faassen E 2000 *Chem. Phys. Lett.* **332** 5
[8] Simon J, André J J, Lehn J M and Rees C W 1985 *Molecular Semiconductors: Photoelectrical Properties and Solar Cells* (Springer)
[9] Roy M, Kumar M, Jaiswal P and Sharma G 2004 *Radiat. Meas.* **38** 205
[10] Samuel M, Menon C and Unnikrishnan N 2005 *J. Phys.: Condens. Matter* **18** 135
[11] Puigdollers J, Voz C, Fonrodona M, Cheylan S, Stella M, Andreu J, Vetter M and Alcubilla R 2006 *J. Non· Cryst. Solids* **352** 1778
[12] Nakato Y, Shioji M and Tsubomura H 1981 *J. Phys. Chem.* **85** 1670
[13] Remaki B, Guillaud G and Mayes D 1998 *Opt. Mater.* **9** 240
[14] El-Nahass M, Zeyada H, Aziz M and El-Ghamaz N 2005 *Solid State Electron.* **49** 1314
[15] El-Nahass M, Farid A, Farag A and Ali H 2006 *Vac.* **81** 8
[16] Antohe S, Tomozeiu N and Gogonea S 1991 *Phys. Status Solidi A* **125** 397
[17] Riad S 2000 *Thin Solid Films* **370** 253
[18] Darwish S 2003 *Egypt. J. Sol.* **26** 55
[19] El-Nahass M, Abd-El-Rahman K, Farag A and Darwish A 2005 *Org. Electron.* **6** 129
[20] El-Nahass M, Zeyada H, Abd-El-Rahman K and Darwish A 2007 *Sol. Energy Mater. Sol. Cells* **91** 1120
[21] Soliman H, Farag A, Khosifan N and El-Nahass M 2008 *Thin Solid Films* **516** 8678
[22] El-Nahass M, Metwally H, El-Sayed H and Hassanien A 2011 *Synth. Met.* **161** 2253
[23] El-Nahass M, Kamel M, Atta A and Huthaily S 2013 *Mater. Chem. Phys.* **137** 716
[24] National Center for Biotechnology Information [Internet], PubChem Database. CID=2735172.
[25] Rhoderick E H and Williams R H 1988 *Metal-Semiconductor Contacts* (Clarendon Press)
[26] Ranuarez J, Sánchez F G a and Ortiz-Conde A 1999 *Solid-State Electron.* **43** 2129
[27] Sze S M 1981 *Physics of Semiconductor Devices* (John Wiley & Sons)
[28] Darwish S, Riad A and Soliman H 1996 *Semicond. Sci. Technol.* **11** 96
[29] Das S and Morris G 1993 *Sol. Energy Mater. Sol. Cells* **28** 305
[30] Darwish A, El-Shazly E, Attia A and El-Rahman K A 2016 *J. Appl. Phys.* **27** 8786
[31] Peumans P, Yakimov A and Forrest S R 2003 *J. Appl. Phys.* **93** 3693
[32] Ganesh V, Manthrammel M A, Shkir M, Yahia I, Zahran H, Yakuphanoglu F and AlFaify S 2018 *Appl. Phys. A* **124** 424
[33] Gould R 1982 *J. Appl. Phys.* **53** 3353
[34] Zeyada H, El-Nahass M and El-Shabaan M 2016 *Synth. Met.* **220** 102
[35] Anthopoulos T and Shafai T 2001 *Phys. Status Solidi A* **186** 89
[36] Riad A 1999 *Physica B Condens Matter.* **270** 148
[37] Malik T A and Abdel-Latif R 1997 *Thin Solid Films* **305** 336
[38] El-Nahass M, Zeyada H and Hendi A 2004 *Eur. Phys. J. Appl. Phys.* **25** 85
[39] Pal U, Samanta D, Ghorai S and Chaudhuri A 1993 *J. Appl. Phys.* **74** 6368
[40] Hamam K J and Alomari M I 2017 *Appl. Nanosci.* **7** 261

IOP Conf. Series: Materials Science and Engineering 1046 (2021) 012011 doi:10.1088/1757-899X/1046/1/012011

Synthesis and characterization of PVA /sPTA proton exchange membranes for fuel cell applications

M M Gomaa*, Y S Elsharkawy, M O Abdel-Hamed and E E Abdel-Hady

Physics Department, Faculty of Science, Minia University, P.O. Box 61519 Minia, Egypt

*Email: mahmoud_gomaa19@mu.edu.eg

Abstract. Using the casting method, proton exchange membranes PEMs for fuel cell were prepared by thermal cross-linking of polyvinyl alcohol (PVA) with sulfophthalic acid (sPTA). The membranes were prepared with different cross-linking temperatures with thickness ranged between 100 to 150 μm. The prepared membranes were characterized by different techniques such as ion exchange capacity, thermal analyses, water uptake and proton conductivity. The prepared membranes showed proton conductivity between 8.46-32.7 mS/cm. Increasing the cross-linking temperature increased the cross-linking density, resulting in an enhancement in the tensile strength and thermal stability of the prepared membranes. The membrane hole sizes were obtained using positron annihilation lifetime spectroscopy (PALS) measurements. Cross-linked PVA/sPTA membranes showed a strong correlation between the free volume hole size with the mechanical stability and methanol permeability of the membranes.

1. Introduction

Proton exchange membrane fuel cells (PEMFC) have attracted attention because of their low operating temperature and fast startup process. One of the essential problems of the fuel cell is the high material costs, in particular, the membrane and the catalysts In order to become commercially available, Low coast and high efficient membranes should be prepared. The standard material used commercially throughout the world is the Nafion. It has several drawbacks because it is unstable at high temperatures and has high fuel permeability, resulting in deterioration of the performance of the fuel cell and a reduction in the life of the cell[1]. Therefore, many researchers are focusing on finding an alternative membrane to Nafion[2-9]. The membrane should have some requirements such as high proton conductivity and high mechanical, thermal and chemical stability[10]. In general, it is believed that the fuel permeability through the PEMs correlates with the structure of the membrane and the open spaces or holes in the PEM material[11-14], so that the structure with less free volume is desirable for PEM. In order to develop the advanced PEM materials with low fuel permeability, It is important to have a deep insight into the hole nanostructures to get more information about the size and distribution of the holes.

The size and distribution of membrane nanopores can be accurately determined using positron annihilation lifetime spectroscopy (PALS), a nondestructive and very sensitive technique. The basic idea of PLAS is to measure the difference between the implantation time and the time of annihilation a positron with an electron. Upon entry into the material, the positron can be directly annihilated with one of the free electrons of the material emitting two gamma rays each of them has energy 511 kev, or it can be located at free volume spaces that form positronium (Ps), which is then annihilation after a longer time (142 ns). Positron annihilation lifetime spectra in polymer usually comprise three or more lifetime components τ_1, τ_2 and $\tau_3$with corresponding intensity I_1, I_2 and I_3, respectively. The third component orthopositronium (o-Ps) is in the range of 1 to 10 ns in the matter and correlates with the radius of the holes in the membranes according to this equation[15]:

$$\tau_3 = 0.5\{1 - \frac{R}{R_0} + \frac{1}{2\pi}\sin(\frac{2\pi R}{R_0})\}^{-1} \quad (1)$$

Here, Ro = R + ΔR is the radius of an infinitely deep square well potential and ΔR = 0.166 nm is the homogeneous electron layer thickness [16]. The volume can be calculated by:

$$V_h = \frac{4}{3}\pi R^3 \quad (2)$$

Modification of the membranes using cross-linking such as chemical cross-linking and thermal cross-linking is a promising method for improving the various properties of the polymers [17-19], including thermal, chemical and mechanical stability. Many researchers are currently focusing on thermal and chemical cross-linking in which the polymer reacts with the cross-linking agent to form a network structure with the aid of a catalyst. In comparison between thermal and chemical cross-linking, Lee et al. [20] concluded that thermal cross-linking can improve the thermal and chemical stability of the polymer membrane more efficiently than chemical cross-linking because of the ability of thermal cross-linking to form a stable three-dimensional network structure after cross-linking.

In this study, PVA was sulfonated with sPTA and then thermally cross-linked at different cross-linking temperatures. The influence of the cross-linking temperature on the various properties is investigated. In addition, the structure of the free volume was studied and correlated with the different properties of the membranes.

2. Experimental

2.1. Materials

Polyvinyl alcohol (PVA, extra pure, Mw =1,25,000 g/mol, from Alpha Chemika), Sulfophthalic acid (sPTA, 50% w/w aqueous solution, from Alfa Aesar, Germany). Sodium chloride (NaCl) and sodium hydroxide (NaOH) were used for titrimetric analysis and purchased from Adwic. Egypt. All chemicals used without further purification.

2.2. Preparation of the Proton Exchange Membranes (PEMs)

The PEM has been prepared using casting technique by heating of (10 % PVA) in an aqueous solution for 6 h at 70°c. After the solution becomes clear, sPTA with concentration 20 wt.% was added to the PVA solution. The PVA/sPTA solution was stirred under the room conditions for 1 day. Then, the solution was poured into a Teflon Petri-dish to avoid the adhesion of the membranes and left for 24 hours to evaporate the excess water. After the evaporation of the excess water, solid membranes were removed from the petri-dish. Finally, the PVA/sPTA membranes were thermally cross-linked for 1 h at different temperatures between 100-120 °c and stored in a desiccator until further testing. The thickness of the prepared membrane ranged from 100 to 150μm.

3. Membrane characterization

3.1. Thermal stability

The thermal stability of pure PVA and PVA / sPTA with different cross-linking temperatures was investigated using a Thermogravimetric Analyzer (TGA) (Model Q50, USA). The TGA spectra were investigated between 35 and 700 ° C at a heating rate of 10 ° C min -1 under a nitrogen atmosphere.

3.2. Water uptake

The water uptake is the tendency of the membranes to absorb water, which can be determined by calculating the percentage weight difference between wet and dry membrane. The water uptake for

PVA/sPTA with different cross-linking temperature was measured at room temperature. The wet membranes were weighted after 24 hours immersion in deionized water and wiped their surface with a tissue paper. The various membranes were then dried in a vacuum oven at 60 ° C for 24 hours and weighted several times under vacuum until the weight became constant to obtain the dry weight of the membranes. The water uptake (WU) of the membranes was calculated according to the following formula:

$$\text{Water uptake } \% = \frac{w_{wet} - w_{dry}}{w_{dry}} \times 100 \quad (3)$$

where w_{wet} and w_{dry} are the weight of the wet and dry membrane

3.3. Ion Exchange Capacity (IEC)
IEC gives information about the concentration of the active groups in the PEMs **(SO_3 groups).** The IEC of the sulfonated membranes was carried out by back-titration technique. The synthesized membranes were weighted after drying at 60 ° C for 1 day under vacuum and then soaking in 30 ml of 1 M NaCl solution for 24 hours at room temperature. During the titration process, H^+ ions in the PEMs are replaced by Na^+ ions. This solution was titrated with 0.05 M NaOH with phenolphthalein as an indicator to determine the concentration of exchanged H+ ions picked up by the membrane. The NaoH was added to the solution until the color changed to pink color. The IEC values, in units of mmole per gram, were calculated according to the following equation [21]:

$$\text{IEC} = \frac{0.05 \times v_{NaoH} \times n}{w_{dry}} \quad \left(\frac{\text{mmol}}{\text{g}}\right) \quad (4)$$

where v_{NaoH} is the NaoH volume used for titrating, n is the quotient of volume the Nacl solution in which the membranes were immersed to the volume of NaoH used for titration andw_{dry} is the dried weight of the membranes.

3.4. Methanol permeability
A homemade side-by-side diffusion cell is used to study the methanol permeability through the PEM[22]. The PEM was sandwiched between two identical glasses reservoirs (A and B). At room temperature, reservoir (A) was filled with 150 ml of 4M methanol while reservoir (B) was filled by distilled water and the two were magnetically stirred at room temperature. Using refractometer, the refractive index of the methanol concentration in the reservoir (B) was measured as a function of time and compared with the calibration curve of the refractometer. Finally, the methanol permeability could be calculated from the concentration versus time curve according to the following equation[23]:

$$P = \frac{K \times V_B \times L}{A \times C_A} \left(\frac{\text{cm}^2}{\text{s}}\right) \quad (5)$$

where C_AK,V_B, L, A and C_A are slope of concentration-time curve, volume of distilled water, thickness of membrane, effective area of permeability and methanol concentration respectively.

3.5. Mechanical properties
The tensile testing of the dried membranes was performed under room conditions (the temperature is about 24 ° C and the relative humidity between 30 and 35%) using a universal material testing machine (Lloyd model LR5K-Plus) at 0.5 mm / min as a constant crosshead speed.

3.6. Proton conductivity
The synthesized membranes were immersed for 24 h in deionized water and were cut into small sections (2×2 cm^2). The ionic conductivity of the cross-linked membranes is obtained at room temperature by the AC impedance spectroscopic technique using HIOKI-3532 LCR Hi-tester over a frequency range from 50 Hz to 1 MHz and an oscillating voltage between 50 and 500 mV. The in-plane ionic conductivity (σ) of the samples was calculated from the respective impedance data according to the following relation:

$$\sigma = \frac{L}{R\times A}\left(\frac{S}{cm}\right) \quad (6)$$

Where L is the distance between the two electrodes, R is the membrane resistance which was taken at the minimum imaginary impedance response in Col-Col plot and A is the cross-sectional area of the membrane.

3.7. Positron Annihilation Lifetime Spectroscopy (PALS)

Positron annihilation lifetime (PAL) measurements are performed using ^{22}NaCl radioactive material as a positron source. Kapton film has enveloped the source. The source activity is approximately 0.297 MBq. The source is placed between 10 layers of the membrane with a total thickness of 1 mm on each side to ensure that the positrons are completely annihilated inside the membrane. A fast-fast coincidence spectrometer with a time resolution of 370 ps was used to collect spectra with more than 2 million counts at room temperature for PVA/sPTA with different cross-linking temperature.

4. Results and discussion

4.1. Thermal Stability

Figure 1 shows the TGA curves for pure PVA and cross-linked PVA/sPTA. The pure PVA showed three weight loss regions with increasing the temperature. The first region between 50–200 °C is caused by the loss of absorbed water molecules while the second one between 200–340 °C is related to the loss of water bound to the polymer matrix. The third region between 340-450 °C is associated with the decomposition and carbonization of the polymer. Contrary, PVA/sPTA showed four regions for weight loss. The first region between 40-150 °C was belonging to the loss of the bounded and free water. The second ranged between (150-210) °C was due to loss of bounded water. The third region (210-450 °C) is attributed to the degradation of the functional groups. Finally above 450 °C, the last loss is the decomposition of the polymeric backbone. It is interesting to find that weight loss values decrease with increasing the cross-linking temperature. Increasing the cross-linking density leads to an increase in the interaction between polymer chains and it is needs more energy to loss. Finally, it is found that increasing the thermal cross-linking temperature for PVA/sPTA is leads to enhancement in thermal stability for the prepared membrane.

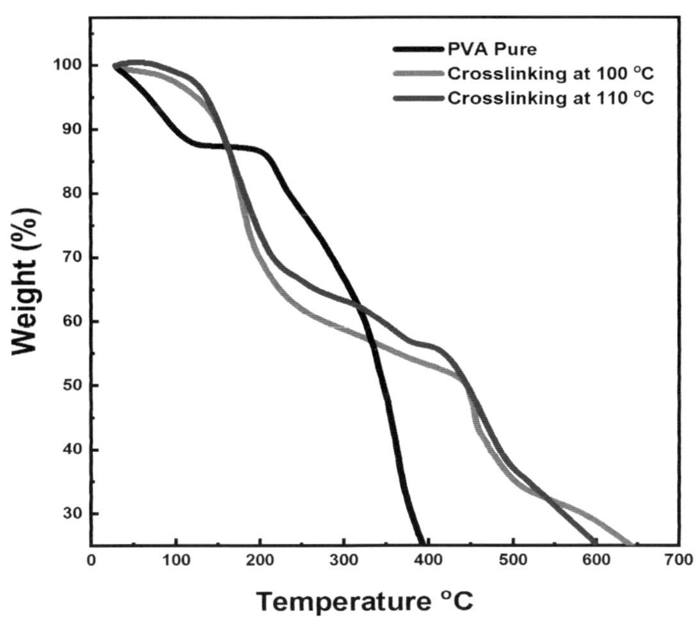

Figure 1. TGA curves for pure PVA and cross-linked PVA/sPTA.

4.2. Water uptake

The hydration of the membrane has a very important function for transporting the protons through the PEM which can reflect the fuel cell performance. There are two types of absorbed water molecules in the membranes. The first type is the free water, which is easy to lose, the second type is bounded water, which interacts with the material itself and requires more energy to lose. The water uptake was determined for the membranes with a different cross-linking temperature of PVA/sPTA and varied between 142 and 40 % , as shown in figure 2. It is obvious that the resulting water absorption values decrease with increasing cross-linking temperature. This behaviour can be attributed to the effect of increasing cross-linking density with increasing the temperature. Hydroxyl groups are responsible for the hydrophilicity of the PVA material, which is known as a highly hydrophilic material due to having a high OH content. Cross-linking limited the concentration of the hydroxyl groups in PVA / sPTA membranes which leads to decreasing the hydrogen bonds interactions, resulting in a decrease in the water uptake of the membranes[7].The thermal cross-linking leads to creating a three-dimensional cross-linking network which restricts the mobility of the chains due to the decrease of the internal free volume space in the structure of the membranes. As a consequence, little water can be trapped in the membrane resulting in lower water uptake.

4.3. Ion Exchange Capacity (IEC)

IEC is an important parameter affects many properties of the PEMs [24] and gives information about the number of functional groups in the membrane that can be considered the way of gaining the membrane the property of proton conductivity [25]. As shown from figure 3, the IEC values of PVA/sPTA decreased with increasing the cross-linking temperatures. This reduction of the IEC is due to forming a hard cross-linking network by thermal cross-linking which reduces the movements of the sulfonic acid molecule and limits the exchange between the ions.

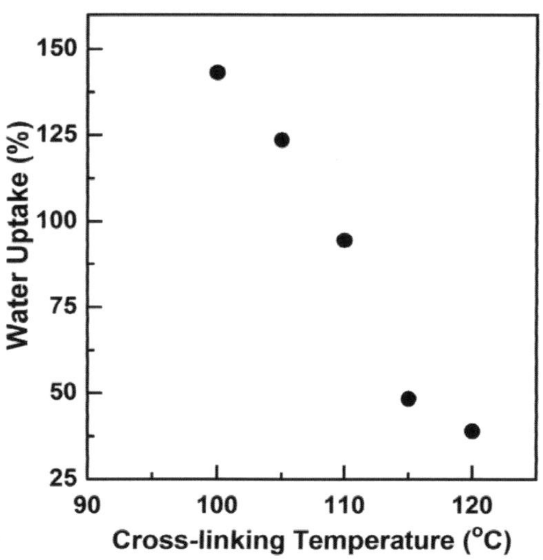

Figure 2. Water uptake of cross-linked PVA/sPTA with different cross-linking temperature.

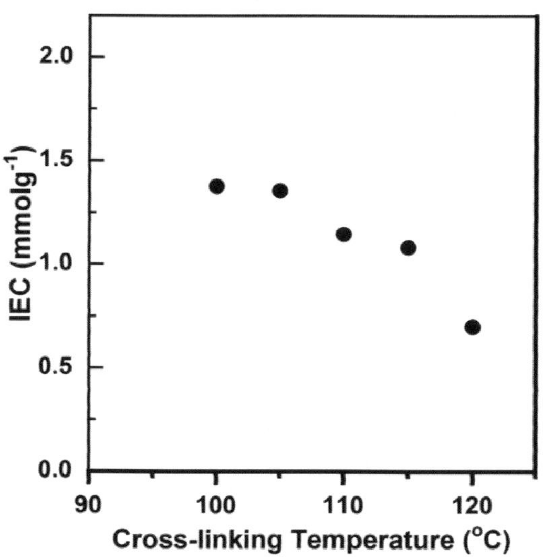

Figure 3. IEC of cross-linked PVA/sPTA with different crosslinking temperature.

4.4. Mechanical Properties

The mechanical stability is one of the requirements of the PEM since it is subjected to a high hot-press pressure to prepare the anode-membrane-cathode which called membrane electrode assembly (MEA)[26]. The membrane should have high mechanical stability to can survive after the hot press process. The tensile testing was performed at room temperature for pure PVA and the PVA/sPTA membranes with the different cross-linking temperature. The obtained results which are presented in figure 4 showed that the cross-linked membranes have higher tensile strength values than that of the pure PVA. The tensile strength values of the prepared membranes ranged between 38 to 80 MPa while it is about 30 MPa for the pure PVA. Moreover, it is clear that with raising the cross-linking temperature, the tensile strength was enhanced. This behaviour can be explained as: when the thermal cross-linking takes place, the polymeric chains connected to each other strongly makes the membrane more rigid. Higher cross-linked temperature leads to stronger polymer chains and lower elongation at break percentages. From the results of PALS, thermal cross-linking leads to a decrease in the dimensions of the holes. The reduction in the size of the holes enhances the interaction between polymer chains and it is become stronger, which is reflected in the increase in tensile strength of the cross-linked membranes. It should be noted that the thermally cross-linking PVA / sPTA membranes exhibited a stronger mechanical property than the commercially available Nafion 117 membrane having tensile strength values of about 10 MPa [27].

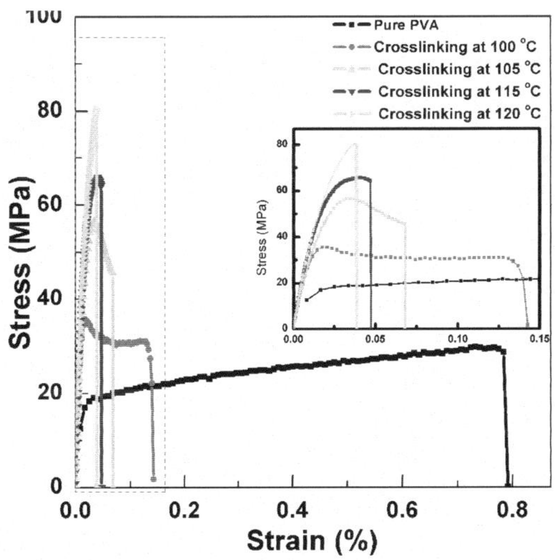

Figure 4. Tensile strength of pure PVA and cross-linked PVA/sPTA with different cross-linking temperature.

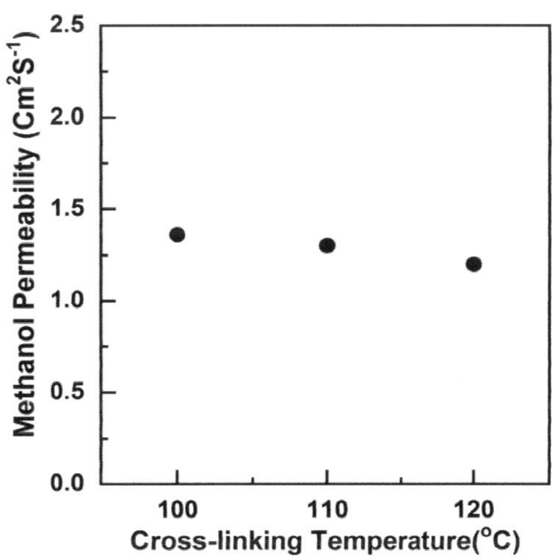

Figure 5. Methanol permeability of cross-linked PVA/sPTA with different cross-linking temperature.

4.5. Methanol permeability

The methanol permeability of PEMs is one of the keys to confirming whether PEMs can be used for long-term operation in fuel cells. High methanol crossover is one of the disadvantages of PEM and affects its properties. It was responsible for reducing the performance of the fuel cell [28]. Figure 5 shows the correlation between the methanol permeability and the cross-linking temperature. It is clear from the figure that increasing the cross-linking temperature of PVA/sPTA membranes leads to decrease the methanol permeability. The methanol permeability values are ranged between (1.2-1.36) x10-6 cm2s-1 as shown in figure 5 which is lower than the permeability of Nafion membrane (1.65 x10-6 cm2s-1)[29]. The results show that the thermal cross-linking reaction can significantly improve the methanol permeability of the membranes, which can cause the polymer to form cross-linking structures. After thermal cross-linking, the membranes have a dense structure that effectively limits the presence of the hydrophilic domains and inhibits swelling of the membranes[30]. In addition, the cross-linking leads to a narrowing of the transport channels of the methanol and plays a strong role as a barrier to methanol permeation. This behaviour is confirmed by reducing the size of the holes in the membrane obtained by PALS results. Decreasing the free volume decrease the interaction between polymers chains leading to enhancement in the tensile strength of the cross-linked membranes.

4.6. The proton conductivity

The proton conductivity of the PEM is the ability of passes the protons from anode to cathode. It is strongly dependent on the water uptake and IEC of the membrane [31]. The proton conductivity was calculated using the in-plane method impedance spectroscopy. The resistance of the membranes obtained from the intersection with the x-axis from the Col-Col diagram (figure 6). The proton conductivity was calculated using Eq. (6) and presented in figure 7 as a function of the cross-linking temperature. It is clear from the figure 7, the proton conductivity of the cross-linked PVA / sPTA membranes decreases with increasing the cross-linking temperature as a behaviour of water absorption and as values of IEC. Decreasing the IEC means decreases the number of active groups (SO_3) in the membrane which is responsible for the conduction of the proton through the membranes. Moreover, a diminution in the size of the free volume and their concentrations with increasing the cross-linking temperature leads to a reduction in the mobility of the charge carrier and is clearly reflected in the reduction of the conductivity values after cross-linking.

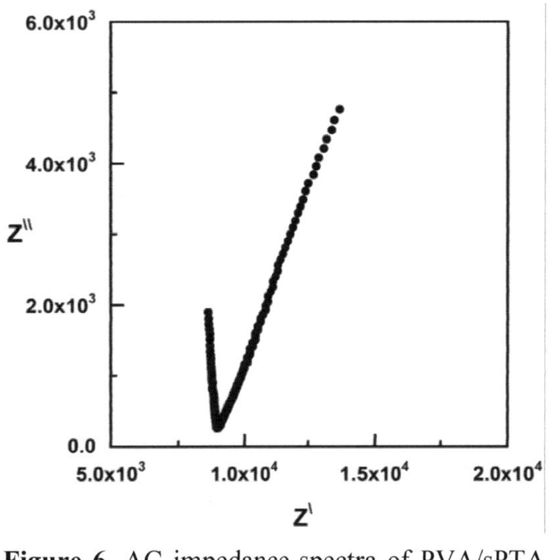

Figure 6. AC impedance spectra of PVA/sPTA with cross-linking at 110 °C.

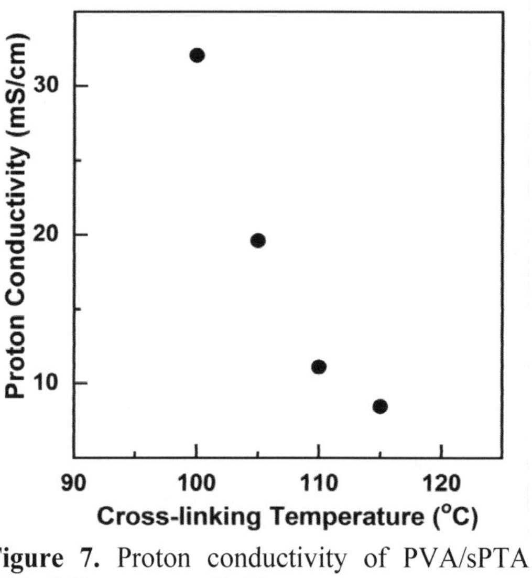

Figure 7. Proton conductivity of PVA/sPTA with different cross-linking temperatures.

4.7. Positron Annihilation Lifetime spectroscopy PALS

The free volume size is studied for cross-linked PVA/sPTA with increasing the cross-linking temperature. Figure 8 shows the *o*-Ps lifetime τ_3 and its intensity I_3 as a function of the cross-linking temperature,including on the right axis of the upper figure , the free volume hole V_h calculated using Eqs. (1) and (2). From figure 8, it is clear that there is no significant change in the o-Ps lifetime, which translates into constant free volume. In contrast, I_3 is dramatically decreased with increasing the cross-linking temperature indicating that the concentration of the holes is decreased. Increasing the cross-linking temperature leads to increasing the density of the cross-linking. After thermal cross-linking, the polymer chains become shorter, which leads to a minimization of the interactions between the polymer molecules, therefore the hole concentration becomes lower.Resulting in an increase in cross-linking density, which reduces the concentrations of the holes without affecting their radius. The reduction in free volume concentrations is reflected in the improvement in thermal and mechanical properties. On the other hand, this leads to a lessening in water absorption and conductivity. It is important to mention that the average size of the holes in Nafion 117 is about 0.18 nm^3[32], which is larger than the cross-linked PVA / sPTA membranes (0.074 nm^3), which have a good correlation with the permeability results.

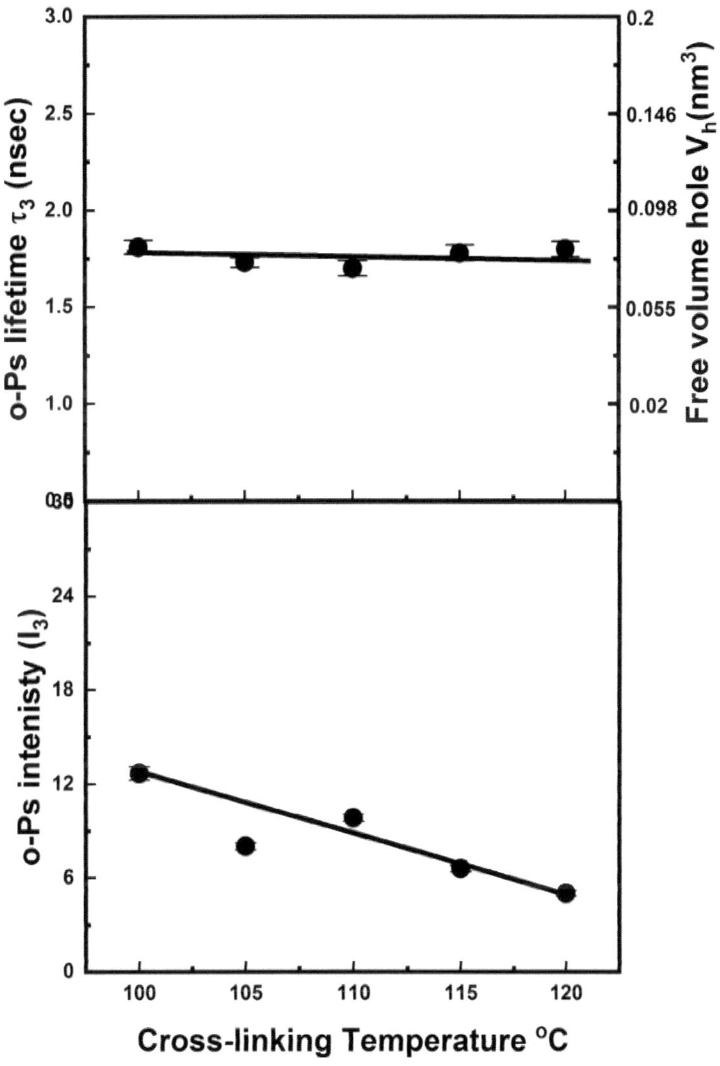

Figure 8. The o-Ps lifetime τ_3 and its intensity I_3 of PVA with concentration 20 wt.% membrane as a function of cross-linking temperature

5. Conclusion

PEM was successfully prepared by thermal cross-linking of PVA /sPTA by casting method. Detailed investigation of PVA/sPTA with different cross-linking temperatures has shown that its thermal and mechanical properties have been improved due to the reduction in the free volume of the membranes with increasing the cross-linking temperature. In addition, the prepared membranes showed methanol permeability less than Nafion membrane. The proton conductivity of the prepared membranes is decreased with increasing the cross-linking temperature and ranged between 8.46-32.7 mS/cm. The increase in the cross-linking density in PVA / sPTA leads to a reduction in the number of holes in the membranes without affecting the size of the holes. It has been observed that there is a strong correlation between water uptake, ionic conductivity, tensile strength, methanol permeability, and free volume holes.

References

[1] Wu J, Yuan X Z, Martin J J, Wang H, Zhang J, Shen J, Wu S and Merida W 2008 *J. Power Sources* **184** 104
[2] Kim H J, An S J, Kim J Y, Moon J K, Cho S Y, Eun Y C, Yoon H K, Park Y, Kweon H J and Shin E M 2004 *Macromol. Rapid Commun.* **25** 1410
[3] Lobato J, Canizares P, Rodrigo M A and Linares J J 2007 *Electrochim. Acta* **52** 3910
[4] Savadogo O and Varela F R 2001 *J. New Mater. Electrochem. Syst.* **4** 93
[5] Abdel-Hady E E, Abdel-Hamed M O and Gomaa M M 2013 *J. Membr. Sci. Tech.* **3** 2
[6] Steele B C and Heinzel A 2011 *Materials for Fuel-Cell Technologies* (Singapore: World Scientific) p 224
[7] Gomaa M M, Hugenschmidt C, Dickmann M, Abdel-Hady E E, Mohamed H F and Abdel-Hamed M O 2018 *Phys. Chem. Chem. Phys.* **20** 28287
[8] Chang Y-N, Lai J-Y and Liu Y-L 2012 *J. Membr. Sci.* **403** 1
[9] Abdel-Hady E E, Mohamed H F M, Abdel-Hamed M O and Gomaa M M 2018 *Adv. Polym. Technol.* **37** 3842
[10] Xing D and Kerres J 2006 *Polym. Adv. Technol.* **17** 591.
[11] Nagel C, Günther-Schade K, Fritsch D, Strunskus T and Faupel F 2002 *Macromolecules* **35** 2071
[12] Kobayashi Y, Haraya K, Hattori S and Sasuga T 1994 *Polym. J.* **35** 925
[13] Mohamed H F M, Ito K, Kobayashi Y, Takimoto N, Takeoka Y and Ohira A 2008 *Polym. J.* **49** 3091.
[14] Gomaa M M, Hugenschmidt C, Dickmann M, Abdel-Hamed M O, Abdel-Hady E E and Mohamed H F M 2017 *Acta Phys. Polon.A* **132** 1519
[15] Tao S 1972 *J. Chem. Phys.* **56** 5499
[16] Nakanishi H 1988 Int. Symp. on Positron Annihilation Studies of Fluids 292.
[17] Kita H, Inada T, Tanaka K and Okamoto K-i 1994 *J. Membr. Sci.* **87** 139
[18] Rezac M E and Schöberl B 1999 *J. Membr. Sci.* **156** 211
[19] Bos A, Pünt I, Wessling M and Strathmann H 1998 *J. Polym. Sci., Part B: Polym. Phys.* **36** 1547
[20] Lee S Y, Kang N R, Shin D W, Lee C H, Lee K-S, Guiver M D, Li N and Lee Y M 2012 *Energy Environ. Sci.* **5** 9795
[21] Rambabu G and Bhat S D 2015 *Electrochim. Acta* **176** 657
[22] Bum Park H, Yong Nam S, Won Rhim J, Min Lee J, Kim S E, Ran Kim J and Moo Lee Y 2002 *J. Appl. Polym. Sci.* **86** 2611
[23] Aricò A S, Sebastian D, Schuster M, Bauer B, D'Urso C, Lufrano F and Baglio V 2015 *Membranes* **5** 793
[24] Chen S-L, Krishnan L, Srinivasan S, Benziger J and Bocarsly A B 2004 *J. Membr. Sci.* **243** 327
[25] Klaysom C, Ladewig B P, Lu G M and Wang L 2011 *J. Membr. Sci.* **368** 48
[26] Therdthianwong A, Manomayidthikarn P and Therdthianwong S 2007 *Energy* **32** 2401
[27] Zheng P, Liu Q, Li Z, Wang D and Liu X 2019 *Polym. J.* **11** 964
[28] Hosseinpour M, Sahoo M, Perez-Page M, Baylis S R, Patel F, and Holmes S M 2019 *Int. J. Hydrog. Energy* **44** 30409
[29] Xu W, Lu T, Liu C and Xing W 2005 *Electrochim. Acta* **50** 3280
[30] Krumova M, Lopez D, Benavente R, Mijangos C and Perena J 2000 *Polymer* **41** 9265
[31] Yamazaki K, Tang Y and Kawakami H 2010 *J. Membr. Sci.* **362** 234
[32] Mohamed H F, Kobayashi Y, Kuroda C and Ohira A 2009 *J. Phys. Chem. B* **113** 2247

Effect of stirring time on ZnO nanoparticles properties and morphology

M A Moghazy*

Environmental Applications of Nanomaterial's Lab., Chemistry Department, Faculty of Science, Aswan University, Aswan, Egypt.

*E-mail: marwa_ch@aswu.edu.eg

Abstract. The impact of stirring time on the ZnO nanoparticles assembly and properties as a novel study was investigated. A precipitation technique has been used for synthesizing ZnO nanoparticles and the impact of stirring time as well as calcination temperature and time were analyzed. X-ray diffraction (XRD), field emission scanning electron microscopy (FE-SEM) and Fourier-transform infrared spectroscopy (FTIR) were used for characterizing the resulting powder. The effect of stirring time, (viz., 0, 2, 4 and 6 h), on the optical properties was investigated using UV-Visible spectroscopy. The XRD analysis shown that all of the acquired powders had a pure form of ZnO nanoparticles with distinct crystallite size. The best circumstances for ZnO nanoparticle synthesis were 2h stirrings time and 400°C for 1 h, as calcination temperature and time, respectively. The FTIR spectra show that ZnO nanoparticles prepared at different stirring times appeared in wavenumber range 435.3 - 458.2 cm^{-1}. SEM pictures reveal different surface distribution with some agglomeration through a variation on stirring time. The bandgap value was identical for 2 and 4 h stirring time with value 3.18 eV while, zero and 6 h stirring showing the same value with slight increase compared with 2 and 4 stirring to be 3.2 eV.

1. Introduction

The multifunctional and morphology ZnO nanoparticles attract many considerations nowadays due to its activity, high binding energy exciton (60 meV) and wide band-gap (3.37 eV) [1]. It is conscious to visible light [2, 3]; nonhazardous material for humans and animals and antibacterial agents because it is resistant to hard processing conditions [4]. Likened to organic materials, inorganic materials like ZnO have excellent thermal stability, greater durability, heat resistance and selectivity [5]. These assets have lead to numerous applications of ZnO in various technological areas, such as gas sensors, photocatalyst, catalysts, varistors, piezoelectric devices, ultraviolet photodiodes, field-emission displays, surface acoustic wave devices, medical and dental materials, sunscreens, rubber, coatings, pigments, concrete, ceramic, optoelectronics, industry, cosmetology, solar cells and cellulose fibers [6-17].

In ZnO nanoparticles synthesis, several synthesis techniques have been used in a pure form with different morphology and crystallite size. Midst the most mentioned techniques are aerosol process [18], sonochemical [19], microemulsions [20], mechanochemical process [21], vapor-liquid-solid growth [22, 23] vapor-phase transport [24, 25], template-assisted systems and vapor-phase epitaxy that extraordinary temperature methods, sol-gel [7], direct and hydrothermal precipitation [26] and spray pyrolysis [27].

Nevertheless, to accomplish the ultimate structure, these approaches usually require multiple stages, complex equipment and high temperature. Of these, the precipitation technique was designated for synthesis ZnO nanoparticles in this study. It is a friendly method, cost-effective, low temperature and forms numerous nanostructures [28, 29].

In supreme synthesis techniques, some variables should be measured and its effect of each on the final product was investigated. For precipitation technique, many parameters as calcination temperature and pH on ZnO nanoparticles size and properties were intensively considered, but the effect of reaction stirring time not considered in literatures till now. The objective of this paper is to synthesize ZnO nanoparticles in pure form and to investigate of the impact of the time-stirring on its optical feature and morphology.

2. Experimental

2.1. Materials
For the preparation of zinc oxide Nanoparticles, zinc acetate and NaOH were used of analytical gr ade (sigma-Alderich) without further purification.

2.2. Synthesis of ZnO nano powder.
Firstly, sodium hydroxide solution (2 M) was added to zinc acetate solution (1 M) with constant stirring for diverse stirring time (0-6 h). The resulted white precipitate was filtered off and washed several times with deionized water, then the obtained precipitate was dried at100 °C. Finally, the resulting powder calcined at different temperature 300 °C-600 °C for different interval times 0.5-4 h.

2.3. Characterization
The resulted powders characterization was set by X-ray diffraction (XRD) BrukerAXSD8 Germany Cu Kα radiation 0.154. FTIR analysis was done by an Agilent Technologies, Cary 630. Field Emission-Scanning Electron Microscopy (FE-SEM) was achieved using (FE-SEM, QUANTAFEG250, Netherlands). Bandgap characterization was done using Jasco V-670 spectrophotometer (ISN-723, Japan).

3. Result and discussion

3.1. XRD Analysis

3.1.1. Effect of calcination temperature. Calcination impact on the ZnO nanoparticles synthesis is an important parameter since it indicates the stability of the substance and its temperature of crystallization. This effect was studied in the range of 100-600°C. As indicated from figure 1, high intensity and complete crystallization of ZnO nanoparticles was obtained at 400°C. As observed from the previous literature and the TGA analysis, the precursor can be completely decomposed to ZnO after annealing at ~ 350°C[16, 30]; whereas, Moharram, et al., found that, the mass leakage ends at about 300 °C which is due to the decomposition of $Zn(OH)_2$. The best improvement of the ZnO NPs after calcination process was at 400°C[31].

The Debye–Scherer approach has been used for considering ZnO nano crystallite size as set out in Eq.(1). Varying calcination temperature resulted in a distinction of the ZnO size. It was observed that, decrease in crystallite size from 82.4 nm to 44.1nm was obtained with scaling up calcination temperature (100-300°C). Further rise in calcined temperature to 600°C, drastically growth in crystallite size was observed. Reduction in crystallite size at low calcined temperature is revealing of the reorganization process; while at calcination temperature >300°C increase in crystallite size is revealing of reinforcement of ZnO phase [32].

$$D = \frac{0.9\lambda}{\beta\ cos\theta} \quad (1)$$

where, D, λ, β and Θ are crystallite size, X-ray source wavelength, the full width at half maximum of the peak and the angle at which diffraction intensity is maximized. From the study it was observed that, the color of obtaining ZnO powder is yellow, which returned to white with cooling. This phenomenon explained by the fact that, ZnO exhibits a thermochromism property wherein it changes its color with increasing temperature. This is probably because of the loss of oxygen which results in the formation of crystal lattice defects [5, 33]

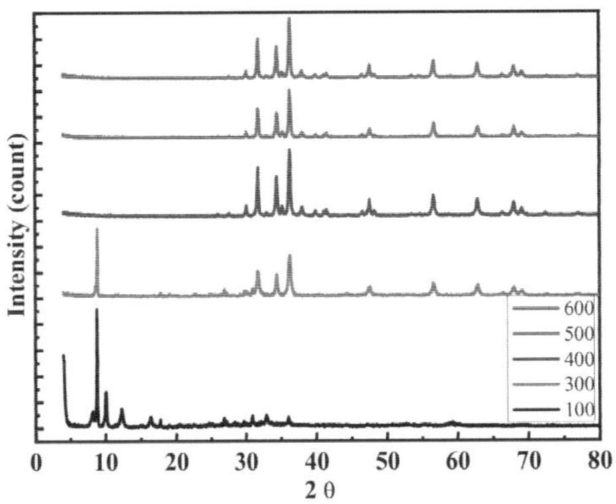

Figure 1. Effect of calcination temperature on the synthesis of ZnO nanopowder at 100, 300, 400, 500 and 600°C.

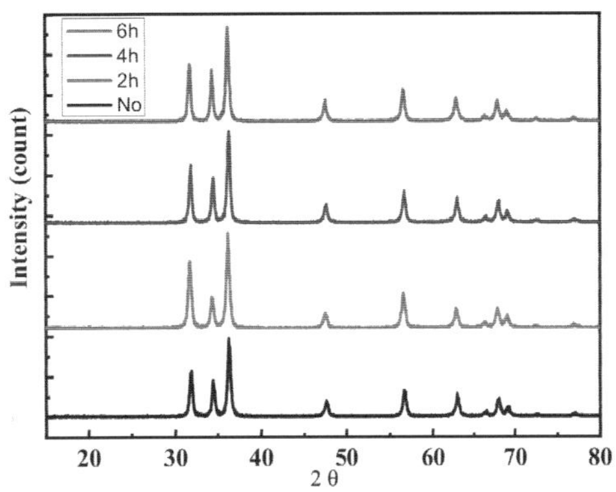

Figure 2. Effect of stirring times on the synthesis of ZnO nanopowder at no stirring, (2 h), (4 h) and (6 h).

3.1.2. Effect of stirring time. The stirring time was analyzed in the range 0-6 h. It was found that all reaction times lead to the pure ZnO formation with different intensity and crystallite size with no second phases appeared. The greatest stirring time is 2 h, as illustrated in figure 2 with the maximum peak intensity and the smallest crystallite size compared with other stirring times. The peaks FWHM increases as stirring time increases. At zero time, figure 3, stirring, the high crystallite size (148 nm) was observed of ZnO nanoparticles as a result of the insufficiency distribution of small particles leading to accumulation. At 2 h stirring time, high crystallization with small 74 nm crystallite size was obtained. With increasing stirring time crystallite size augmented. This may explain as small particles acquired energy from the stirring process leading to an increase in the collision rate. Consequently, the combination of small particles into larger particles and hence aggregation resulted in the grain size enlarges [34, 35]. Surabhi Siva Kumar, et al., [16] synthesized nano ZnO for 12 h in dynamic stirring.

3.1.3. Effect of calcination time. Calcination time was varied in the range 0.5-4 h. From XRD pattern, figure 4, it was clear that, as calcination time increased the peak intensity and crystallite size gradually increased to be maxima at 4h. Not any diffraction peaks of second phases were perceived. At the short time, (0.5h), high intensity and 123nm crystallization size was obtained. The phenomenon of scale up crystallite size with calcination time qualified to the growth of the crystalline volume to the surface ratio, which happened due to particle size enlargement [36].

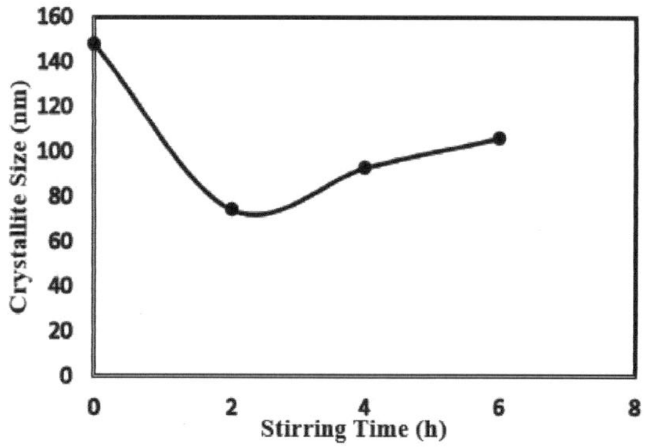

Figure 3. Effect of stirring times on the ZnO nanopowder crystallite size.

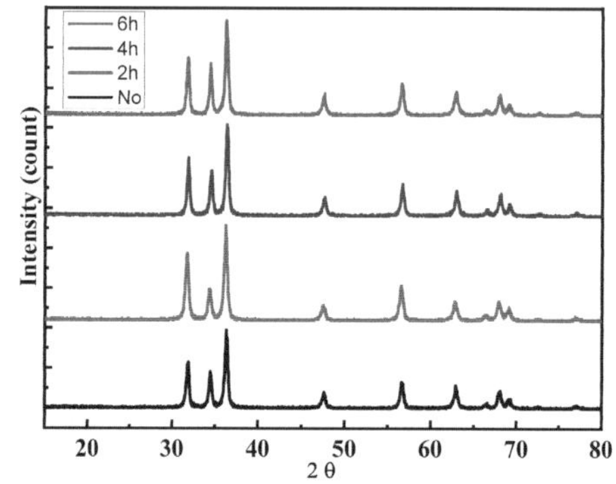

Figure 4. Effect of different stirring times on the synthesis of ZnO nanopowder.

3.2. FE-SEM analysis

The stirring periods impact on the ZnO nano powder morphology has been evaluated by mean of the field emission scanning electron microscope (FE-SEM). The FE-SEM images of nano ZnO with stirring time: zero, 2, 4 and 6 h were represented figures 5. All images demonstrate spherical natures at high magnification level with particle size in nano scale (smaller than 100 nm).

At high magnification, spherical shapes were observed for all stirring times as illustrated from figure 5(a, c, e, g) with a tendency of agglomeration growth for zero and 6 h stirring times. Some spherical cages (raspberry-like) for 4 h stirring time appeared. The particle size was found to be 20-92, 25-46, 32-72 and 45-82 nm for zero, 2, 4 and 6 h stirring times, respectively. At low magnification, the surface topography was diverse with respect to stirring time. For zero time, figure 5(b), some plate-like morphology appeared plus the spherical particles, whereas the surface is a flat and the well-ordered morphology in the form of a flake-like identity with uniform surface topography was obtained in case of 2h stirring as shown in figure 5(d). No agglomeration appeared was obtained. Stirring for 4 h, figure 5(f), gives surface morphology of a sponge-like shape. Irregular platelets also observed. The structure for 6h stirring appeared in figure 5 (h) many agglomerations and the particle arranged to form a sheet-like surface.

3.3. FT-IR analysis

Figure 6 exemplifies the FTIR bands of the produced nano ZnO, stirred at zero, 2, 4, and 6 h. The synthesized ZnO nanoparticles peaks were observed between 435.3 and 458.2 cm^{-1} [37, 38]. The peak area scaled up as stirring time increased, indicating increasing crystallite size. The C=O stretching was observed for all stirring times between 1622 until 1634cm^{-1} [39], which gradually increased with increasing stirring time. Peaks related to O-H were observed in 3435cm^{-1} [40] which diminished regularly as stirring time raised. This may relate to the more consuming OH^- reacted with zinc salt as time increased [41].

3.4. Optical properties

By measuring the diffuse spectroscopy of reflectance, the clarification of optical properties was achieved. The diffuse reflectance (R) was considered against wavelength as displayed in figure 7 for the altered stirring times of ZnO. The spectra were done in the wavelength range 200-1200 nm. The bandgap was calculated using the Kubelka-Munk function F(R) given in Eq. (2) [1, 42]. The gap energy was estimated through drawing the $F(Rh\nu)^2$ Kubelka-Munk function versus energy ($h\nu$) as exhibited in figure 8. The straight line

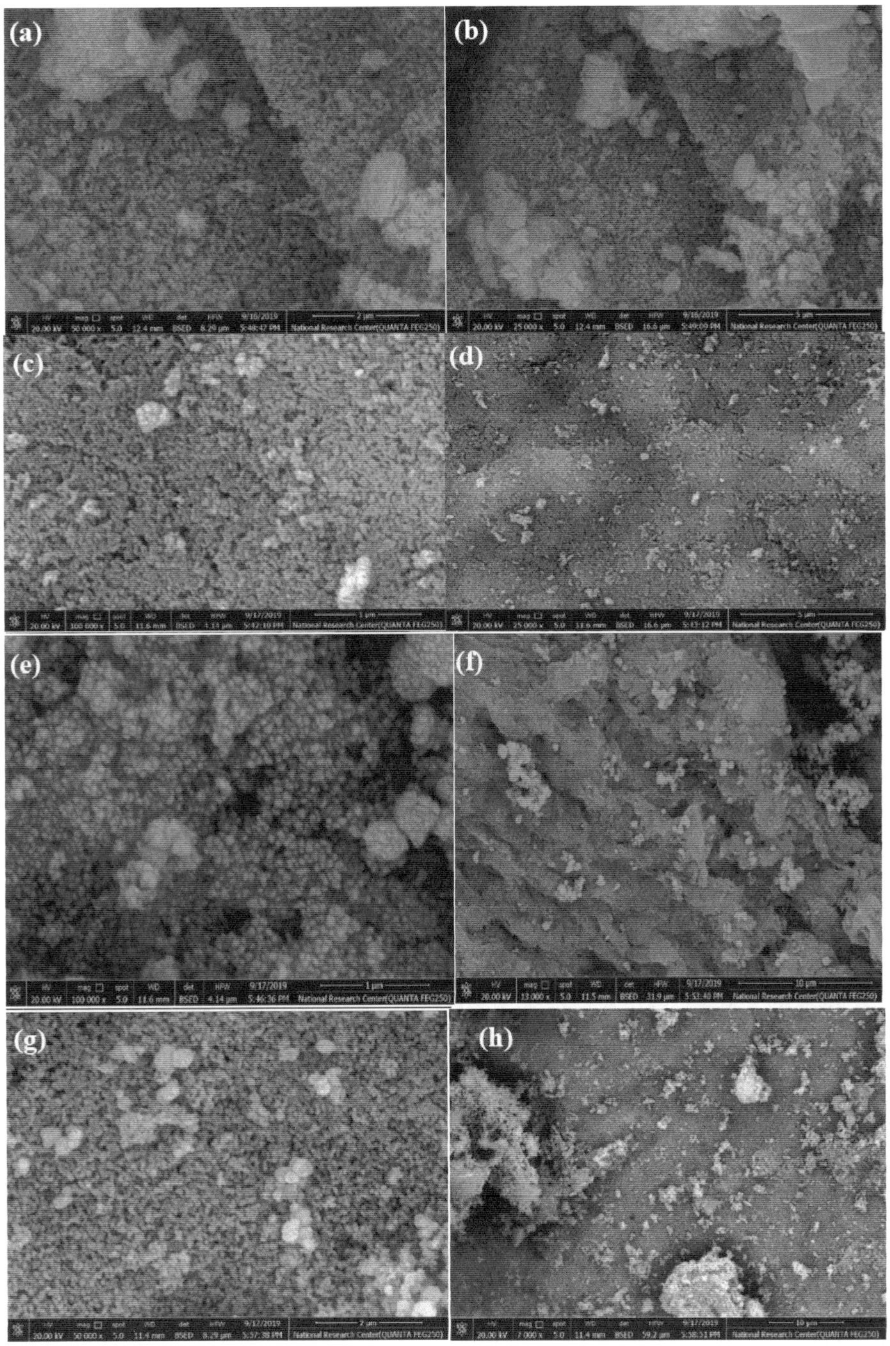

Figure 5. Effect of stirring times on the morphology of ZnO nanopowder at stirring time: (a,b) zero, (c,d) 2, (e,f) 4 and (g,h) 6h at high (2 μm) and low (10 μm) magnification.

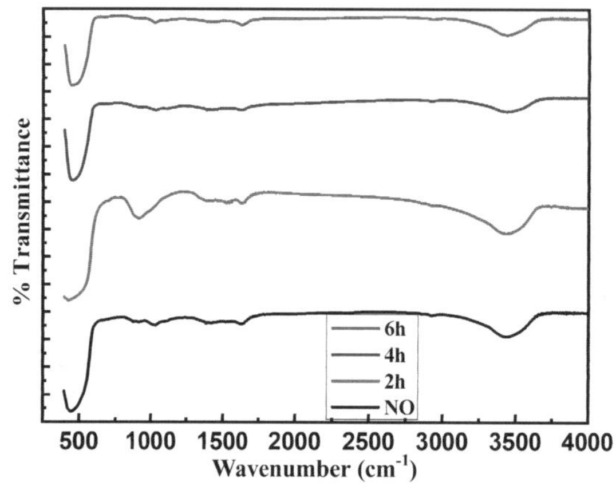

Figure 6. FT-IR spectrum of ZnO nanoparticle synthesis at stirring time 0, 2, 4 and 6h.

of the plot was extrapolated to the energy axis. It was evident that the gap energy of zero and 6 h stirring times have to be 3.2ev, while for 2 and 4h stirring times the gap energy calculated to be 3.18 ev.

$$F(R) = \frac{(1-R)^2}{2R} \quad (2)$$

The difference in gap energy values associated with the variation in particle size, morphologies and the agglomeration formation of the nano ZnO [43]. For zero and 6 h stirring time, high particle size compared to 2 and 4 h stirring time, it has been observed that the bandgap for zero and 6 h stirring are higher than other stirring time. The slight difference in band gap is related to structural variation and the appearance of some agglomerations which resulted from alteration of stirring times [1, 41]. Previous literature determined bandgap of ZnO nanoparticle to be 3.41 eV[32], 3.63 eV[44], 3.10-3.37eV [45], and 3.35 eV [46].

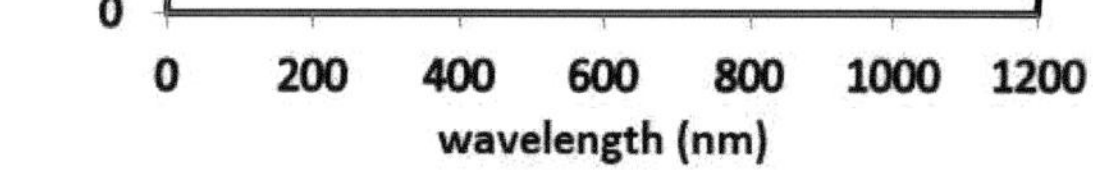

Figure 7. Reflectance spectra of ZnO nanopowder for samples synthesized at different stirring times.

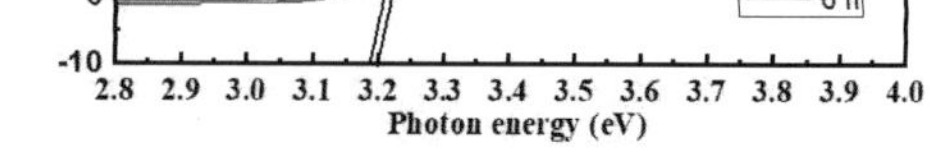

Figure 8. Bandgap of ZnO nanopowder using the Kubelka-Munk function equation for nano ZnO synthesized at different stirring times.

4. Conclusion

Nano ZnO was well prepared under numerous synthesis conditions through the precipitation process. The XRD and the FTIR confirm the great crystalline and pure form of nano ZnO nanoparticles in each synthesized powder. The particle size relies on calcined temperature and time. It crystallized with scaling up the calcination temperature as well as time. Nano ZnO synthesis at 400 °C for 1 h calcination time and 2 h stirring time show the best feature and minimum crystallite and particle size. The stirring period influence on the structure and optical feature of prepared nano ZnO. Spherical assembly with a variety of particle sizes was found whereas, the surface organization for each stirring time was different. The bandgap increase from 3.18 to 3.2 ev with stirring time increment from 2 to 6 h.

Acknowledgements
Special thanks and gratefully acknowledge to professor G. M. Taha for his support and valuable discussion.

References
[1] Prakoso S P 2011 *World J. Condens. Matter Phys.* **01** 130
[2] Wang Z L 2004 *J. Phys. Condens. Matter.* **16** R829
[3] Herrera-Rivera R, Olvera M d L and Maldonado A 2017 *J. Nanomater.* **2017** 1
[4] Cloutier M, Mantovani D and Rosei F 2015 *Trends Biotechnol.* **33** 637
[5] Dimapilis E A S, Hsu C, Mendoza R M O and Lu M 2018 *Sustain. Environ. Res.* **28** 47
[6] Xu J, Pan Q, Shun Y and Tian Z 2000 *Sens. Actuators B Chem.* **66** 277
[7] Suchorska-Woźniak P, Nawrot W, Rac O, Fiedot M and Teterycz H 2016 *IOP Conf. Ser. Mater.* Sci. *Eng.* **104** 012030
[8] Bai S, Liu X, Li D, Chen S, Luo R and Chen A 2011 *Sens. Actuators B Chem.* **153** 110
[9] Phuruangrat A, Yayapao O, Thongtem S and Thongtem T 2016 *Russ. J. Phys. Chem.* . **90** 949
[10] Gupta T K 1990 *J. Am. Ceram. Soc.* **73** 1817
[11] Koa S C, Kimb Y C, Leeb S S, Choic S H and Kimd S R 2003 *Sens. Actuators A Phys.* **103** 130
[12] Soniaa S, Kumarib H L J, Ruckmanib K and Sivakumara M 2017 *Mater. Sci. Eng. C* **79** 581
[13] Pulit-Prociak J, Chwastowski J, Kucharski A and Banach M 2016 *Appl. Surf. Sci.* **385** 543
[14] El-Feky O M, Hassan E A, Fadel S M and Hassan M L 2014 *J. Cult. Herit.* **15** 165
[15] Arefi M R and Rezaei-Zarchi S 2012 *Int. J. Mol. Sci.* **13** 4340
[16] Salahuddin N A, El-Kemary M and Ibrahim E M 2015 *Nanosci. Nanotechnol.* **5(4)** 82
[17] Varaprasad K, Raghavendra G M, Jayaramudu T and Seo J 2016 *Carbohydr. Polym.* **135** 349
[18] Kodas T T 1989 *Angew. Chem.* **101** 814
[19] Phuruangrat A, Yayapao O, Thongtem S and Thongtem T 2016 *Russ. J. Phys. Chem.* . **90** 949
[20] Yıldırım Ö A and Durucan C 2010 *J. Alloys Compd.* **506** 944
[21] Shen L, Bao N, Yanagisawa K, Domen K, Gupta A and Grimes C A 2006 *Nanotechnology* **17** 5117
[22] Kong Y C, Yu D P, Zhang B, Fang W and Feng S Q 2001 *Appl. Phys. Lett.* **78** 407
[23] Heo Y, Varadarajan V, Kaufman M, Kim K, Norton D, Ren F and Fleming P 2002 *Appl. Phys. Lett.* **81** 3046
[24] Huang M H, Wu Y, Tran H F N, Weber E and Yang P 2001 *Adv. Mater.* **13** 113
[25] Lu J G, Ye Z Z, Huang J Y, Zhu L P, Zhao B H, Wang Z L and Fujita S Z 2006 *Appl. Phys. Lett.* **88** 063110
[26] Wang J, Ge L, Li Z, Li L, Guo Q and Li J 2016 *Ceram. Int.* **42** 8545
[27] Messing G L, Zhang S and Jayanthi G V 1993 *J. Am. Ceram. Soc.* **76** 2707
[28] Kahouli M, Barhoumi A, Bouzid A, Al-Hajry A and Guermazi S 2015 *Superlattices Microstruct.* **85** 7
[29] Sepulveda-Guzman S, Reeja-Jayan B, de la Rosa E, Torres-Castro A, Gonzalez-Gonzalez V and Jose-Yacaman M 2009 *Mater. Chem. Phys.* **115** 172
[30] Raoufi D 2013 *Renew. Energy.* **50** 932
[31] Moharram A H, Mansour S A, Hussein M A and Rashad M 2014 *J. Nanomater.* **2014** 1
[32] Ashraf R, Riaz S, Kayani Z N and Naseem S 2015 *Mater. Today.* **2** 5468

[33] Weidenkaff A, Reller A W, Wokaun A and Steinfeld A 2000 *Thermochim. Acta.* **359** 69
[34] Singh N, Dhruvashi, Kaur D, Mehra R M and Kapoor A 2012 *Open Renew. Energy J.* **5** 15-18
[35] Xavier C S, Sczancoski J C, Cavalcante L S, Paiva-Santos C O, Varela J A, Longo E and Li M S 2009 *Solid State Sci.* **11** 2173
[36] Alibe I M, Matori K A, Sidek H A A, Yaakob Y, Rashid U, Alibe A M, Zaid M H M and Khiri M Z A 2018 *Molecules.* **23** 1
[37] Moussawi R N and Patra D 2016 *RSC Adv.* **6** 17256
[38] Handore K, Bhavsar S, Horne A, Chhattise P, Mohite K, Ambekar J, Pande N and Chabukswar V 2014 *J. Macromol. Sci. A.* **51** 941
[39] Fernandes D M, Silva R, Hechenleitner A A W, Radovanovic E, Melo M A C and Pineda E A G 2009 *Mater. Chem. Phys.* **115** 110
[40] Meshram J V, Koli V B, Phadatare M R and Pawar S H 2017 *Mater. Sci. Eng. C* **73** 257
[41] Kayani Z N, Saleemi F and Batool I 2015 *Appl. Phys. A.* **119** 713
[42] Naeem M, Hasanain S K and Mumtaz A 2008 *J. Phys. Condens. Matter.* **20** 1
[43] Baruwati B, Kumar D K and Manorama S V 2006 *Sens. Actuators B Chem.* **119** 676
[44] Debanath M K and Karmakar S 2013 *Mater. Lett.* **111** 116
[45] Davis K, Yarbrough R, Froeschle M, White J and Rathnayake H 2019 *RSC Adv.* **9** 14638
[46] Singh J, Kumar P, Hui K S, Hui K N, Ramam K, Tiwari R S and Srivastava O N 2012 *CrystEngComm* **14** 5898

Microstructural changes and fracture resistance of nano-crystalline monolithic zirconia restorations upon aging

M H Kamel[1,*], H Zaghloul[2], T Salah[3]and L Ghanem[4]

[1]Department of fixed prosthodontics, Faculty of Oral and Dental Medicine, Misr International University, KM 28 Cairo – Ismailia Road Ahmed Orabi District, Cairo, EG. ORCID ID 0000-0003-2198-3301.
[2]Department of fixed prosthodontics, Faculty of Oral and Dental Medicine, Misr International University, KM 28 Cairo – Ismailia Road Ahmed Orabi District, Cairo, EG . ORCID ID 0000-0002-3271-4441.
[3]Department of crown and bridge, Faculty of Dentistry, Ain Shams UniversityOrganization of African Unity St, El-Qobba Bridge, Al Waili, Cairo Governorate,EG.
[4]Department of fixed prosthodontics ,Faculty of Oral and Dental Medicine, Misr International University, KM 28 Cairo – Ismailia Road Ahmed Orabi District, Cairo, EG .

*E-mail: Mostafa.hussin@miuegypt.edu.eg

Abstract. To assess the effect of nano-modification of two CAD/CAM monolithic Zirconia systems (wet and dry-milled) on microstructural changes and fracture resistance upon aging. Twenty monolithic Zirconia discs (10mm x1.5mm) were divided into 2 groups (n=10) according to fabrication system;wet-milled (Incoris TZI) and dry-milled (CeramillZolid). Each group was subdivided into two subgroups (n=5), whether subjected to aging or not. Aging proceduresincluded accelerated hydrothermal aging and cyclic loading. Microstructural changes were assessed quantitatively and qualitatively using x-ray diffraction (XRD) and scanning electron microscope (SEM) respectively. Discs were subjected to fracture resistance test usinguniversal testing machine. Meansandstandard deviations were recorded for phase transformation (t-m)of Zirconia and repeated ANOVA was used to correlate phase transformation with fracture resistance results ($p \leq 0.05$). For aged subgroups, XRD analysis revealed significantly higher phase transformation of wet-milled than dry-milled ($p \leq 0.05$) and SEM analysis revealed wide zone phase transformation for wet-milled zirconia and minimal zone for dry-milled. For both systems, no significant difference between the mean flexural strength values ofnon-agedand aged zirconia ($p \leq 0.05$).Nanocrystalline microstructure of monolithic zirconia improved its resistance to aging. The fracture resistance of the two systems was not affected by aging.

IOP Conf. Series: Materials Science and Engineering 1046 (2021) 012013 doi:10.1088/1757-899X/1046/1/012013

1.Introduction

Esthetics is becoming a major demand in modern dentistry. The advances in recent dental ceramics are concerned with providing materials with high esthetic parameters, together with function to provide long term success for the restorations [1].Zirconia is one of the materials has the potential to substitute the metal-ceramic fixed dental prosthesis (FDP) due to their high biocompatibility [2] and similar mechanical properties[3][4] with those of metal-ceramic restorations. The superior mechanical properties are due to the finer grain size and the tetragonal-monoclinic transformation toughening mechanism (t-m) of yttria-stabilized Zirocnia (3Y-TZP) which leads to compressive stresses in the material and results in reduced crack propagation [5].

On the other hand, Zirconia shows inferior optical properties than those of natural tooth structure due to its inherent opacity [6]. This opacity required veneering of zirconia with more translucent ceramic to simulate the natural esthetic appearance of the tooth structure. The veneering process had a negative impact on the strength and longevity of the zirconia-based restorations, [7], [8][9].

The oral cavity is an inhospitable environment due to the presence of moisture, temperature and PH fluctuations and biting forces that induce zirconia degradation, hence a t→m phase transformation [10][11]. This phenomenon is known as low temperature degradation (LTD) [12].One of the hypotheses currently most favored is that the filling of oxygen vacancies by “water derived species” (hydroxyl, oxygen, or hydrogen ions) probably leads to accumulation of internal tensile stresses in the grains in contact with water. (t-m) transformation is accompanied by surface uplift of grains and large stresses that can provoke the creation of cracks along the grain boundaries [13][14]. Previous studies [15][16], [17]discussed the relation between (t-m) transformation of zirconia and the reduction of its mechanical properties. Long term exposure to hydrothermal aging, has produced significant reduction in its mechanical parameters, including flexural strength and Young’s modulus of elasticity. CAD/CAM machined Zirconia showed initial high resistance to hydrothermal aging but deteriorated at a faster rate upon prolonged exposure to oral conditions[18][19].

Many studies [20][21][22], [23]clarified the role of nano technology in improving the properties of zirconia, regarding to the optical properties and resistance to aging. Reducing the grain size to nano scale (70 – 82 nm) together with high density and eliminating the internal porosity had been proposed as a solution for enhancement of aging resistance[24][25]. These nanomodifications of Zirconia had led to introduction of monolithic translucent zirconia which in turn overcomes the drawbacks of Zirconia-based systems. By reviewing the literature, the impact of the nanocrystalline microstructure on the fracture and aging resistance of 3Y-TZP requires furtherer investigations. So, the aim of the present study was to assess the effect of nano-modification of two CAD/CAM monolithic Zirconia systems (wet and dry) on microstructural changes and fracture resistance upon aging. The null hypothesis of this study was that nano-modification of two CAD/CAM monolithic Zirconia systems (wet and dry) had no effect on the microstructural changes and the fracture resistance upon aging.

2.Materials and Methods

2.1. Sample grouping

Twenty monolithic Zirconia discs (10 mm diameter x 1.5 mm thickness) were divided into 2 groups (n=10 each) according to the system used for fabrication. Group I; fabricated from wet-milled monolithic nanocrystalline Zirconia system (incoris TZI, Dentpsly Sirona, Germany), Group II; fabricated from dry-milled monolithic nanocrystalline Zirconia (CeramillZolid, Ammangirbach, Austria). Each group was further subdivided into 2 subgroups (n=5) a: not subjected to aging and b: subjected to aging.

2.2. Discs Fabrication

For the purpose of standardization of monolithic Zirconia discs, a copper mould was designed with diameter of 10 mm and thickness of 1.5 mm. For the wet-milled monolithic zirconia discs; the mould was scanned by an Intraoral scanner (Omnicam, Dentsply Sirona, Germany) and discs were designed using Inlab 3Dsoftware (V4.2) (Dentsply Sirona, Germany). Zirconia discs were milled using InLab MCX5 machine (Dentsply Sirona, Benseheim, Germany). Sintering process was carried out for all the discs inFire HTC speed(Dentsply Sirona, Germany) following the manufacturer's recommendations. While for Dry-milled monolithic Zirconia discs, scanning of the mouldwas carried out with digital optical scanner (Ceramillmap 400, Ammangirbach, Austria), designing of the disc with CAD software (Ceramillmind, Ammangirbach, Austria), milling using Ceramill motion (Ammangirbach, Austria) and finally sintering process was carried out in furnace (Ceramiltherm, Ammangirbach, Austria) following the manufacturer's recommendations.

2.3. Aging Procedures

Zirconia discs of group Ib and IIb were subjected to 2 types of aging to simulate the conditions existing in the oral cavity (hydrothermal aging and cyclic loading). For hydrothermal aging, discs were placed in a steam autoclave (Hydra,Carlo Di Georgi, Italy) at temperature 1210C and 1 bar pressure for 5 hours to simulate hydrothermal aging for 5 years inside the oral cavity **Error! Reference source not found.**.To simulate cyclic loading, a custom-made machine loading device was designed (figure 1). The machine delivers 1,250,000 cycles at right angle to the discs at frequency of 300 cycle/ min using load of 50 N forces[26].

2.4. X- ray diffraction analysis (XRD)

XRD analysis was used to detect the crystalline phases inside the monolithic Zirconia samples. XRD analysis were performed with a powder diffractometer (Xpert pro, PANalytical B.V, Netherlands), using Cr Ka radiation (40 mA, 40 kV), a scan range of 0–90(2h), a step size of 0.02 and a scan time per step of 4 s. Zirconia fractions were determined with software support. The percentage of Zirconia transformation (t-m) was calculated using Gravie- Nicholson equation [27].

$$Xm = Im\,(-1,1,1) + Im\,(1,1,1) / Im\,(-1,1,1) + Im\,(1,1,1) + It\,(1,0,1) \quad (1)$$

where Xm is the amount of transformed monoclinic phase, Im (-1,1,1) is the peak recorded at 2θ angle 28.40^0, Im (1,1,1) is the peak recorded at 2θ 31.41^0 and It (1,0,1) is the tetragonal peak recorded at 2θ 30.34^0.

2.5. Fracture testing

Discs sample strength was determined using biaxial flexural test in a universal mechanical testing machine (Instron, University Ave, Norwood, USA) at a cross-head speed of 1mm/min. Custom made sample holder was designed for flexural strength testing (figure 2). Samples were successively placed on the testing platform on the universal testing machine to remain stable during load application and the maximum breaking loads were recorded in Newton (N).

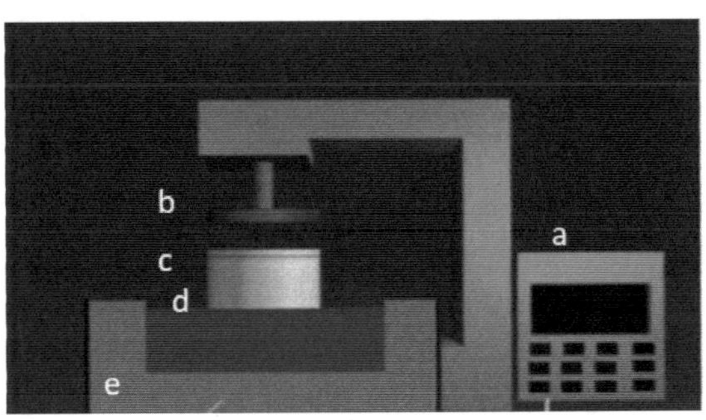

Figure1. Diagram showing custom made cyclic loading machine.: (a) digital panel to control the number ofload cycles and load intensity, (b) attachment for cyclic load application, (c) acrylic base to hold the disc, (d) monolithic Zirconia disc and (e) machine platform to hold acrylic base – zirconia disc assembly.

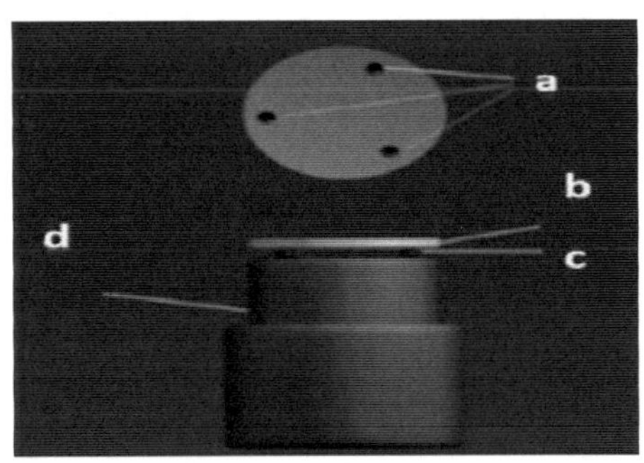

Figure 2. Custom made holder for zirconia discs: (a) top view with 3 equidistant stainless-steel balls (b) Zirconia disc (c) Stainless steel balls projecting from the base (d) Stainless steel base of (12mm in diameter).

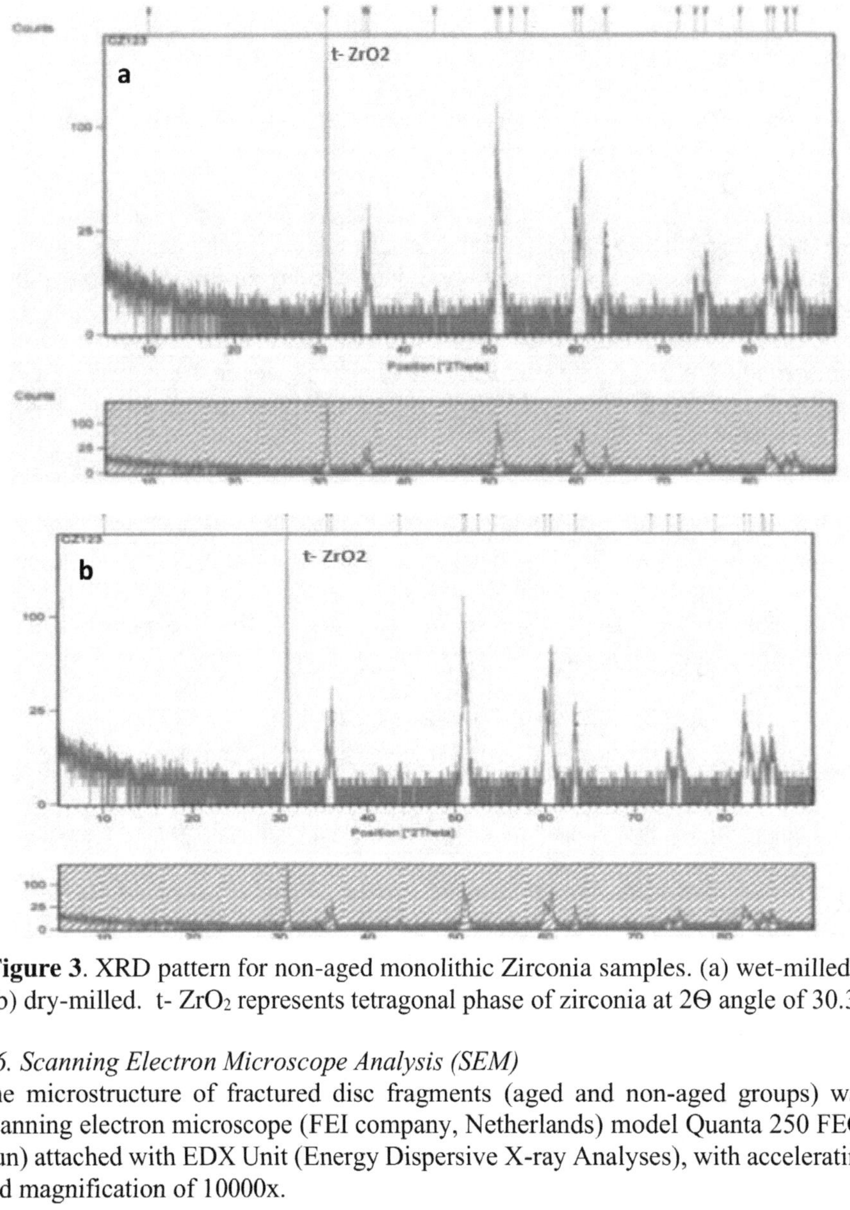

Figure 3. XRD pattern for non-aged monolithic Zirconia samples. (a) wet-milled and (b) dry-milled. t- ZrO_2 represents tetragonal phase of zirconia at 2Θ angle of 30.3^0.

2.6. Scanning Electron Microscope Analysis (SEM)

The microstructure of fractured disc fragments (aged and non-aged groups) was assessed using Scanning electron microscope (FEI company, Netherlands) model Quanta 250 FEG (Field Emission Gun) attached with EDX Unit (Energy Dispersive X-ray Analyses), with accelerating voltage 30 K.V and magnification of 10000x.

2.6. Statistical Analysis

Quantitative data were presented as mean and standard deviation (SD). Data were explored for normality by checking the data distribution, calculating the mean and median values and using Kolmogorov-Smirnov and Shapiro-Wilk tests. Flexural strength data showed parametric distribution. For parametric data; repeated measures Analysis of Variance (ANOVA) was used in testing the effect of milling system, aging and their interactions on mean flexural strength. Turkey's post-hoc test was used for pair-wise comparison between the groups when ANOVA test is significant. Whereas for the values of XRD measurements, mean and standard deviation (SD).Theamount of (t-m) transformation were recorded for aged and non-aged groups. The significance level was set at$P \leq 0.05$.

3. Results

3.1.XRD Results

XRD pattern for wet-milled and dry-milled monolithic nanocrystalline Zirconia systems were recordedfornon-agedsamples (figure 3 a and b). Peak for tetragonal Zirconia was recorded at 2Θ angle of 30.3. No peaks for monoclinic Zirconia were recorded at 2Θ angle of 28. The amount of each phase was measured by counts for each phase or element.

In samples subjected to aging, peak for monoclinic Zirconia was recorded at an angle of 28 for both systems (figure 4). The monoclinic peek in case of dry-milled Zirconia was smaller than that of wet-milled Zirconia.

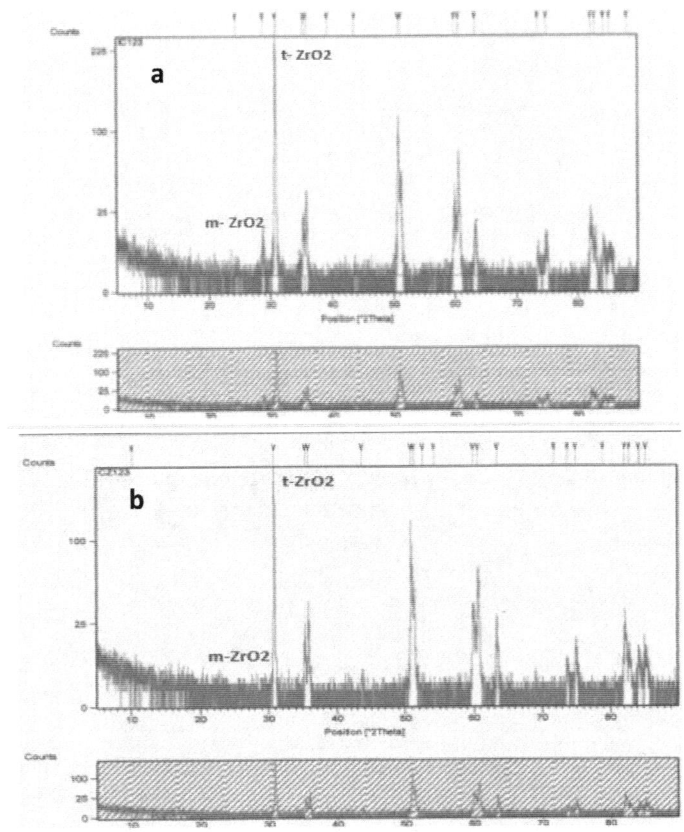

Figure 4. XRD pattern for aged monolithic Zirconia. (a) wet-milled and (b) dry-milled. t- ZrO_2 represents tetragonal phase at 2Θ angle of 30.3^0, m- ZrO_2 represent monoclinic phase at 2Θ angle of 28.4^0.

The mean and the standard deviations of the amount of phase transformation (t-m) calculated by Gravie- Nicholson equation shown that wet-milled system showed a statistically significantly higher mean amount of transformation of (t-m) (0.101±0.006) than dry-milled system (0.027±0.005) (figure 5).

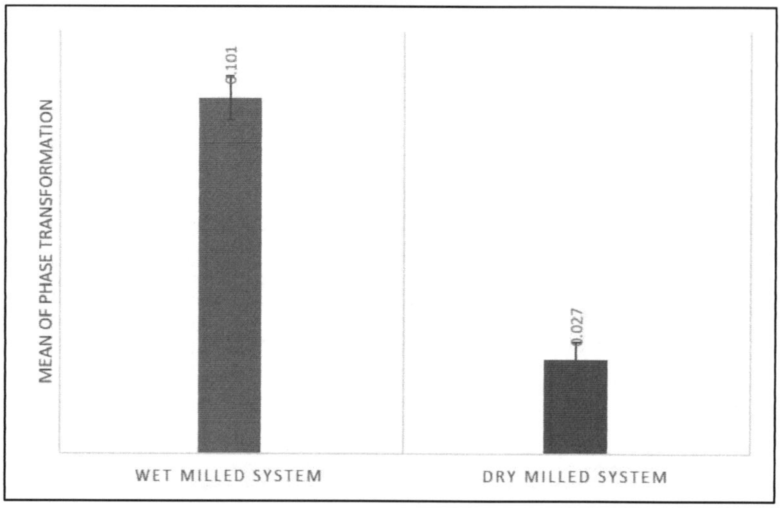

Figure 5. Bar chart representing mean values for comparison between amounts of transformation of (t-m) in the aged two milling techniques.

3.2. Fracture Resistance results

Regardless of aging; wet-milled system showed statistically significantly lower mean flexural strength value (1326.4±171.0) than dry-milled system (1949.8±76.4) (figure 6).

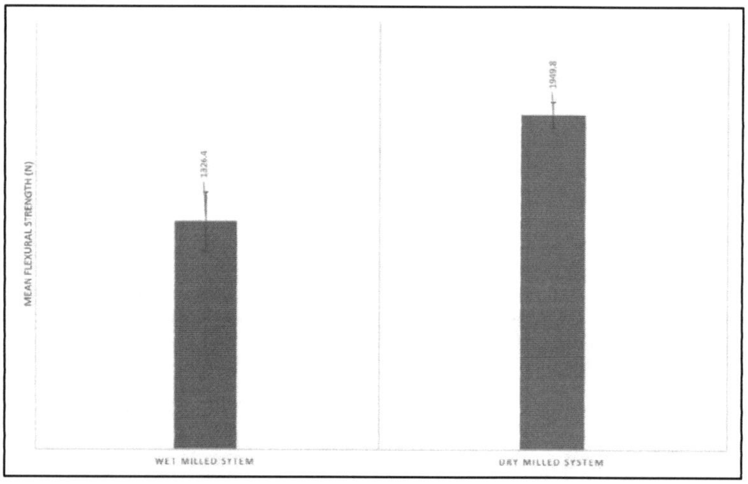

Figure 6. Bar chart representing mean values for comparison between wet-milled and dry-milled Zirconia systems

Flexural strength results have shown that, either with wet or dry milling systems, there was no statistically significant difference between the mean flexural strengthvaluesofnon-aged (wet-milled1382.1±209.6N,dry-milled 1946.2±98.9N) andaged samples (wet-milled1270.7±118.9N,dry-milled 1953.4±57.6N) (figure 7).

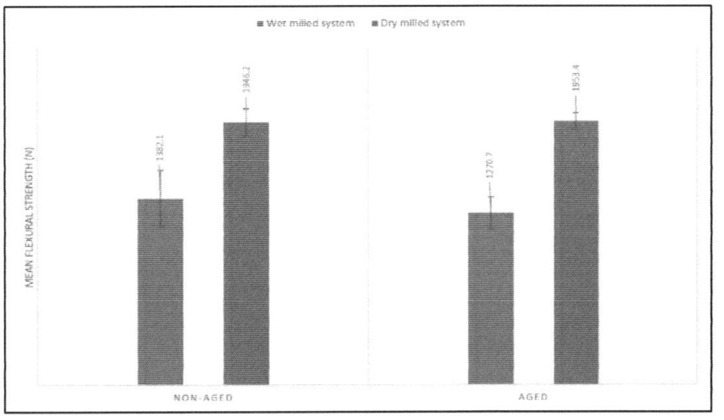

Figure 7. Bar chart representing mean values for comparison between flexural strength of milling systems for non-aged and aged samples.

3.3. Scanning Electron Microscope (SEM) Analysis

For non-aged samples, the SEM photomicrograph of wet-milleddisc(Incoris TZI) showed areas of subsurface porosities, no surface cracks and very narrow zone of transformation below the surface (figure 8a). While, the SEM photomicrograph of dry-milleddisc(CeramilZoild)showed homogeneous microstructure, no zone of transformation and /or subsurface porosities (figure 8c). Aged, the SEM photomicrograph of wet-milleddisc(Incoris TZI) showed an area of transformation below the surface, subsurface porosities and/or grain pull-out at the surface (figure 8b). While, the SEM photomicrograph of dry-milled disc (CeramilZoild) showed minimal zone of subsurface transformation, no subsurface porosities and/or grain pull-out from the surface (figure 8d).

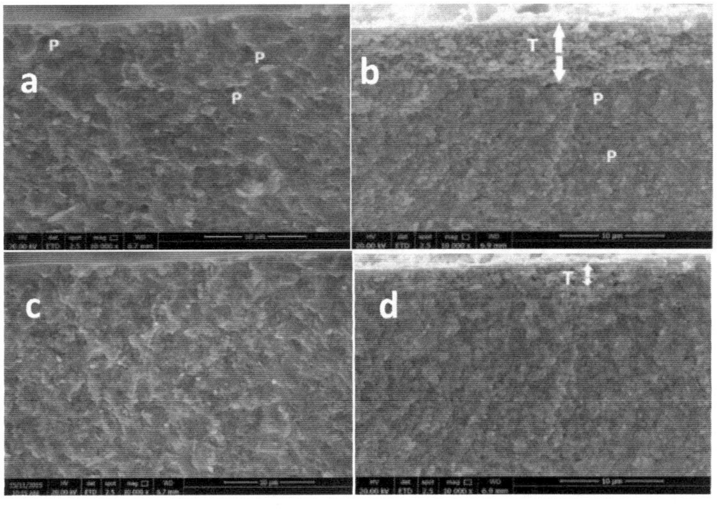

Figure 8. SEM photomicrographs of monolithic zirconia systems; a;non-agedwet-milled disc (Incoris TZI). b; aged wet-milled disc (Incoris TZI). c;non-ageddry-milled disc (CeramilZoild). D; aged dry-milled disc (CeramilZoild). P, represents subsurface porosity.T, represents transformation zone (magnification 10000x).

4. Discussion

The nano technological evolution stood behind the development of a new class of nano ceramic materials that nas been remarkably and intensively used in dentistry [28,29]. Previous studies [30],31] summarized the benefits connected to the use of nanostructured ceramics with reduced grain size that leads to an improvement of the strength; since finer grain size (≤200nm or ≤0.4 μm) [31,32] leads to stronger materials, while other study Hallman et al. [33] has shown critical grain size of ≤0.36 μm to resist LTD. So, the aim of the present study was to assess the impact of the nano microstructure of two monolithic Zirconia systems (wet and dry-milled) on its aging resistance, (t-m) transformation and finally its flexural strength. The null hypothesis of the study was partially accepted, as the wet milling system induced significant (t-m) phase transformation, yet this transformation of nano crystalline Zirconia was below the level of inducing reduction of flexural strength.

In the present study, accelerated hydrothermal aging was selected and simulated extraorally by subjecting samples to steam autoclaving for 5 hours, as 1 hour of steam autoclaving at 1210C at 1 bar pressure simulates one year of hydrothermal aging in the oral cavity. Cyclic loading mechanical stress in the oral cavity was simulated by custom made machine delivering 1250,000 cycles at frequency of 300 cycle/ min using load of 50 N force. This load was found to simulate 5 years inside the oral cavity [26].

XRD has been chosen as method for the assessment of microstructural changes. XRD has the advantage of revealing the phase's count of the same element or crystalline structure inside the material based on its volumetric change [34]35]. The same condition which occurs within the structure of Zirconia due to phase transformation of tetragonal to monoclinic phase accompanied by volumetric change of crystal sizes.

In the present study, XRD analysis for non-aged samples of both tested systems didn't show any peaks of monoclinic phase indicating (t-m) transformation (figure 3 a &b) which appeared as homogeneous microstructure, no zone of transformation and /or subsurface porosities in the SEM photomicrograph of dry milling system CeramilZoild (figure7 c) and very narrow zone of transformation below the surface in the SEM photomicrograph of wet-milled system Incoris TZI (figure 7a). WhileXRD analysis of aged samples showed the appearance of monoclinic peaks in both wet and dry-milled system which appeared as an area of transformation below the surface, subsurface porosities and/or grain pull-out at the surface in the SEM photomicrograph of wet milling system (Incoris TZI) (figure 6 b). While, the SEM photomicrograph of dry milling system (CeramilZoild) showed minimal zone of subsurface transformation, no subsurface porosities and/or grain pull-out from the surface (figure 6d). The results may be attributed to LTD under the effect of moisture of aging procedure. These results were in accordance with many studies [10], [11] .They studied the impact of oral conditions on aging resistance of Zirconia. They claimed that phase transformation and appearance of monoclinic phase occurred under the effect of moisture and cyclic loading.

Further description for the role of water in the process of Zirconia degradation has been described in Lange's thermodynamic model [39], Where the chemical free energy associated with the transformed particles provoked the creation of cracks along the grain boundaries [13]].

In the present study, the amount of phase transformation of aged two monolithic Zirconia systems (Incoris TZI (0.101±0.006) & CeramillZolid (0.027±0.005)] were less than those of other studies[35-37].They found increased in the amount of phase transformation of monolithic nano-crystalline ultra-translucent Zirconia following aging process. This contrast in the results may be attributed to the difference in the microstructure modification. The presence of cubic phase of ultra-translucent Zirconia was responsible for reduction of its strength and overall aging resistance compared to the tetragonal nano-crystalline tetragonal translucent Zirconia tested in the present study.

The impact of microstructural changes (phase transformation) on the function was assessed by measuring the flexural strength of monolithic Zirconia. Yet non-aged dry-milled system showed significantly higher strength values (1946.2±98.9N) than that of wet-milled system (1382.1±209.6 N).This can be attributed to the effect of water during the milling process as previously discussed.

Aging process didn't significantly affect the flexural strength of both monolithic Zirconia systems (wet milling 1270.7±118.9N, dry milling 1953.4±57.6N). These results may be attributed to the beneficial effect of nano modification of the grain size [31-33], as both Zirconia systems (wet and dry-milled) have grain size less than 400nm [32,33].The results of the present study were consistent with those of Zhang et al [40]. They claimed that densification of microstructure enhanced the aging resistance of monolithic Zirconia, as it minimized the effect of moisture on acceleration the aging process overtime. Also, Kim et al. [41] clarified that the flexural strength started to decrease significantly after monoclinic phase reached 12 to 54 %. This explanation was behind the non-significant effect of aging and phase transformation on the flexural strength of both monolithic Zirconia systems.

Limitations of this study were the inability to simulate all the aging conditions in the oral cavity. So, further future studies are required to study the effect of different aging conditions within the oral cavity on the performance of monolithic Zirconia for longer periods of time.

5. Conclusions

Within the limitations of this study, the following can be concluded: 1) Nanocrystalline microstructure of monolithic Zirconia positively improved its resistance to aging. 2) Dry-milled monolithic Zirconia system is more resistant to microstructural changes than wet-milled Zirconia. 3) Microstructural changes of both systems did not negatively affect the fracture resistance.

References:

[1] Shenoy A and Shenoy N 2010 *J. Conserv. Dent.* **13**195
[2] Piconi C and Maccauro G 1999 *J. Appl. Biomater. Func.* **20**1
[3] Filser F, Kocher P, Weibel F, Luthy H, Scharer P and Gauckler L J 2001 *Int. J. Comput. Dent.* **4**89
[4] Quinn J B, Sundar V, Parry E E and Quin G D 2010 *Dent. Mater.* **26**13
[5] Christel P S 1989 *Bull. Hosp. Jt. Dis. Orthop. Inst.* **49**170
[6] Herrernan M J, Aquilino S A, Diaz-Arnold A M, Haselton D R, Stanford C M and Vargas M A 2002 *J. Prosthet. Dent.* **88**4
[7] Dittmer M P, Borchers L, Stiesch K P 2009 *Acta Biomater.* **5**3231
[8] Kohorst P, Brinkmann H, Dittmer M P, Borchers L M 2010 *J. Oral Rehabil.* **37**283
[9] Torabi K, Vojdani M, Giti R, Taghva M and Pardis S 2015 *J. Adv. Prosthodont.* **7**233
[10] Khamverdi Z and Moshiri Z 2012 *J. Dent. Res.* **4**1
[11] Schubert H and Frey F 2005 *J. Eurp. Ceram. Soc.* **25**1597
[12] Kobayashi K, Kuwajima H and Masaki T 1980 *Solid State Ion.* **3**489
[13] Guo X and Schober T 2004 *J. Am. Ceram. Soc.* **87**746
[14] Chevalier J 2006 *Biomaterials.* **27**535
[15] Brochers L, Stiesch M, Bach F W, Buhl J C, Hubsch C, Kellner T, Kohorst P and Jendras M 2010 *Acta Biomater.* **6**4547
[16] Munoz-Tabarez J A, Jimenez-Pique E and Anglada M 2011 *Acta Biomater.* **5**9473
[17] Cattani-Lorente M, Scherrer S, Ammannb P, Jobin M and Wiskott H W 2011 *Acta Biomater.* **7**858
[18] Sindel J, Petschelt A, Grellner F, Dierken C and Greil P 1998 *J. Mater. Sci. Mater. Med.* **9**291
[19] Kim J, Covel N and Guess P 2010 *J. Dent. Res.* **89**91
[20] Zhang Y 2014 *J. Dent. Mater.* **30**1195
[21] Tuncel I, Turp I and Üşümez A 2016 *J.Adv. prosthdont.***8**181
[22] Attachoo S and Juntavee N 2019 *J. Clin. Exp. Dent.* **11** 146
[23] Limsamai N, Kyaw P P, Pongpruaksa P and Suputtamongkol K 2019 *M. Dent. J.* **39**99
[24] Hirano M 1992 *British Ceram. Trans. J.***91**139
[25] Bajraktarova-Valjakova E, Korunoska-Stevkovska E, Kapusevska B, Gigovski N, Bajraktarova-Misevska C and Grozdanov A 2018 *J. Med. Sci.* **6**1742
[26] Ozcan M 2016 *J. Prosthdont.* **27** 364
[27] Garvie R and Nicholson P 1972 *J. Am. Ceram. Soc.* **55**303
[28] Ceren N, Turp V, Emir F, Akgüngör G, Ayyildiz S and Şen D 2016 *Aydin Dent. J.* **3**55
[29] Guth J F, Zuch T, Zwinge S, Engels J, Stimmelmayr M and Edelhoff D 2013 *Dent. Mater.* **32** 865

[30] Jue J F, Chen J and Virkar A V **1991** *J. Am. Ceram. Soc.* **74**1811
[31] Munoz-Saldana J, Balmori-Ramirez H, Jaramillo-Vigueras D, T and Schneider GA 2003 *J. Mater. Res.* **18**2415
[32] Li J F and Watanabe R 1998 *J. Am. Ceram. Soc.* **81**2687
[33] Hallmann L, Ulmer P, Reusser E, Louvel M and Hämmerlea C H F **2012** *J. Eur. Ceram. Soc.* **32** 4091
[34] Chevalier J, Gremillard L, Virkar A V and Clarke D R 2009 *J. Am. Ceram. Soc.* **92**1901
[35] Gauna M R, Conconi M S, Gomez S , Suárez G, Aglietti F and Rendtorff N M 2015 *Ceram. Silikáty* **59** 318
[36] Hannink R H J, Kelly P M and Muddle B C 2000 *J. Am. Ceram. Soc.* **83**61
[37] Denry I and Kelly J R 2008 *Dent. Mater. J.* **24**299
[38] Flinn B D, Raigrodski A J, Singh A and Mancl L A 2014 *J. Prosthet. Dent.* **112**1377
[39] Lange, F 1982 *J. Mater. Sci.* **17**225
[40] Zhang H B, Li Z P, Kim B N, Morita K, Yoshida H and Hiraga K 2012 *J. NanoMater.* **41** 5
[41] Kim J W, Covel N S, Guess P C, Rekow E D and Zhang Y 2010 *J. Dent. Res.***89**91

IOP Conf. Series: Materials Science and Engineering 1046 (2021) 012014 doi:10.1088/1757-899X/1046/1/012014

Optical properties of one-dimensional photonic crystal and light absorption enhancement in planar a-Si:H solar cell

B Sadoun*[,1], S Mouetsi[1], A Hocini[2] and A Hocini[1]

[1]Laboratory of Electronics and New Technology (LENT),
Department of electrical engineering, Faculty of Sciences and Applied Sciences,
University of Larbi ben M'hidi, Oum Elbouaghi, Ain Beida, 04201, Algeria.
[2]Laboratory of Signal and Systems Analysis (LSSA),
University of Mohamed Boudiaf, BP.166,Ichebilia Road, M'sila, 28000, Algeria.

Email: sadoun.boumediene@univ-oeb.dz

Abstract. The generation of the photovoltaic has intensified over the last decade, moving from the most basic applications based on elementary devices to one of the most important applications of energy. Photonic Crystals (Ph.Cs) are very promising systems for applications in the field of electromagnetic waves and for real achievements in microwaves field. This particular property of the Ph.Cs offers the possibility of the control of light propagation in an increased way, and thus makes possible to consider many applications in nanotechnology and photovoltaic field. The a:Si-H is a very attractive material for simulation and experimental applications, when deposited in thin layers, it has an efficient refractive index. The Plane Wave Expansion (PWE) method was used to assess the band gap of the structure, the Rigorous Coupled-Wave Analysis (RCWA) method integrated in RSoft CAD Software was used to calculate the optical properties, integrates with thin film structure of the solar cell with Ph.Cs. The one-dimensional optical properties of photonic crystal with the absorbent layer of photovoltaic cell on both hydrogenated amorphous silicon (a:Si-H) and silicon (Si) are studied in this paper in order to enhance the yield of the absorption of light on a solar cell with an efficiency geometrical parameters.

1. Introduction

Advanced optical engineering is an important technology in providing new concepts and designs for photovoltaic cells with thin films. Its aim is to reduce the reflection of light at her top surface, as well as to strengthen the photonic path within the absorbed material [1]. This latter, enhances the absorption of photons, and enables the conversion efficiency of photovoltaic cells from photovoltaic engineering. In addition, the use of thin layers reduces the path of photons within the absorbed silicon, which tends to reduce the amount of light absorbed. For example, the use of photovoltaic cells for non-crystalline silicon layer with a total thickness of the absorbed substances is about 100nm [2], even if an anti-reflection coating (ARc) is used. This latter, makes a combination of optoelectronics loses because the conversion efficiency is up to~10% [3]. The optical path-way required to achieve large absorption in long waves exceeds the thickness of the silicon layer. In particular, Bermale P. et al [4] have performed a diffraction grating and reflection on the backside of the silicon solar cells [5], they have realized that it is immature with a thickness of 100 nm.

There is another way to strengthen the light collection and the absorption, this can be done by using a set of golden nano-particles on the front side of a newly introduced silicon in the photovoltaic cell, and this is approved by Matheu P. et al [6]. Before beginning the search, we proposed a planar light surface to collect incident light, and to multiply it with the slow-wave of the guided wave that stands inside the absorbent layer [7]. These echoes have been successfully used to achieve photovoltaic surface devices such as micro-controllers or broadband reflectors. Actually, most economically accessible photovoltaic cells are produced using mono-crystalline or poly-crystalline silicon[8]. These cells can revive photovoltaic yields of around 24.9% with non-concentrated light [9]. Silicon (Si) thin film technology offers many possibilities of production (modules of 1 m² or more). In addition, the moderate temperatures (150-300°C), required for the production of Si thin films induce low power consumption during industrial processes, and offer the possibility of using a wide variety of inexpensive substrates, such as glass, stainless steel or plastics [10].

Our work aims to demonstrate the concept of this interest in the special case of solar cells, based on the a-Si:H and Si layers with a thickness of up to 100 nm. It is not only for economic reasons that the thickness of the absorption thin layer is limited, but also because of the low propagation length of the photovoltaic contained in this material [11]. A One-Dimension Photonic Crystal (1DPh.C), obtained with a full a-Si:H and Si coatings can enhance the optical absorption of this layer [12] and it is therefore expected to enhance the efficiency of photovoltaic cells and increase the yield, provided they are at the lowest cost. Also, such calculations are made for a specific incidence angle and certain absorption, reflection power of a patterned or an un-patterned layers of a-Si:H and Si.

Table. Physical properties of the Si and a-Si:H.

	Si	a-Si:H
Band gab [eV]	1.1	1.7-1.8
Photo-conductivity [s/cm]	$10^{-2} - 10^{-4}$	$10^{-4} - 10^{-5}$
Boiling point [°C]	3265	2900
Density [g/cm^3]	2.33	2.285

We consider the infrastructure as described in figure 1(a) and figure 1(b). The first layer consists of an un-patterned absorbent thin layer on a substrate of glass, while the second absorbent layer consists of a patterned absorbent thin layer on the 1D Ph.C, respectively.

The light emitted at this resonance can then be locked at the wavelength associated with the parameters of the two structures [13]. The lattice parameter (Period) is the same in the two directions of space.

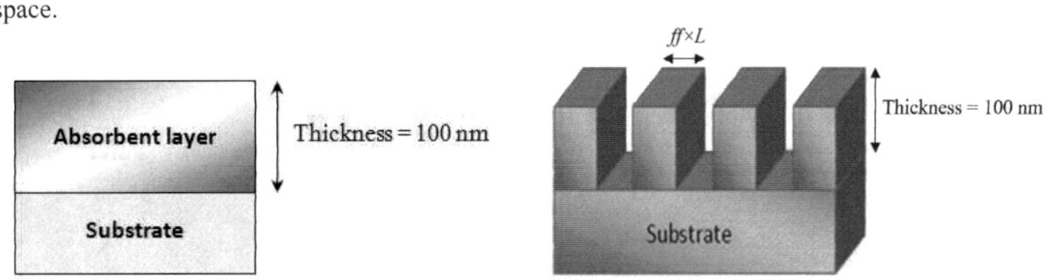

(a) (b)

Figure 1.(a) Thin film solar cell structure made by un-patterned layer on a glass substrate;(b) The patterned layer with 1D Ph.C.

The main parameters of the Ph.Cs are then lattice parameter (L), and Filling Factor (*ff*) known as the diameter of the air holes during this period, *ff* is then given by (D/L), D is the width of the layer. The thicknesses of the two configurations are considered the same; they are equal to 100nm [2].

Our study, focused on the use of Si because it is cheaper in the market and most industries are based on it, because of its technological importance in the manufacture of solar chips and photovoltaic production [14], we have also used a:Si-H because it is more sensitive than Si[15], which is a hydrogen atom, it is not present in the chemical bond gave, which make it more sensitive and more reactive to the sunlight compared with Si [16].

Both design and digital simulation are closely related. In our numerical simulations, we used the RCWA Method, which we used to study changes in the shape of the solar cell for its ability to absorb high in the visible field of optical changes [17], ranging from 350nm to 750nm (visible spectrum).

2. Method

The Rigorous Coupled-Wave Analysis Method (RCWA), it is a semi-analytical method used in computational electromagnetic, this method is most often applied to solve the diffusion from periodic dielectric structures [18].It is a Fourier space method in which the devices and the fields are represented by a sum of spatial harmonics [19]. This method allows solving Maxwell's equations, we start with the calculation of Reflectance (R), the Transmittance (T) and thus the Absorption (A) [11].

$$A=1-R-T \quad (1)$$

The absorbance of a material, denoted by A, is given by [11] :

$$A = \log_{10}\frac{I_0}{I} = -\log_{10}T \quad (2)$$

where: I_0 is the transmitted radiant flux by the material; I is the received radiant flux by the material; T is the transmittance of the material.

We calculate the normalized absorption efficiency η, defined as follows [20] :

$$\eta = \frac{\int_{350}^{\lambda g} I(\lambda)A(\lambda)\frac{\lambda}{\lambda g}d\lambda}{\int_{350}^{\lambda g} I(\lambda)\frac{\lambda}{\lambda g}d\lambda} \quad (3)$$

where: λg is the photon wavelength corresponding to the band gap; $I(\lambda)$ is the solar spectral irradiance. We chose the spectral range from 350 nm to λg in the integration (visible spectrum). In a wave fall incident on a planar structure ,as well as how to distribute the electromagnetic field in the structure[21], we suggest some of the optical properties contribute to our understanding of how the optical proliferation in the planar structure. Measuring the absorbance throughout [22], the structure means to take into account the reduced area of the active layer due to the patterning of thin layer[23].

3. Results and Discussion

3.1. The absorption with the patterned and the un-patterned thin layers

Spectral properties of a new set of parameters used are shown in figure 2.We use the RSoft simulation software [24], to simulate the two a patterned and an un-patterned thin layer structures. The absorption was monitored as the wavelength varies, we found that the initial values of *ff* are fixed at 60%.All of our studies are in the visible spectrum that's in between ~(350nm-750nm)[25]. With RSoft CAD software, we put the lattice on L=500nm for 1D Ph.C to calculate the ideal parameters. The structure of the integrated absorption of 1D Ph.C thin layer is first optimized by varying the L for the thickness of 100 nm, after that the absorption is compared with the un-patterned thin layer. We noticed that the spectra corresponding to the layer of 1D Ph.C differ from the spectrum of the an un-patterned thin layer.

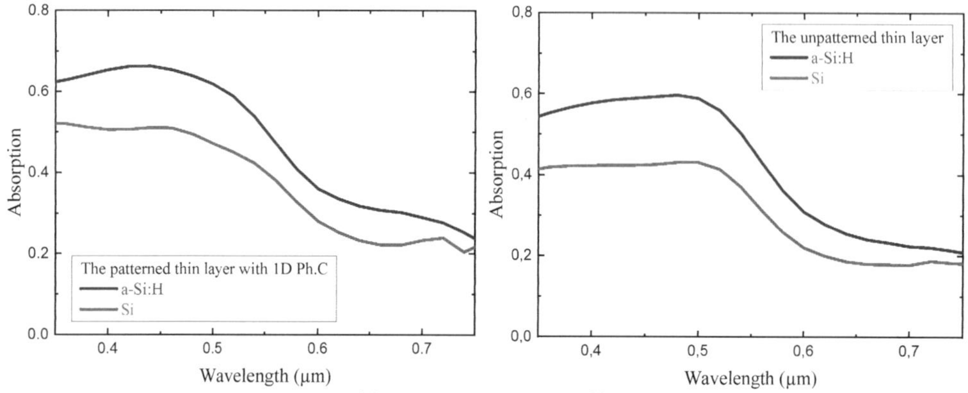

(a) (b)

Figure 2. The variation of the absorption power of a:Si-H and Si, on (a) Thin layer patterned Spectral structure of the 1D Ph.C. (b) thin layer unpatterned spectral structure.

For the 1D Ph.C layer in a-Si:H, the highest absorption ratio of 62% reaches between ~(350 nm-500 nm), and the absorption up to ~(500 nm-600nm) begins to decrease due to the extinction coefficient (*k*)[26], until the ratio reaches 32% between the ends of the wave of 610 nm and 750 nm. This is for a-Si:H, but for Si, the amount of absorption up to about 52% between the ends of the wave 350 nm and 450 nm, which is less knowledgeable than a-Si:H being more efficient than it. They also start decreasing ranges between (450 nm-610 nm) until they reach 20% ~(610 nm-750 nm).

In the second curve, an un-patterned thin film solar cell (without 1D Ph.C), we can observe; as for a-Si:H, the absorption ratio reaches 54% ~(350 nm-520 nm) and is also declining due to the effect of extinction coefficient and the efficiency of the solar cell until it reaches 20% ~ (620 nm-750 nm) As for the Si, up to about 41% ~ (350 nm- 520 nm) and then to decrease to reach the ratio of 15% ~(630 nm-750 nm).As a primary result, the absorption in the solar cells and increase the yield ratio is better by using photonic crystals that have the ability to do so [27], and the quantity of the Si absorption is little compared with a-Si:H.

3.2. The lattices parameters variation

The dimension was chosen in order to maintain as much of the optical band gap as possible [28]. The simulation of certain values of parameter L (lattice parameter) was repeated. Thus, for the used process, we determine the absorption of different values for the lattice parameter L=400 nm, L=500 nm and L= 600 nm respectively. In addition, we calculate the absorption in the two structures with the same thickness = 100 nm.

The two curves cross the 1D Ph.C patterned layer for the purpose of studying the role change in two different a-Si-H and Si in the absorbent layer of the solar cell.

In the a-Si:H patterned layer 1D Ph.C, the absorption ratio up to about 70% at L=600 nm, and reaches the lowest drop between 580 nm and 750 nm. With the lattices varying, in L=400 nm, we notice that the absorption ratio reached about 55%, which is lower compared to L=600 nm. On the other hand, with the Si absorbent layer 1DPh.C, we can observe that the absorption rate in thin layer solar cell, reached 52% at L=500 nm and L=600 nm, but starting from 400 nm, increases the ratio of L=600 nm to 55% to start decreasing ~(550 nm-750 nm). As for L=400 nm, it is less than 48% until it reaches 32% between 600nm and 750nm.

When we conducted a study of the variation of lattice parameter constant (L), we found that the better one is L= 600 nm in both curves. Which can be deduced here, is that the (L) is larger, the absorption rate increased especially between 350nm and 500nm. That is, the (L) is a powerful factor for increasing yield in thin-film solar cells and in nanotechnology for nanoparticle and patterning [29].

In fact, the absorption is usually higher regardless of the wavelength used. For example, the absorption between 350 nm and 500 nm for the a-Si:H patterned thin layer (Figure (3.a)), approximately two times greater than the Si patterned thin layer (Figure (3.b)).

(a) (b)

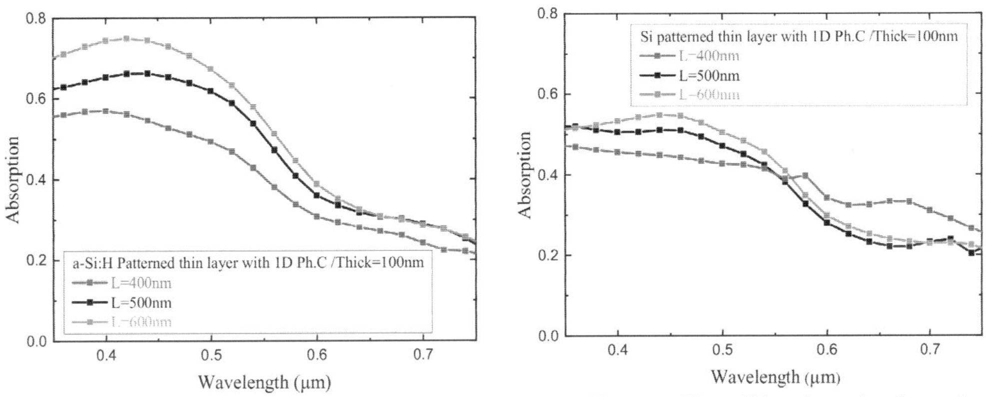

Figure 3. Variation of lattice parameter (L = 400 nm, L = 500 nm and L = 600 nm) on the absorption power of a:Si-H (a) and Si (b) with a thickness of layer 100 nm.

The absorption can be improved globally by a combination of the effect of slow lighting patterns and the thorough regression of reflection [30], probably because of the high average index in the case of 1D Ph.C The absorption of the patterned structure becomes two times greater in the wavelength range between 350nm and 500nm in the a-Si:H patterned thin layer.

3.3. The incidence angle

Figure 4 shows the adoption of the absorption spectra angle in many solar cell structures. The patterned thin film photovoltaic cell is characterized by the diffraction of the conversion efficiency on the angle of the incident light occurrence [31].

(a) (b)

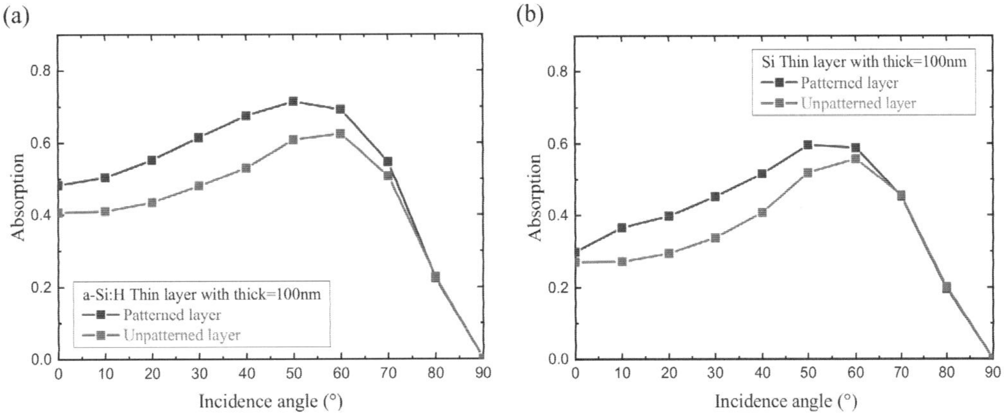

Figure 4. The variation of the absorption of: (a) a-Si:H patterned and un-patterned thin layers,(b) Si patterned and un-patterned thin layers, in terms of the incidence angle (°).

In the two curves, the red line represents the absorbance process during the polarization state in the patterned thin layer, while the blue corresponds to the absorption in the un-patterned thin layer. The patterned thin layer is more sensitive to the incidence angle [32].

In the a-Si:H thin layer, the absorption process starts increasing ranges between(49%-67%) for the patterned structure (with 1D Ph.C), beginning from the vertical position of the angle to reach 50°, which in turn recorded the largest absorption rate at this angle, which reaches ~70%, which is very high in terms of studying the impact of incidence angles. Then, they begin to gradually decrease to the absence at a 90°. As for the un-patterned thin layer, they begin to increase gradually between 0° and 50° until they reach ~58% at 60°, which in turn records the largest absorption rate in the layer is not engraved with a-Si:H, then gradually decrease until it reaches the 90°. Therefore, their variability with respect to the incidence angle is considered, which may account for the high global absorption decline in the case of 1D Ph.C.

As for the incidence angle using Si, it is the lowest percentage of a-Si:H, which is an active and most sensitive material in photovoltaic. Between 0° and 50°, we notice an increase in the absorption ratio, to reach ~58.5% of the patterned thin layer and ~50% for the un-patterned thin layer at 60°. And then, the gradual extinction, as well as with the effect of the coefficient (*k*), Begin to gradually decrease until the angle 90°.Furthermore such as mentioned above, the slow light pattern is affected by a relatively low refractive index, therefore the spatial spacing between places of diffusion of light is limited [33], and the incidence angle is increasing.

The manufacture of these solar photovoltaic cells requires the need to verify the uniformity of absorption with respect to the change of topographic parameters [34]. After that, we need to know the processing techniques that are usually used, to develop new patterns of Sub-Wavelength (SbW) on a planar surface[35].After the comparative study of solar cells influencing the optical properties, we found that the improvement in light absorption was between ~(350 nm-580 nm), which means an increase in yield. That is to say, that there is a qualitative improvement through the choice of material quality is a:Si-H compared with Si[36]. We explain chemically that the loss of hydrogen in a:Si-H compared with Si, has caused a high agitation and a high sensitivity to this substance, which in turn causes more agitation of the sun to the lattice parameter in the absorbent layer of the solar cell [37], which in turn contributes to a high yield. Furthermore, Ouanoughi A.et al [38] have done the study of improving the absorption of light in the solar cell, In our work we tried to develop the idea by studying the vertical position of the incidence angle in the solar cell, which in turn contributed to the enhancement of the absorption of light in the solar cell using photonic crystals.

We concludethe difference between silicon (Si) and Hydrogenated amorphous silicon (a-Si:H) in terms of cost and ease of fabrication is probably, that the Si was condition to rapidly cooled in a vacuum environment [39], and the a-Si:H was rapidly cooled in an aerated (in the present of H_2) environment[40]. In addition, the Si would cost more to fabricate in a comparison to the a-Si:H[41].

4. Conclusion

In this work, we have shown some properties related to both the absorption of photons lifetime in photo-voltaic cells using one-dimensional photonic crystal and the absorption of light in the hard cells to enhance the performance of the yield. The coupling of emitted light greatly increases the absorption of the a-Si:H layer, using a simple one-dimensional photonic crystal layer, which makes it possible to find out how to enhance the absorption of light in absorbent thin layer of the solar cell. In addition, the absorption offers good stability for the incident angle. The proposed design is completed tolerant in terms of malfunctions during the manufacturing process, which make it possible to fabricate with low cost for nanotechnology such as the three-dimensional lithography or a nanoparticle. The cost of this additional technological step should be overcome by the advantages of using a higher absorbent layer in the solar cell, with optimal absorption of sunlight.

Acknowledgment

For acknowledgment, we are honored in this work, to thank the Laboratory of Signal and Systems Analysis (LASS) at the University of Mohamed Boudiaf, M'sila, Algeria.

IOP Conf. Series: Materials Science and Engineering 1046 (2021) 012014 doi:10.1088/1757-899X/1046/1/012014

References

[1] Park Y, Drouard E, El Daif O, Letartre X, Viktorovitch P, FaveA, Kaminski A, Lemiti M and Seassal C 2009 *Opt. Express* **17** 14312
[2] Zeng L, Yi Y, Hong C, Liu J, N Feng, Duan X and Kimerling L C 2006 *Appl. Phys. Lett.* **89** 111111
[3] Meier J, Kroll U, Vallat-Sauvain E, Spitznagel J, Graf U and Shah A 2004 *J. Sol. Energy***77** 983
[4] Bermel P, Luo C, Zeng L, Kimerling L C and Joannopoulos J D 2007 *Opt. Express* **15** 16986
[5] Baird W E, Moharam M G and Gaylord T K 1983 *Appl. Phys. B-Lasers O* **32** 15
[6] Matheu P, Lim S H, Derkacs D, McPheeters C and Yu E T 2008 *Appl. Phys. Lett.* **93** 113108
[7] Duché D, Escoubas L, Simon J J, Torchio P, Vervisch W and Flory F 2008 *Appl. Phys. Lett.* **92** 193310
[8] Drouiche N, Djouadi-Belkada F, Ouslimane T, Kefaifi A, Fathi J and Ahmetovic E 2013*Desalin. Water Treat.* **51** 5965
[9] Green M A,Emery K,King D L, Hishikawa Y and Warta W 2006 *Prog. Photovolt. Res. Appl.***14** 455
[10] Deng X andSchiff EA2003 *Handbook of Photovoltaic Science and Engineering* (England: John Wiley & Sons Ltd) pp505
[11] Ferry V E, Marc A V, Hongbo B T L, Ruud E I S, Harry A A and Polman A 2009 *Appl. Phys. Lett.* **95** 183503
[12] Moharam M G and Gaylord T K 1981 *J. Opt. Soc. Am.* **71** 811
[13] Wu M S, Vail E C, Li G S, Yuen W and Chang-Hasnain C J 1996 *IEEE Photon. Technol. Lett.* **8** 98
[14] Alsema E and Wild M J 2005 *Mater. Res. Soc. Symp. Proc.* **895** 0895
[15] Amer N M and Jackson W B 1984 *Semicond. Semimet.* **21** 83
[16] PeartonS J, Stavola M and Corbett J W 1991 Mater. Sci. Forum 38 25
[17] Chopra K L, Paulson P D and Dutta V 2004 *Prog. Photovolt.***12** 69
[18] Semenikhin I, Zanuccoli M, Benzi M, Vyurkov V, Sangiorgi E and Fiegna C 2012 *Opt. Quant. Elect*.**44** 149
[19] Callens MK, Yokoyama D and Neyts K 2015 *Opt. Express* **23** 21128
[20] Cao L, White J S, Park J S, Schuller J A, Clemens, B M and Brongersma M L 2009 *Nat. Mater.* **8** 643
[21] Kosyachenko L A, Lytvynenko V Y and Maslyanchuk, O L 2016 *J. Semicond.* **50** 508
[22] Fujiwara H and Kondo M 2005 *Appl. Phys. Lett.* **86** 032112
[23] Bozzola A, Liscidini M and Andreani L C 2012 *Opt. Express* **20** A224
[24] RSoft Design Group, Inc. 200 Executive Blvd. Ossining, NY 10562.
[25] Li Y and ZouY 2008 *Adv. Mater.* **20** 2952
[26] Lipinski M, Zieba P, Jonas S, Kluska S, Sokolowski M and Czternasetk H 2004 *Opto-Electron. Rev.* **12** 41
[27] Priolo F, Gregorkiewicz T, Galli M and Krauss T F 2014 *Nat. Nanotechnol.* **9** 19
[28] Reineix A and JeckoB 1996 *Ann. Telecommun.***51**656
[29] Moulé A J and Meerholz K 2009 *Adv. Funct. Mater.* **19** 3028
[30] Huang Y F, et al.2007 *Nat. Nanotechnol.* **2** 770
[31] Forberich K, Dennler G, Scharber M C, Hingerl K, Fromherz T and Brabec C J 2008 *Thin Solid Films* **516** 7167
[32] Smith H I 1974 *Proc. IEEE***62** 1361
[33] Green M A, Jianhua Zhao Wang A and Wenham S R 1999 *IEEE Trans. Electron Devices* **46** 1940
[34] Tenent R C, Barnes T M, Bergeson J D, Ferguson A J, To B, Gedvilas L M, Heben M J and Blackburn J L 2009 *Adv. Mater.* **21** 3210
[35] Zhao G, Jiao Y C, Zhang F and Yang X 2010 *J. Electromagnet. Wave.* **24** 33
[36] Vaillant F and Jousse D 1986 *Phys. Rev. B Condens. Matter.* **34** 4088

[37] Zhao S, Gao Z, Chen C, Wang G, Zhang B, Chen Y, Zhang J, Li X and Qin Y2016 *Carbon* **98** 196
[38] Ouanoughi A, Hocini A and Khedrouche D 2016 *Front. Optoelectron.* **9** 93
[39] Rose M J, Snell A J, Lecomber P G, Hajto J, Fitzgerald A G and Owen A E *1992 Mater. Res. Soc. Symp. Proc.* **258** 1075
[40] Bansal A, Li X, Yi S I, Weinberg W H and Lewis N S 2001 *J. Phys. Chem. B* **105** 10266
[41] Donolato C 1985 *Appl. Phys. Lett.* **46** 270

Design of high sensitive glucose concentration sensor of photonic crystal cavity

A Hocini[1,*], S Mouetsi[1], A Hocini[2], B Saadoune[1]

[1]Department of Electrical Engineering, Faculty of Sciences and Applied Sciences, Pole Ain Beida, University of Oum El Bouaghi, 04201. Algeria.
[2] Laboratoire d'Analyse des Signaux et Systèmes, Department of Electronics, University of M'Sila BP.166, Route Ichebilia, M'Sila, 28000, Algeria.

*Email: raoufho28@gmail.com; hocini.abderraouf@univ-oeb.dz

Abstract. Photonic sensors, in light of proposed structures comprised of rhombus–molded gaps matched with two waveguides, have been developed as another enthusiasm for specialists in science and technology .In this paper, it has been demonstrated that by utilizing the rhombus–formed openings examination, they got outcomes which are brilliant inside the depression locale. Consequently, the quality factor (Q) and the affectability (S) can be essentially improved for the proposed structure. This investigation are performed by reproduction utilizing limited contrast time area (FDTD). Different affectability are acquired as 494.15 nm/RIU and 445.75 nm/RIU and 482.84 nm/RIU and 483.63 nm/RIU and quality factor as high as 5739.5 and 1.17×10^4, 8.7×10^4 and 3489.9 have been accomplished, deducing in a location breaking points of 5.52×10^{-5}, 2.84×10^{-5}, 3.67×10^{-6} and 9.14×10^{-4}. This blend angle gives the proposed structure an extra exhibition as a component to play out a name free biosensing for the biomedical visualization.

1. Introduction

Optical sensors, that offer prompt revelation and measurement of organic investigation, have pulled in an extraordinary consideration owing to their promising highlights, for example, security in a burnable and hazardous environment, assurance from electromagnetic impedances, snappy reaction and the separated web based detecting capacity. Photonic gems (PhC) are occasional dielectric components with the capacity to screen and control light age and stream. The periodicity can be shaped by embedded and punctured openings in a dielectric component, making a photonic band hole (PBG) area [1–2]. PhC have been utilized in few applications, as in twist waveguides, channels, sensitizers, lasers, amplifiers, and resonators. Among those optical innovations, receptive photonic precious stone (PhC) built sensors have gotten a lot of consideration because of their microcosmic size, their high affectability, negligible model preparing denied of fluorescence classifying and the chance of participate in MEMS (Micro Electro-Mechanical Systems) [3]. More often than not, it is about physical and compound/biochemical challenges, the PhC sensors offer an incredibly solid narrow mindedness inside the substance to be broke down, because of the effect of the photonic taboo band [4].The light can be packed in an extremely little volume, communicating with a colossal fluid light [5]. This wonder leaves the sensors to be very sensitive to minor refractive list (IR) factors that are shaped by the natural medium, with limits on the dividers of the PhC test. The examination is completed continuously and doesn't require the utilization of markers (bio-location without name). The acquired discoveries can be improved via doing advancements in the sensor frameworks, for instance a warmth balance, the movement of the coupling and the topological streamlining of the geometry (shape, opening size, thickness and Si level). At this premise, an assortment of IR sensor proposition have been broken down and acknowledged by means of various sorts of PhC structures, for example, smaller scale depressions [6-7-8], waveguides [9-10], Slotted waveguides [11-12-13-14], hetero-structures [15], from which critical advancements in detecting stages have been arranged. A large portion of these arranged sensors work as a solitary thought or sensor and the quantity of brands that can be veiled on one event is very little. So, as to ease this lack, the fundamental thought is to build up a framework to permit blending a few research thoughts on a colossal substrate. That is the reason PhC-based sensors have been imagined and tried. A few photonic gem topologies have been proposed in the writing for structure of biosensors, contrasted and photonic precious stone smaller scale hole which has more deformity modes and optical restriction impact. So, as to improve the affectability of photonic precious stone miniaturized scale hole sensor, it is important to propose a sensor structure with colossal optical confinement impact [16-17-18-19].

In this paper, we depict the system and the reproduction procedure of the proposed PhC biosensor that is speculatively utilized for evaluating the glucose's focus. Considering the silicon-on-encasing (SOI) innovation, the planned structure is molded by two waveguides with ring-formed openings pit framework. The detecting standard depends on the move of reverberation frequency λ0, which occurs because of the adjustment in RI of the sensor when the PhC's air gaps are involved by a homogenous demonized water (n = 1.33) or other fluid (distilled water n = 1.3147, methanol n = 1.316, acetone n = 1.3445, isopropyl alcohol n = 1.363 [20]).For the glucose recognition, it has been affirmed that the resounding frequency design changes its unearthly position following a direct demonstration when the glucose fixation somewhere in the range of 0% and 60% is applied. The properties of the sensor are reproduced utilizing the limited distinction time-space (FDTD) calculation (RSoft CAD). In theory and statistics, light concentration in photonic crystals is estimated by resolving Maxwell's calculations in a varied dielectric medium, stated in equations (1) and (2); E and H are the electric and magnetic field intensities.

$$\nabla\times E(r,t)=-\frac{\partial}{\partial t}\left[\mu_0.H(r,t)\right] \quad (1)$$

$$\nabla\times H(r,t)=-\frac{\partial}{\partial t}\left[\varepsilon.\varepsilon_0(r).E(r,t)\right] \quad (2)$$

μ is the permeability which equals to μ0 as the measured material is non-magnetic,
ε :symbolizes the permittivity which is habitually written as $\varepsilon=\varepsilon_0.\varepsilon(r)$, where ε_0 is the permittivity of vacuum and ε(r) is the relative permittivity of the material.

Equation (3) is obtained by solving Equation (1) and inserting it into the time derivative of Equation (2):

$$\nabla\times\left[\frac{1}{\varepsilon(r)}\nabla\times H(r,t)\right]=-\frac{\partial^2}{\partial t}\left[\mu_0.H(r,t)\right] \quad (3)$$

Assuming harmonic time dependence for the magnetic field with angular frequency ω.
H(r, t) = H(r) e-iωt, we obtain the master equation for the magnetic field:

$$\nabla\times\left[\frac{1}{\varepsilon(\vec{r})}\nabla\times H(\vec{r})\right]=\left(\frac{\omega}{c}\right)^2 H(\vec{r}) \quad (4)$$

The spatial and temporal steps are related through this equation:

$$\Delta t\leq\frac{1}{c}\left(\Delta x^{-2}+\Delta y^{-2}\right)^{1/2} \quad (5)$$

Where c is the speed of light in vacuum,

Δx and Δy are spatial stages in the $x$and $y$directions, respectively.

In order to terminate the computational field, the perfectly matched layers (PMLs) are employed [21].

2. Materials and theoretical model

The proposed plan depends on a 2D-PhC structure. It comprises of a periodical triangular grid of air-expanded gaps punctured in a silicon (Si) piece (refractive list n_{si}=3.45).The range of the air openings is r=0.38a, with a cross section steady a=550 nm, the thickness of the PhC chunk is set as h=230 nm. The Silicon dioxide (SiO_2) layer with a thickness of 1500 nm and low refractive record is utilized as a supporting layer to the high filed Silicon based chunk, to keep light to the pit center, forestalling optical misfortunes in the lower substrate, and the light detainment in the vertical course is then protected by full interior reflection. So as to diminish the computational time which may happen in 3D structures, the PhC is changed into a 2D framework with the dielectric medium having the viable refractive file of 2.868, which relates to the powerful file of the focal guided TE mode. For the accompanied numerical reenactments, we have applied the powerful file approach given in [22], in blend with two different techniques specifically the 2D-PWE and the 2D-FDTD from the RSoft programming bundle. The dispersal properties of the deliberate PhC structure have been investigated utilizing the 2D-PWE technique in Band SOLVE programming. For the TE polarization, the photonic band hole (PBG) covers the scope of ω_1=0.278(a/λ) to ω_2=0.4 (a/λ) comparing to the wide frequency goes of 1323.7>λ>1908.1 nm.

Subsequently, the viable working frequency extend is adequate to satisfy the detecting need, where the full mode creates a wide frequency move. Figure 1 shows the schematic of the planned biosensor. This last comprises of two waveguide couplers and one cavity. The two waveguides are acquired by expelling one column of air-expanded gaps in the x course, to couple light source in and out the Silicon section punctured ring-molded gaps PhC depression.

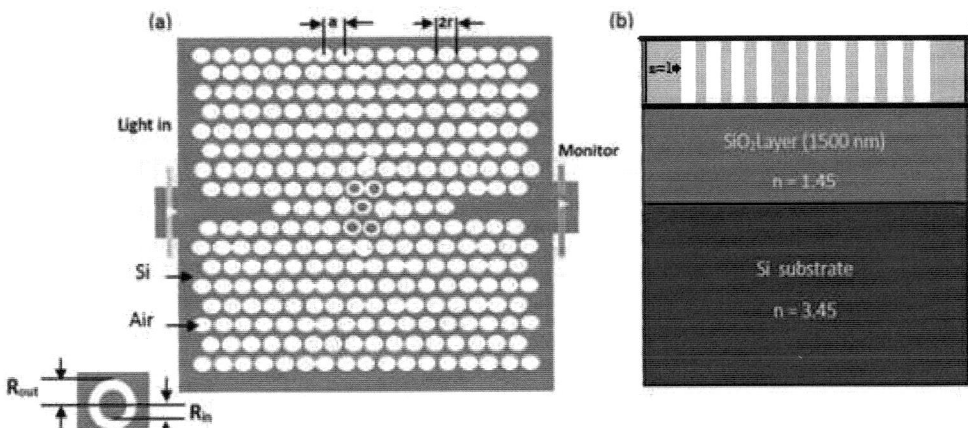

Figure 1. (a) Schematic diagram of the proposed rhombus-shaped photonic crystalcavity coupled to an input and output the PC waveguides.
(b) Schematic side and sectional view of the proposed PC biosensor structure in SOI substrate.

It is important to note that the ring-formed gaps arrangement doesn't just give greater adaptability as far as planning new structures in correlation with to roundabout molded ones, but also to improve the optical field containment, thus light-liquid cooperation inside the ring of low-dielectric material territory [23-24].

The ring-formed gap parameters are characterized by their internal and external range Rin=121nm (0.22a) and Rout=247.5 nm (0.45a) individually. In this manner the ringed air-expanded district is given by Rout−Rin=126.5 nm. The entire depression framework is isolated from the info and yield waveguides by three PhC gaps.

Two-dimensional limited contrast time-area (2D-FDTD) technique is utilized to gauge the structure's usefulness, considering the Perfect Matched Layers (PML) conditions in computations, to guarantee no back appearance in the broke down locale's breaking point.

In reproduction process, a TE energized Gaussian optical heartbeat, covering the entire recurrence scope of intrigue, is propelled at the information port to energize the hole modes. A force screen was put toward the finish of the yield waveguide to gauge the transmitted sign, and to improve the reproduction's precision, FDTD investigation is completed with a lattice size of 0.01 nm.

3. Sensing properties of the designed structure

So, for showing the working head of the PhC pit based RI biosensor, a progression of FDTD reproductions have been led. The starter detecting examination is finished by assuming the neighborhood penetration of de-ionized (DI) water in the hole detecting zone, this compares to the change in refractive file (RI) of the ring molded gaps from 1.33 (DI-water) to 1.332305450 of (0 g/l to 60 g/l).

The relation between both the glucose's RI n and concentration C, can be expressed as follows [25]:

$$n = 0.00011889.C + 1.33230545 \tag{6}$$

For a neighborhood invasion, an exact system dependent on small scale penetration innovation by means of empty submicron size pipettes has been tentatively shown in [26]. This strategy permits to control a fluid affidavit inside an ideal PhC gap, and not influencing the nearby ones.

For assessing the usefulness of the structure, for instance fluid, should be in contact with the sensor, and the reproduction is done by changing the RI of the gaps from 1.3147 (refined water) to 1.363 (isopropyl liquor). This prompts a move in the yield transmission range that relates to an adjustment in the RI.

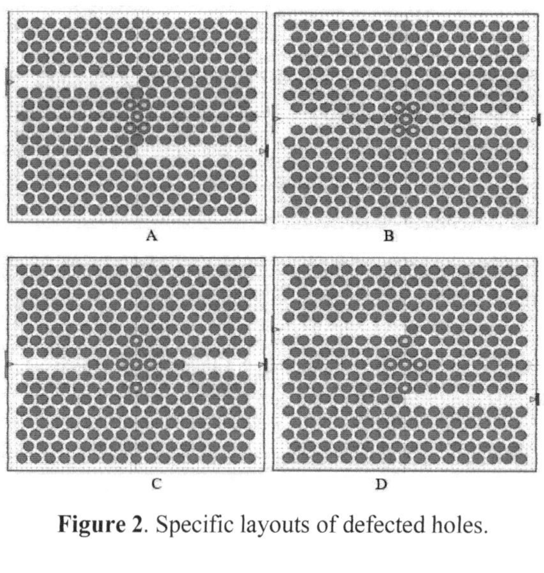

Figure 2. Specific layouts of defected holes.

IOP Conf. Series: Materials Science and Engineering 1046 (2021) 012015 doi:10.1088/1757-899X/1046/1/012015

Figure 2 presents the specific layouts of defected holes. As can be seen from figure 3, the quality factor, sensitivity and transmission spectrum of resonant mode will change with N of (Design-A, Design-B, Design-C and Design-D).

Normalized transmission spectrum of the PC cavity as a function of wavelength for different numbers N (DesignA, DesignB, DesignC and DesignD) is also illustrated in figure 3. According to 0g/l to 60 g/l with different numbers of functionalized circular, we can find that the normalized transmittance remains the same and can reach to 84.5. It can be seen in this figure that when the number of functionalized holes increases, the resonant wavelength shifts toward larger amounts. Figure 3 is in their lower values, which can explain the choice of the local infiltration instead of the total infiltration.

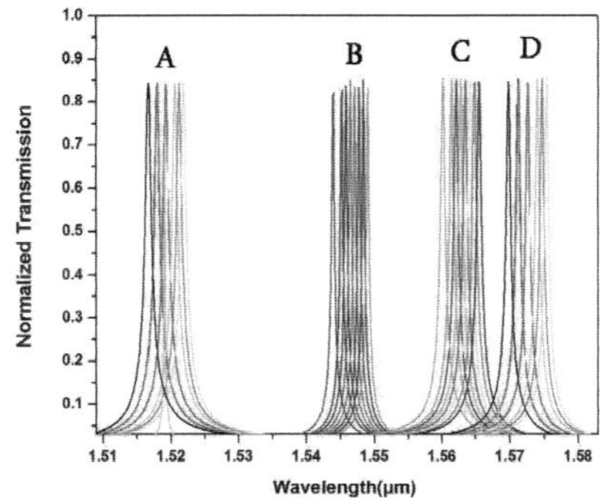

Figure 3. Transmission spectra.

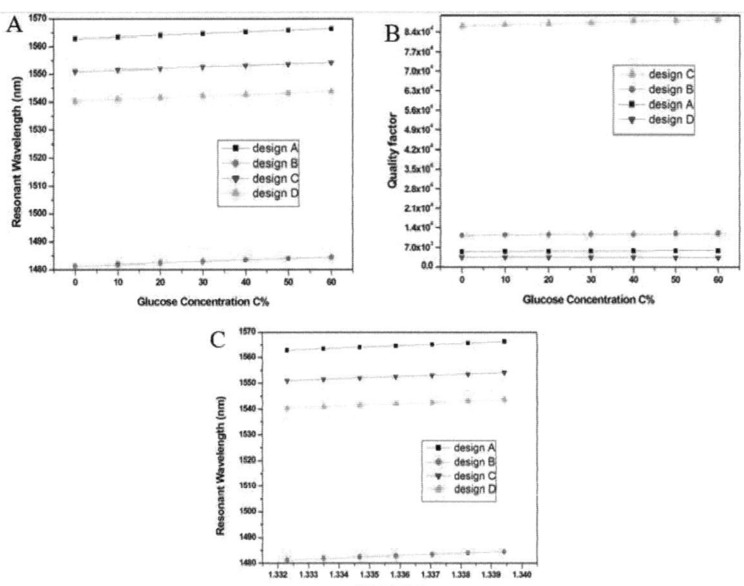

Figure4. (a) The resonant wavelength variations according to the change in the glucose concentration. (b)The quality factorvariations according to the change of the glucose concentration. (c) Resonant wavelength shift as a function of the refractive index n.

As it tends to be found in figure 4(a), when the glucose fixation changes as per various quantities of N (Design A, Design B, Design C and Design D) there is a particular field at frequencies 1519.11nm to 1575.39nm, individually.

As indicated by figure 4(b) the best quality factor has a place with the structure N=0 by a position of 106. It is seen in the bend that affectability in all structures has significant and Figure 4(c) shows the frequency move of the plans A-D as a component of progress in RI. Great qualities from the relating estimations of sensitivities are 520, 537, 589, 604 nm/RIU for Design-A, Design-B, Design-C and Design-D, individually.

Reproduction results have been outlined in table 1. It is clear that the movements in resounding frequencies have been gotten in any event, for the slightest change in the example RI, subsequently; unique glucose concentrations in the infiltrated test could be all around identified. It can likewise be noted that the quality factor esteem increments relatively to its underlying worth before the tests invasion. This is because of the expansion in the surrounding refractive file, which diminishes the PhC differentiate [30]. In this manner, we can reason that the glucose biosensor based PhC-RSH pit presents high exactness and it is exceptionally touchy to the arrangement RI variations. A 0.2% RIU change in the RI arrangement may prompt the resonance top moving almost 1.1 nm. Moreover, high sensor exhibitions require additionally a low estimation of as far as possible (DL) which is conversely relative to Q and S. The identification breaking point of refractive record change can be communicated as follows [27-28]:

$$DL = \frac{\lambda_0}{10.Q.S} \tag{7}$$

Table 1. Comparison of performance's parameters of the proposed biosensor with various similar PC-based designs.

Sensing structure	Sensitivity (nm/RIU)	Q factor	detection limit (RIU)
[28]RI biosensor formed by two waveguide and one micro-cavity.	330	3.82×10^{3}	-
[23]RI biosensor consists of H2-type nano-cavity and broadband W1 waveguide.	131.70	2.96×10^{3}	-
[25]RI sensor based on a ring–slot cavity coupled to an input and output waveguides.	160	10^{7}	8.75×10^{-5}
[29]RI sensor consists of ring resonant cavity coupled to an in and out line defect PhCstructure.	330	8.2×10^{6}	1.24×10^{-5}
[30]RI sensor consists of ring cavity coupled to an opt fluidic slow-light waveguide in a PhC platform.	293	950	-
[31]RI biosensor based ring-shaped holes cavity coupled to an input and output waveguides.	462.61	1.112×10^{5}	3.03×10^{-6}
This work for design A	494.15	5739.5	5.52×10^{-5}
B	445.79	1.17×10^{4}	2.84×10^{-5}
C	482.24	8.7×10^{4}	3.67×10^{-6}
D	483.63	3489.9	9.14×10^{-4}

The table displays the performance parameters such as sensitivity, quality factor and detection limit of the designed biosensorsbasedon PC-RSH cavity andalso shows a comparison with several previously reported PC-based biosensors. The proposed biosensor exhibited significantly higher sensitivitydue to the increase in light matter interaction within the sensingarea. Additionally, the quality factor value and the detection limit of thedesigned device were favorably comparable to the reported works.

4. Conclusions

In this paper, we have proposed a novel design of refractive index optical sensor based on 2D photonic crystal. The structure consists of a 2D triangular array of silicon substrate. The proposed sensor is formed with an glucose concentration widthmodulated line-defect microcavity. In order to enhance the sensitivity of our sensor, we have tuned some structural parameters, such as the position, diameter of holes and introducing an air-slot. The principle of detection is based on the measurement of wavelength due to change in RI. In the present work, the variation in the analyte glucose concentration is analyzed by using FDTD simulations. Various sensitivities and high Quality factor have been attained , concluding very small detection limit respectively.These four structures are considered as basic structures for the performance of the detection process .In this case, the sensitivity is achieved to 482.82 nm/RIU with a detection limit of 3.67×10^{-6} RIU. Which can be important in many applications and be widely used in the detection of organisms, nanoscale sensors and explosive environmental.

References

[1] Liedberg B, Lundström I and Stenberg E 1993 *Sens. Actuators. B Chem.* **11** 6372
[2] Heideman R G and Lambeck P V 1999 *Sens. Actuators. B Chem.* **61** 100127
[3] Scherer A, Painter O, Vuckovic J, Loncar M and Yoshie T. 2002 *IEEE T. Nanotechnol.* **8** 411
[4] Mortensen N A, Xiao S and Pedersen J 2007 *Microfluid. Nanofluidics* **4** 117127
[5] Caër C, Serna-Otálvaro S F, Zhang W, Le Roux X and Cassan E 2014 *Opt. Lett.* **39** 57925794
[6] Benmerkhi A, Bouchemat M and Bouchemat T 2016 *Optik* **127** 56825687
[7] Hocini A, Moukhtari R, Khedrouche D, Kahlouche A and Zamani M 2017 *Opt. Commun.* **384** 111117
[8] Dutta H S and Pal S 2013 *Opt. Quant. Electron.* **45** 907917
[9] Bougriou F, Bouchemat T, Bouchemat M and Paraire N 2013 *Eur Phys J-Appl Phys* **62** 11201
[10] Zhang Y, Zhao Y and Wang Q 2013 *Sens. Actuators. B Chem.* **184** 179188
[11] Lai W C, Chakravarty S, Wang X, Lin C and Chen R T 2011 *Opt. Lett.* **36** 984986
[12] Harhouz A and Hocini A 2015 *J. Electromagn. Waves Appl.* **29** 659667
[13] Hocini A and Harhouz A 2016 *Nanophotonics* **10** 016007016010
[14] Di Falco A, O'Faolain L and Krauss T F.2009 *Appl. Phys. Lett.* **94** 06350310635033
[15] Säynätjoki A, Mulot M, Vynck K, Cassagne D, Ahopelto J and Lipsanen H. 2008 *Photonics Nanostruct.* **6** 4246
[16] Qiu M 2002 *Appl. Phys. Lett.* **81** 1163
[17] Kafesaki M, Soukoulis C M and Agio M.2002 *Appl. Phys.* **96** 40334038
[18] Dutta H S, Goyal A K and Pal S 2014 *Nanophotonics* **8** 083088083088
[19] Rigamonti G, Bello V and Merlo S 2018 *Sensors* **18** 603
[20] Yeh Y L 2008 *Opt. Lasers Eng.* **46** 667670
[21] Dorfner D F, Hurlimann T, Zabel T,Frandsen L H, Abstreiter G and Finley J J 2008 *Appl. Phys. Lett.* **93**. 181103181106
[22] Faraon A, Englund D, Fushman I, Vučković J, Stolz N and Petroff P 2007 *Appl. Phys. Lett.* **90** 213110213113
[23] Loncar M and Scherer A 2004 *Optical Microcavities* (World Scientific Publishing)
[24] Zhou J, Tian H, Yang D, Liu Q and Ji Y 2014 *Opt. Commun.* **330** 175183

[25] Benmerkhi A, Bouchemat M and Bouchemat T 2016 *Photonics Nanostruct.* **20** 717
[26] Huang L, Tian H, Zhou J, Liu Q, Zhang P and Ji Y 2015 *Opt. Commun.* **335** 7377
[27] Arafa S, Bouchemat M, Bouchemat T, Benmerkhi A and Hocini A 2017 *Opt. Commun.* **384** 93100
[28] Benelarbi D, Bouchemat T and Bouchemat M 2017 *Opt. Quant. Electron* **49** 347
[29] Arafa S, Bouchemat M, Bouchemat T and Benmerkhi A 2017 *Optik* **131** 4957
[30] White I M and Fan X 2008 *Opt. Express* **16** 10201028
[31] Dominguez-Juarez J L, Kozyreff G and Martorell J 2011 *Nat. Commun.* **2** 18

IOP Conf. Series: Materials Science and Engineering 1046 (2021) 012016 doi:10.1088/1757-899X/1046/1/012016

Diagnosing different types of skin carcinoma based on their optical properties: A Monte-Carlo implementation

Omnia Hamdy* and Ibrahim Abdelhalim

Engineering Applications of Laser Department, National Institute of Laser Enhanced Sciences "NILES", Cairo University, Giza, Egypt.

*E-mail: omnia@niles.edu.eg

Abstract. Skin cancer is a very common and serious type of cancers worldwide. Among many kinds of non-melanoma skin cancers, Basel Cell and Squamous Cell Carcinoma are highly treatable in case of early detection. Various Diagnosing techniques are employed to detect skin cancer, such as dermoscopy, OCT, biopsy and physical examination according to the medical case. However, the non-invasive optical methods are gaining validity due to their competitive advantages including safety and functionality. In addition, they are painless and high sensitive to the examined tissue metabolic changes. The propagation of light in any biological tissue is controlled be its optical absorption and scattering properties that highly depend on the wavelength of the utilized light. Monte-Carlo simulation is a forward numerical method used to describe light propagation in biological tissues depending on their optical parameters. In this work, Monte-Carlo simulation method was implemented to characterize the light propagation in normal dermis, Infiltrative Basal Cell Carcinoma, Nodular Basal Cell Carcinoma, and Squamous Cell Carcinomas in order to differentiate healthy from cancerous tissues. The obtained results provided information about the amount of light reflectance, transmittance, absorbed fraction and fluence rate distribution in the examined tissues showing different values at each condition over a wide range of wavelengths, which provide a simple, safe and functional tool for diagnosing these categories of skin carcinoma.

1. Introduction

Among different diseases affecting the human skin, skin cancer is enrolled as the most dangerous diseases. It mainly occurs because of the uncontrolled cell growth on the skin layers. The most problem of these abnormal cells is their ability to move or extent to other tissues and body organs [1,2]. Basal Cell Carcinoma "BCC" and Squamous Cell Carcinoma "SCC" are very common types of skin cancer, BCC commonly appears in the face but can also be found in neck, ears, head and shoulders and rarely causes a metastasis [3,4], however, it is not very dangerous if it is early detected. While, SCC occurs in the cells that form the upper layer of the skin and can extend to deeper layers. It mostly appears in body parts that is exposed to UV light such as legs or feet. If not early detected, SCC can reach a very large size and may cause metastasis [4].

In the field of medical diagnosis, the physician can determine the type of skin cancer through many techniques such as optical coherence tomography "OCT", reflectance confocal microscopy, biopsy and dermoscopy [5]. High frequency ultrasound and Doppler sonography are non-invasive techniques implemented to characterize the skin according to the backscattered echo resulting from pulsed ultrasound [6]. While, in biopsy, a skin sample from the affected part is detached from the body to be examined [6]. OCT is also noninvasive imaging technique utilized to show the linear characteristics of the skin such as scattering, absorption birefringence and refractive index with acceptable resolution images [7,8] . As a minimally invasive technique, Raman spectroscopy is considered a powerful tool for clinical diagnosis of skin malignancies and in vivo skin cancer screening that almost doesn't require sample preparation [8,9].

In addition to the conventional practical methods, Monte-Carlo (MC) simulation is a numerical method utilized to validate the theoretical basis of many skin cancer detection techniques [10–13]. MC method can simulate the diffuse propagation of light in single and/or multilayer tissue sample based on its optical absorption and scattering characteristics at a specific wavelength.

In this work, the Monte-Carlo simulation method was utilized to distinguish between the normal and diseased tissues. The differentiation criteria depended on the variations in optical absorption and scattering properties between healthy and cancerous skin tissues at laser wavelengths starts from 470 nm to 970 nm. The obtained reflectance, transmittance, absorbed fraction profiles of Infiltrative Basal Cell Carcinoma, Nodular Basal Cell Carcinoma, and Squamous Cell Carcinomas provided a promising diagnosing method for these types of skin carcinoma.

2. Methods

2.1. Monte-Carlo Implementation

Monte Carlo stimulation for photon propagation in biological tissue is a statistical method based on the tracing the propagation of a large bundle of photons generated by a computer. It stimulates photon propagation in a highly scattering medium such as biological tissue by applying photon propagation rules including probability distribution and angle of deflection [14,15].

Photons on Monte Carlo stimulation are not treated by wave phenomena; however, the phase and polarization are ignored. The stimulation process starts with photon weight equals to unity, then, after the photon hits the surface of the tissue, a portion of photon will reflect and escape while the rest will internally reflect or scatter and propagate continuously. This process is repeated until the weight of photon become less than a specified threshold. The number of photons used in the stimulation is a matter of desired precision and the required spatial resolution. To implement Monte-Carlo, main optical information of the examined tissue must be known including, absorption coefficient, scattering coefficient, anisotropy, refractive index and the thickness of the sample[16].

In the present study, common types of skin cancers with previously published optical absorption and scattering parameters [17] have been examined. The selected tissue types are; Infiltrative Basal Cell Carcinoma, Nodular Basal Cell Carcinoma, and Squamous Cell Carcinoma with a sample thickness equals to 1 mm. Number of photons in MC implementation was 50000 photons.

2.2. Tissue optical parameters

Tissue absorption and scattering parameters can be determined from diffuse light measurements including reflection and transmission. These measurements are collected using either integrating spheres [18,19] or distant detectors [20–22] based techniques. Diffuse reflection and transmission are then utilized in suitable mathematical model to reconstruct the optical parameters. Inverse Monte-Carlo [23], Kubellka-Munk [24]and inverse adding doubling [25] are the most common mathematical models used for this purpose.

The optical parameters of many normal and diseased biological tissue types including human skin have been calculated over a wide range of wavelengths in [26]. Some of these published values have been introduced to our MCML implementation to differentiate between Infiltrative Basal Cell Carcinoma, Nodular Basal Cell Carcinoma, Squamous Cell Carcinoma and normal dermis.

2.3. Determining the Fluence rate distribution

Recent advances in describing laser transfer energy in biological tissues are based on transport theory of light propagation that utilizes the radiative transport equation RTE to represent this propagation [27]. The radiative transport equation can be written as. [28]

$$s.\nabla L(r,s) = -(\mu_a + \mu_s)L(r,s) + \mu_s \int_{4\pi} p(s,s')d\omega' \tag{1}$$

where L(r, s) is the radiance of light at position r traveling in a direction of the unit vector s and $d\dot{\omega}$ is the differential solid angle in the direction $\dot{s}$. μa is the tissue absorption coefficient and μs is the scattering coefficient.

When a laser beam is incident on a turbid medium as biological tissue, the radiance can be divided into a coherent and diffuse term. The diffuse radiance is the main problem which transport theory has to deal with, since scattered photons do not follow a determined path. Therefore, many approximations and statistical approaches were introduced to the transport equation depending on the value of the albedo of the medium [29].

Using radiative equation to calculate the light distribution in tissue requires the knowledge of the absorption and scattering coefficients as well as the phase function. In order to estimate these parameters, one must have a solution of the radiative transport equation. The diffusion equation (2) is an approximation to the radiative transport equation in which the radiance is assumed to be nearly isotropic, it is a partial differential equation that is significantly easier to be solved than the radiative transport equation [30].

$$\frac{\partial\phi(\vec{r},t)}{c\partial t} + \mu a\phi(\vec{r},t) - \nabla.[\mathrm{D}\nabla\phi(\vec{r},t)] = S(\vec{r},t) \tag{2}$$

where the constant $\mathrm{D} = \frac{1}{3(\mu_a + \mu'_s)}$ is the diffusion coefficient and $\mu'_s = (1-g)\mu_s$ is the reduced scattering coefficient and g is the anisotropy factor. $S(\vec{r},t)$ represents the source term and $\phi(\vec{r},t)$ is the fluence rate.

In this study, the diffusion equation was solved using the finite element method [31,32] under the environment of COMSOL Multiphysics 5.2 software [33]. Helmholtz Equation (3) in COMSOL is employed to model the diffusion equation presented in (3) in the steady state.

$$\nabla(-C\nabla u) + au = f \tag{3}$$

Comparing the parameters in equation (3) with those in (2) we get, $u = \phi$, $C = D$, $a = \mu_a$, and $f = S$. A 1 cm circular model was created with a point source at the center of the model as presented in figure 1(a). The implemented mesh (figure 1(b)) has element size ranges from 0.000057 to 0.02 cm. Based on that predefined finite element model; the fluence rate (W/cm^2) distribution at the surface of the various selected tissues was obtained.

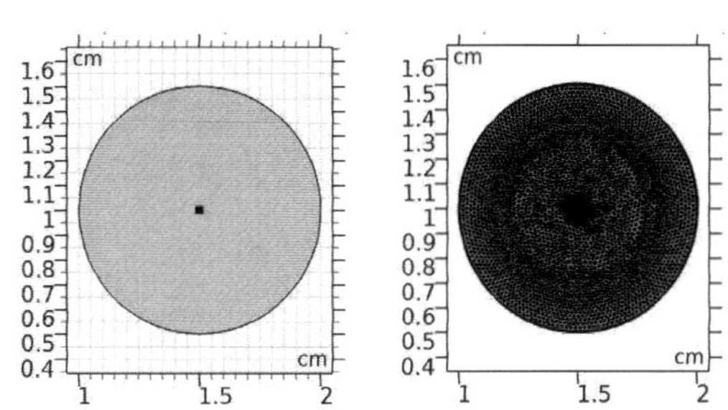

Figure 1. The implemented finite element (a) model, (b) mesh.

IOP Conf. Series: Materials Science and Engineering 1046 (2021) 012016 doi:10.1088/1757-899X/1046/1/012016

3. Results and discussion

Using Monte-Carlo simulation, the spatially resolved steady state diffuse reflectance profiles for normal dermis, infiltrative, nodular, and squamous cell carcinoma were obtained at 470, 570,670,770,870,970 nm laser irradiation of 2.7 mm spot size as presented in figure 2. As illustrated from the figure, Normal dermis shows almost maximum reflectance values over the considered wavelengths. However, at 470 and 570 nm, the maximum reflectance was noticed in infiltrative basal cell carcinoma. Moreover, the lowest reflectance was observed in squamous cell carcinoma over the all wavelengths. Reflectance of nodular basal cell carcinoma always intermediates squamous cell carcinoma and infiltrative basal cell carcinoma.

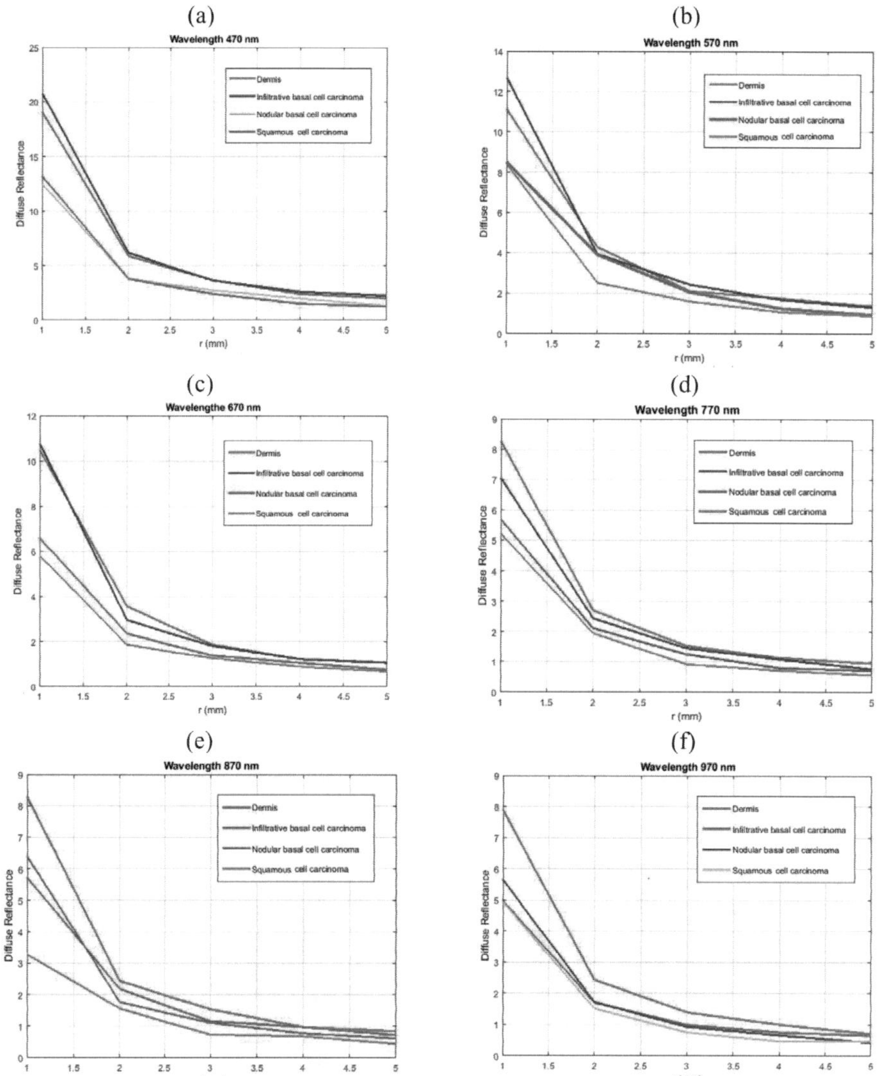

Figure 2. Diffuse reflectance profile of normal dermis, infiltrative, nodular, and squamous basal cell carcinomas at (a) 470 nm, (b) 570 nm, (c) 670 nm, (d) 770 nm, (e) 870 nm, (f) 970 nm.

The transmission and absorbed fraction in each studied case at selected wavelengths were also determined. Figure 3(a) and (b) illustrates the ttransmittance and absorption profiles for normal dermis, infiltrative, nodular, and squamous cell carcinomas respectively.

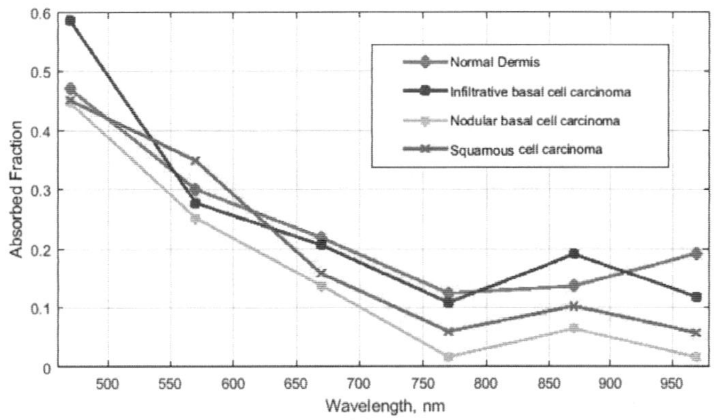

(a)

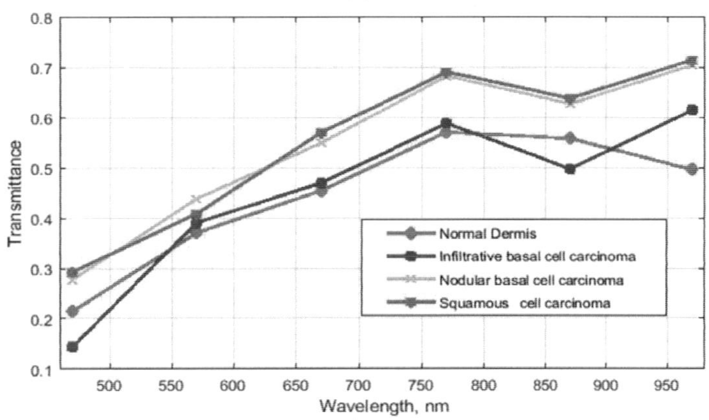

(b)

Figure 3. (a) transmission profile, (b) absorbed fraction

As conducted from figure 3, infiltrative basal cell, nodular basal cell, and squamous cell carcinoma almost take the same regime over the studied wavelengths. Unlike the normal dermis which takes the opposite behaviour from 750 to 950 nm for both transmission and absorption profiles. Moreover, infiltrative basal cell carcinoma has almost the maximum absorbed fraction and minimum transmission rather than other types of carcinoma. As more investigations are required for accurate differentiation, the fluence rate distribution at the surface of the examined tissues is obtained as presented in figure 4.

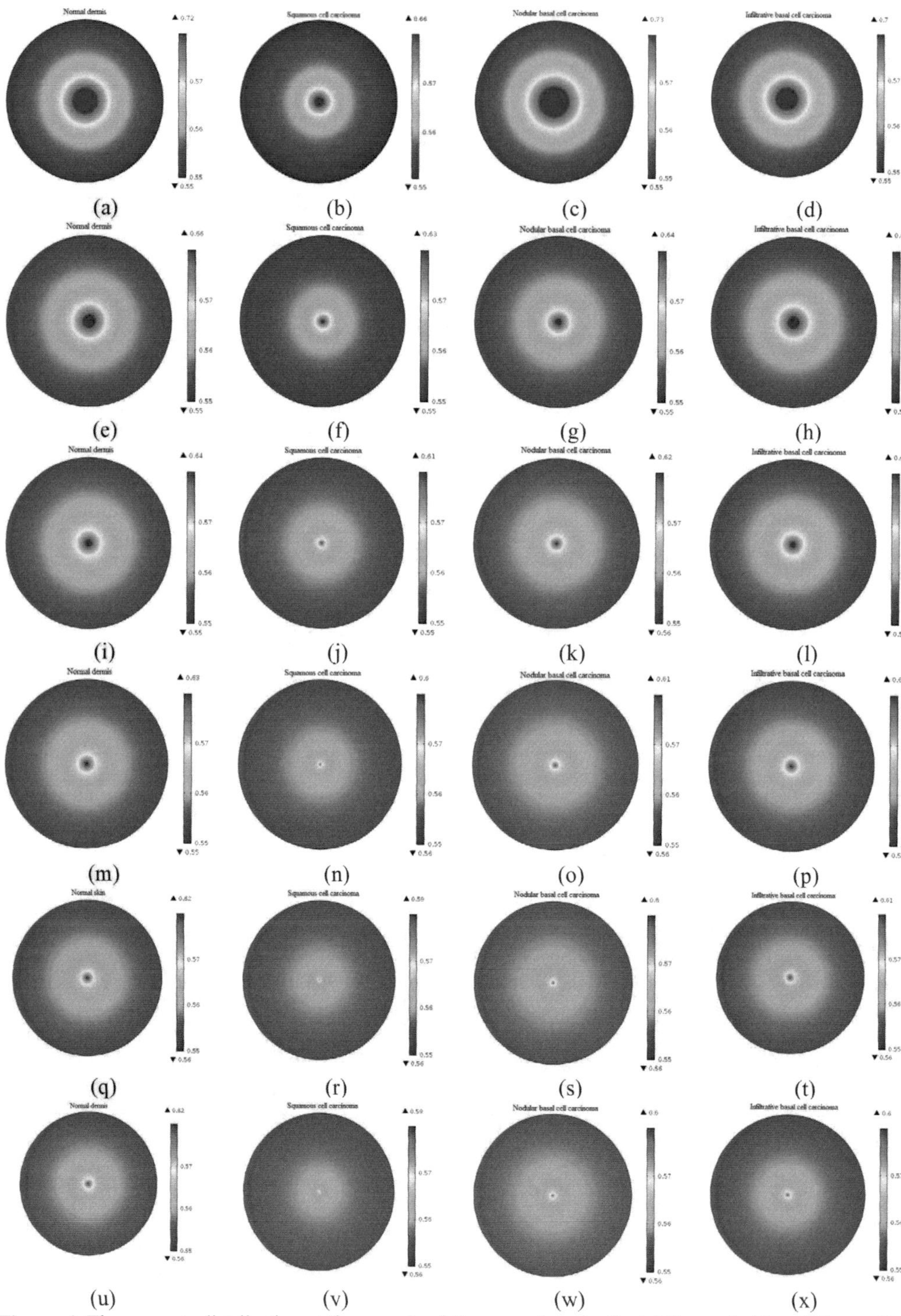

Figure 4. Fluence rate distribution at the examined tissue surface (a-d) at 470 nm, (e-h) at 570 nm, (i-l) at 670, (m-p) at 770, (q-t) at 870, (u-x) at 970

According to the optical fluence rate images of figure 4 and the curves presented in figure 5, squamous cell carcinoma gives the minimum fluence rate along the all used wavelengths followed by nodular basal cell. However, infiltrative basal cell almost gives the same fluence rate as the normal dermis. Therefore, these results can help in differentiating squamous cell and nodular basal cell carcinoma from the normal skin.

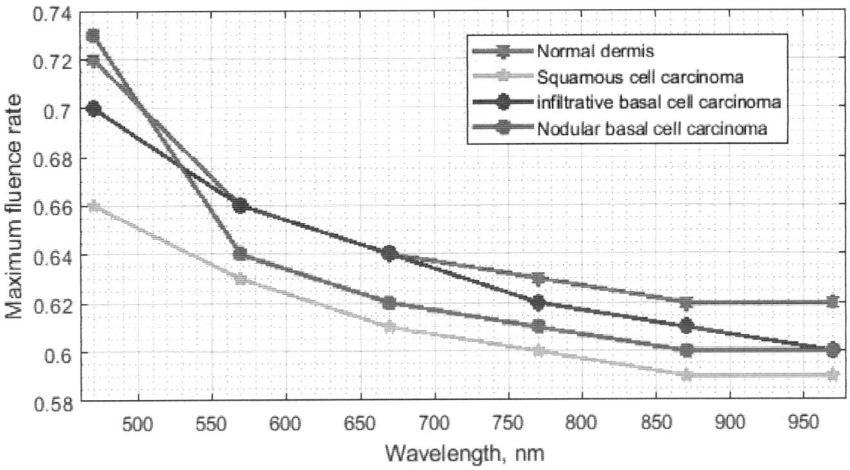

Figure 5. Maximum Fluence rate at different laser wavelengths

The obtained variations in diffuse reflectance profiles, transmittance, absorbed fraction and fluence rate distribution between the different tissue types are mainly due to the difference in the tissue's absorption and scattering properties. These properties are wavelength dependent and highly related to tissue chromophores concentrations including haemoglobin, melanin, lipids and water contents [34]. Tissue diffuse reflectance is considered as a finger print for each tissue type and condition as it changes upon appearing any abnormality in the tissue [21]. Therefore, both tissue optical parameters and diffuse reflectance profile can powerfully assist in the process of medical diagnosis and therapy monitoring [35–38]. Therefore, our proposed results can offer additional diagnostic information compared to other traditional techniques.

4. Conclusion

Three types of skin carcinoma infiltrative, nodular and squamous basal cell have been compared with normal dermis using Monte Carlo stimulation for light propagation in biological tissue at different laser wavelengths. This study showed maximum reflectance for normal dermis at all employed wavelengths except for 470 nm and 570 nm, the maximum reflectance was for infiltrative basal cell carcinoma which can be used to differentiate between normal and cancerous skin. Moreover, light transmittance and absorbance have been studied. Absorbance and transmittance showed significant values variation for normal rather than cancerous tissues especially at wavelength range 750 nm to 950 nm out of this range values are very close.

Acknowledgment

The authors would like to thank the oregon medical laser center "omlc" for sharing Monte-Carlo simulation source code "MCML" free on the web.

References

[1] Eide M J, Asgari M M, Fletcher S W, Li L, Alexander G L and Altschuler A 2013 *J. Am. Board Fam. Med.* **26** 648

[2] Pyne J H, Fishburn P, Dicker A and David M 2015 *Dermatol Pr. Concept* **5** 21
[3] Bichakjian C, Armstrong A, Baum C, Bordeaux J S, Brown M, J.Busam K, Eisen D B, Iyengar V, Lober C, Margolis D J, Moyer J, Olencki T and Rodgers P 2018 *J. Am. Acad. Dermatol.* **78** 540
[4] Ogden S and Telfer N R 2009 *Medicine* **37** 305
[5] Bakos R M, Blumetti T P, Roldán-Marín3 R and Salerni G 2018 *Am. J. Clin. Dermatol.* **19** s3
[6] Mogensen M and Jemec G B E 2007 *Dermatologic Surg.* **33** 1158
[7] Calin M A, Parasca S V, Savastru R, Calin M R and Dontu S 2013 *J Cancer Res. Clin. Oncol.* **139** 1083
[8] di Ruffano L F, Dinnes J, Deeks J J, Chuchu N, Bayliss S E, Davenport C, Takwoingi Y, Godfrey K, O'Sullivan C, Matin R N and Tehrani H 2018 *Cochrane Database Syst. Rev.* **12** CD013189
[9] Feng X, Moy A J, Nguyen H T M, Zhang Y, Zhang J, Fox M C, Sebastian K R, Reichenberg J S, Markey M K and Tunnell J W 2018 *J. Biomed. Opt.* **23** 057002
[10] Mowla A, Taimre T, Leng Y, Bertling K, Wilson S J, Prow T W, Soyer H P and Raki A D 2016 *Proc. SPIE* **9887** 98870T
[11] Dolganova I N, Neganova A S, Kudrin K G, Zaytsev K I and Reshetov I V 2016 *J. Phys. Conf. Ser.* **673** 012014
[12] Hamdy O, Youssef D and El-azab J 2018 *9th Cairo International Biomedical Engineering Conference (CIBEC)* 53–56
[13] Mowla A, Du B W, Thomas Taimre 2 Karl Bertling, Wilson S, Soyer H P and Raki A D 2017 *Biomed. Opt.* **8** 4037
[14] Zhu C and Liu Q 2013 *J. Biomed. Opt.* **18** 050902
[15] Wang L, Jacquesa S L and Zheng L 1995 *Comput. Methods Programs Biomed.* **47** 131
[16] Hamdy O and Ismail T 2019 *Novel Intelligent and Leading Emerging Sciences Conference (NILES)* 21–24
[17] Boas D A, Pitris C and Ramanujam N 2011 *Handbook of Biomedical Optics* (New York: CRC press)
[18] Hamdy O, Fathy M, Al-Saeed T A, El-Azab J and Solouma N H 2017 *Optik* **140** 1004
[19] Hamdy O, Abdel-salam Z and Abdel-harith M 2020 *Anal. Methods* **12** 2093
[20] Hamdy O, El-Azab J, Solouma N H, Fathy M and Al-Saeed T A 2017 The use of optical fluence rate distribution for the differentiation of biological tissues *2016 8th Cairo International Biomedical Engineering Conference, CIBEC 2016*
[21] Hamdy O, El-Azab J, Al-Saeed T A, Hassan M F and Solouma N H 2017 *Materials* **10** 1
[22] Abuelmakarem H S, Hamdy O, Sliem M A, El-azab J, M.A O-H and Ahmed W A 2020 *J. Phys. Conf. Ser.* **1472** 012001
[23] Honda N, Ishii K, Kajimoto Y, Kuroiwa T and Awazu K 2019 *J. Biomed. Opt.* **23** 075006
[24] Wei H-J, Xing D, Wu G-Y, Jin Y and Gu H-M 2003 *World J. Gastroenterol* **9** 2068
[25] Prahl S A, van Gemert M J and Welch A J 1993 *Appl. Opt.* **32** 559
[26] Tuchin V V 2007 *Tissue optics: light scattering methods and instruments for medical diagnosis* (Washington USA:SPIE Press)
[27] Kim A D 2004 *J. Opt. Soc. Am. A* **21** 820
[28] Markolf N H 2007 *Laser-Tissue Interactions: Fundamentals and Applications* (Berlin: Springer)
[29] ISHIMARU A 1977 *Proceeding of the IEEE* **65** 1030
[30] Ishimaru A 1996 *Wave Propagation and Scattering in Random Media* (IEEE Press)
[31] Saouli A and Mansour K 2013 *J. Intense Pulsed Lasers Appl. Adv. Phys.* **3** 41
[32] Lehtikangas O, Tarvainen T, Kim A D and Arridge S R 2015 *J. Comput. Phys.* **282** 345
[33] Singh R and Jose I 2011 Diffuse Optical Tomography : A Simulation Based Approach *The proceeding of the 2011 COMSOL conference* (Bangalore) pp 1–6
[34] Salomatina E, Jiang B, Novak J and Yaroslavsky A N 2006 *J. Appl. Anim. Res.* **11** 1

[35] O'Sullivan T D, Cerussi A E, Cuccia D J and Tromberg B J 2012 *J. Biomed. Opt.* **17** 71311
[36] Wilson B C and Jacques S L 1990 *IEEE J. Quantum. Electron.* **26** 2186
[37] Kumar D and Singh M 2003 *IEEE Trans. Biomed. Eng.* **50** 1012
[38] Serafetinides A A, Makropoulou M and Drakaki E 2008 *Proc. SPIE* **7027** 1

A DFT investigation of anion-π interactions between halogen oxyanions (XO_3^-, X = Cl, Br, I) and aromatic π-systems (benzene and coronene)

S M Elgengehi[1,*], Kh E El-Kelany[2,3] and S El-Taher[1]

[1]Department of Chemistry, Faculty of Science, Cairo University, 12613 Giza, Egypt

[2]Institute of Nanoscience and Nanotechnology, Kafrelsheikh University, 33516 Kafrelskeikh, Egypt

[3]CompChem lab, Chemistry Department, Faculty of Science, Minia University, 61519 Minia, Egypt

*E-mail: smahmoud@sci.cu.edu.eg

Abstract. Anion-π interactions have proved to play a key role in many chemical and biochemical phenomena. In order to investigate the nature and the applications of these interactions, the anion-π interactions of the chlorate (ClO_3^-), bromate (BrO_3^-), and iodate (IO_3^-) oxyanions with benzene and coronene have been studied using density functional theory (DFT). The structures, adsorption energies, and electronic properties of XO_3^- (X = Cl, Br, I) adsorbed on benzene and coronene have been explored using dispersion corrected DFT hybrid functional, M06-2X-D3, with the def2-TZVP basis set. Benzene and coronene have been respectively used as a building block and a simple flake of graphene. The interaction of these anions with benzene is found to be unexpectedly favorable. Their binding strength is further increased when the anions are adsorbed on coronene. Moreover, energy decomposition analysis based on symmetry adapted perturbation theory (SAPT) has been performed and the nature of interaction is characterized using noncovalent interaction analysis (NCI) based on reduced density gradient (RDG) method. Generally, the favorable binding of these anions with benzene or coronene is mainly ascribed to dispersion interactions. These anion-π interactions with graphene flakes or with benzene can be utilized in the design of anions receptors and detectors.

1. Introduction

Noncovalent interactions have proved to be of great influence in various chemical and biological applications [1-4]. One of these interactions is the anion-π interaction that has attracted significant attention theoretically [5-10] and experimentally [11-13]. Because of the essential role of anions in biological and chemical systems and their connection to environmental pollution, anion recognition has grown as an important research area. Since there are many ways to bind anions (hydrogen bonding, electrostatic, etc.) [14], the design of neutral anion receptors using aromatic rings is a recent promising research field [15-16].

The previous theoretical studies have been focused on understanding the nature of interaction between anions and simple electron-deficient aromatic systems (e.g., substituted benzene). Generally, it has been believed that electrostatic interactions between aromatics with positive quadrupole moment (Q_{zz}) and anions, as well as induction are the dominant forces in these types of interactions [8]. In another study, it has been demonstrated that the binding of anions is mainly due to the interactions with the dipoles induced by the ring substituents [9-10]. Despite what is expected of the repulsion between anions and electron-rich aromatic systems (with negative Q_{zz}), it has been reported that the polarizability of the benzene substituents causes a favorable interaction between halides and substituted benzene [17-19]. Because the previous studies were more concentrated on the interaction between halides and substituted benzene, here in this study the interaction of polyatomic oxyanions namely: chlorate (ClO_3^-), bromate (BrO_3^-), and iodate (IO_3^-) with unsubstituted benzene is investigated, where their binding is unexpectedly found favorable. Furthermore, a new adsorbent; coronene (which is a simple flake of graphene) is introduced to improve the anion-π binding by making use of its large unoccupied π orbitals. In recent years, graphene has been recognized as an emerging nanomaterial with exceptional properties and wide applications in technology [20-23]. These flakes can be used in the removal of toxic anions like arsenite, arsenate, phosphate, etc. as expected by previous experimental results [24-26]. Also, several theoretical studies [27-37] have been performed to investigate the adsorption of anions on graphene or graphene-based materials.

To the best of our knowledge, the adsorption of the trigonal pyramidal anions (ClO_3^-, BrO_3^-, and IO_3^-) on benzene and coronene has never been studied so far. However, these oxyanions are toxic and unsafe for living organisms [38-39]. For instance, bromate is a possible carcinogen, highly soluble in water and it is difficult to be removed using conventional methods [40]. So, it is important to design a receptor for these kind of anions. In this study, we aim to carefully study the interactions of the above mentioned anions with benzene and with the larger surface area coronene. Understanding the nature of anion-π interaction has helped in utilizing them in anion recognition [41-42], and in the design of anion receptors [43-45].

2. Computational details

In order to investigate the adsorption of the 3 oxyanions; chlorate (ClO_3^-), bromate (BrO_3^-), and iodate (IO_3^-) on benzene and coronene surfaces, M06-2X-D3/def2-TZVP level of DFT theory is used. It was found that M06-2X has broader accuracy than other popular DFT functionals when studying noncovalent interactions [46]. Although M06-2X has medium-range electron correlation to some extent [47], adding the D3 version of Grimme's dispersion correction [48] to account for long-range electron correlation has proven to improve its results in noncovalent interactions [49]. It has also been verified that def2-TZVP [50] basis set provides accurate results for noncovalent interactions involving anions without the need of adding diffuse functions [51]. Effective potentials were used for the core electrons of the iodine atom [52]. Geometry optimization and frequency calculations were done using Gaussian09 package [53]. All molecular structures were optimized without any symmetry constraints and with the default convergence parameters. Vibrational frequency analysis was carried out at the same level of theory to verify the stability of each complex. Depending on the position and orientation of each oxyanion on every adsorbent, the complexes change from local minima to first or even second-order saddle points. Adsorption energy E_{ads} is obtained as the difference between the total calculated energies of the complex and its interacting species using the following expression:

$$E_{ads} = E_{complex} - E_{\pi\text{-system}} - E_{XO3^-} \quad (1)$$

where $E_{complex}$ is the total energy of the optimized adsorbed system, $E_{\pi\text{-system}}$ is the energy of the π-system (C_6H_6 and $C_{24}H_{12}$), and E_{XO3^-} is the energy of the isolated anion where (X = Cl, Br, I). Due to the resulted basis set superposition error (BSSE), all calculated adsorption energies were corrected using the counterpoise method [54].

To analyze and understand the nature of the intermolecular orbital interactions between the oxyanions and the studied adsorbents, natural bond orbital (NBO) analysis was employed [55-56]. Furthermore, interaction energy decomposition analysis of the low-energy complexes of chlorate and bromate was carried out using the symmetry-adapted perturbation theory (SAPT) [57-59]. The total interaction energy is decomposed into electrostatic (E_{elst}), exchange-repulsion (E_{exch}), induction (E_{ind}), and dispersion (E_{disp}) components. sSAPT0 has been chosen along with the def2-TZVP basis set. sSAPT0 is a wavefunction-based SAPT method that provides the best error by unit computational effort [60]. SAPT calculations were performed using PSI4 software [61]. Moreover, NCI analysis based on the reduced density gradient (RDG) method [62] has been carried out using Multiwfn program [63]. This analysis has helped in visualizing and characterizing the studied anion-π interactions.

3. Results and discussion

3.1. Structures and adsorption energies of C_6H_6–XO_3^- and $C_{24}H_{12}$–XO_3^- (X = Cl, Br, I) complexes

Here, the adsorption of the three halogen oxyanions (ClO_3^-, BrO_3^-, IO_3^-) on benzene and coronene is carefully explored. The calculated geometrical structures of XO_3^- (X = Cl, Br, I) in gas phase and in adsorbed phase are listed in table 1. It is worth noting that the M06-2X-D3 bond lengths ($r_{X\text{–}O}$) and bond angles (∠O–X–O) are in reasonable agreement with the corresponding previously reported experimental values [64-66]. As shown in table 1, the bond angles are 0.99%–3.74% larger than the experimental parameters while the bond lengths are approximately similar to the experimental values. As in experiment, the M06-2X-D3 optimized geometries show that $r_{X\text{–}O}$ bond length gets longer when going from ClO_3^- (1.480 Å) to IO_3^- (1.785 Å). On adsorption, the oxyanions are not geometrically changed significantly, which is consistent with physisorption. For example, the change in the bond angles upon adsorption is ~ 0.07-0.33 (deg) with respect to the calculated parameters in the gas phase.

Table 1. Bond lengths and bond angles of XO_3^- (X = Cl, Br, I) anions in gas phase and adsorbed phase.

Anions	Gas phase		Adsorbed phase	
	Experimental	This work	C_6H_6	$C_{24}H_{12}$
ClO_3^-				
$r_{Cl\text{–}O}$ (Å)	1.480 [64]	1.480	1.481	1.480
∠O–Cl–O (deg)	106.60 [64]	107.66	107.73	107.80
BrO_3^-				
$r_{Br\text{–}O}$ (Å)	1.640 [65]	1.637	1.638	1.637
∠O–Br–O (deg)	105.5 [65]	106.72	106.81	106.65
IO_3^-				
$R_{I\text{–}O}$ (Å)	1.790 [66]	1.785	1.786	1.785
∠O–I–O (deg)	102.0 [66]	105.81	105.71	105.48

The optimized structures of the low-lying energy conformers of C_6H_6–XO_3^- complexes are shown in figure 1, together with the corresponding adsorption energies (E_{ads}) in kJ/mol and the equilibrium distance between the X-atom of the anion and the ring centroid (z), $R_{X\text{–}z}$. Because our present goal is to analyze the nature of the interaction between the halogen oxyanions and the π-system of the electron-rich benzene

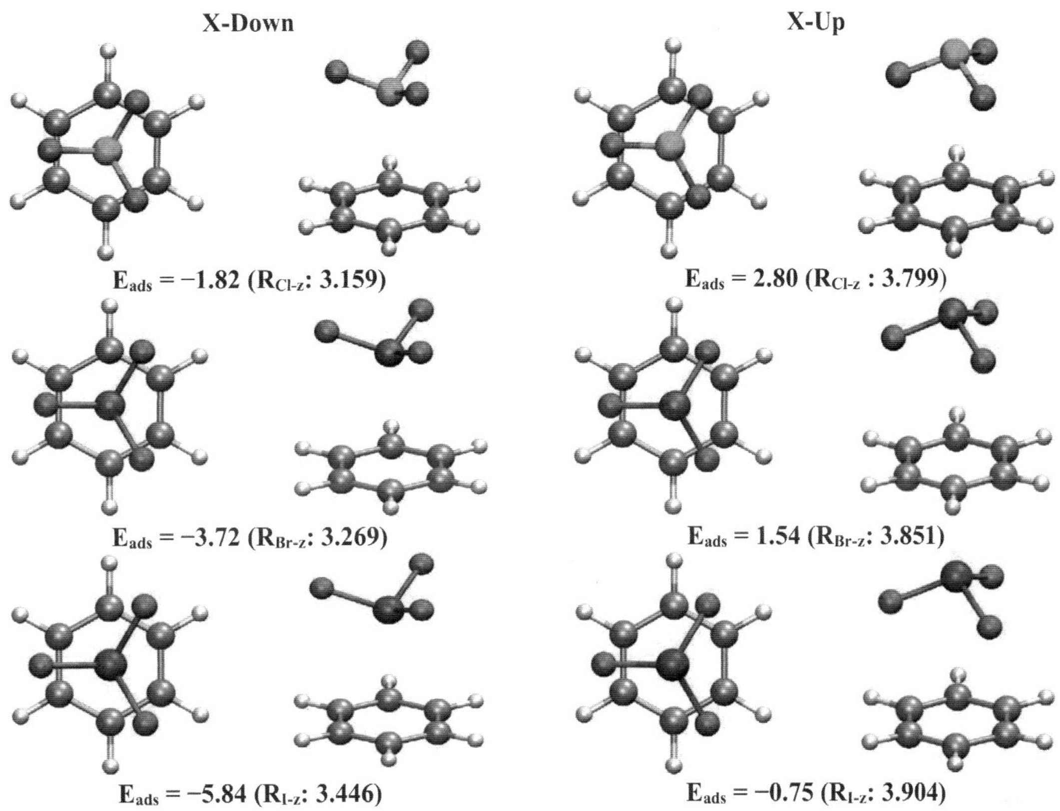

Figure 1. Optimized structures of the $C_6H_6–XO_3^-$ (X = Cl, Br, I) complexes calculated at the M06-2X-D3/def2-TZVP level with BSSE correction. Down and Up mean downward and upward orientations of the halogen atom. Adsorption energies (E_{ads} in kJ/mol) and R_{X-z} equilibrium distances (Å) between the ring centroid z and the anion halogen atom X are given.

and coronene rings, we only considered two low-lying energy conformers of the $C_6H_6–XO_3^-$ complexes indicated by X–Up and X–Down as depicted in figure 1. In both conformers, the anions are located along the z-axis of the benzene ring with the halogen atoms perpendicularly located above the benzene centroid. The oxygen atoms in each of these conformers can have staggered tripod orientation in which they are stacked on the midpoint of the C–C bonds (∠O-X-z-C dihedral angle d = 30°) or they can have an eclipsed orientation in which the oxygen atoms are stacked on the C atoms (d = 0°). Both orientations of oxygen atoms are nearly isoenergetic, so only the staggered conformations are considered here. All benzene complexes are characterized by second-order saddle points with two or three imaginary frequencies (ranging from ~20.0*i* to 60.0*i*), which originate from the X–O swings toward the H-atoms as the anions prefer hydrogen bonding rather than binding to the π-ring [67]. Careful inspection of the data given in figure 1 shows that the X- Down conformers are more favorable than the corresponding X-Up ones. The adsorption energies of the X-Down (X-Up) conformers are −1.82 (2.80), −3.72 (1.54), and −5.84 kJ/mol (−0.75 kJ/mol) for ClO_3^-, BrO_3^-, and IO_3^-, respectively. This unfavorable binding of the X-Up conformers can be explained on the basis that the negatively charged oxygen atoms in these conformers are pointed toward the benzene ring, and hence the repulsive electrostatic interaction between the oxyanions and the negative quadrupole moment of benzene ($Q_{zz} = -8.7 \pm 0.5$ DÅ [68]) is maximized. As noticed, there is an increase in the binding strength as going from ClO_3^- to IO_3^-, which can be related to the corresponding decrease in the anions electronegativity. However, the equilibrium distances ($R_{X\text{-}z}$) increase when going from ClO_3^- (3.159 Å) to BrO_3^- (3.269 Å) and to IO_3^- (3.446 Å). This behavior can be assigned to the steric effect that increases as the size of the halogen atom increases and thus the oxyanion is pushed a little farther from the benzene surface.

Different orientations and positions were studied for the trigonal pyramidal anions on coronene and the conformers are shown in figure 2, with geometrical and energetic properties summarized in table 2. Three adsorption sites were considered: Top (T), Bridge (B), and Hollow (H) sites with respect to the halogen atom.

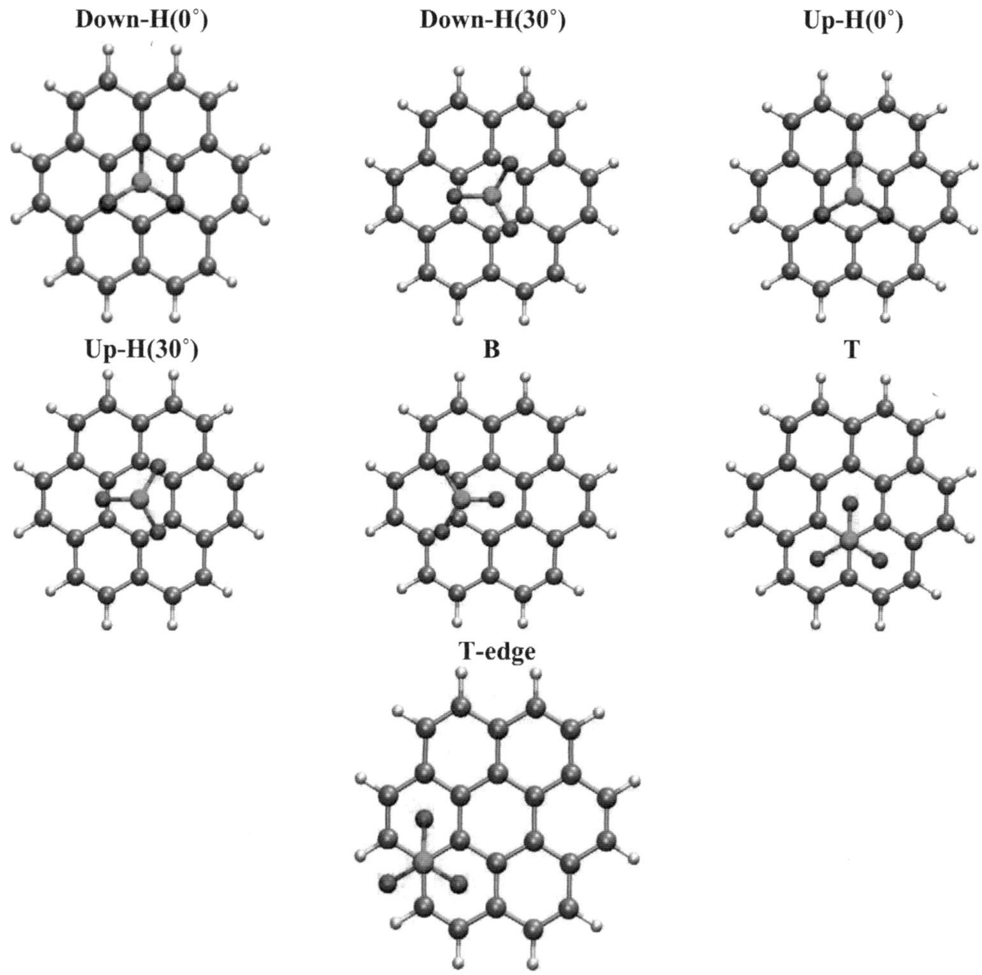

Figure 2. Optimized structures of the different conformers of $C_{24}H_{12}$–ClO_3^- complexes calculated at the M06-2X-D3/def2-TZVP level. The same conformers are studied for both $C_{24}H_{12}$–BrO_3^- and $C_{24}H_{12}$–IO_3^- complexes.

Table 2. Calculated adsorption energies (E_{ads}, in kJ/mol), and equilibrium distances ($R_{X\text{-}z}$, in Å) for the different conformers of $C_{24}H_{12}$–XO_3^- (X = Cl, Br, I) complexes. All energies are BSSE corrected.

Adsorption site	ClO_3^-		BrO_3^-		IO_3^-	
	E_{ads}	$R_{Cl\text{-}z}$	E_{ads}	$R_{Br\text{-}z}$	E_{ads}	$R_{I\text{-}z}$
Down-H(0°)	−3.81	2.943	−0.27	3.102	−0.46	3.373
Down-H(30°)	−3.95	2.949	−0.36	3.102	−0.51	3.373
Up-H(0°)	−13.72	3.506	−14.54	3.576	−14.56	3.667
Up-H(30°)	−13.29	3.507	−14.01	3.576	−14.32	3.660
B	−18.25	3.524	−19.02	3.603	−19.68	3.709
T	−21.92	3.481	−21.93	3.563	−21.48	3.663
T-edge	−25.11	3.456	−26.83	3.522	−28.01	3.608

Since there are a lot of possibilities for the different orientations on each position, only the stable orientations are considered here. For example, the dihedral angle (d) ∠O–X–C–C in T site can range from 0° to 60°, and it is found that the optimization of all T-conformers with the different d values, whether the halogen atoms are up or down, will lead to only one conformer: X-Up with d = 60°, so this conformer is referred to as T site. There is also only one X-Up orientation of B site, as shown in figure 2. However, there are more than one stable conformer for the H site: Down-H(0°), Down-H(30°), Up-H(0°), and Up-H(30°). In contrast to C_6H_6–XO_3^-, X-Up orientations are more stable than the X-Down orientations, see table 2. This finding suggests that there is an interaction between the oxygen atoms and the π-system of coronene. It was also reported previously that the Up tripod geometry of ClO_4^- is more stable than the Down geometry when adsorbed on electron deficient aromatic rings [45, 69]. Among the Up conformers, T site is the most stable one with oxygen atoms interacting with the centers of the rings, where there is depletion in the electron density. Moreover, XO_3^- (X = Cl, Br, I) anions form much more stable complexes when adsorbed on T site at the edge of the coronene with a difference of 3.19, 4.90, and 6.53 kJ/mol with respect to the corresponding T site (center) complexes, respectively. All the optimized geometries of the T-conformers are minima, while those of the B-conformers have one small imaginary frequency that corresponds to a transition state of the corresponding T-conformer. On the other hand, H-conformer geometries have 2 or 3 imaginary frequencies (< 40*i*) for Up-H(0°) and Up-H(30°), given that Up-H(0°) is slightly more stable than Up-H(30°) by 0.24-0.53 kJ/mol. Overall, the stability of the studied complexes depends on the sites and orientations the adsorbed anions adapt, which follows the following order: Down-H(0°) < Down-H(30°) < Up-H(30°) < Up-H(0°) < B < T < T-edge. Similar to benzene complexes, the binding strength increases as the halogen electronegativity decreases and the equilibrium distances increases with increasing the halogen size. The adsorption energies for the most stable conformers are −25.11 (ClO_3^-), −26.83 (BrO_3^-), and −28.01 kJ/mol (IO_3^-).

In order to understand the nature of the anion-π interaction, NBO analysis was performed on all the local minima complexes at the same level of calculation. The NBO charge transfer (ΔQ) from the anions to the coronene is calculated and compiled in table 3 as well as the second order perturbation energies (E^2) resulting from the second order perturbation analysis of Fock matrix in the NBO basis.

The analysis shows a stabilization due to the interaction between the lone pair orbitals (LP) of the oxygen atoms in the oxyanions and the anti-bonding π^* C–C orbitals. However, ΔQ and E^2 are relatively lower in C_6H_6–XO_3^- complexes, compared to $C_{24}H_{12}$–XO_3^-. This finding may conclude that the favourable binding in oxyanions-benzene complexes lack of any contribution of donor-acceptor interaction and is mainly dispersion in nature as will be discussed later in the SAPT section. The reported E^2 values in table 3 are the sum of the energies from individual contributions related to the dominant interaction LP (O) → π^* C–C. Inspection of the results in table 3 reveals that as the distance between the oxygen atoms of the interacting oxyanions and the surface of coronene becomes shorter, the LP (O) → π^* C–C interaction increases, causing a subsequent increase in the charge transfer and the donor-acceptor nature, as indicated by the largest ΔQ and E^2 values for $C_{24}H_{12}$–ClO_3^-. A comparison between the results of E_{ads} in table 2 and E^2 in table 3 shows that there may be forces other than charge transfer causes the interaction energy to increase when going from ClO_3^- to IO_3^-, these forces may be of electrostatic and/or dispersion nature as will be shown later. Furthermore, the enhancement in the stability when the oxyanion moves to the edge may be ascribed to an increase in the attractive electrostatic interaction, since the charge transfer decreases at the T-edge for each anion as shown in table 3. Therefore, charge transfer takes part in stabilizing the adsorption process of the oxyanions on coronene.

Table 3. NBO charge transfer (ΔQ) from the donor oxyanion molecule to the acceptor graphene flake (coronene). Second order perturbation energies (E^2) for the intermolecular oxygen lone-pair (LP) and antibonding π^* C-C orbital interactions are also given. Charges are given in atomic units and energies in kJ/mol.

Complex	T site		T-edge site	
	ΔQ	E^2 [LP (O) → π^* (C–C)]	ΔQ	E^2 [LP (O) → π^* (C–C)]
$C_{24}H_{12}$–ClO_3^-	0.031	10.92	0.027	10.42
$C_{24}H_{12}$-BrO_3^-	0.025	9.67	0.022	9.41
$C_{24}H_{12}$–IO_3^-	0.028	9.50	0.023	9.79

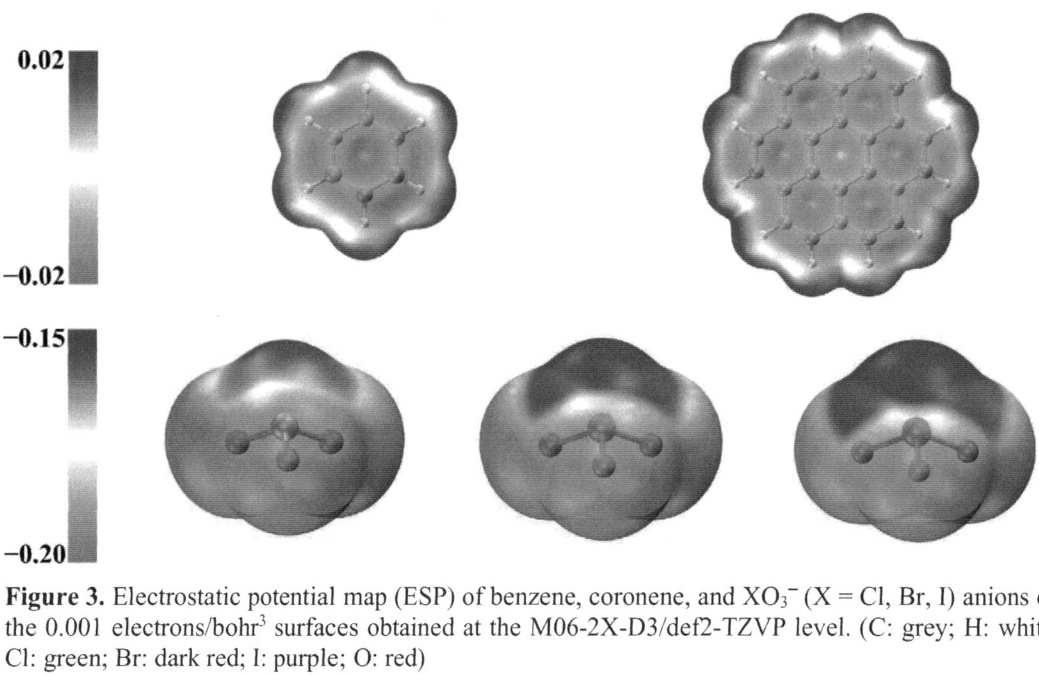

Figure 3. Electrostatic potential map (ESP) of benzene, coronene, and XO_3^- (X = Cl, Br, I) anions on the 0.001 electrons/bohr^3 surfaces obtained at the M06-2X-D3/def2-TZVP level. (C: grey; H: white; Cl: green; Br: dark red; I: purple; O: red)

Inspection of the electrostatic potential map (ESP) of coronene with respect to benzene (figure 3) reveals that there is some relative electron deficiency in the centers of the rings of coronene as represented by the decrease in the intensity of the red color. This verifies that there is a difference between the anions interaction with coronene and with benzene [31]. In addition, the increase in the relative positive charge density around the halogen atom when going from ClO_3^- to BrO_3^- and to IO_3^- in figure 3 is obvious. According to NBO analysis, the charges on Cl, Br, and I in their isolated anions are respectively +1.819, +2.077, and +2.521, while the negative charges on oxygen atoms are −0.940, −1.026, and −1.174. In agreement with the ESP maps of benzene and coronene, the NBO negative charges on carbons are decreased from −0.212 in benzene to −0.008 (−0.180) in the central ring (edges) of coronene. Accordingly, the binding strength of the studied halogen oxyanions increases when adsorbed on coronene, compared to their adsorption on benzene.

3.2. SAPT analysis of C_6H_6–XO_3^- and $C_{24}H_{12}$–XO_3^- (X = Cl, Br) complexes

Table 4 lists the SAPT interaction energies components of the Down complexes of benzene and all the local minima complexes of coronene. This energy decomposition analysis helps in determining the forces that are responsible for the anion-π interactions. As shown in table 4, the calculated interaction energies (E_{int}) at M06-2X-D3/def2-TZVP are approximately in agreement with the sSAPT0 total interaction energies (E_{tot}). It is found that the electrostatic interaction (E_{elst}) in benzene complexes is favourable. Also, E_{elst} increases in C_6H_6–BrO_3^- by ~ 7.5 kJ/mol, compared to C_6H_6–ClO_3^- complex because of the increase in the nuclear positive charge acquired by the bromine atom in BrO_3^-. The most dominant interaction in benzene complexes is due to the dispersion force (E_{disp}), which becomes larger when BrO_3^- is adsorbed due to its larger size (larger polarizability). Careful inspection shows that the anion-induced polarization (E_{ind}) has relatively small favorable contributions to the interaction energies of both ClO_3^- and BrO_3^- complexes with benzene Therefore, trigonal pyramidal anions can interact favorably with electron-rich benzene ring especially when the oxygens are oriented upwards because of the favorable dispersion, induction, and electrostatic forces that compensate for the repulsive exchange forces.

Table 4. sSAPT0/def2-TZVP components for the conformers of C_6H_6–XO_3^- and $C_{24}H_{12}$–XO_3^- (X = Cl, Br). The M06-2X-D3/def2-TZVP interaction energies E_{int} are also given. All energies are in kJ/mol.

Complex	Conformer	E_{elst}	E_{exch}	E_{ind}	E_{disp}	E_{tot}	E_{int}
C_6H_6–ClO_3^-	X-Down	−0.88	31.53	−8.57	−25.49	−3.41	−2.68
C_6H_6–BrO_3^-	X-Down	−8.35	34.76	−5.72	−26.55	−5.86	−4.52
$C_{24}H_{12}$–ClO_3^-	T	11.63	66.77	−45.99	−62.53	−30.12	−22.34
	T-edge	2.91	63.21	−44.67	−51.60	−30.15	−26.53
$C_{24}H_{12}$–BrO_3^-	T	11.43	70.18	−46.25	−68.67	−33.31	−22.64
	T-edge	1.18	65.67	−44.39	−56.39	−33.93	−28.33

In the case of coronene, the negatively charged oxygen atoms are facing the electron cloud of the flake causing repulsive E_{elst} values. There is a relatively small decrease in the repulsive E_{elst} when going from $C_{24}H_{12}$–ClO_3^- to $C_{24}H_{12}$–BrO_3^- complex. However, the electrostatic energy decreases substantially to 2.91 (1.18) kJ/mol from 11.63 (11.43) kJ/mol for ClO_3^- (BrO_3^-) when they are adsorbed at the edge of coronene. This lowering in E_{elst} is the main reason for the increase in the stability of the anions when adsorbed at the edge, compared to their adsorption in the coronene centroid. The results in table 4 show that the dispersion interaction comes to be the first driving force in the stabilization of these anion-π interactions. The E_{disp} values of ClO_3^- (BrO_3^-) complexes with coronene at the T and T-edge sites are −62.53 (−68.67) and −51.60 kJ/mol (-56.39 kJ/mol), respectively, which are more than two-fold larger compared with those of benzene, −25.49 (−26.55 kJ/mol). This dominant contribution of dispersion energy is accompanied by a considerable contribution of the induction force (E_{ind} =~46.0 kJ/mol). This substantial increase in either E_{disp} or E_{ind} may be explained on the basis that the dispersion and induction interactions depend on the polarizabilities of the interacting partners, which, in turn, correlate with the flake size as well as the oxyanion size. Overall, one may predict that the anion-π interaction can be favorable even for the electron-rich benzene ring when the dispersion and induction energies are large enough to compensate for the repulsive electrostatic and exchange energies.

3.3. Non-covalent interaction (NCI) – reduced density gradient (RDG) analysis

The NCI analysis which is based on the reduced density gradient method (RDG) is very important in the characterization of weak interactions such as van der Waals interactions [62]. The RDG function is defined as:

$$\mathrm{RDG(r)} = \frac{1}{2\,(3\pi^2)^{1/3}} \frac{|\nabla\rho(\mathrm{r})|}{\rho(\mathrm{r})^{4/3}} \tag{2}$$

where ρ(r) is the electron density. By using Bader AIM theory, critical points where electron density is aggregated or depleted can be distinguished by the sign of the second largest eigenvalue of the Hessian matrix of electron density (λ_2). According to NCI analysis, weak interactions are found when $\lambda_2 \sim 0$, strong attraction when $\lambda_2 < 0$, and strong repulsion when $\lambda_2 > 0$. Therefore, the sign of λ_2 can give information about the nature of the interaction. Besides, the relative strength of the interaction correlate with the electron density ρ(r), for example: weak interactions have very small ρ. Finally, defining a real space function sign(λ_2) ρ(r) and mapping it according to the RDG values gives us a live picture of the interaction type, figure 4.

Figure 4 (a) displays the NCI surface of the Down complexes of C_6H_6–XO_3^- and the most stable complexes of $C_{24}H_{12}$–XO_3^- (X = Cl, Br, I), while figure 4 (b) represents their scatter plots (RDG versus sign(λ_2)ρ). In the scatter plots, the regions where sign(λ_2) ρ(r) values are around 0.005 a.u correspond to dispersion interactions, while those of sign(λ_2) ρ(r) > 0.005 a.u correspond to strong repulsion (*e.g.* steric effect) and sign(λ_2) ρ(r) < −0.005 a.u values indicate strong attraction (*e.g.* hydrogen bonding). Careful inspection of Figure 4 reveals the absence of any hydrogen bonding in these types of interaction since there are no blue-colored surface in figure 4 (a) or blue spikes in figure 4 (b). Also, it verifies the significant role of dispersion interactions in oxyanion-π binding, the dispersion forces take place above the π-electron density of benzene and coronene (green color in figure 4 (a)). The plots show 2 green-colored spikes at the region of weak interactions. At last there are regions of red color that show areas of steric effect (repulsion) at the center of the rings due to the nonbonding overlap.

4. Conclusion

Based on this study, one may conclude the following: 1) The study demonstrates the presence of unexpected anion-π interactions between the oxyanions (ClO_3^-, BrO_3^-, and IO_3^-) and two of electron-rich π-systems (benzene and coronene). The origin of the binding in these complexes can be largely attributed to dispersion and induction effects, which, in turn, correlate with the increase in the polarizability of the interacting partners when going from ClO_3^- to IO_3^- and from benzene to coronene. 2) The increase in the adsorbent surface area when going from benzene to coronene introduces large unoccupied π-orbitals that results in a parallel increase of the binding strength due to the existence of charge transfer between the lone pairs of the anion oxygens and the π^* orbitals of the coronene. This is why the Up orientations of the oxyanions on coronene are more stable than the Down orientations. 3) Moreover, it has been demonstrated that edge adsorption has the most stable binding energy which may be ascribed to the lowering of the repulsive electrostatic force when the anion is moved from the center to the edge. 4) Overall, it has been noticed that the increase in the polarizability and the electropositivity of the anions results in a parallel increase in the E_{ads} when going from ClO_3^- to BrO_3^- and to IO_3^-. 5) These findings are hoped to help in further understanding of the different mechanisms in the binding of polyatomic anions with small or large π-systems and provide a potential application in the design of neutral anion receptors.

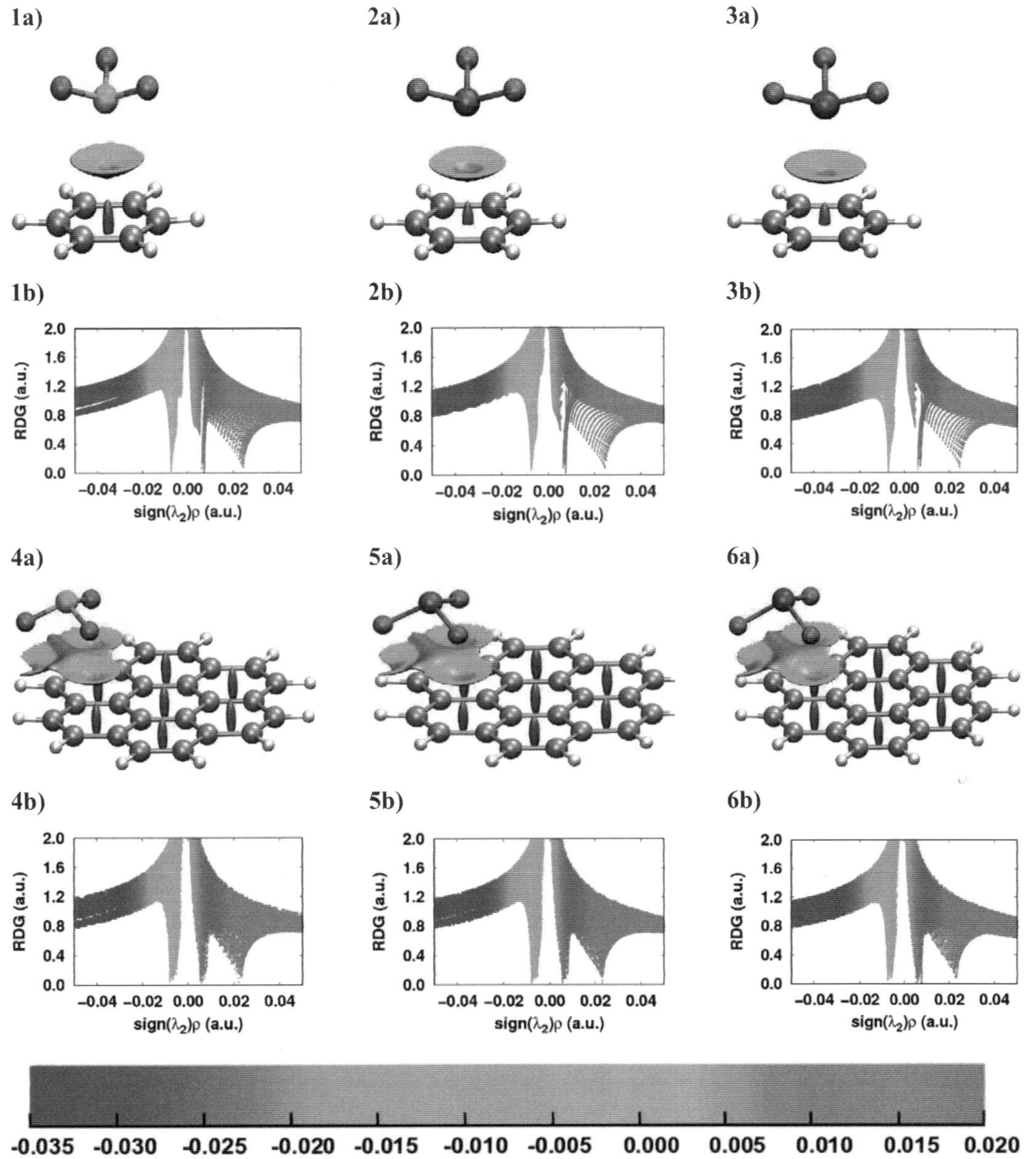

Figure 4. a) NCI surfaces (isovalue = 0.5 au) of C_6H_6 (1-3) and $C_{24}H_{12}$ (4-6) complexes at the M06-2X-D3/def2-TZVP level. The surfaces are colored on a blue-green-red scale according to values of sign(λ_2)ρ, ranging from -0.035 to 0.02 au. b) The corresponding scatter plots of reduced density gradient (RDG) versus the electron density ρ multiplied by the sign of the second Hessian eigenvalue (λ_2) for each complex. (Cl: green; Br: dark red; I: purple).

Acknowledgement
The authors acknowledge Compute Ontario (Graham) in partner with Compute Canada (www.computecanada.ca) for providing the necessary computational facilities.

References

[1] Meyer E A, Castellano R K and Diederich F 2003 *Angew. Chem. Int. Ed.* **42** 1210
[2] Salonen L M, Ellermann M and Diederich F 2011 *Angew. Chem. Int. Ed.* **50** 4808
[3] Molina P, Zapata F and Caballero A 2017 *Chem. Rev.* **117** 9907
[4] Schneider H J 2009 *Angew. Chem. Int. Ed.* **48** 3924
[5] Quiñonero D, Garau C, Rotger C, Frontera A, Ballester P, Costa A and Deyà P M 2002 *Angew. Chem. Int. Ed.* **41** 3389
[6] Mascal M, Armstrong A and Bartberger M D 2002 *J. Am. Chem. Soc.* **124** 6274
[7] Alkorta I, Rozas I and Elguero J 2002 *J. Am. Chem. Soc.* **124** 8593
[8] Kim D, Tarakeshwar P and Kim K S 2004 *J. Phys. Chem. A* **108** 1250
[9] Wheeler S E and Houk K N 2010 *J. Phys. Chem. A* **114** 8658
[10] Wheeler S E and Bloom J W G 2014 *J. Phys. Chem. A* **118** 6133
[11] Dawson R E *et al.* 2010 *Nat. Chem.* **2** 533
[12] Giese M, Albrecht M and Rissanen K 2016 *Chem. Commun.* **52** 1778
[13] Demeshko S, Dechert S and Meyer F 2004 *J. Am. Chem. Soc.* **126** 4508
[14] Beer P D and Gale P A 2001 *Angew. Chem. Int. Ed.* **40** 486
[15] Garau C, Frontera A, Ballester P, Quiñonero D, Costa A and Deyà Pere M 2005 *Eur. J. Org. Chem.* **2005** 179
[16] Frontera A, Gamez P, Mascal M, Mooibroek T J and Reedijk J 2011 *Angew. Chem. Int. Ed.* **50** 9564
[17] Clements A and Lewis M 2006 *J. Phys. Chem. A* **110** 12705
[18] Geronimo I, Singh N J and Kim K S 2011 *J. Chem. Theory Comput.* **7** 825
[19] Xu Z, Singh N J, Kim S K, Spring D R, Kim K S and Yoon J 2011 *Chem. Eur. J.* **17** 1163
[20] Geim A K and Novoselov K S 2007 *Nat. Mater.* **6** 183
[21] Geim A K 2009 *Science* **324** 1530
[22] Georgakilas V, Tiwari J N, Kemp K C, Perman J A, Bourlinos A B, Kim K S and Zboril R 2016 *Chem. Rev.* **116** 5464
[23] Gong X, Liu G, Li Y, Yu D Y W and Teoh W Y 2016 *Chem. Mater.* **28** 8082
[24] Yu F, Sun S, Ma J and Han S 2015 *Phys. Chem. Chem. Phys.* **17** 4388
[25] Vasudevan S and Lakshmi J 2012 *RSC Adv.* **2** 5234
[26] Li L, Fan L, Sun M, Qiu H, Li X, Duan H and Luo C 2013 *Colloids Surf. B. Biointerfaces* **107** 76
[27] Zhu C and Yang G 2016 *ChemPhysChem* **17** 2482
[28] Patra N, Esan D A and Král P 2013 *J. Phys. Chem. C* **117** 10750
[29] Schyman P and Jorgensen W L 2013 *J. Phys. Chem. Lett.* **4** 468
[30] Foroutan-Nejad C and Marek R 2014 *Phys. Chem. Chem. Phys.* **16** 2508
[31] Shi G, Ding Y and Fang H 2012 *J. Comput. Chem.* **33** 1328
[32] Farajpour E, Sohrabi B and Beheshtian J 2016 *Phys. Chem. Chem. Phys.* **18** 7293
[33] Kunisada Y, Kishi H, Dimas F, David M Y, Nakanishi H, Kasai H, Asari T and Hayashi S 2010 *Jpn. J. Appl. Phys.* **49** 02BB04
[34] Cortés-Arriagada D and Toro-Labbé A 2016 *RSC Adv.* **6** 28500
[35] Cortés-Arriagada D and Toro-Labbé A 2016 *Appl. Surf. Sci.* **386** 84
[36] Hizhnyi Y, Nedilko S, Borysiuk V and Shyichuk A 2017 *Nanoscale Res. Lett.* **12** 71
[37] Saha B and Bhattacharyya P K 2019 *Comput. Theor. Chem.* **1147** 62

[38] Snyder S A, Vanderford B J and Rexing D J 2005 *Environ. Sci. Technol.* **39** 4586
[39] Bürgi H, Schaffner T and Seiler J P 2001 *Thyroid* **11** 449
[40] Butler R A Y, Godley A, Lytton L and Cartmell E 2005 *Crit. Rev. Env. Sci. Tec.* **35** 193
[41] Gamez P, Mooibroek T J, Teat S J and Reedijk J 2007 *Acc. Chem. Res.* **40** 435
[42] Lucas X, Bauzá A, Frontera A and Quiñonero D 2016 *Chem. Sci.* **7** 1038
[43] Frontera A 2013 *Coord. Chem. Rev.* **257** 1716
[44] Tuo D-H, Liu W, Wang X-Y, Wang X-D, Ao Y-F, Wang Q-Q, Li Z-Y and Wang D-X 2019 *J. Am. Chem. Soc.* **141** 1118
[45] Kim D Y, Singh N J and Kim K S 2008 *J. Chem. Theory Comput.* **4** 1401
[46] Zhao Y and Truhlar D G 2011 *Chem. Phys. Lett.* **502** 1
[47] Zhao Y and Truhlar D G 2008 *Theor. Chem. Acc.* **120** 215
[48] Grimme S, Antony J, Ehrlich S and Krieg H 2010 *J. Chem. Phys.* **132** 154104
[49] Goerigk L and Grimme S 2011 *Phys. Chem. Chem. Phys.* **13** 6670
[50] Weigend F and Ahlrichs R 2005 *Phys. Chem. Chem. Phys.* **7** 3297
[51] Bauza A, Quinonero D, Deya P M and Frontera A 2013 *J. Phys. Chem. A* **117** 2651
[52] Peterson K A, Figgen D, Goll E, Stoll H and Dolg M 2003 *J. Chem. Phys.* **119** 11113
[53] Frisch M J *et al.* 2009 Gaussian 09 (revision E.01) Gaussian Inc.Wallingford, CT
[54] Boys S F and Bernardi F 1970 *Mol. Phys.* **19** 553
[55] Reed A E, Weinstock R B and Weinhold F 1985 *J. Chem. Phys.* **83** 735
[56] Reed A E, Curtiss L A and Weinhold F 1988 *Chem. Rev.* **88** 899
[57] Jeziorski B, Moszynski R and Szalewicz K 1994 *Chem. Rev.* **94** 1887
[58] Hohenstein E G and Sherrill C D 2010 *J. Chem. Phys.* **133** 104107
[59] Parrish R M, Hohenstein E G and Sherrill C D 2013 *J. Chem. Phys.* **139** 174102
[60] Parker T M, Burns L A, Parrish R M, Ryno A G and Sherrill C D 2014 *J. Chem. Phys.* **140** 094106
[61] Turney J M *et al.* 2012 *WIREs Comput. Mol. Sci.* **2** 556
[62] Johnson E R, Keinan S, Mori-Sánchez P, Contreras-García J, Cohen A J and Yang W 2010 *J. Am. Chem. Soc.* **132** 6498
[63] Lu T and Chen F 2012 *J. Comput. Chem.* **33** 580
[64] Sowa H 1995 *J. Solid State Chem.* **118** 378
[65] Santoro A and Siegel S 1960 *Acta Crystallogr.* vol 13 (Copenhagen: Munksgaard Int. Publ. Ltd.) pp. 1017
[66] Schulz H 1973 *Acta Crystallogr. B* **29** 2285
[67] Schneider H, Vogelhuber K M, Schinle F and Weber J M 2007 *J. Am. Chem. Soc.* **129** 13022
[68] Battaglia M R, Buckingham A D and Williams J H 1981 *Chem. Phys. Lett.* **78** 421
[69] Kim D Y, Geronimo I, Singh N J, Lee H M and Kim K S 2012 *J. Chem. Theory Comput.* **8** 274

Bond strength of hybrid ceramics: the effect of nanotechnology

T Anwar[1,*], A Mokhtar[2] and H Zaghloul[1]

[1] Fixed Prosthodontics Department, Misr International University. Km 28 Misr-Ismailia Road, Cairo, Egypt
[2] Fixed Prosthodontics Department, Cairo University. 1 Gamaa Street, Giza, Egypt

*Email: tarek.anwar@miuegypt.edu.eg

Abstract. To evaluate the effect of nanotechnology on bond strength of hybrid ceramics. Twelve CAD/CAM ceramic slices were divided into 3 groups (n=4) (Lava Ultimate [LU], Vita Enamic [VE], and IPS e.max [EM]). Each was subdivided into 2 subgroups (n=2) (hydrofluoric acid etching [HF] and Sandblasting [SB]). Each was subdivided into 2 subdivisions (n=1) (Scotchbond Universal [SbU] and Futurabond U [FbU]). Silane was applied for 1 minute. Five micro-cylinders of RelyX Ultimate were cemented on each slice. 5000 thermocycles were done. Micro-shear bond strength (μSBS) was performed using the universal testing machine. SEM analysis was used to identify the mode of failure and the effect of surface treatment. Surface roughness (Ra) was measured by contact profilometer. Three-way ANOVA was used to study the effect of ceramic material, surface treatment, and universal adhesive on μSBS. Bonferroni's post-hoc test was used when ANOVA test was significant ($P \leq .05$). Two-way ANOVA was used to study the effect of ceramic microstructure and surface treatment on surface roughness. LU showed the highest μSBS and the least Ra values. The nanostructure of LU enhanced its bonding properties and decreased its surface roughness. HF and silane application could be considered a reliable surface treatment for bonding of hybrid ceramics. The presence of silane in SbU, in addition to the separate silane application, was enhanced the μSBS of the sandblasted LU.

1. Introduction

Dental glass-ceramics attain high esthetics qualities, color stability and good biocompatibility to the oral tissues. However, they are brittle, liable to fracture and cause wear to the opposing enamel [1-3]. On the other hand, resin composite restorations are soft, compliant and do not cause wear to the opposing dentition but are subjected to their own wear and discoloration [1,2].

Hybrid ceramics are a recent class of dental ceramics having the advantages of both ceramics and resin composites, known as ceramic-reinforced polymer systems [4]. These hybrid materials are believed to be of higher flexibility and better fracture toughness while retaining less brittleness, rigidity and hardness than the original materials [1,5]. Hybrid ceramics could be considered as promising materials for indirect restorations. Not only for the resiliency, proper marginal integrity, and fewer laboratory steps, but also for their bonding performance.

Two main types of the hybrid ceramics are present in the dental market: polymer-infiltrated ceramic network material (PICN) and the resin nanoceramic (RNC), which is based on the nanotechnology [5]. Based on the literature, nanotechnology has played an important role in dental materials where the decrease of particles' size has enhanced several properties such as bonding, strength and color [3,5].

Computer-aided design/ Computer-aided manufacture (CAD/CAM) technology has produced those hybrid materials for different indications; inlays, onlays, laminate veneers, single crowns and fixed dental prostheses (FDPs) [6]. One of the criteria for success of these CAD/CAM materials is their strong and durable bond to the underlying substrates, whether enamel, dentin or a restorative material [4]. To obtain such a reliable bond, the literature recommends using resin cements [2,4-9]. Bonding by resin cements helps in increasing retention, allows better marginal adaptation and fracture resistance and decreases microleakage [9].

Several previous studie [9-11] researched the bond strength of resin cements to glass ceramics using different surface treatments. It has been widely approved that the most suitable surface treatment for glass ceramics is HF, followed by silane application then an adhesive bonding agent [9-14]. HF acid etching removes the glassy matrix of the ceramics network in order to expose the crystalline structure [13,15,16]. The silane-coupling agent is a bifunctional molecule that enhances a chemical reaction between the resin cement and the ceramic. It increases wettability and improves contact angle, which enhances adhesion [9,11,12,14,17].

Universal Adhesives contain all components required for the adhesive bonding in one product, including etchant, primer (such as 10-methacryloyloxydecyl dihydrogen phosphate (10-MDP)) and bonding resin. Some universal adhesives may contain silane in their composition [6,18,19].

Minimum information is available regarding the effect of nanotechnology on the bond strength of hybrid ceramics. Thus, the purpose of this study was to evaluate the effect of nanotechnology on bond strength of hybrid ceramics using two surface treatment protocols and two universal adhesives on the μSBS of CAD/CAM hybrid ceramics compared to a lithium disilicate-based glass ceramic. The research hypotheses were that 1) LU would have better μSBS values than VE. 2) HF would be a better surface treatment protocol than SB, in terms of μSBS of CAD/CAM ceramics. 3) The MDP-containing universal adhesive would improve bond strength of CAD/CAM ceramics more than the MDP-free universal adhesive.

2. Materials and methods

All materials used in the present study are listed in table 1. The power analysis was for a 3 x 2 x 2 fixed effects analysis of variance; the first factor (ceramic type) included 3 levels, the second factor (surface treatment protocol) included 2 levels and the third factor (universal adhesive) included 2 levels. Based upon the results of Wahsh MM *et al.* [3], the effect size for the μSBS was found to be (0.6), using alpha (α) level of (5%) and Beta (β) level of (20%) i.e. power = 80%; the study included 5 samples per cell for a total of 60 samples. Sample size calculation was performed using IBM® SPSS® Sample Power® Release 3.0.1.

2.1. Sample grouping

Twelve CAD/CAM ceramic specimens were divided into 3 equal groups (n=4) according to the ceramic material (EM, VE and LU). Each group was further subdivided into two equal subgroups (n=2) according to the surface treatment protocol (HF and SB). Each subgroup was further subdivided into two equal subdivisions (n=1) according to the universal adhesive (SbU and FbU).

Table 1. List of brand names, material descriptions, manufacturers, and lot numbers used in this study.

Brand Name	Material Description	Manufacturer	Composition	Lot Number
IPS e.max CAD	Lithium disilicate glass ceramic CAD/CAM block HT A2/C14	Ivoclar Vivadent AG, Schaan, Liechtenstein	SiO_2, Li_2O, K_2O, P_2O_5, ZrO_2, ZnO, other oxides, coloring oxides	S50380
Vita Enamic	Polymer-infiltrated ceramic network CAD/CAM block 1M1-T/ EM-14	Vita Zahnfabrik, Bad Säckingen, Germany	86 wt% feldspathic-based ceramic network, 14 wt% acrylate polymer network (infiltrated into feldspathic-based ceramic network)	41720
Lava Ultimate	Resin-nano ceramic CAD/CAM block A2-HT/14L	3M ESPE, St. Paul, MN, USA	Bis-GMA, UDMA, bis-EMA, TEG-DMA, 80% SiO_2 (20nm), ZrO_2 (4-11) nm, aggregated ZrO_2/ SiO_2 cluster	N666576
Scotchbond Universal	Universal Adhesive	3M ESPE, St. Paul, MN, USA	10-MDP, dimethacrylate resins, HEMA, methacrylate-modified polyalkenoic acid copolymer, nanofiller, ethanol, water, initiators, silane	598837
Futurabond U	Universal Adhesive	Voco, Cuxhaven, Germany	HEMA, bis-GMA, HEDMA, acidic adhesive monomer, urethane dimethylacrylate, catalyst, silica nanoparticles, ethanol	1544437
RelyX Ultimate Clicker	Adhesive Resin Cement	3M ESPE, St. Paul, MN, USA	10-MDP, dimethacrylate resins, HEMA, Vitrebond Copolymer, Filler, Ethanol, Water, Initiators, Silane	591508

Bis-GMA: bisphenol A-glycidyl dimethacrylate; TEGDMA, triethyleneglycol-dimethacrylate; UDMA: urethane dimethacrylate. MDP: 10-methacryloyloxydecyl dihydrogen phosphate; HEMA: hydroxyethyl methacrylate

2.2. Ceramic specimen preparation

Three CAD/CAM ceramic blocks were used in this study. Each was cut into 4 rectangular specimens of 2 mm thickness using a low-speed cutting saw (Isomet 4000 Linear Precision Saw, Buehler) under continuous water-cooling [20] and was confirmed using a digital caliper (Mitutoyo, Kanagawa Prefecture). The EM group was fired following the crystallization program recommended by the manufacturer. For ease of handling and support, each ceramic specimen was embedded in a block of cold cure acrylic resin (Acrostone, Egypt). 9% HF (Ultradent Porcelain Etch, Ultradent) was applied on the HF subgroup for 60 s, rinsed for 60 s, and dried for 30 s [21]. The surfaces of the SB subgroup were abraded with 50-µm aluminum oxide (Al_2O_3) particles in a dental sandblasting unit (Renfert, Germany) at a 2 bar pressure (30 psi) at a distance of 10 mm for 15 s each [14]. The nozzle of the sandblaster was fixed at a 90° angle to the ceramic surface and applied in a circular motion, until the entire surface appeared matte [22]. All sandblasted specimens were ultrasonically cleaned (Toption Digital Ultrasonic Cleaner, Shaanxi, China) with distilled water for 180 s, and dried for 30 s [23]. Both subgroups received an application of silane coupling agent (Ultradent Silane, Ultradent) with a brush and left to dry for 60 s, following the manufacturer's recommendation. Each universal adhesive was applied by a microbrush, scrubbed for 20 s covering the entire ceramic surface, and dried for 5 s, following the manufacturer's recommendation.

For μSBS test, five transparent polyethylene microtubules were cut from a 6 FG nelaton catheter using a sharp blade, of internal diameter of 0.9 mm and 1 mm height [3]. They were placed on each ceramic surface giving a total of 60 microtubules. RelyX Ultimate resin cement was packed into each microtubule until it was completely filled. A glass microscopic slide was gently pressed on their top surfaces to remove excess cement and to obtain a flat surface. All resin cement-filled microtubules were light cured with a dental light cure unit of 1000 mW/cm^2 (Woodpecker LED D Curing Light, Woodpecker, Guilin, China) for 3 s of tack curing following the manufacturer's recommendation. A sharp dental explorer was used to remove excess cement around each microtubule. Then, they were light cured for 40 s, following the manufacturer's recommendation [24]. After light curing, the microtubules were cut using a surgical blade size 15 leaving the resin cement micro-rods properly bonded on the ceramic surface. The same operator performed all procedures using dental loupes of magnification 2.5X (Univet, Italy).

A pilot study demonstrated that air bubbles might be incorporated in the microtubules during resin cement packing. So all specimens were examined under stereomicroscope (Nikon Eclipse MA100, Nikon, Tokyo, Japan) at magnification of 10X to identify if any air bubbles occurred. Only one micro-cylinder was found to be defective and was replaced by another.

2.3. Thermocycling

All specimens were subjected to thermocycling in distilled water in a thermocycling unit (Julabo FT200, Julabo, Seelbach, Germany) for 5000 cycles between -5°C and 55°C with a dwell time of 20 s and transfer time of 7 s [25].

2.4. Micro-shear bond strength test

Each acrylic block was mounted on the lower jig of the universal testing machine (Instron 3345, Instron Corporation, MA, USA). An orthodontic wire (0.2 mm diameter) was used to loop around the ceramic surface/resin cement micro-rod interface (figure 1). Micro-shear force was applied on each micro-rod, with the crosshead speed at 0.5mm/min until fracture occurred. μSBS was calculated by dividing the maximum load at failure (N) over the bonded surface area (mm^2) and recorded in megapascals (MPa).

2.5. Scanning electron microscopy (SEM) analysis

2.5.1. Mode of failure analysis. After μSBS test, the fractured interfaces of the specimens were analyzed using SEM at 100X Magnification to determine the mode of failure. The failure mode was classified as cohesive failure (within resin cement or ceramic), adhesive failure (at resin cement/ceramic interface), or mixed failure (both cohesive and adhesive).

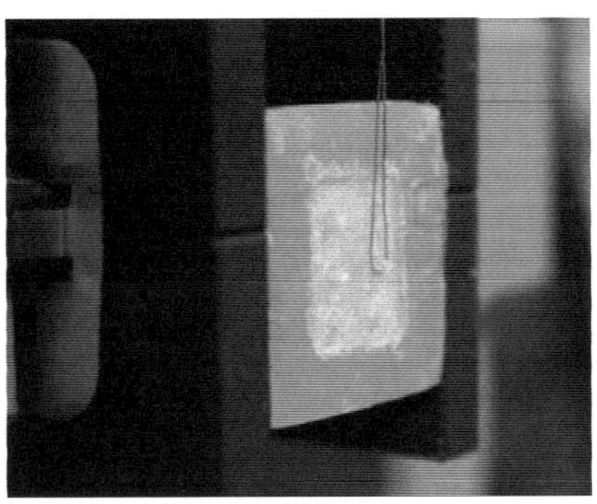

Figure 1. Representative ceramic specimen embedded in acrylic block mounted on universal testing machine and having orthodontic wire looping around resin cement micro-cylinder.

2.5.2. Effect of surface treatment. To identify the effect of surface treatment protocols on the ceramic surfaces, three ceramic specimens of each group were subjected to Scanning Electron Microscopy (SEM) (Quanta 250 FEG, ThermoFisher Scientific, MA, USA), after fine gold-sputtering in a sputter coater (Emitech K550X Sputter Coater, Quorum Technologies, Kent, UK) [24]. The SEM analysis was carried out before and after different surface treatment protocols at magnification 2000x.

2.5.3. Surface roughness analysis. Three representative ceramic specimens of each material were used to measure the surface roughness (Ra) before and after different surface treatment protocols using contact profilometer (Form Talysurf 60, Taylor Hobson, Ametek, Berwyn, PA, USA). The mechanical stylus moved in a linear motion for 5 mm distance at 3 random areas on each ceramic surface. The profiler was calibrated after every 3 specimens by testing the Ra of a standard pretested specimen [26].

2.5.4. Statistical analysis. Data were explored for normality by checking the data distribution and using Kolmogorov-Smirnov and Shapiro-Wilk tests. Data showed parametric distribution.

Three-way analysis of variance (ANOVA) was used to study the effect of ceramic type, surface treatment protocol, universal adhesive and their interaction on mean μSBS. Bonferroni's post-hoc test was used for pair-wise comparisons when ANOVA test is significant. Two-way ANOVA was used to study the effect of ceramic type, surface treatment protocol and their interaction on mean surface roughness (Ra). The significance level was set at $P \leq .05$. Statistical analysis was performed with IBM® (IBM Corporation, NY, USA) SPSS®* Statistics Version 20 for Windows (SPSS, Inc., an IBM Company).

3. Results

3.1. Micro-shear bond strength results

Three-way ANOVA results (table 2) showed that ceramic type, surface treatment and universal adhesive had a statistically significant (P <0.001) effect on the mean μSBS values. The interaction between the three variables had no statistically significant effect on the mean μSBS values (P=0.910).

The mean, standard deviation (SD) values and results of three-way ANOVA test for comparison between μSBS of different interactions are present in table 3. The results of the present study showed that LU had the highest statistically significant μSBS values (P<0.001), either with HF or SB and using SbU or FbU, followed by VE and EM. Also, HF showed the highest statistically significant μSBS values (P<0.001). SbU showed a higher statistically significant μSBS value than FbU with the sandblasted LU subgroup only (P=0.007). While no significant difference was evident between both universal adhesives for the remaining subgroups (P>0.05).

Table 2. Three-way ANOVA results for the effect of different variables on mean micro-shear bond strength.

Source of variation	Sum of Squares	df	Mean Square	*F*-value	*P*-value	*Partial Eta Squared (Effect size)*
Ceramic type	1079.1	2	539.6	148.6	<0.001*	0.861
Surface treatment	481.6	1	481.6	132.6	<0.001*	0.734
Universal adhesive	15.8	1	15.8	4.3	0.043*	0.083
Ceramic type x Surface treatment x Universal adhesive interaction	0.7	2	0.3	0.1	0.910	0.004

*df: degrees of freedom = (n-1), *: Significant at P ≤ 0.05*

Table 3. The mean, standard deviation (SD) values and results of three-way ANOVA test for comparison between micro-shear bond strength of the different interactions.

Surface treatment	Universal Adhesive	Vita Enamic		Lava Ultimate		IPS e.max		*P*-value *(Between ceramic types*
		Mean	SD	Mean	SD	Mean	SD	
Sandblasting	Scotchbond	8.08 ^{BG}	2.67	12.53^{ADG}	2.68	0.49 ^{CG}	0.25	<0.001*
	Futurabond U	6.39 ^{BG}	0.60	9.10 ^{AEG}	3.17	0.25 ^{CG}	0.08	<0.001*
	P-value *(Between Universal Adhesives)*	0.165		0.007*		0.842		
Hydrofluoric acid	Scotchbond	10.63 ^{BF}	1.42	18.29 ^{AF}	1.81	6.89 ^{CF}	1.88	<0.001*
	Futurabond U	10.57 ^{BF}	1.70	16.85 ^{AF}	2.49	7.60 ^{CF}	0.92	<0.001*
	P-value *(Between Universal Adhesives)*	0.961		0.238		0.558		
P-value *(Between surface treatments)*	Scotchbond	0.040*		<0.001*		<0.001*		
	Futurabond U	0.001*		<0.001*		<0.001*		

**: Significant at $P \leq 0.05$*
A, B, C superscripts in the same row indicate significant differences between ceramic types.
D, E superscripts in the same column indicate significant differences between universal adhesives
F, G superscripts in the same column indicate significant differences between surface treatments

3.2. SEM Analysis

3.2.1. Mode of Failure Analysis. The percentages of failure modes of all the tested groups are presented in table 4. Adhesive failure was predominant in all ceramic specimens (100%), except for the sandblasted LU, which showed 80 % adhesive failure and 20 % mixed failure (figure 2). Figure 3 shows representative SEM images of the mode of failure of the tested ceramic surfaces after μSBS tests.

Table 4. Mode and Frequency (Percentage) of Failure.

Surface Treatment	Universal Adhesive	IPS e.max (n=5)		Vita Enamic (n=5)		Lava Ultimate (n=5)	
Hydrofluoric acid	Scotchbond	A	5 (100)	A	5 (100)	A	5 (100)
	Futurabond U	A	5 (100)	A	5 (100)	A	5 (100)
Sandblasting	Scotchbond	A	5 (100)	A	5 (100)	A	4 (80)
						M	1 (20)
	Futurabond U	A	5 (100)	A	5 (100)	A	4 (80)
						M	1 (20)

A: Adhesive failure. M: Mixed failure.

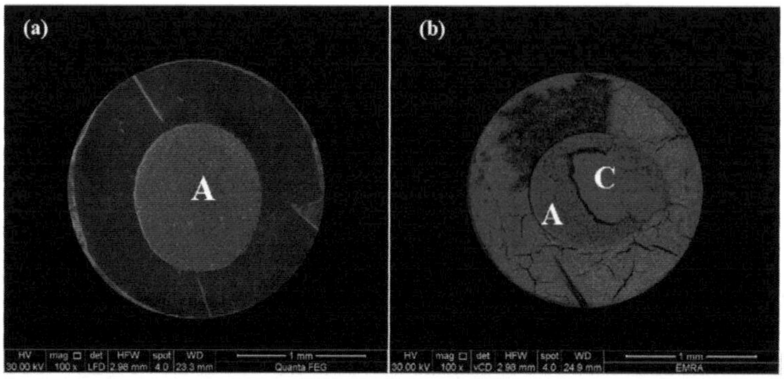

Figure 2. Representative SEM images showing the mode of failure. (a): adhesive failure. (b): mixed failure. *A: adhesive layer. C: Cement.* At 100X Magnification

3.2.2. Effect of Surface Treatment. The SEM analysis of the effect of surface treatment protocols on different ceramic materials is presented in figure 3. SEM analysis of different ceramic surfaces under different surface treatments shows the impact of both the HF and SB. It displays the controlled effect of smooth and uniform action of HF, while on the other side; SB shows haphazard and random bombardment on all ceramic surfaces creating cracks and irregularities (figure 3).

3.3. Surface roughness (Ra) analysis

Two-way ANOVA results (table 5) showed that ceramic type, surface treatment, and the interaction between the two variables had a statistically significant (P<0.001) effect on mean (Ra) values. Since the interaction between the variables is statistically significant, so the variables are dependent upon each other.

Table 5. Two-way ANOVA results for the effect of different variables on mean (Ra)

Source of variation	Sum of Squares	df	Mean Square	*F*-value	*P*-value	*Partial Eta Squared (Effect size)*
Ceramic type	0.549	2	0.274	50.416	<0.001*	0.691
Surface treatment	0.232	2	0.116	21.331	<0.001*	0.487
Ceramic type x Surface treatment interaction	0.539	4	0.135	24.766	<0.001*	0.688

*df: degrees of freedom = (n-1), *: Significant at P ≤ 0.05*

Table 6. The mean, standard deviation (SD) values and results of two-way ANOVA test for comparison between (Ra) of the different interactions

Surface treatment	Vita Enamic		Lava Ultimate		PS e.max		*P*-value *(Between ceramic types)*
	Mean	SD	Mean	SD	Mean	SD	
No treatment	0.43^{BE}	0.02	0.28^{C}	0.03	0.63^{AD}	0.12	<0.001*
Sandblasting	0.49^{E}	0.08	0.40	0.06	0.46^{E}	0.10	0.069
Hydrofluoric acid	0.87^{AD}	0.06	0.33^{C}	0.02	0.56^{BD}	0.06	<0.001*
P-value *(Between surface treatments)*	<0.001*		0.080		0.001*		

**: Significant at P ≤ 0.05*

A, B, C superscripts in the same row indicate significant differences between ceramic types.

D, E, F superscripts in the same column indicate significant differences between surface treatments

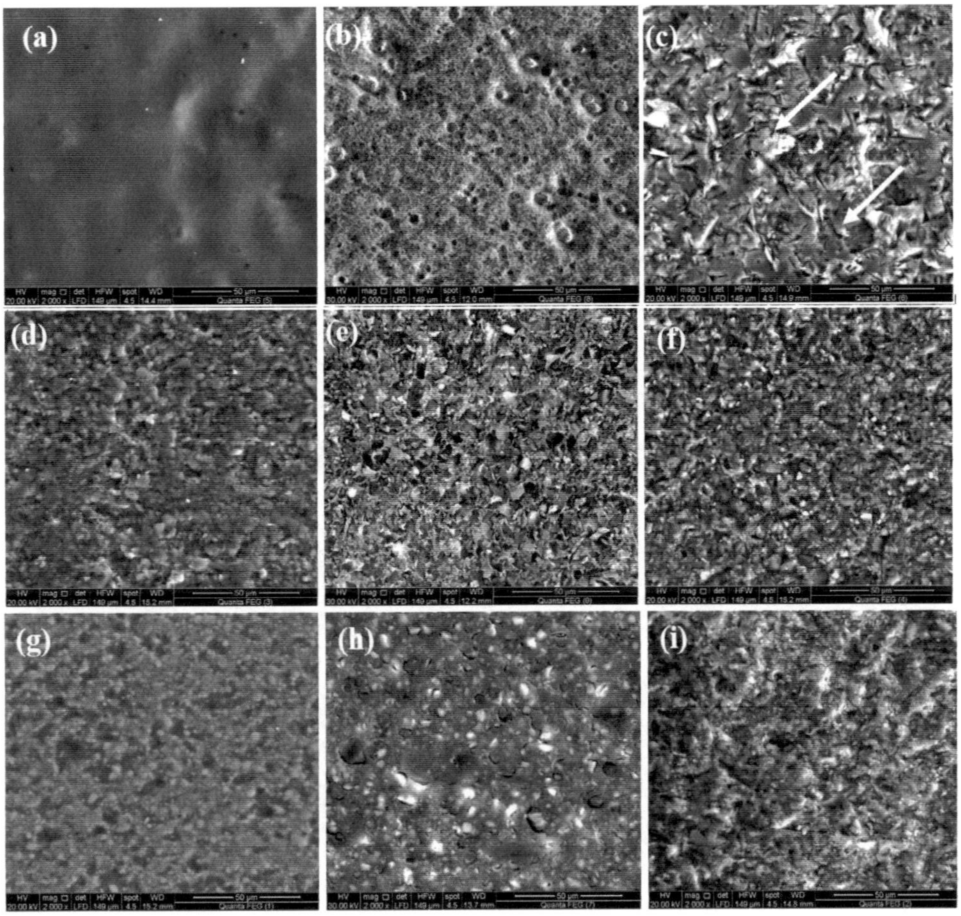

Figure 3. SEM images of different ceramics under different surface treatments at 2000X. *(a-c):* EM, *(a)*: No treatment. Note the homogeneity and uniformity of glass matrix. *(b)*: 9% HF. Note the uniformly distributed needle-like shaped crystals and porous glass matrix. *(c)*: SB. Note the more persistent crystalline fillers. Matrix is more affected with the present micro-cracks and flaws (white arrows). *(d-f):* VE, *(d)*: No treatment. Note the parallel oriented prismatic-shaped crystals. *(e):* 9% HF. Note the non-uniform distribution of crystalline phase. *(f):* SB Note the various sizes of prismatic-shaped crystals, surface affected by sandblasting. *(g-i):* LU, *(g):* No treatment. Note the well-distributed nanoparticles in the polymeric matrix. *(h):* 9% HF. Note the silica (dark) and the zirconia (light) nanoparticles. Flaws are present on the surface. *(i):* SB Note the zirconia crystals in the irregular and rough polymeric matrix (dark).

The mean, standard deviation (SD) values and results of two-way ANOVA test for comparison between (Ra) of the different interactions are presented in table 6. It was shown that LU had the lowest statistical significant ($P<0.05$) mean Ra value for both groups: no surface treatment and HF. While for the SB subgroup, there was no statistically significant difference ($P= 0.069$) between the ceramic materials. For the surface treatment protocols, HF showed the highest statistical significant ($P<0.05$) mean Ra value for VE and EM subgroups. While for the LU, there was no statistically significant difference between surface treatments ($P= 0.080$). Surface roughness analysis using contact profilometer is shown in figure 4 representing the effect of different surface treatments on each ceramic material.

(a)
(b)
(c)
(d)
(e)
(f)
(g)

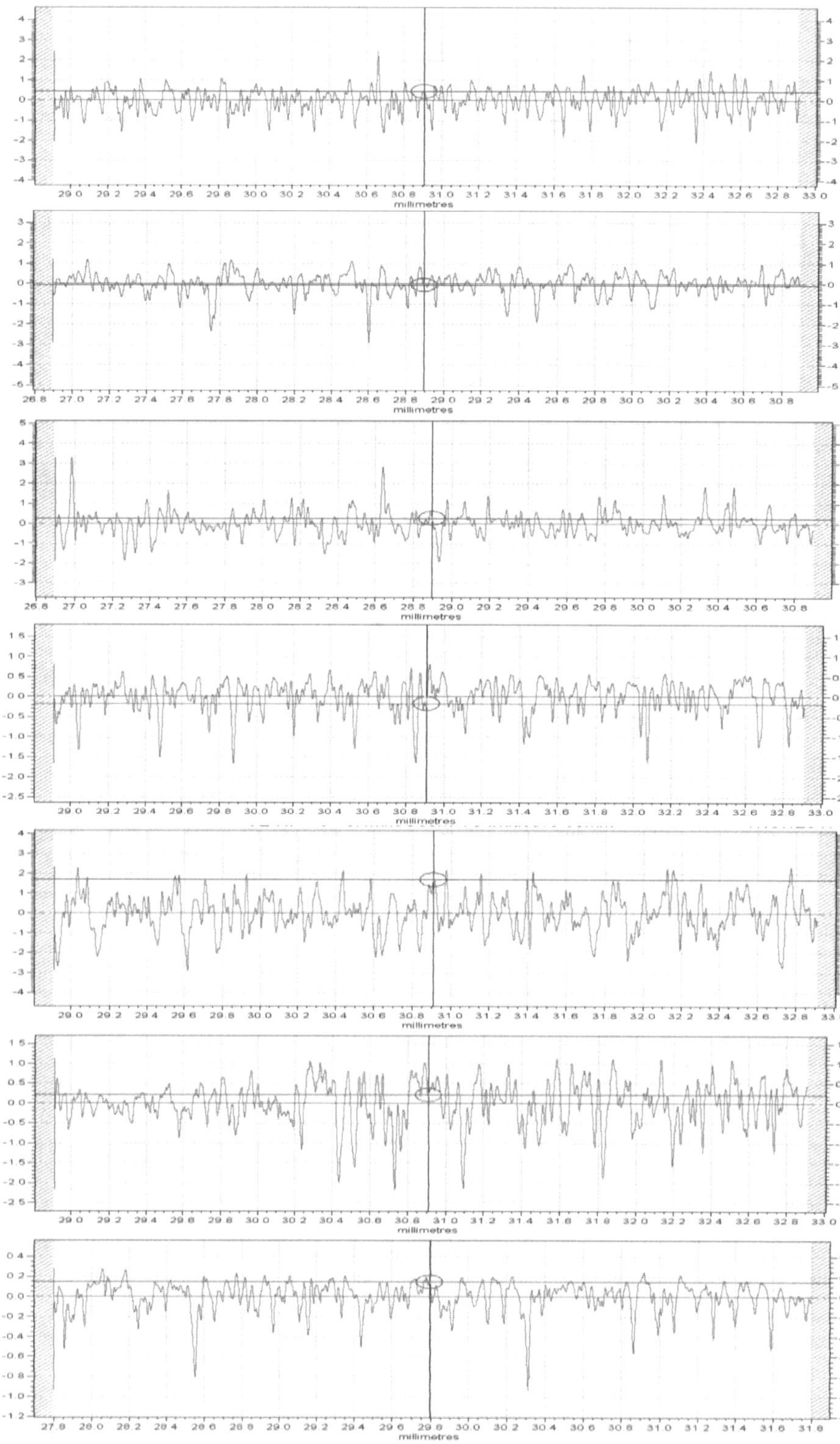

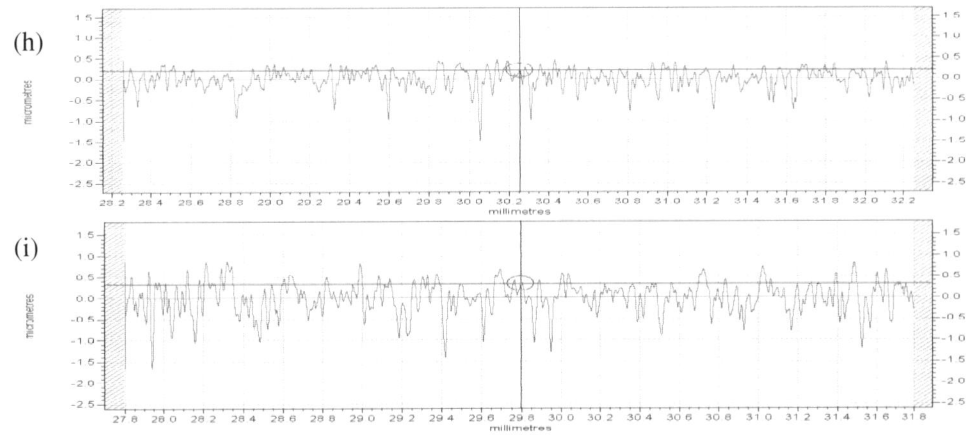

Figure 4. Surface roughness analysis using contact profilometer of different ceramics surfaces after different surface treatments. *a:* EM without surface treatment. *b:* EM after HF. *c:* EM after SB. *d:* VE without surface treatment. *e:* VE after HF. *f:* VE after SB. *g:* LU wtihout surface treatment. *h:* LU after HF. *i:* LU after SB.

4. Discussion

Nano-technology, adhesive dentistry and CAD/CAM era have opened doors for construction of new materials, restorations and introduction of minimal invasive dentistry [27-29]. Resin-ceramic bond increases fracture resistance of both tooth and restoration itself, decreases micro-leakages, overcomes the internal flaws of the material and provides further retention and durability [9,30]

The aim of the present study was to evaluate the μSBS of CAD/CAM ceramics, focusing on different surface treatment protocols and universal adhesives. According to the results obtained, the first and second hypotheses of the present study were accepted. The nanotechnology enhanced bonding properties of LU than VE. Both surface treatment protocols were statistically significant, where HF showed higher μSBS values than SB for all specimens, the third hypothesis was partially accepted as there was no statistical significance between the two tested universal adhesives except for sandblasted LU, the MDP-containing SbU showed statistically significant higher μSBS values than the non-MDP containing FbU.

For the ceramic type, LU showed significantly the highest mean μSBS value, followed by VE and EM. This finding was in accordance with those of *Al-Harbi et al* [26] and *Fasbinder and Neiva* [31]. They concluded that the shock-absorbing effect present in hybrid ceramics (LU and VE) was an advantage over the glass ceramic (EM). It could be postulated that the nanotechnology of the microstructure of LU played an important role in its bonding performance. The zirconia-silica nanoclusters, present in the highly cross-linked resinous matrix, were able to distribute stresses equally beside the resilient effect of the resin matrix [26,31].

In the present study, SB presented a significantly lower mean μSBS value than HF. This came in agreement with those of *Valian and Moravej-Salehi* [22], *Campos et al* [32] and *Menees et al* [33]. They clarified that HF formed homogenous porosities of uniform size. On the other hand, SB led to surface roughness on the ceramic surface, but could also lead to increased stress concentrations due to rough and sharp peaks, voids formation, and inhibit wettability between the ceramic and the bonding agent, besides micro-fractures at the surface [22,32,33]. The results of the present study were verified by SEM analysis for specimens treated with HF (figures 4-6).

The manufacturer of LU does not recommend HF of 5%, as it is not sufficient to roughen the surface. Interestingly, in the present study, LU showed higher μSBS values with HF rather than the recommended SB protocol. A possible justification for this result is the presence of nanomer silica particles and nanoclusters of silica/zirconia within the LU, which can be etched if a high concentration of HF is used for adequate duration as 60 seconds. This was in agreement with *Elsaka* [5], who found the same result in her study on the surface treatment effect on micro-tensile bond strength of CAD/CAM hybrid ceramics.

However, this result came in disagreement with that of *Duzyol et al* [34], who studied the effect of different surface treatments on the micro-tensile bond strength between LU, EM and CEREC blocks. They found that SB gave better bond strength values than HF. This might have been due to the low concentration of 5% of HF etch used in that study.[64] This point requires further investigations regarding the effect of HF on mechanical properties, as fracture resistance, flexural strength, and physical properties as color, of LU.

In the present study, sandblasted EM specimens showed the least μSBS values compared to the other groups. This came in accordance with that of *Ozdemir and Aladag* [35]. They found that SB caused damaging effect on the ceramic surface. Their SEM analysis showed increasing irregularities and micro-flaws on the surface, which could decrease bond strength values [35]. This coincided with the SEM analysis of EM surface after SB in the present study. It revealed micro-cracks and internal flaws formation due to the bombardment of SB. Irregular crystalline fillers were more persistent with various sizes (figures 4-6).

Several authors [6,23,25,36] have studied the performance of different universal adhesives on bond strengths of different CAD/CAM ceramics. They found that the MDP functional monomer might not be able to interact with the crystalline phase of the glass ceramics to produce stable bonds [36]. This explains the non-significant difference of μSBS values between the universal adhesives of the present study with the glass ceramic. They also concluded that the presence of silane could possibly enhance the chemical adhesion between the adhesive and the ceramic, which will accordingly improve the bond strength [23,25,36]

In the current study, in addition to the separate silane application, SbU contains silane incorporated in its composition, which could increase chemical adhesion between LU, which also contains silane in its micro-structure, and the resin cement and hence, increase the μSBS [23,25,36]. Also, the 10-MDP functional monomer found in SbU might have bonded to the zirconia nanomers and zirconia-silica nanoclusters found in LU, owing to the high affinity of zirconia to the phosphate ester of 10-MDP leading to its higher μSBS values than the MDP-free FbU.

The failure mode results of the present study showed predominance of adhesive failure. This came in accordance with the results of a review by *Sirisha et al.* [37], who stated that the nature of the μSBST commonly results in majority of adhesive failures. The result of the present study was also in agreement with a study conducted by *Sundfeld et al.* [38], who also found predominance of adhesive failure in most of the specimens. They stated that adhesive failures do not always indicate poor quality bond, but instead, they imply that the interfacial bond strength has been truly and accurately assessed [38].

The surface roughness results of the current study showed that LU had the least significant surface roughness. This came in agreement with the results of *Elsaka* [5] and *Fasbinder and Neiva* [31] who compared the surface roughness between LU and VE using different surface treatments. Both studies claimed that the nanosize filler particles of LU aided in the low inherent Ra values compared to other ceramics. Surface roughness depended greatly on the inherent crystalline structure of the material [5,31].

The findings also came in accordance with several studies [15,32,38,39,40] that demonstrated the positive correlation between HF and surface roughness of ceramics. These studies concluded that HF was able to modify the ceramic surfaces by forming irregularities in a controlled manner, selectively removing the glass matrix and exposing the crystalline fillers [15,32,38,39]. On the other hand, SB could have possibly flattened the pre-existing irregularities present on the unpolished CAD/CAM surfaces, leading to the non-significant difference with untreated specimens [40] Surprisingly, LU showed statistically the lowest significant mean (Ra) value, yet it showed statistically the highest significant mean μSBS value. This came in agreement with studies conducted by *Elsaka* [5], *Iorizzo et al.* [41] and *Kallio et al.* [42]. Their results showed that as Ra increased, this could impede wettability, inhibit intimate contact and increase stress concentrations. Finally, they concluded that high Ra values do not necessitate high bond strength values [5,41,42].

Micro-shear bond strength test was used in the present study due to the small bonded surface area to be tested [37]. Also, it is considered a more economical method than other tests. Two main reasons behind the use of μSBS are that first; the specimens are only stressed during the mold removal by a sharp surgical blade. Second; each ceramic specimen yields more than one micro-cylinder, which will provide simultaneous and similar factors on each one yet are considered different readings. In the present study, five micro-cylinders were bonded on each ceramic specimen (n=12) giving a total of 60 different readings. Orthodontic wire loop was used in the test, since a knife-edge chisel can lead to more severe stress concentration at the load application area [43].

Several limitations could be found in this study. This is an in-vitro study, which does not explain or resemble the exact situation clinically. Yet, it gives an idea of the behavior of the tested CAD/CAM materials with respect to different surface treatment protocols and universal adhesives tested. Only one group of HF was used with no variations in either the concentration or the duration of etching. Last, there were no tested specimens before thermocycling to evaluate the effect of thermal aging on decreased bond strength values.

Further studies are needed to assess the effect of the tested different surface treatment protocols on the mechanical and physical properties of CAD/CAM hybrid ceramics. Also, variations in universal adhesives and resin cements could be beneficial in future studies to understand the action of each material on the bond strength obtained. Last but not least, more clinical studies are required and needed to reach a proper surface treatment protocol and suitable choice of universal adhesives for the newly introduced CAD/CAM ceramics.

5. Conclusions

Within the parameters used and the limitations of this study, the following could be concluded: the nanostructure of LU enhanced its bonding properties and surface roughness. HF and silane application could be considered as a reliable surface treatment protocol for bonding of CAD/CAM hybrid ceramics. The presence of silane in SbU, in addition to the separate silane application, was able to enhance the μSBS of the sandblasted LU.

References

[1] Coldea A, Swain M V and Thiel N 2013 *Dent. Mater.* **29** 419
[2] Mante F K, Ozer F, Walter R, Atlas A M, Saleh N, Dietschi D and Blatz M B 2013 *Compend. Contin. Educ. Dent.* **34** 2
[3] Wahsh M M and Ghallab O H 2015 *Tanta. Dent. J.* **12** 178
[4] Duarte S, Sartori N, Cascione D and Phark J 2014 *Quintessence Dent. Technol.* 36
[5] Elsaka S E 2014 *J. Adhes. Dent.* **16** 531
[6] Zaghloul H, Elkassas D W and Haridy M F 2014 *Eur. J. Dent.* **8** 44
[7] Technical and Scientific Documentation. Vita Enamic. 2013
[8] 3M ESPE. Technical Product Profile: Lava Ultimate. 2011
[9] Blatz M B, Sadan A and Kern M 2003 *J. Prosthet. Dent.* **89** 268
[10] Nagai T, Kawamoto Y, Kakehashi Y and Matsumura H 2005 *J. Oral. Rehabil.* **32** 598
[11] Spitznagel F A, Vuck A, Gierthmuehlen P C, Blatz M B and Horvath S D 2016 *Compend. Contin. Educ. Dent.* **37** 630
[12] Mainjot A K, Dupont N M, Oudkerk J C, Dewael T Y and Sadoun M J 2016 *J. Dent. Res.* **95** 487
[13] Borges G A, Sophr A M, De Goes M F, Sobrinho L C and Chan D C N 2003 *J. Prosthet. Dent.* **89** 479
[14] Park J-H and Choi Y-S 2016 *J. Adv. Prosthodont.* **8** 275
[15] Zogheib L V, Della B A, Kimpara E T and Mccabe J F 2011 *Braz. Dent. J.* **22** 45
[16] Torres S, Borges G, Spohr A, Cury A, Yadav S and Platt J 2009 *Oper. Dent.* **34** 399
[17] Matinlinna J P, Chem I and Özcan M 2004 *Int. J. Prosthodont.* **17** 155
[18] Isolan C P, Valente L L, Münchow E A, Basso G R, Pimentel A H, Schwantz J K, da Silva A V and Moraes R R 2014 *Appl. Adhes Sci.* **2** 25

[19] Alex G 2015 *Compend. Contin. Educ. Dent.* **36** 15
[20] Lise D, Perdigão J, Van Ende A, Zidan O and Lopes G 2015 *Oper. Dent.* **40** 524
[21] Güngör M B, Nemli S K, Bal B T, Ünver S and Doğan A 2016 *J. Adv. Prosthodont.* **8** 259
[22] Valian A and Moravej-Salehi E 2014 *J. Adv. Prosthodont.* **6** 387
[23] Siqueira F, Millán Cárdenas A, Gutiérrez Reyes M, Malaquias P, Hass V, Reis A, Loguercio A D, Perdigao J. 2016 *J. Adhes. Dent.* **18** 1
[24] Cekic-Nagas I, Ergun G, Egilmez F, Vallittu P K and Lassila L V J 2016 *J. Prosthodont. Res.* **60** 265
[25] Stawarczyk B, Krawczuk A and Ilie N 2015 *Clin. Oral. Investig.* **19** 299
[26] Al-Harbi F A, Ayad N M, ArRejaie A S, Bahgat H A and Baba NZ 2017 *J. Prosthodont.* **26** 432
[27] Davidowitz G and Kotick P G 2011 *Dent. Clin. North. Am.* **55** 559
[28] Fasbinder D J 2012 *Comp. Cont. Educ. Dent.* **33** 50
[29] Beuer F, Schweiger J and Edelhoff D 2008 *Br. Dent. J.* **204** 505
[30] Spitznagel F A, Horvath S D, Guess P C and Blatz M B 2014 *J. Esthet. Restor. Dent.* **26** 382
[31] Fasbinder D J and Neiva G F 2016 *J. Esthet. Restor. Dent.* **28** 56
[32] Campos F, Almeida C, Rippe M, de Melo R, Valandro L and Bottino M 2015 *Oper. Dent.* **40** E83
[33] Menees T S, Lawson N C, Beck P R and Burgess J O 2014 *J. Prosthet. Dent.* **112** 1164
[34] Duzyol M, Sagsoz O, Polat S N, Akgul N and Yildiz M 2016 *J. Prosthodont.* **25** 466
[35] Özdemir H and Aladağ L I 2017 *Biotechnol. Biotechnol. Equip.* **31** 815
[36] Cardenas A M, Siqueira F, Hass V, Malaquias P, Gutierrez M F, Reis A, Perdigao J and Loguercio A *J. Adhes. Dent.* **19** 1
[37] Sirisha K, Ravishankar Y, Ravikumar P and Rambabu T 2014 *J. Conserv. Dent.* **17** 420
[38] Sundfeld D, Correr-Sobrinho L, Pini N I, Costa A R, Sundfeld R H, Pfeifer C S, Martins L R 2016 *Braz. Dent. J.* **27** 727
[39] Sundfeld Neto D, Naves L Z, Costa A R, Correr A B, Consani S, Borges G A, Correr-Sobrinho L *Oper. Dent.* **40** 470
[40] Brum R, Mazur R, Almeida J, Borges G and Caldas D 2011 *Oper. Dent.* **36** 478
[41] Iorizzo GM, Prete F, Mazzanti B, Timellini G, Scotti R and Baldissara P 2014 *Dent. Mater.* **30** e24
[42] Kallio T T, Lastumäki T M, Lassila L V J and Vallittu P K 2014 *Acta. Odontol. Scand.* **72** 202
[43] Sirisha K, Rambabu T, Shankar Y and Ravikumar P 2014 *J. Conserv. Dent.* **17** 305

IOP Conf. Series: Materials Science and Engineering 1046 (2021) 012019 doi:10.1088/1757-899X/1046/1/012019

A novel merging solar parabolic collector with thermoelectric generator using geothermal energy

A E Atta[1], Nabila Shehata[2,*], Hamdy F M Mohamed[1,3] and Mohamed R. O. Ali[4]

[1]Renewable Energy Science & Engineering Department, Faculty of Postgraduate Studies for Advanced Science (PSAS), Beni-Suef University, P.O. Box 62511 Beni-Suef, Egypt
[2]Environmental Science and Industrial Development Department, Faculty of Postgraduate Studies for Advanced Science (PSAS), Beni-Suef University, P.O. Box 62511 Beni-Suef, Egypt. E-mail: nabila.shehata@psas.bsu.edu.eg
[3]Physics Department, Faculty of Science, Minia University, P.O. Box 61519 Minia, Egypt
[4]Mechanical Power Engineering& Energy Department, Faculty of Engineering, Minia University, Minia, Egypt

*E-mail: Nabila.shehata@psas.bsu.edu.eg

Abstract. Geothermal energy is probably the biggest wellspring of sustainable power sources. Compared with sunlight and wind power the geothermal power has numerous focal points, including being impenetrable to climate changes, having a steady base burden, and high heat productivity (for the high-temperature geothermal assets). In any case, the total capacity of installed geothermal energy plants falls behind sun based and wind. In this work, the geothermal power has been converged with solar energy. A test model concentrator thermoelectric generator utilizing an explanatory dish concentrator was created and tried. A sunlight-based gatherer has been utilized as the hotspot for the hot side of the thermoelectric module and the geothermal power has been applied as the cold wellspring for the cold side for the thermoelectric module. An electric flow as per the Seebeck hypothesis has been produced, as another wellspring of the sustainable power source. The capability of utilizing concentrated sunlight-based power age framework dependent on the thermoelectric module and geothermal power has been examined. The maximum voltage 11.7 V was obtained in the present work, when the maximum temperature for the hot side 93 °C and cold side 21°C.

1. Introduction

The recent energy crisis, the decay of petroleum derivatives the resulting environmental burdens, have gotten progressively critical, with solid consideration being energized the utilization of sustainable energy sources, for example, sunlight based, hydroelectric, wind and geothermal vitality. The International Energy Outlook 2016 (IEO2016) recorded significant development in worldwide energy requests over the 28-year time frame from 2012 to 2040. Worldwide requirement for vitality increments from 549 quadrillion British warm units (Btu) in 2012 to 629 quadrillion Btu in 2020 up to 815 quadrillion Btu in 2040, with 48% development 2040 compared to 2012 [1].

Sustainable energy sources are developing exceptionally quick as a wellspring of vitality for power age, with yearly expanding normal 2.9% from 2012 to 2040 [2]. So as to meet the future quickly developing power request and the natural issues of utilizing fossils fuels, enormous endeavors must be done in the improvement for sustainability of power supplies including wind, hydropower, sun-oriented and geothermal sources. These inexhaustible sources are foreseen to be the fastest creating to produce electric power. Their contribution to the worldwide mix is estimated to grow from 19% in 2008 to 23% in 2035. Over 82% of increment in hydroelectric and wind power in a similar timeframe. In perspective on the increasingly more stringent need to substitute fossil and nuclear fuels, it is especially critical to advance additionally sustainable power sources like solar, geothermal, thermoelectric module and wind energy [3].

Geothermal energy is probably the biggest wellspring of sustainable power source as per the World Energy Assessment [4]. Compared with solar and wind systems geothermal power has numerous points of interest, including being not to rely upon the climate, having a stable baseload, and high heat productivity if the geothermal asset has high-temperature. However, the maximum limit of installed geothermal power comes behind solar power and wind. In 2011, the introduced intensity of photovoltaic (PV) and wind were 70 and 240 GW, respectively. According to the Geothermal Energy Association (GEA), the all-out geothermal power introduced on the planet was about 11.2 GW as of May 2012. The normal yearly development pace of geothermal power is about 2%, while that of PV is about 58% during a similar time of 2006-2011 [5].

The geothermal energy has been utilized in bygone eras for washing, bathing, and for space heating. These days, it is generally utilized industrially for both creating power and direct uses. Worldwide, geothermal power plants generated about 10 GW of electricity in 2007, and in practice supply 0.3% of global electricity need. An extra 28 GW of direct geothermal power limit is utilized, hot spas, space warming, industrial processes, agricultural, and desalination applications [6]. The innovation utilized for producing power by utilizing the direct use of geothermal energy called the Organic Rankine Cycle (ORC). A remarkable example is the 250 kW ORC plant in Chena Hot Springs, Alaska, which produces electricity from very low temperature (74 °C) geothermal resource [7].

Sunlight based collector is one of the heat exchangers types that change sun-based radiation energy to inside heat control by using a medium. The most noteworthy portion of solar system is the sun-based collector. It is the part which absorbs incoming solar beam, converts it into heat, and moves this heat to a medium (usually water, air, or oil) flowing through the collector. The solar collectors are normally sorted into two classes as per focus proportions; non-focusing - and focusing - collectors [8]. Concentrating solar power (CSP) is a power age innovation that utilizations mirrors, lenses or focal points to focus the sun's beams and, in the vast majority of the present concentrating solar power systems, which it heats a liquid and produces steam. The steam drives a turbine and creates control similarly to ordinary power plants. Various thoughts are being examined and not all future CSP plants will in a general sense use a steam cycle [9]. A non-concentrating solar collector has the equivalent catching territory as its retaining zone while concentrated solar collector depends on the sun's tracking. It usually has inward reflecting surfaces to capture and center the sun-oriented light to a lot littler getting zone, resulting in an increased heat flux with the goal that the thermodynamic cycle can accomplish higher effectiveness when working under higher temperatures [8].

Most current thermal power generation technologies rely on the conversion of thermal energy into mechanical work before electricity is produced. A thermoelectric power module has no moving parts and is conservative, quiet, trustworthy, and clean. Thermoelectric generators (TEGs) productivity can be improved by expanding the temperature gap between cold and hot sides. The thermoelectric power cycle, with charge transporters (electrons) go about as the working medium, follows the fundamental laws of thermodynamics and firmly like the power cycle of a traditional heat engine [10]. Thermoelectric generators (TEGs) are semiconductor gadgets [11]. The benefits of TEGs are various like direct power transformation, not at all like many heat motors that first convert heat power into mechanical power and afterward convert this mechanical power into power utilizing an alternator. There is neither moving parts nor working liquids inside the TEG, consequently no maintenance and no additional costs. The TEG has a long-life expectancy, particularly when working with consistent heat sources. The TEG can be utilized for small scale age in exceptionally restricted spaces or to deliver kilowatts (no scale impact). Close to quiet activities, it appreciates any working position [12].

The low development pace of geothermal power was researched for the principle factors include high initial investment, high exploration risk, long payback, and construction time, trouble to evaluate asset, and trouble to modularize [13]. The main advantage of utilizing geothermal energy for direct use in the low temperature extends to moderate is more common and exist at economic drilling depths [14]. Also, portions of a geothermal could be utilized for instance as an underground heat power stockpiling system [15]. Despite TEGs efficiency is about 5–10%, taking into consideration the reuse of waste energy gain, their efficiency cannot be ignored [16].

2. State of art

Since Seebeck's discovery, various materials have been considered worthy of generating thermoelectricity. The first TEGs were built on electrical conductors and semiconductors, such as iron, bismuth, lead, antimony, zinc, copper and different alloys, Later, in the 20th century, numerous other thermoelectric materials (TMs) were created: composites, ceramics, etc. Though, the updated semiconductors still the basic TMs for the production of thermoelectric effects [17]. The effectiveness of a thermoelectric module relying on the temperature gaping between both gadget sides [18]. In present-day TEGs, at least two kinds of materials are typically utilized in one leg, to increase the proficiency of the TEG [19]. In solar thermoelectric generator (STEG), the hot side for TEG is the sun absorbs and the cold side is the warmth sink. The devices that utilized in sun-powered thermoelectric power age comprise of two sections. The first is the collector, which ingests sunlight-based radiation as thermal power. The second is a thermoelectric generator that utilizes the consumed thermal power to create flow current [20]. Geothermal resources are classified dependent on their lake temperatures alone [21]. The classic classification of the geothermal energy store temperature is low-enthalpy resource matches repository temperature under 150 °C and high-enthalpy resources are available if the temperature outperforms 150 °C [22].

Thermoelectric power generators have been considered as a good alternative clean technology because of their special advantages. Thermoelectric power gives a decent application in the immediate change of waste heat energy into electric power where it isn't essential to take the heat energy input cost into consider. The use of this innovation in changing over waste heat energy legitimately into electric power can likewise expand the general efficiencies of power transformation systems [10]. A small power generation system comprised of a sunlight-based concentrator (parabolic dish) and thermoelectric module. The sun-powered heat is gathered by the parabolic dish and concentrated on one side of the thermoelectric module to become (hot side). While the opposite side's temperature is kept lower as (cool side). The temperature gap is created between the hot side and the cool side produces electrical energy by the thermoelectric generator. This power will be put away into the battery. In this system, the most extreme temperature contrast came to around 202 °C and the maximum open voltage was about 3.8 V [23]. The productivity of a small power generation system comprising of thermoelectric module and sun oriented parabolic dish gatherer together to create the power was improved utilizing. The TEG performance was enhanced by the perfectly insulated heat sink and acrylic cover that encloses the receiver plate. There is a 2.11% expansion in all-out effectiveness for TEG with spread. The greatest voltage of the thermoelectric module was 4 V, which is 10.75% more than TEG without cover for the same sunlight radiation [24].

A small scale solar parabolic dish thermoelectric generator was investigated. Solar parabolic dish collector is made of satellite dish antenna fitted with a polished aluminum sheet as concentrator surface. It is made of satellite dish reception apparatus fitted with a cleaned aluminum sheet as concentrator surface. The thermoelectric generator is put on the focal plane with manual tracking for the parabolic dish. The concentrated sunlight-based radiation and the water that cooled heat sink was the driving potential to produce power. The operating factors such as receiver plate temperature, power output, and conversion efficiency are studied with respect to solar radiation. They inferred that beneficiary plate temperature, power output, and change proficiency are shifting legitimately with the sun-powered radiation [25]. A power generator manufactured which was provided by the thermostatic water circulator, imitation the geothermal. The cooling cycle was made out of a water pump and water holder.

Thermal couples for temperature measurements were installed at the input and output pipes. The values of voltage, current, and power were measured by electrical multimeter. The TEG system could produce more than 1 KW at a temperatures gap of around 120 °C. The prompt effectiveness of the TEG system arrived at 4.5% at an input temperature of around 95 °C on the hot side and a temperature of 30 °C on the cold side. This proficiency increment exponentially increases the input temperature [26]. Afterward, this module was improved by investigating the TEG temperature and mechanical structure. The output power was improved by 34.6% and the efficiency raised from 4.5 to 6.5% [27].

Fan et al. [28] created a model of concentrator TEG utilizing sun parabolic dish 180 cm diameter. The amassed system as the CTEG was able to produce electric of up to 5.9 W when the temperature distinction was 35 °C. A high open voltage was acquired from the thermoelectric generator utilizing sunlight power (57 V DC) when the temperature gapping was 67 °C. The system comprised of 8 emptied tube sun-oriented gatherers (4-series and 2 parallel) and 24 TEG associated arrangement. The absolute yield voltage is a total of TEG singular voltage [29]. Electric power was created from sunlight-based lake utilizing a blend of thermoelectric modules and thermosyphon. The short current and the open-circuit voltage esteems were 0.4 A and 26 V, respectively when the temperature gapping of 75 °C was kept up maintained across 16 (3D) thermoelectric cells. The TEG comprised of 3 layers (2 cool sides and one hot side). This type of TEG was more effective [30].

A small power generation system has been built manufactured utilizing joined the system, thermoelectric modules, and thermosyphon for producing power from low heat sources like a solar lake. The temperature gapping between the upper convective zone and the lower convective zone of sun-oriented lake is 20 °C produces 4.6 V and 0.12 A [31]. Various methods for utilizing thermoelectric power systems were checked on. The most effective path for improving thermoelectric power system execution is to use it with hybrid systems. The thermoelectric module can be utilized with, parabolic collectors and flat plate collectors or evacuated tube collectors to generate heat and electricity. Such a hybrid system increases the overall performance and efficiency of the thermoelectric power generation system which can be economic device [32]. The aim of this study is to develop and test a hybrid system based on geothermal energy for the purpose of increasing its efficiency. Three different models will be used by changing the hot and cool sides to improve the effectiveness of each side.

Electricity from solar energy was produced using a thermoelectric generator (TEG) system. Two-phase closed thermosiphon type (TPCT) was powered by the solar system. Heat a pipe is used to heat the hot surface of the TEG, and a 1200 cc capacity passive water-cooled system was implemented to cool the cold side using five TEGs. The system was operated by solar power under three different conditions: non-reflector, semi-reflector and full-reflector. Electrical and thermal data was recorded in a computer from 08:00 to 15:00. The highest difference in temperature between the hot- and cold-side was 88.9 °C. The highest open circuit voltage values in total were 4.38V, 7.53V and 8.20V [33]. Mismatch of thermoelectric materials is considered a reason for the low efficiency of conversion. Subsequently, choosing the material and optimizing the models is a significant step before implementation of large scale production. Comsol-Multiphysics (5.3) software was applied for this purpose. The results showed that the maximum simulated conversion efficiency was 9.2% at temperature gradient 500 K [34].

An analysis of a point-of-use thermoelectric generator was presented. The design, implementation and performance of the generator for powering electronic monitoring devices and charging batteries was discussed. using six Laird thermoelectric modules (Laird PB23 Series, HT8, 12). It was revealed that placing a hot side for TEG on a 160°C steam pipe with a 30°C ambient environment for a cold side for TEG (Δ T of 130°C), TEG produced 31.2 volts (V) open circuit and 0.89 amperes (A) short circuit [35]. Optimization model has been developed to simulate TEG based power generation system. TEG module has been chosen from the commercially available modules with efficiency of 1.91% for the targeted low-grade waste heat temperature of Th=90 °C and Tc=14 °C. Predicted maximum net power was 99 W from 50 TEG modules [36].

Zhang et al [37] found the peak output voltage and power values of the device in single-output mode range from 183.1 to 370.7 mV and 33.5 to 137.2 mW, respectively.

In this work, we aim to test a combination between a thermoelectric module with geothermal energy (low Temperature) as a cold side (GTEM), and connected it to a solar collector (parabolic dish) as a hot side (STEM). And to optimize this system.

3. Experimental set up

The geothermal system pilot-scale consists of the following components: The solar parabolic dish collector: A concentrate solar thermal energy on a focal line made of copper pipe filled with hot fluid (Therminol oil PV-1 or water) and connected with the hot side insulated storage tank a shown in figure 1(a). Two major internal tanks as shown in figure 1(b), the first tank contains the hot source and the other one contains the cooled water and geothermal system. Thermoelectric modules consist of 12 TEG, (BI2Te3) connected in series and located in the middle of the two separated insulated tanks TEGs connected with AVO-Meter to measure the output open voltage as shown in figure 1(c).

A geothermal system, using the ground as a source heat exchanger designed to cool the cooling medium (water) circulated in 2 parallel pipes made of PVC, buried in 1 m depth under sand soil and connected to the cold insulated tank a shown in figure 2. Ten calibrated Thermo-Couples are used to measure temperatures in the main points along the experiment starting with inputs of hot fluid that passing through the copper pipe in solar parabolic dish ending with cooling water passing through the soil temperatures. The temperature of these thermo-couples was measured by a digital data logger. Both solar radiation intensity and wind speed were also measured by (Digital LCD Wind Speed Meter Pyranomete). The experimental part of this work was carried in the Heat Engines Laboratory of Mechanical Power Engineering and Energy of premises of the Faculty of Engineering, Minia University, Egypt.

Figure 1. The main movable components of the system; (a) parabolic trough, (b) tanks assembly, and (c) TEG fixed on the bottom of the tank.

Figure 2. Photos of the geothermal system pilot scale.

IOP Conf. Series: Materials Science and Engineering 1046 (2021) 012019 doi:10.1088/1757-899X/1046/1/012019

The merged SPD/TEG/GHP system is built, flushed and cleaned as shown in figure 3. The ambient temperature is measured by a calibrated digital thermometer. The terminal oil is pumped into SPD for heating. The solar oil streams into the protected cylinder in the solar parabolic dish to fall into a tank to represent to the hot side of the TEG. On the other side, the tap is opened to the flow of water into the geothermal system to be cooled and then the water enters the insulated tube into the reservoir which represents the cold side for TEG. The measurement was recorded for a sunshine period 6 h from 9 a.m. till 3 p.m. Ten calibrated thermocouples are used to measure temperatures in the main points (input oil for SPD, output oil from SPD, Tank for hot side, Tank for cold side, input water for geothermal system, soil temperature for geothermal system and output water). These measurements are done by data Logger GL 240 every 30 min.

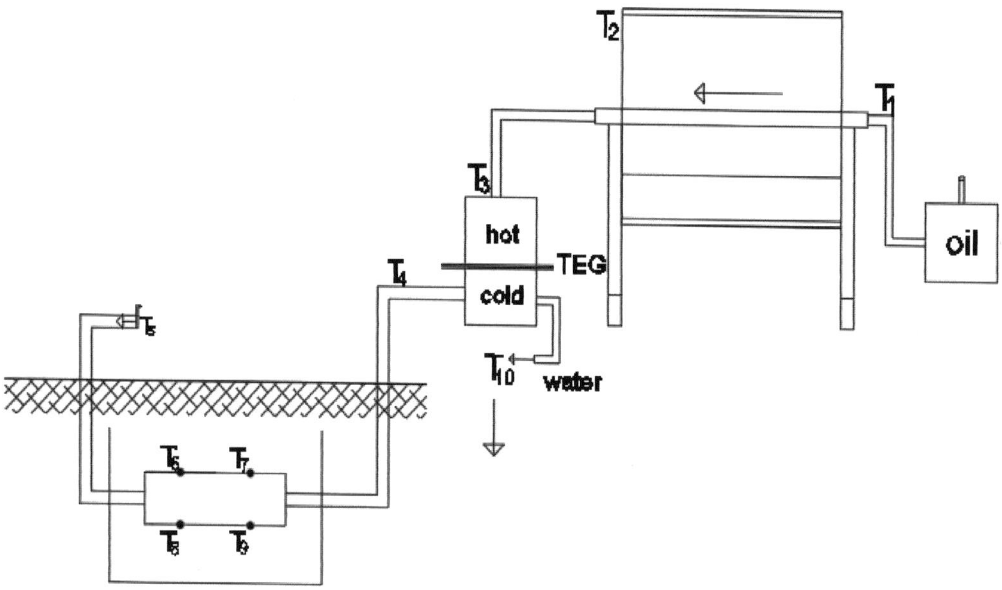

Figure 3. Diagrammatic sketch for the test rig.

The experiments were done in three models to gauge the effectiveness of the system in each model. In the model (1), water is used in the solar collector to be used as a hot Side for TEG. In the model (2), water is replaced with solar oil to measure its impact on the efficiency of the system. In the model (3), the soil is dampened instead of dried to improve the effectiveness of the cooling system on the cold side for TEG in the geothermal system.

4. Results and Discussion

Figure 4 shows the solar beam radiation intensity as a function of the time during the days in the model (1). It is clear that the solar intensity increases with time up to 12 o'clock and then it decreases with time. The maximum solar intensity is at 12 o'clock because the sun is perpendicular to the earth around this time. The temperature of both hot and cold sides as a function of the time in the model (1) using the water for three different runs at 15, 17 and 20 August is shown in figure 5. The behaviors of the three runs are almost the same indicating that there are no many differences during one week in August. It is clear that the cold side temperature did not change with the time indicating the temperature stability in the geothermal system. On the other hand, the hot side temperature of the TEG increases over time until it reaches its maximum value at 12 o'clock and then it decreases. This conduct is nearly equivalent to the power of solar radiation conduct (figure 4) which states that high sun-based radiation power prompts increment the temperature of the hot side.

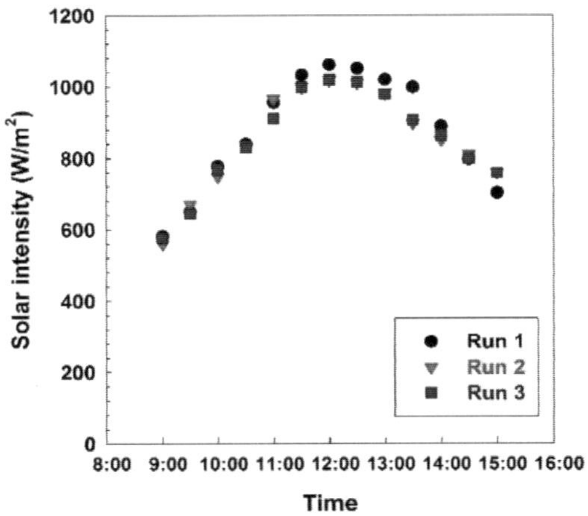

Figure 4. Solar intensity vs. time for model (1).

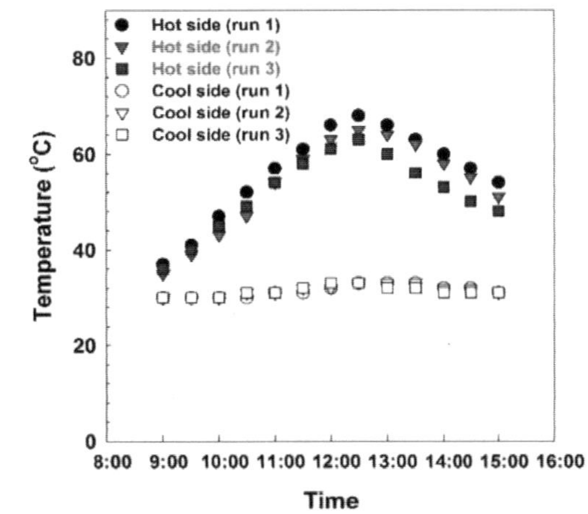

Figure 5. The temperatures for both sides of TEG vs. time for model (1).

Using solar collector to heat the water which used for the hot side and the geothermal system was used as the cold side for TEG. The open voltage as a function of time for these three days is shown in figure 6. It is clear from the figure that, the open voltage increases by increasing the time until it reaches its maximum value at 12 o'clock and then it decreases. This increases on the open voltage connected to increase the temperature difference between the hot and cold side [$\Delta T=T_{hot}-T_{cold}$] of the TEG at 12 o'clock. The maximum open voltage is about 5.9 V when the solar radiation intensity is maximum at noon where ΔT is 35 °C and the solar radiation is 1062 W/m^2.

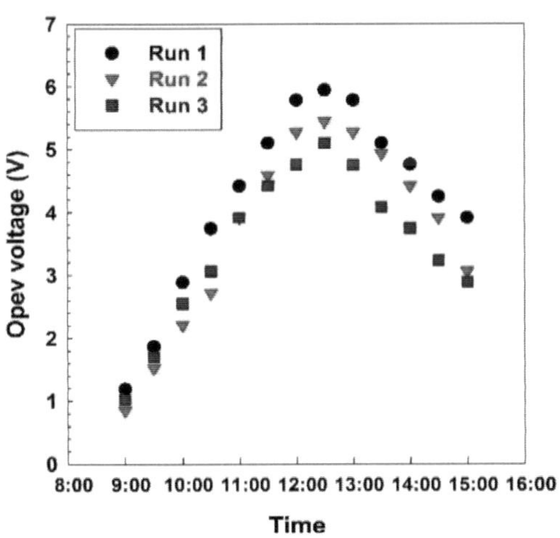

Figure 6. Output voltage vs. time for model (1).

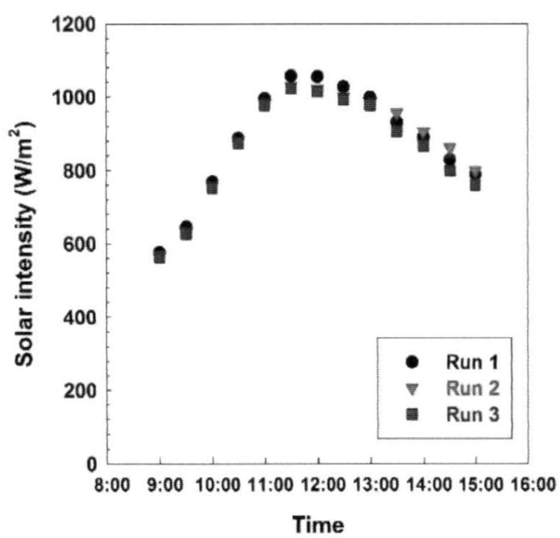

Figure 7. The solar intensity as a function of the time during 3 runs for model (2).

The solar beam radiation intensity as a function of the time during the days in the model (2) is shown in figure 7. It is clear that the solar intensity increases with time up to 11:30 am and then it decreases with time. The maximum solar intensity is at 11:30 am because the sun is perpendicular to the earth around this time. Figure 8 shows the temperature of both hot and cold sides as a function of the time in the model (2) using the therminol vp–1 oil for three different runs at 23, 24 and 25 August. The behaviors of the three runs are almost the same indicating the stability of the system. Clearly that the temperature of the hot side of the TEM in the second model higher than that of the hot side in the first model. The maximum value of the temperature of the hot side in the second model 96 °C while in the first model

65 °C. On the other hand, the cold side temperature did not change with the time indicating the temperature stability in the geothermal system.

The open voltage readings as a function of time for three days (Three runs) is shown in figure 9. The open voltage increments by expanding the temperature gapping between the hot and cold side [$\Delta T = T_{hot} - T_{cold}$] of the TEG. The maximum open voltage in model (2) is 10.2 V when the solar radiation intensity is maximum at noon where ΔT is 59 °C and the solar radiation is 1051 W/m^2.

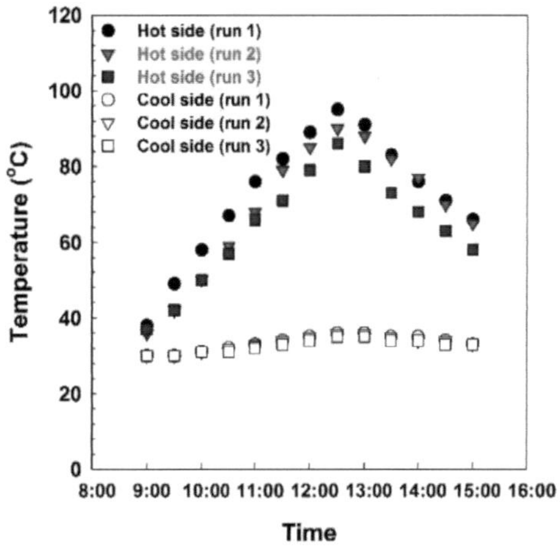

Figure 8. The temperatures at both sides of TEG vs. time for model (2).

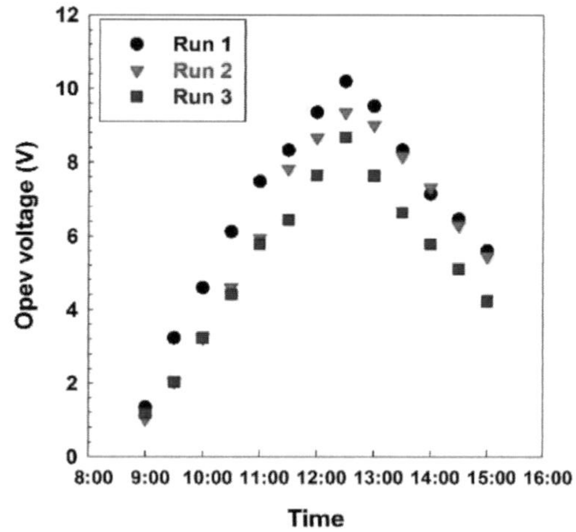

Figure 9. Output voltage vs. time for model (2).

According to Özbas (2019), 8.2 volt was produced using 5TEG when the temperature difference on both sides of the TEG was 89 °C. In this work, 10.2 volt were obtained using 12 TEG when the temperature difference was 59 °C. we can conclude that our system is effective although it applies higher number of TEG where the difference in temperature between the two models is 30 °C [33].

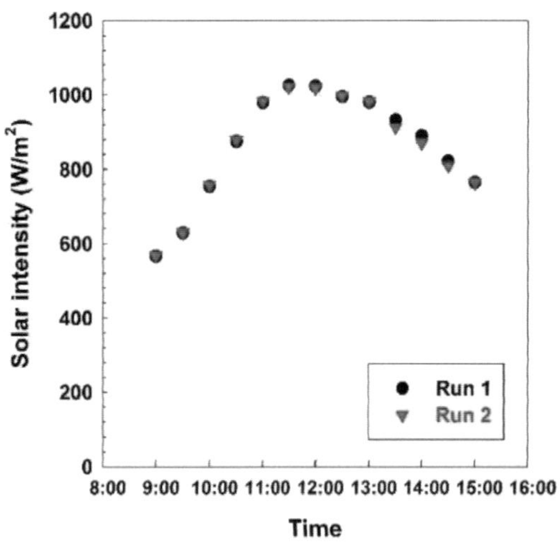

Figure 10. The solar intensity as a function of the time during 2 runs for model (3).

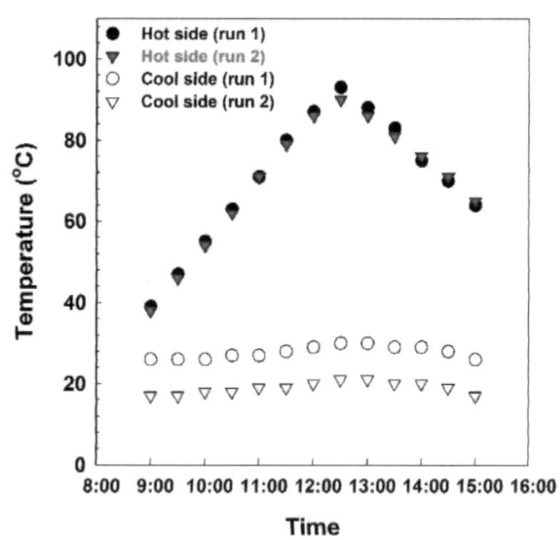

Figure 11. The temperatures for both sides of TEG vs. time for model (3).

The solar beam radiation intensity as a function of the time during the days in the model (3) is shown in figure 10. It is also clear that the solar intensity increases with time up to 11:30 am and then it decreases with time. The maximum solar intensity is at 11:30 am because the sun is perpendicular to the earth around this time. Figure 11 shows the temperature of both hot and cold sides as a function of the time in the model (3) using the therminol vp–1 oil for one run on 27 August. In this run, water was added to the soil to become moist soil. This was a major cause of hypothermia on the cold side of the TEG. On the other hand, the hot side temperatures didn't change from the previous model. On other hand, figure 12 shows the open voltage as a function of time for the model (3). It was discovered that diminishing the cold side temperatures increment the gapping in temperature between the hot side and the cold side, which prompts increment of the open voltage outside from the TEG. In this case, the maximum open voltage value is 11.7 V when ΔT is 72 °C and the solar radiation is 1056 W/m^2.

Our results are lower than that achieved in other studies. This difference arises from the TEG number or efficiency. For example, 57 V (DC) was generated using 24 TEG at Δt = 67 ° C [29] while in this work, we use only 12 TEG. Our results also less than Ref. [30]. due to the difference in efficiencies of the TEG.

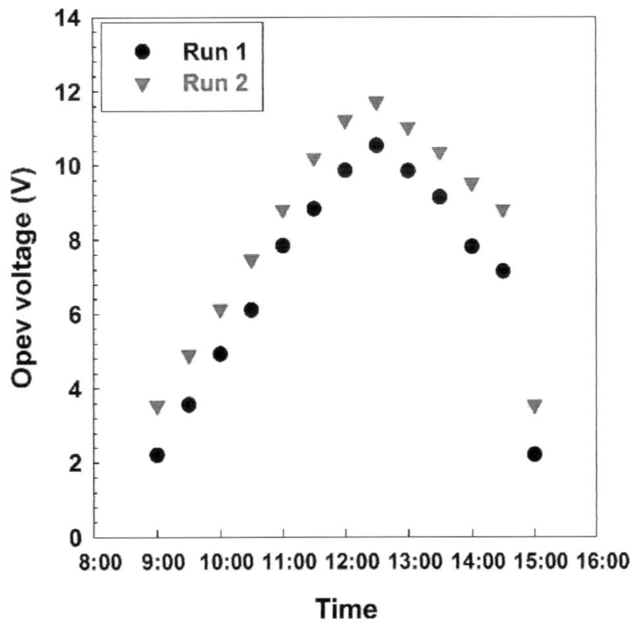

Figure 12. Output voltage vs. time for model (3).

5. Conclusions

The most proficient route for improving the thermoelectric generator is to utilize it in hybrid systems. In this system, the thermoelectric module was combined with geothermal energy (low Temperature) as a cold side (GTEM), and connected to a solar collector (parabolic dish) as a hot side (STEM). In addition, any increase in the temperature difference will increase system efficiency. Nowadays, this innovation is presently being developed stage, and need improvement to build its effectiveness by the research of new thermoelectric semiconductor materials.

References

[1] Doman L E 2016 *International Energy Outlook 2016 With Projections to 2040* (Washington: https://www.eia.gov/outlooks/ieo/)
[2] Doman L E 2013 *International Energy Outlook 2013 with Projections to 2040* (Washington: https://www.eia.gov/outlooks/ieo/)
[3] Teske S, Crespo L and Richter C 2016 *Solar Thermal Electricity: Global outlook 2016* (Netherlands: Ottho Heldringstraat) (Belgium : Rue de l'Industrie 10, B-1000 Brussels) (Spain: Apartado 39E-04200 Tabern) P12

[4] Brown M M, Desai N and Doucet G 2000 *Energy and the Challenge of Sustainability. UNDP* (New York : NY 10017) P165
[5] Kalogirou S A 2004 *Prog. Energy Combust. Sci.* **30** 231
[6] Panzera F, Sicali S, Lombardo G, Imposa S, Gresta S and Amico S D 2016 *Environ. Earth. Sci.* **75**1140
[7] Erkan K, Holdman G, Blackwell D, and Benoit W 2007 *Thirty-Second Workshop on Geothermal Reservoir Engineering* Stanford University, Stanford, California, January 22-24, 2007 SGP-TR-183.
[8] Tian Y, Zhao C Y 2013 *Appl. Energy* **104** 538
[9] Gielen D 2012 *Renewable Energy Technology Cost Analysis Series* (Germany: IRENA) 2
[10] Ismail B I, Ahmed W H 2009 *Recent Pat. Eng.***2** 27
[11] Doms I, Merken P and Van Hoof C 2007 *IEEE* **44** 17530
[12] Champier D 2017 *Energy Convers. Manag.***140** 167
[13] Kewen L, Changwei L and Pingyun C 2013 *Thirty-Eighth Workshop on Geothermal Reservoir Engineering* (Stanford, California)
[14] Lund W J, Boyd L T 2016 *Geothermics* **60** 66
[15] Andersson O 2007 A*quifer Thermal Energy Storage (ATES)* (Paksoy: Springer) 155
[16] Mamur H and Ahıska R 2014 *Int. J. Renew. Energy Environ. Eng.* **4** 128
[17] Polozine A, Sirotinskaya S, and Schaeffer L 2014 *JMR* **17** 1260
[18] Aydogan H, Ozcelik E A, Acaroglu M and IŞIK H 2014 *Appl. Mech. Mater.* **446** 858
[19] Karpe S 2016 *Int. J. Curr. Eng. Technol.* **4** 144
[20] Fan H, Singh and Akbarzadeh A 2011 *J. Electron. Mater.* **40** 1311
[21] Muffler P and Cataldi R 1978 *Geothermics* **7** 53
[22] Chandrasekharam D and Bundschuh J 2008 *Low-Enthalpy Geothermal Resources for Power Generation* (London : Taylor & Francis Group) 149
[23] Zhu N, Matsuura T, Suzuki R and Tsuchiya T 2014 *Energy Procedia* **52** 651
[24] Muthu G, Shanmugam S and Veerappan R A 2014 *Energy Procedia* **54** 2
[25] Eswaramoorthy M, Shanmugam S and Veerappan R A 2013 *Int. J. Energy. Eng.* **3** 62
[26] Liu C, Chen P, and Li K 2014 Thirty-Ninth Workshop on Geothermal Reservoir Engineering Stanford, California, P 24-26
[27] Chen J , Li K, Liu C, Li M, Jia L and Jiang S 2017 *Energies* **10** 1329
[28] Lertsatitthanakorn C, Jamradloedluk J and Rungsiyopas M 2014 *Energy Procedia* **52**150
[29] Faraji A Y, Date A , Singh R and Akbarzadeh A 2014 *Energy Procedia* **57** 2112
[30] Singh M, Nirapure S and Mishra A 2015*IOSR J. Mech. Civ. Eng.* **12** 40
[31] Tundeea S, Srihajonga N and Charmongkolpradita S 2014 *Energy Procedia* **48** 453
[32] Sathawane N S and Walke P V 2014 *Int. J. Adv. Res. Sci. Eng.* **3** 36
[33] Özbaş E 2019 *Experimental Investigation of Passive Water Cooling in Solar Heating Thermoelectric Generator* (Turkey: Gazi University)
[34] Yusuf A and Ballikaya S 2020 *Appl. Sci.* **10** 1
[35] Dell R, Petralia M T, Pokharel A and Unnthorsson R 2019 *Thermoelectric Generator Using Passive Cooling* (London : intechopen)
[36] Rana S, Date A, Iqbal A and Akbarzadeh A 2019 *Energy Procedia* **160** 723
[37] Zhang Z, Wu Y, Li W and Xu D 2020 *Energies* **13** 947

IOP Conf. Series: Materials Science and Engineering 1046 (2021) 012020 doi:10.1088/1757-899X/1046/1/012020

Monitoring of acoustic emission behaviour during early-age cement paste hydration

M A Shawkey[1], M A Abdelzaher[1], H M Mahmoud[2] and M M Rashad[3]

[1]Department of Environmental Science and Industrial Development, Faculty of Postgraduate Studies for Advanced Sciences, Beni-Suef University, Egypt.
[2]Faculty of Science, Beni-Suef University, Egypt.
[3]Department of Electronic and Magnetic Materials, CMRDI, Helwan, Egypt.

*E-mail: moustafa.shawkey@psas.bsu.edu.eg

Abstract. Valuation of Portland cement hydration using non-destructive techniques at early age is of great interest for predicting its thermal and durability characteristics. The acoustic emission (AE) technique is verified to be a high-quality non-destructive tool for real-time monitoring of structural micro-changes during the lifetime of Cement-based material composites. In this study, AE was applied to study the relationship between recorded signal and mechanisms that were happened during cement paste hydration. By using AE sensors, Portland cement paste samples with w/c ratio of 0.33 and 0.4 were explored during the first 72 h of casting. The AE resulting signals in each stage of hydration were studied to describe the various mechanisms that were running throughout the hydration process.

1. Introduction

Cement is a composite material that normally used in structural buildings. The early age properties of cement composites control the strength of their entire development. Furthermore, cement composites setting and hardening characteristics are critical factors, affecting the behaviors of these composite during their life span [1-3]. Therefore, applying non-dangerous testing at the early age of these composites is of interest to monitor early-age cracking, which is a critical and fundamental issue for their durability [4].

The hardening stages of concrete or cement-based materials are considered as the most crucial period during their life span and it is necessary to have solid data in this stage for a deeper understanding of their early age characterization [5, 6]. Water reacts with different phases of cement that hydrates and forms a new mixture [7]. Topo-chemical theory as well as through-solution reactions are the two principal mechanisms that clarify the C-S-H gel formation since the beginning of cement hydration [8-10].

One of the key issues with such materials is to be capable to predict their mechanical characterization and behaviors. The long-age behavior of concrete is intently connected with its state at an early age and in precise its hydration process [11]. Since this period is the core, it has been widely studied over the course of recent years [12, 13]. The know-how of their mechanical behavior at an early age becomes imperative to forecast their overall performance in service. To describe the hydration of cement paste at an early age, from a few minutes to a few hours after casting, distinctive strategies have been developed [14].

Understanding the mechanism that runs throughout hydration of cement composites persists with the enhancement in state-of-the-art strategies to represent the integrated multidisciplinary science [15]. However, the transition of concrete from a fresh fluid state to harden stable state is nevertheless the considerable event, which desires to be addressed. Besides, the dependence of thermal and mechanical properties of cement composites on the early levels of hydration initiates interest in hydration study [16-19]. Hence, a clear understanding of the microstructural observations occurring during the successive phases of the cement hydration technique will trigger adequate information to describe its durability and thermal features [20, 21].

2. Material and methods

In order to study the AE responses during early age of cement composites and check out relevant of hydration stages, testing was carried out on two different cement paste samples. By applying the AE test, we could explore the characteristics of wave propagating during cement paste samples.

In this study, the passive AE technique was applied; AE sensors were used for continuously exploring the elastic waves that were developed as a result of any microstructural changes during the hydration. Besides passive testing, temperature change was measured using the thermocouple to associate the AE parameters outcomes with temperature change in both samples . AE signal clustering was characterized in order to understand the relationship between AE waves and the corresponding explanations associated with various stages of cement hydration.

Cement paste specimens in this study were prepared using a Portland-Limestone Cement CEM II/A-L 32,5R, Blue Circle General Purpose Cement, manufactured by Tarmac Ltd, UK [22, 23] to match with the physical property and chemical composition requirements released in the British Standards Institution (BS EN 197-1:2011) [24]. Table 1 displays the main constituent properties of cement used in this study. Cement and normal tap water were used to form plain cement paste with water to cement ratio (w/c) 0.33 and 0.4. Cubic moulds with an edge length of 100 mm were made to form cement paste samples. The AE recording started approximately after 20 min of casting cement paste samples until 72 h.

Table 1. Main constituent properties of cement [24, 25].

Property	Cement (Quantity, %)
Clinker (K)	80-94
Limestone (L)	6-20
Minor Additional Constituents	0-5

SAEU25 acoustic emission system used in this study is consists of a sensor data acquisition system and broadband sensors. This system permits to capture the main parameters that such as event, hit, rise time, duration of event, count to peak, amplitude (in dB). Figure 1 describes different AE parameters captured from the signal waveform. The plug-in filter of this system has bandpass from 20 to 400 kHz and the preamplifiers provide 40 dB gain. The bandpass of these preamplifiers is from 20 to 100 kHz. The cement paste samples were mixed and poured in moulds in an open area and then they were moved into the lab room at a controlled temperature of 22 ± 3 °C. Sensors were placed over the top of the cement paste specimens, and other sensor was attached over the floor to act as a control sensor. This sensor was used to refine AE results from unwanted signals that identified with surrounding noise. In addition, k-type thermocouple was used to measure temperature. Temperature recording was captured continuously at every 4 min using HH-520 4-channel thermocouple data logger. Figure 2 displays a drawing of the experimental setup.

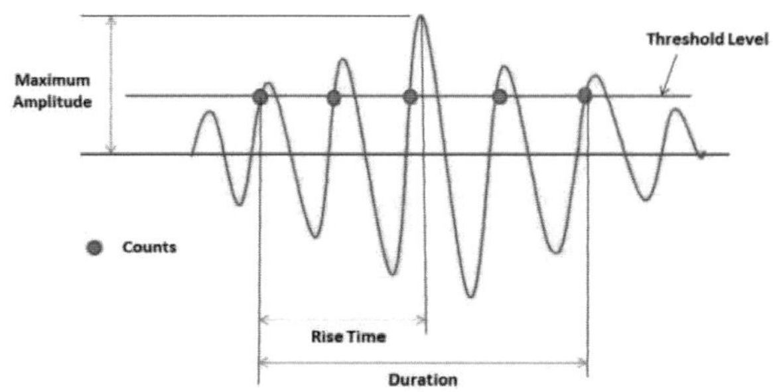

Figure 1. Typical AE waveform with related parameters.

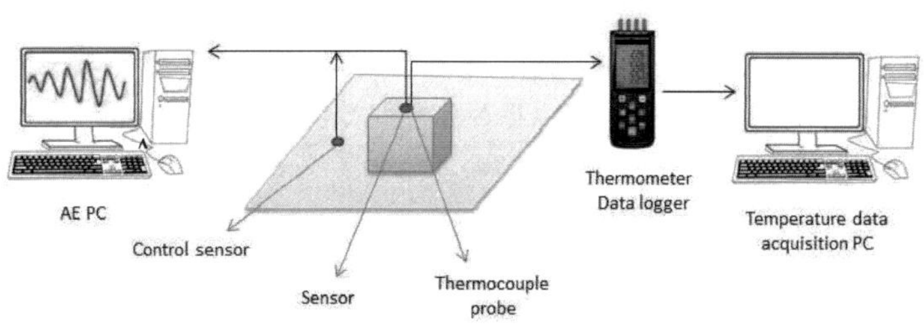

Figure 2. Experimental setup.

3. Results and Discussion

In this study, experimental results obtained from the acoustic emission experiments conducted on cement paste at different water/cement ratios were introduced and evaluated. The acoustic emission data (hits, events, amplitude, duration, energy, rise time, rise counts) versus time and heat of cement hydration measurement were classified to explore relationships between the AE data signals and temperature during first 72 h of hydration time. An operation for AE activity clustering was discussed to discover potential stages of cement hydration and correlate them to the classified AE signals.

Figure 3 shows the AE cumulative counts and change in temperature as a function of time in cement pastes 1 and 2. It is obvious that the cumulative number of counts for cement 1 is much higher than for cement 2 (About twice). The feature of a high acoustic activity occurring in these samples expresses the physical and chemical responses happening during hydration. Another perception concerns the pace of variety that is more prominent on account of cement 1 than cement 2. An explanation was implemented on both AE clustering signals and temperature data of the cement paste samples in order to analyses the distinctions in AE activity happening during the hydration.

The internal temperature measured in cement samples was regarded as an indication of hydration progress. We could expect that the exothermic reaction represented by a high rise of the internal released temperature was firmly related to the precipitation of hydrates. In addition, we noticed that an enormous exothermic phenomenon began sooner than the AE activity (AE cumulative counts). Through the first 35 h of cement hydration, the temperature of the two cement paste samples raised and then reduced. A notable change in the temperature rate happened from the beginning of the experiment until the time when the temperature reached their optimum values (T_{max} = 27.2 °C at t= 14.2 h, T_{max} = 26.6 °C at t = 18.2 h) for cement 1 and 2, respectively. The AE activity started just before the maximum heat of hydration. It first slightly increased which we suggest that acoustic activity generation came by contact interactions between cement particles.

IOP Conf. Series: Materials Science and Engineering 1046 (2021) 012020 doi:10.1088/1757-899X/1046/1/012020

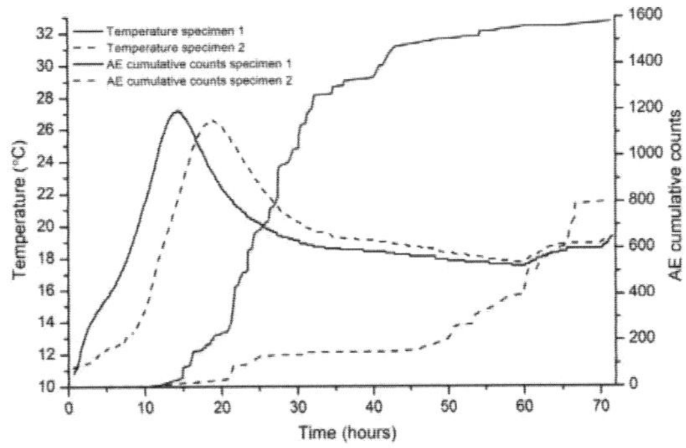

Figure 3. AE cumulative counts and internal temperature as a function of time.

Figure 4(a, b) displays AE counts and hydration temperature with respect to time for cement pastes 1 and 2, respectively. From figure 4(a) AE counts are seen as prevalently dynamic during the period between 12 and 45 h while Majority of AE counts in figure 4(b) are registered for sample 2 appears in 42–69 h period.

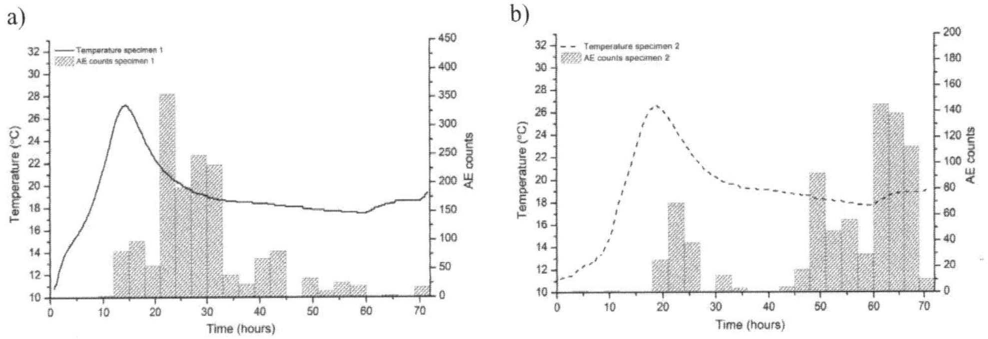

Figure 4. AE cumulative counts and internal temperature of cement paste specimens as a function of time: (a) for w/c 0.33 and (b) for w/c 0.4.

As explained in figure 5(a, b), values of duration and amplitude are directly proportional to numbers of counts. Moreover, it is shown also that by getting higher values of amplitude and longer values of duration, larger values of energy were captured.

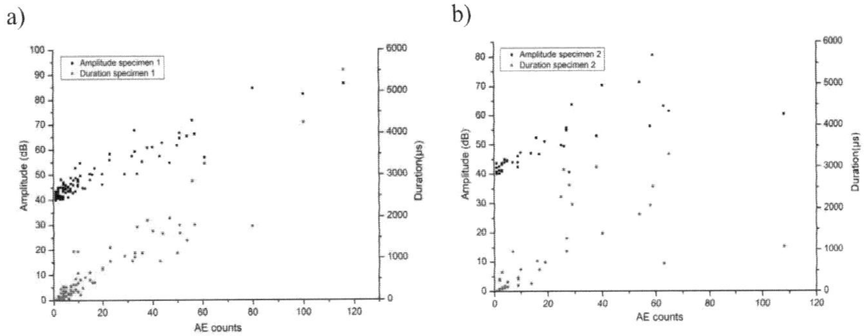

Figure 5. Amplitude and duration of cement paste specimens as a function of AE counts: (a) for w/c 0.33 and (b) for w/c 0.4.

Figures 6 and 7 show duration and energy of the AE signals obtained during the first 72 h of cement hydration, respectively. Intense presence of duration and high energy is clearly visible at the early age for cement 1 while it is observed in sample 2 that high energy activity with shorter duration happens at a later age.

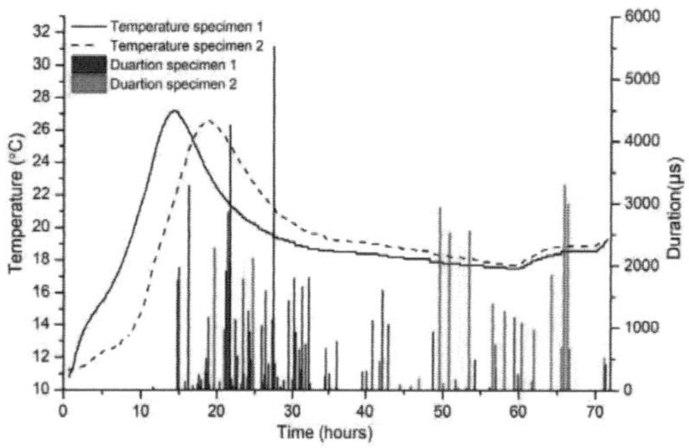

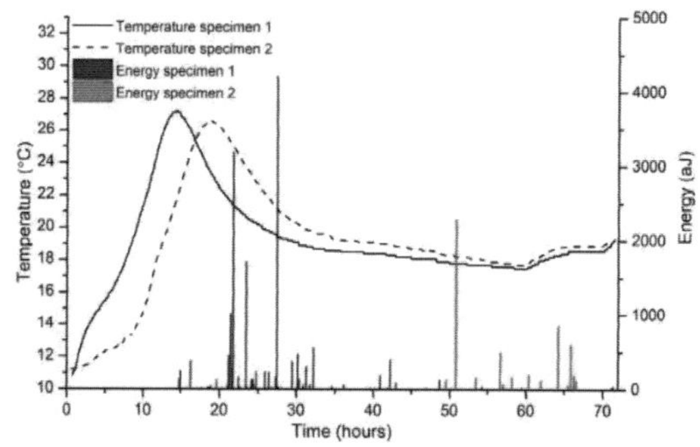

Figure 6. Duration of AE activity during first 72 h. **Figure 7.** Energy of AE activity during first 72 h.

Three-dimensional graphs, hits versus time and amplitude, were depicted in figure 8(a, b) for cement 1 and 2, respectively. It is observed the majority of hits which own amplitude of 40–50 dB at the early age for specimen 1 and at a later age for specimen 2. Cement grains that were started to mix with water subjected to intense chemical reaction led to gathering between particles (acceleration stage).

We noticed also a rapid increase of AE hits with amplitude 40–60 dB during the middle stage in specimens 1 and 2, which evinced the progression of hydration reaction (deceleration stage). At the final stage, it was suggested that nearly all the particles were completely reacted and accumulation of AE hits continued to grow during further stages of hydration that expressed the hardening state (steady state stage).

a) b)

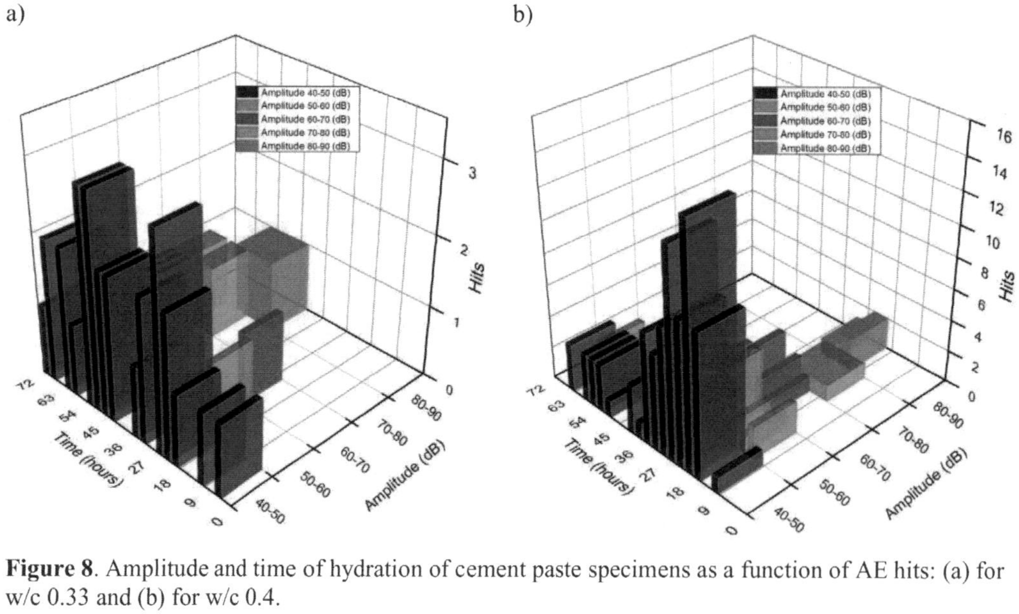

Figure 8. Amplitude and time of hydration of cement paste specimens as a function of AE hits: (a) for w/c 0.33 and (b) for w/c 0.4.

4. Conclusion

In this study, the results of experimental testing carried out to explore the hydration process in cement paste samples, at different water/ cement ratios, were obtained during the first 72 h of cement hydration. The analysis of hydration stages based on the development of AE activity showed vividly that development of AE events has a good correlation to the changes in the microstructure in the cement samples, therefore, providing a better understanding of the connection between results of acoustic emission and hydration process. The results are summarized as follows: the AE cumulative counts for cement sample 1 (w/c = 0.33) are nearly twice as high as for cement sample 2 (w/c = 0.4), these counts were captured as a result the physical and chemical response happened during hydration. A notable exothermic phenomenon, observed by thermocouple data logger data, started to appear sooner than the beginning of AE activity. In addition, the reach of sample 1 to a higher optimum temperature, the end of acceleration stage, in nearer hydration time (T_{max} = 27.2 °C at t= 14.2 h) compared to sample 2 (T_{max} = 26.6 °C at t = 18.2 h) gave an indication of the hydration rate as well as the development of the evolution of AE events. AE counts were displayed obviously in sample 1 over the period of 33 hours between 12 h and 45 h whereas the substantial existence of AE counts for sample 2 appeared in 42–69 h period. A clear presence of duration and high energy for cement 1 was noted in early age while high energy activity with a shorter duration for sample 2 happened at a later age. Majority of hits own amplitude of 40–50 dB were obtained at the early age for specimen 1 and at the later age for specimen 2. Moreover, the dramatic increase of AE hits with amplitude 40–60 dB throughout the middle stage in samples 1& 2 reflected the progression of hydration reaction (deceleration stage). The continuous growing of AE cumulative hits during further stages of hydration expressed that nearly all the particles are completely reacted as well as indicated the reach of the hardening state (steady state stage).

References

[1] Pazdera L, Topolar L, Korenska M, Smutny J and Bilek V 2014 *Advanced Analysis of Acoustic Emission Parameters during the Concrete Hardening for Long Time. In 11th European Conference on Non-Destructive Testing (pp. 1–8).*
[2] Abo-EL-ENein S A, Abdel-Gawwad H A, Hodhod O A, El-Kattan I M 2016 *J. Am. Sci.* **12** 51
[3] Pazdera L, Topolar L, Korenska M, Vymazal T, Smutny J and Bilek V 2015 *Mater Tehnol.* **49** 703
[4] Balboul B A, Abdelzaher M, Hamouda A S and Zaki A H 2019 *Egypt. J. Chem.* **62** 1105
[5] Qin L, Ren H W, Dong B Q and Xing F 2014 *Materials* **7** 6908
[6] Abdel-Gawwad H A, Hassan H S, Vásquez-García S R, Israde-Alcántara I, Ding Y C, Martinez-Cinco M A and El-Kattan I M 2020 *J. Clean. Prod.* **252** 119875.
[7] Lu Y, Zhang J and Li Z 2013 *Constr. Build. Mater.* **46** 183
[8] Abdelzaher M A, Hamouda A S, Ismail I M and El-Sheikh M A 2018 *Key Engineer. Mater.* **786** 248
[9] El-Kattan I M, Abdelzaher M A and Farghali A A 2020. *J. Mater. Res.Tech.* https://doi.org/10.1016/j.jmrt.2020.05.087
[10] Chotard T J, Smith A, Boncoeur M P, Fargeot D and Gault C 2003 *J. Eur. Ceram. Soc.* **23** 2211
[11] Tantawy M A, El-Roudi A M, Abdalla E M and Abdelzaher M A 2013 *J. Engineer.* **2013** 1
[12] Chotard T, Rotureau D and Smith A 2005 *J. Eur. Ceram. Soc.* **25** 3523
[13] Assi L, Soltangharaei V, Anay R, Ziehl P and Matta F 2018 *Cem. Concr. Res.* **103** 216
[14] Chotard T J, Barthelemy J, Smith A, Gimet-Breat N, Huger M, Fargeot D and Gault C 2001 *J. Mater. Sci.* **20** 667
[15] Qin L, Lu Y and Li Z 2010 *J. Mater. Civil Eng.* **22** 1323
[16] Thirumalaiselvi A and Sasmal S 2019 *Constr. Build. Mater.* **196** 411
[17] Ahmed M A, Hasaneen Y A, Assaf K A, El-dek S E and Shawkey M A 2015 *Microelectron.*

Eng. **146** 53
[18] Abdel-Gawwad H A, El-Enein A, Heikal M, Abd El-Aleem S, Amer A A and El-Kattan I M 2018 *Constr. Build Mater.* **176** 676
[19] Ahmed M A, Hasaneen Y A, Assaf K A and Shawkey M A 2015 *Open J. Civ. Eng.* **5** 131
[20] Chotard T J, Smith A, Rotureau D, Fargeot D and Gault C 2003 *J. Eur. Ceram. Soc.* **23** 387
[21] Lura P, Couch J, Jensen O M and Weiss J 2009 *Cem. Concr. Res.* **39** 861
[22] Tarmac 2017 *Cement Declaration of Performance Sheet* (Online) (Available at: https://dop.asp-bd.co.uk/DOP_FILES/562355_Dunbar_mastercrete.pdf).
[23] Tarmac 2018 *Blue Circle Cement: Data Sheet* (Online) (Available at: https://tarmac-bluecircle.co.uk/wp-content/uploads/2018/09/General-Purpose-Cement-approved-04-09-2018.pdf).
[24] BS-EN197-1 2011 *Cement Part 1: Composition, Specifications and Conformity Criteria for Common Cements* (British Standard)
[25] Keihani R, Bahadori-jahromi A and Clacy T J 2019 *Int. J. Civ. Eng.* **13** 588

IOP Conf. Series: Materials Science and Engineering 1046 (2021) 012021 doi:10.1088/1757-899X/1046/1/012021

Recent advances and challenges in management of urea wastewater: A mini review

A Zaher and Nabila Shehata*

Environmental Science and Industrial Development Department, Faculty of Postgraduate Studies for Advanced Science (PSAS), Beni-Suef University, P.O. Box 62511 Beni-Suef, Egypt.

*E-mail: nabila@psas.bsu.edu.eg

Abstract. Treating the wastewater with high level of urea and ammonia-nitrogen is one of the problems faced by different plants in the world. The basic methods of urea removal of wastewater streams are electrochemical oxidation, biological treatment, adsorption and hydrolysis using thermal treatment or enzymes and decomposition with the aid of catalyst, strong oxidant. This work reviewed recent of different nanosized and nanostructured particles in different techniques for urea removal from wastewater and the challenges for treatments to draw the attention of researchers and scientists toward novel research ideas in this area.

1. Introduction

Urea is abundant in the environment due to its presence as a major product in human and industrial plants effluents. In addition to enormous amount of urea in different industrial plants, huge amounts are leaching from agro-breeding farms and fields (figure 1). The major problem of urea as pollutants is its presence in toxic form and even if it hydrolyzes, it forms another toxic pollutant (ammonia gas). The later will decompose to yield additional pollutants such as NO_x (x=1, 2 and 3), finally, these nitrogen oxides cause eutrophication. Most of the treatment methods are still under developing or being applied in small scale. These methods are mainly electrochemical oxidation (EO), decomposition (using catalysts, strong oxidants, biological film, adsorption and hydrolysis (with or without the aid of enzyme). Because urea is cheap, widely available, non-toxic, stable, it can be transported and stored easily, urea, was investigated as a source of hydrogen in fuel cell [1]. Ewelina Urban et al. reviewed techniques used for urea removal from wastewater [2]. Ljubica Matijašević et al. suggested some preferable modifications in the current urea industrial plants. They analyzed status of the current plant and developed a computer simulation based on the real process parameters. Finally, a scenario for minimizing the negative impact on the environment was proposed [3]. However, due to the rapid modernization, water shortage, the urea should be managed efficiently in the wastewater. Several researches investigated urea management via different systems from lab scale to full scale. Figure 1 concludes the main processes and products of urea management.

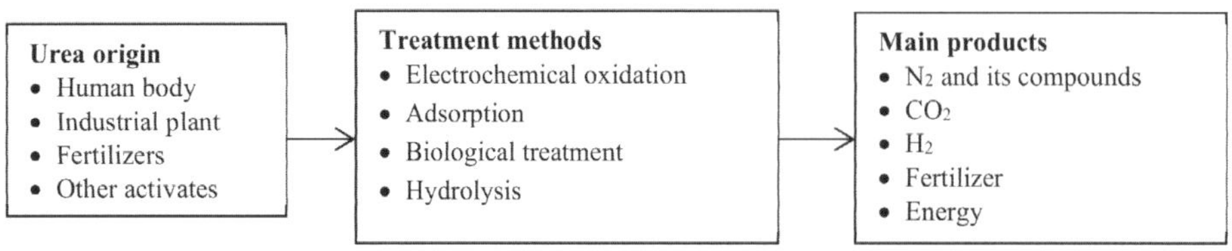

Figure 1. A schematic flow diagram for urea management.

2. Methods of urea management in wastewater

2.1. Electrochemical oxidation

Electrochemical oxidation of urea involves destroying/decomposing of urea in oxidation reaction where it losses one or more electrons followed by reaction in solution and transformation to another compound according to the following reactions [4],[5]:

Anode: $CO(NH_2)_2+6OH^- \rightarrow N_2+5H_2O+CO_2+6e^-$

Cathode: $6H_2O+6e^- \rightarrow 3H_2+6OH^-$

The overall reaction: $CO\ (NH_2)_2+H_2O \rightarrow N_2+3H_2+CO_2$

Bolzan and Iwasita investigated electrochemical oxidation of urea on Pt electrode. They detected production of nitrogen oxides, CO_2 and H_2 [6]. Hernlem (2005) investigated decomposition of urea (up to 750 mg/l) in electrolysis unit. Formation of byproducts, effect of temperature and initial concentration on urea conversion were studied [7]. A lot of studies have suggested modifications and have investigated different techniques and catalysts. Nickel and nickel based materials have drawn the attention of numerous researchers (figure 2). In the next sections, some of these investigations.

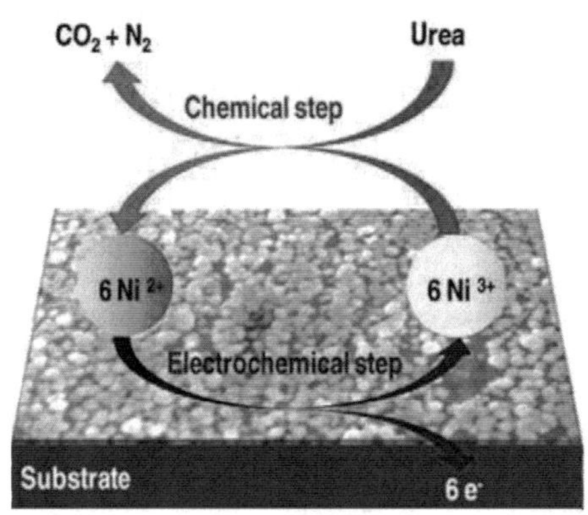

Figure 2. The mechanism of urea decomposition over Ni electro-catalyst surface [8].

2.1.1. Nickel based catalysts. Nickel was preferred as electrode for urea removal due it its highly boosting of the interfacial electroactive sites, charging transfer rates for urea electro-oxidation [9], and high catalytic activity [10]. Wang, et al. synthesized electro-catalysts based on nickel hydroxide nano-sheets for urea EO. This electrode decreases the over potential and enhance the current density [11]. The nickel catalyst was also suggested by King and Botte for the oxidation of urea. They introduced Rh metal to the catalysts. The results showed that the later reduces both the surface blockage and the over potential for urea EO, increases catalyst stability and improves the current density [12]. Yan et al.

compared binary and tertiary composites based on Ni. They found that the tertiary one (ZnCo@Ni) lowered the urea onset potentials by 80 mV [13].

Ding et al. compared mesoporous electrode ($NiCo_2O_4$) with cobalt oxide. They showed that the former is much stable and more active than the later. This electrode delivered 136 mA cm^{-2} mg^{-1} of current density and 0.7 V of oxidation potential when the initial concentration of urea was 0.33 M [9]. A screen printed electrode for urea EO was modified using multi-walled carbon nanotubes-nickel oxide nanoparticles composite [14]. Ewelina Urbańczyk et al. investigated three electrodes as EO cell anodes (Ni, Ni-Pt and Ti-Pt electrodes). Based on their results, the sintered Ni–Pt electrodes were more active than others [15]. Ye et al. synthesized low cost Ni@carbon sponge electrode (figure 3). The produced electrode was characterized and investigated for urea oxidation in alkaline solution. The electrode achieved an onset oxidation potential of 0.24 V, the peak current density in NaOH (5 M) and urea (0.1 M) solutions was 0.29 A/cm^2. They attributed these promising results to the highly porous network (0.40%) and the presence of enormous Ni active sites. The alkaline medium improved the electo-catalytic activity of the electrode due to the increase in current density of the anode > 0.35 V (vs. Ag/AgCl) [16]. Szpyrkowicz (2000) studied EO of urea wastewater from texture plant. The best EO results were obtained with Ti/Pt@Ir anode. They attributed the good results of the electrode catalytic activity to the Cl^- electro-oxidation into Cl_2 [17].

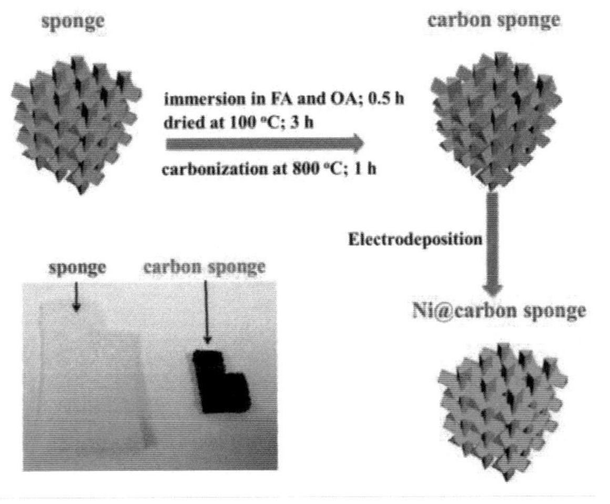

Figure 3. Synthesis of Ni@carbon sponge electrode [16].

Nanotechnology have been used to modify the catalyst via increasing the specific surface area (internal and external) (figure 4). They used the doping technique to introduce active/new surface functional groups to the catalyst. Guo et al. prepared Ni nanowire electrode using electrodepositing technique. They tested the efficiency of this electrode for EO of ammonia using electrochemical impedance spectroscopy, cyclic voltammetry and chronoamperometry tests. The onset oxidation potential of this electrode vs. Ag/AgCl was 0.25 V and peak current density and urea in KOH (5M) were 0.16 A/cm^2 and 0.33 M, respectively [18]. Yang et al. (2019) investigated the effect of nanorods form of Ni/MO oxide on urea oxidation. They discovered that the ratio of Ni and MO oxides play a significant role in the overall process efficiency. A good urea catalytic stability was recorded at current density 0.096 A/cm^2 [19].

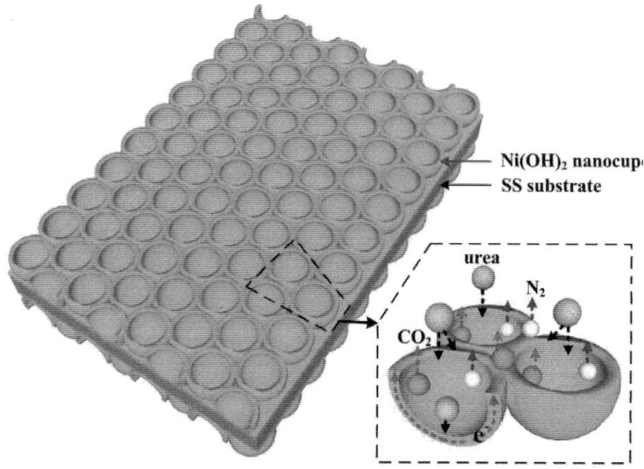

Figure 4. A schematic illustration of nickel hydroxide electrode for enhanced urea electrolysis [20].

Barakat et al. (2017) used doping technique to introduce N_2 to nanofibers of Ni-C. Then, they calcinated the nanofiber under temperature 750 °C in the presence of a noble gas (Ar). The proposed technique saved the energy required for electrolysis by saving significantly the urea onset EO potential [21]. Tammam et al. used nickel oxide nanoparticles to modify carbon electrode, the electrode was tested for EO of urea in alkaline medium. Also, it was found that the reticulated structure is much better than the planar and the maximum conversion is dependent on the loading extent [22]. Tungsten was suggested as a co-catalyst to increase the activity of Ni/C nanostructures in the urea EO, the nanofibers that containing 35 wt% tungsten showed the best performance (figure 5).

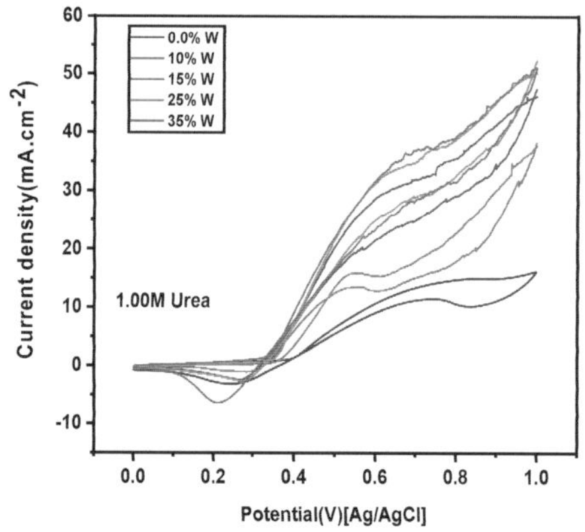

Figure 5. Influence of tungsten content on the electro oxidation activity of the NiW-C nanofibers for a 1.0 M urea EO in solution of KOH (1 M) [23].

Zeng et al. prepared NiCo@LDH as an efficient and highly durable catalyst for urea EO. They concluded that the spacing in the LDH interlayer structure played a control role for the EO of urea [24]. Li et al. developed Ni/N@graphene as a catalyst for urea EO. After modification with CeO_2, the oxidation current density of the catalyst was doubled. Introducing CeO_2 to the catalyst improved its activity, by improvement the pore volume and structure. This resulting in increase in NiO% and improve in catalyst mass transfer [25]. Alajami et al. prepared NiMn-incorporated carbon nanofibers and studied the composition, morphology, and synthesis-temperature on EO of urea (2M). The results represented that the best values of current density and onset potential were achieved using nanofibers of NiMn@carbon, with manganese acetate (10 wt%) and calcination temperature 850 °C [26]. Rahimpour et al. proposed a membrane for decomposition of NH_3 on Ni–Al_2O_3 catalyst. They suggested that more than 4500 tons of hydrogen can produced yearly from the urea plant. They modeled the treatment loop of urea in wastewater. They concluded that any increase in temperature or pressure and membrane reactor diameter yielded high urea management. And the countercurrent flow was better than co-current one [27].

2.1.2. Other catalysts. Shuguang Shen et al. investigated different alumina catalysts loaded on ceramic particles in a fixed bed reactor to hydrolysis urea in wastewater. The order of activity was α-Al_2O_3<γ-Al_2O_3<η-Al_2O_3. The results proved that the alumina catalysts depend mainly on temperature and it may depend also on the basic strength. The η-Al_2O_3 showed the best results concerned with hydrolysis temperature and the concentration of urea in the effluent (4.96 mg/l). They concluded that η-Al_2O_3 is a promising catalyst to replace current hydrolysis process [28]. Hernlem applied current (0.6- 2.5 A) using oxide-coated titanium electrodes in a bench scale electrolysis device for treating urea in stimulated solution. Increasing the temperature or even the urea initial concentrations increased the energy efficiency of urea decomposition [7]. Amstutz et al. (2012) compared EO of fresh and old urea synthetic solution using IrO_2 anodes. It was found that urea could be managed efficiently in fresh solution in contrast to the old solution where the carbonate resulted from hydrolysis of urea by the bacteria hinder ammonia removal [29]. Mahmoud et al. studied graphite electrodes bench-scale electrochemical cell for decomposition of urea. They attributed the increasing in urea removal rates (0.78 g/h) to the increase in current density and the decrease in the current efficiency to 82%. In another study, graphite electrode was used for oxidation of urea (initial concentration 50-100 mg/l), they studied different parameters like energy consumptions, current efficiency and the rate of urea removal. They recorded a proportional relationship between the rate at which urea was removed and the current density rate, also the increasing in power consuming. They reached a maximum removal rate 0.33g/h for initial concentration of human urine 100 mg/l. The lowest energy consumption was 196x103 Wh/kg [30]. Jayishnu Singla et al. (2019) tested quaternary mixed metals oxides (Pt/Ru/Ir/Ti@titanium) anode for EO of urea. The optimum conditions for photo-EO to achieve >94.78% or urea oxidation was pH 4, initial concentration of NaCl 1.45g/l, and current density 0.018 A/cm^2 and reaction time 135 min. They attributed the high urea removal % to the development of reactive OH• and Cl_2 species. They estimated the cost of operating for this process to be 0.78 $\$/m^3$. The durability tests showed that the developed electrode could work for 90 cycles without reduction in its efficiency. They recommended it for on-site urine treatment [31]. Simka et al. (2009) compared between Ti/(RuO_2–TiO_2) 40:60 and Ti/Pt electrodes for the decomposition of urea. They studied the process kinetics, and other parameters. It was found that in case of Ti/Pt electrode, the urea decomposed in the solution while in case of Ti/(RuO_2–TiO_2) electrode, urea decomposed on the electrode [32]. Mahmoud et al. tested porous graphite electrodes in electrochemical cell. The effect of salt strength, the rate and concentration at which urea flow along with the current density at the anode on urea management were deeply discussed. The results proved that when the current efficiency decreased and its density increased, the urea management increased. Finally, the energy power consumption increased, and the maximum urea removal rate was 0.78 g/h [33]. Rahimpour et al. introduced a membrane contain Pd that destruct urea in wastewater to H_2. The effect of temperature, pressure, thickness of catalyst layer and configuration of flow were examined. The membrane reactor and urea treatment were modeled mathematically [27].

A lot of studies focused on Nickel in special way and also other catalysts like palladium, graphite, etc. The main challenge of urea management via EO process is the price of catalyst which increase the total cost of the process. The future studies should focus on finding low-cost electrodes, and searching for natural catalysts. Another trend is to improve the efficiency of the existing catalysts to compensate the high cost. This will be done through introducing new functional groups to the catalyst or/and improving the specific surface area, structure and pore volume of the catalyst.

2.1.3. Fuel cell. Urine can be utilized as fertilizer, in struvite form, but this technology consumes intense energy. Microbial Fuel Cell (MFC) consider a promising technology with added value where urea wastewater will be managed effectively and electricity will be produced. Because urea is safer and less toxic than hydrazine and ammonia, it was investigated as a hydrogen-storage material for low-temperature fuel cells. The low-power density resulting from the sluggish anode reaction is still the main challenge for applying urea in fuel cells. Another drawback is that microorganisms are only able to metabolize the organic substrates rather than urea in urine. Wei Xu et al. reviewed urea electrocatalysis in numerous fuel cells [34]. Jiseon You maximized urine utilization in production of energy and struvite. At the beginning, MFC increases the urea hydrolysis rate in the same time of producing electricity which improved struvite formation. Later, the initial effluent was mixed with Mg, and additional reduction in COD was achieved. The system produced 14.32 W/m^3 of power [35]. Barakat et al. introduced nanoparticles of NiMn@activated carbon as new effective onset potential electrocatalyst for urea oxidation in fuel cell. This system delivered 130 mA/cm^2 of current density and -85 mV of onset potential combined with manganese up to 10 % [36]. Doping technique was suggested by Abdelkareem et al. to increase the overall efficiency of urea fuel cell. They found that introduction of cadmium to nickel@carbon (nano-fibers) worked as a co-catalyst. Cadmium binds the carbon nano-fibers, and improved the active site for urea oxidation [37]. Wang et al. studied urea EO in fuel cell to generate power. This fuel cell delivered 78% of coulombic efficiency when urea initial concentration was high as 3940 mg/L. In addition to remove more than 80% of nitrogen of initial concentration 2630 mg/L [38]. Wei Xu et al. prepared NiCo/C bimetallic nanoparticles with various weight % of Co (0- 40 %) in direct urea fuel cell (DUFC). The maximum power density (1.6 mW/cm^2) delivered at Co 10 %. The oxidant was O_2 at temperature 60°C and urea (0.33 M) was the fuel. The effects of temperature and urea initial concentration on DUFC electocatalytic activity were also studied [10]. Rong Lan and Shanwen Tao (2011) developed Ni nano-catalyst to be used in fuel cell for oxidation of urea/urine. The optimum condition to achieve a power of density 14 mW/cm^2 was urea (1M), as fuel, at room temperature. When the temperature increased to 60 °C, the power density reached 4 mW/cm^2 [39]. C. Santoro et al. (2013) investigated fresh urine as a fuel in single chamber microbial fuel cell. The power generation was doubled when Pt was introduced to the cathode (figure 6). After 4 d, 75% of COD was reduced [40]. Carpenter and Stuve (2019) suggested urea oxidation to be paired with co-current reduction of water to produce H_2 for energy storage, or reduction of O_2 to directly produce energy in a DUFC. The challenge is that urea concentration in their study was 0.33 M to simulate human urine concentration in contrast to diluted urea concentrations in wastewater, this resulting in reduction in the overall urea oxidation rate. However, the urea oxidation occurred at potentials that favor generation of Ni^{3+} which encourage Ni^{2+}/Ni^{3+} reversible redox equation [41]. The human urine was directly fed to MFCs (batch and continuous modes) as a fuel [42, 43]. Barakat et al. applied NiSn@carbon nanofibers as anode for urea EO in DUFC. Measurements of electrochemical process shown that applying a co-catalyst (Tin 15%) that was calcinated at temperature 850 °C enhanced significantly the electro-catalytic activity [44].

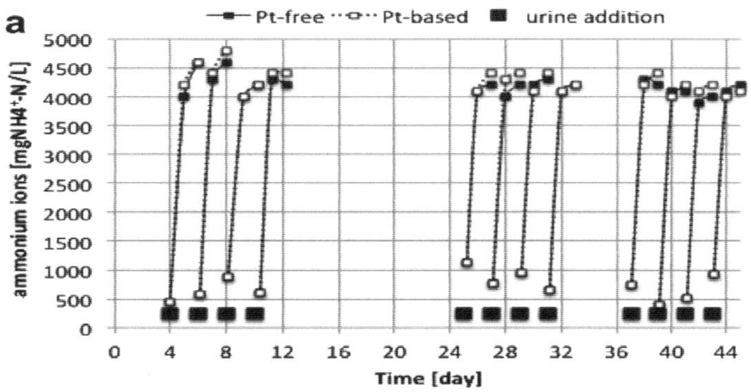

Figure 6. Ammonium concentration profile in SCMFCs at 2d cycles [40].

However, the challenges facing feasibility of EO using DUFC technique is the catalyst material, morphology, cost and efficiency. The researchers should search for lowering the cost and increasing the efficiency of the catalyst via increasing the specific surface area or modification using different techniques like doping, mixing in composites, etc. It can be seen also, that the researchers focused on investigating the current density, onset potential to track the urea management and generate energy. They concluded that increasing temperature would increase the catalyst activity. The targeted modification of catalyst surface improves its efficiency. Another challenge in this research area is the fuel cell itself. This recent technique is still under development, production of energy in the fuel cell is very low, there is still much research for maximizing the power production and also increase the treatment efficiency. However, fabrication of fuel cell on the large scale will promote management of urea in the wastewater. Subsequently, research of urea management in wastewater will be developed parallel with research in fuel cell development.

However, the future research should focus on urea management in DUFC, optimization of electrochemical oxidation process in the cell and lowering the cost of operation. As this technique is clean, the waste will be water or energy (H_2) and both of them are clean and safe for the environment.

2.2. Biological treatment

Biological treatment of urea referred to contact the urea wastewater with microorganisms in biofilter where the microbes will catabolize the urea to final products/byproducts. Campos et al. (2003) utilized urea wastewater in both a biofilm airlift nitrifier reactor and an anoxic sludge blanket up-flow reactor for both urea hydrolysis and nitrification. The maximum nitrogen removal percentages were obtained at 1.0 (formaldehyde/N-NH^{4+}). The efficiency of N removal decreased when this percentage reached 1.5 due to increase in the formaldehyde quantity which decreased the nitrification [45]. Converti (2006) used cultivations of Spirulina platensis in batch system. The maximum nitrogen concentration of 1.7 mM sustained the growth of these microalgae without inhibition. Transferring NH_3 into the cell was controlled by hydrolysis of urea by urease enzyme. The pulse-feeding of urea nitrogen sources can sustain the long term-cultivation of S. platensis [46]. Yan Lilong et al. (2013) studied biological aerated filter to manage urea of highly concentrations in sewage from a chemical plant. They stated that the heterotrophic bacteria were the main bacteria among different communities responsible for the urea hydrolysis. Nitrobacteria and nitrosobacteria also contributed to some extent in the hydrolysis process [47]. Dahlan et al. applied Fenton oxidation process on urea from fertilizer plant using P. Fluorescens bacteria. The study estimated the inhibition concentration (723,219 mg/l). No observed effect of concentration (2533,658 ppm) and lowest observed effect of concentration (393,992 mg/l) versus the P. Fluorescens cell accumulation for 96 h. The highest average percentage of ammonium reduction was 94.50% at 2500 ppm of ammonium-nitrogen [48]. Yavari et al. treated urea wastewater from industrial wood (timber) plant using teak phytoremediation. A constructed bench-scale was studied in term of plant biomass, nitrogen recovery and influent flow rate for 56 d. The teak showed a considerable nitrogen uptake [49].

Here, the challenge is the sensitivity of the microorganisms, complexation of reaction and difficulty in the control. The process is costly due to intensive control of the process parameters (temperature, pH, inhibitors, initial concentration and flowrate of urea). The researchers should investigate resistance of microorganisms in sever conditions of pH, temperature and high initial urea concentrations. Another challenge in this technique is the production of undesirable products which act as additional burden to the environment. The configuration of the treatment system plays important role, any fault in the design could cause leakage of the urea with the downstream, overloads on the system, and failure in the process.

2.3. Adsorption

Adsorption technique for wastewater management is widely used [50]–[54]. It involves contact the wastewater (synthetic or real) with a definite amount of adsorbent. The maximum capacity of adsorption attributed to the adsorbent surface area or/and the surface functional groups. Adsorption of urea from wastewater onto nanoporous and nonporous adsorbent was investigated. In 1975, Roger Parsons investigated adsorption of urea (initial concentration 0.1-4 M) onto mercury. They suggested that the urea molecule oriented parallel to mercury. The data was modelled using Langmuir equation [55]. Climent et al. investigated urea management using adsorption onto Pt electrode. They stated that only nitrogen atom is responsible for the adsorption mechanism [56]. Abd El-Lateef et al. investigated the adsorption of urea onto nano-iron phosphate as a function of temperature, agitation speed, reaction time, catalyst mass and solution pH. The highest capacity of adsorption recorded was 102 mg/g at pH (7-9). The kinetic data showed that film diffusion model controls the process at the beginning and then the matrix diffusion model was the predominant. The thermodynamic parameters have been also determined and analyzed. The nanoparticles were stable and the adsorbent recyclability reached 10 cycles [57]. Kameda et al. studied management of urea using ordinary activated carbon. The urea attraction mechanism was the linkage via dipole–dipole reaction between amino and carbonyl/hydroxyl groups of urea and activated carbon, respectively. Then the adsorbed molecule interacted with another one via –C=O group of the former and–NH_2 group of the later which form a multilayer physical adsorption [58]. Simha et al. investigated fly ash as adsorbent for urea from urine. They stated the optimum condition of pH 6, temperature 30 °C, mass of ash 1.9 and urea initial concentration 13.5 g/L maximize the adsorption capacity to 410 mg/g. The reaction mechanism was described to be spontaneous physical adsorption, they attributed the physical nature of adsorption to Van der Waal's force and electrostatic interactions. The film diffusion played important role in controlling that adsorption process. Further, they derived a model to minimize the fly ash amount required for the maximum adsorption capacity to design a multistage batch adsorber [59]. Ooi et al. transformed oil palm to activated carbon fibers (ACF). They explained that the ionized electronegative -OH groups on the adsorbent surface in the alkaline urea solution caused ionic repulsion towards partially anionic urea. The strength of this ionic repulsion increases in case of increasing OH functional groups onto ACF [60]. Wang et al. successfully deposited MgAl@LDH on η-Al_2O_3 to improve urea hydrolysis. They concluded that the highest urea degradation rate, resulted from the high basicity attributed to the active sites of Mg–OH. The urea concentration reached 8.162 mg/L at 165 °C after 120 min which is below the effluent standard. The most important finding of that study is that the prepared adsorbent could efficiently recycled. The slight reduction in activity was referred to the decreasing in basicity [61]. Ooi et al. investigated urea removal by adsorption onto empty fruit bunch based ACF, mesoporous silica, palm kernel shell based granular activated carbon. The porous and nano-sized adsorbents present high affinity for urea. They attributed this to the presence of carboxyl and amine functional groups on the adsorbents surfaces that facilitated the urea adsorption via hydrogen bonding [62]. Liu et al. prepared chitosan/cupper membrane for urea removal. The maximum capacity of urea adsorption was 78.8 mg/g. This value was highly depending on Cu % in the membrane matrix. This was owned to that the presence of four coordinated sites of Cu(II), two of them attached Cu to the chitosan membrane through $-NH_2$ groups and the other two sites are available for bending urea [63]. Abidin et al. suggested oxidized starch nanoparticles as adsorbent for urea. The developed adsorbent showed a promising adsorption capacity (185.2 mg/g). High removal percentage (95 %) was achieved in a relatively short time (4 h). Finally, 91.6 % of the adsorbent was recovered or reused [64]. Ganesapillai and Simha evaluated activated carbon from bamboo for urea adsorption in terms of reaction time, temperature and urea initial concentration. The maximum capacity of urea adsorption reached 146.1 mg/g. The process was described as spontaneous physical reversible exothermic process. And both of intraparticle diffusion and surface adsorption control the kinetics of urea adsorption onto the surface of the adsorbent [65]. Ganesapillai et al. evaluated walnut shells, microwave activated carbonized based on bamboo shoots, and coconut shells for the adsorption of urea in terms of agitation speed, temperature, adsorbent dose, and urea initial concentration. The kinetic studies proved that pseudo 2nd order kinetic model is the best to describe adsorption process of urea onto microwave activated carbon and walnut shells while pseudo 1st order model is fitted coconut shells [66]. Safwat et al. compared activated alumina with activated carbon for the removal of urea from wastewater. The management efficiency was recorded as function of pH strength. The optimum pH to achieve maximum efficiencies of activated alumina (24 %) and activated carbon (31%) was 9. Pseudo 2nd order model expresses the kinetic results for urea adsorption onto activated alumina. Equilibrium studies stated that Redlich-Peterson isotherm was the preferable to describe ureas/alumina adsorption system while Temkin isotherm was the best to describe the other [67]. The major materials used for the adsorption of urea from different systems were activated carbon, zeolite [68,69] (figure 7) and silica and its based materials [70].

Figure 7. SEM images of powdered zeolites [69].

However, there are many research gaps in urea adsorption studies such as: the adsorption technique is more effective in dealing with low urea concentration, the reaction is also sensitive to pH or/and temperature. Additionally, the adsorption technique is implemented on small volumes of wastewater, scaling up the process is under development. Regeneration of the adsorbents is another challenge where the adsorbents losses percentage of their mass and activity during regeneration process and even the regeneration process is costly. Also, the reaction resulted sludge (wasted adsorbents) which is consider as additional burden on the environment. The upcoming studies should search for recycling this sludge in the process or reusing it in anther application like road pavement, fillers, etc. Using biodegradable adsorbents is preferable. And finally, the impact of this sludge on the environment should be accessed.

2.4. Hydrolysis

Zhigang et al. investigated both urea hydrolysis and nutrients (phosphorus and nitrogen) recovery. They studied the effect of temperature, pH and stored/fresh urea ratio on hydrolysis process. It was found that high temperature (> 20 °C), low pH (< 8) and stored urea enhanced the hydrolysis process. Under optimum conditions, the recovery of PO_4^{3-}-P and NH^{4+}-N reached > 85% and 95%, respectively [71]. Rahimpour et al. investigated urea management in wastewater using a cascade of hydrolyser/desorber. They targeted to approach 0 ppm of urea in the effluent. They stated that the removal efficiency of urea and ammonia could be improved by increasing the inlet wastewater temperature, steam flow rate and decreasing the reflux ratio [72]. Rahimpour et al. developed and validated a model for different ranges of urea concentrations in thermal hydrolyser/desorber loops. They solved the model and suggested desorbers/reactors height, flow rate and temperature [73]. Rahimpour and Mottaghi compared countercurrent and cocurrent modes in thermal hydrolyzer for urea decomposition, they applied equilibrium-stage model to the two modes. The simulation data showed that countercurrent mode was superior than concurrent in achieving high removal efficiency and new environmental standards [74].

2.5. Other techniques

Rychen Philippe et al. (2016) suggested ion exchanging column for removing urea from water by adding nitrite to the influent before passing through a strong cation exchange resin. The NO_2 could be added in different forms (nitrous acid, nitrogen dioxide gas or sodium nitrite salt). Regeneration of the column is conducted using an acidic solution [75].

Machdar et al. (2018) studied adding $MgCl_2$, KH_2PO_4, and KOH to a fertilizer plant wastewater. This resulting an added-value, where the urea in the wastewater is treated and slow-release fertilizer is produced. The effects of molar ratios and solution pH were discussed. It was found that, the highest ammonium removal efficiency was 94.7% at the ratios 1.2/1/1 of $Mg^{2+}/NH^{4+}/PO_4^{3-}$ at pH 9 [76].

Eiroa et al. analyzed thermal decomposition of urea. They stated that the production of ammelide, ammeline and cyanuric acid begin at temperature > 190 °C and completed at 250 °C. Urea was hydrolyzed under anoxic environment. The efficiency of conversion reached 100 % even in the presence of formaldehyde in continuous reactor [77]. Lundström et al. (2011) investigated the impact of TiO_2, Fe-Beta, and γ-Alumina on the thermal destruction of urea. The best results dedicated to TiO_2, also the humid condition was preferable [78].

Although there are numerous techniques for urea management in wastewater, but still most of them are achieved in small scale. This is attributed to the high cost of reactants. Much efforts are still required to minimizing the cost of the materials, systems and the overall process.

Barmaki et al. modeled the thermal hydrolysis of urea from industrial wastewater. They used three models to provides the hydrolyser height, concentrations distribution and temperature for numerous components. They examined the wastewater flowrate, steam and temperatures. They concluded that any increase in temperature would increase the urea removal efficiency [79].

3. Recommendations for future research on urea wastewater management

It can be seen that discharging urea to water bodies cause negative impact on human and the environment. Subsequently, urea should be managed efficiently in wastewater. Different techniques were discussed in this review. Every technique has its advantages and challenges. One of the main challenges in all these techniques is the possibility to produce further wastes. So, it is highly recommended that the future studies in this area should investigate green materials. The researchers could study biodegradable materials to reduce their negative impacts on water bodies and also, they can work on industrial wastes to add values to these wastes and protect the environment from additional wastes.

Implementation of nanotechnology in urea management is highly recommended to improve the efficiency of the treatment process. The nanotechnology increase the materials performance through improving the mechanical, physical and chemical characteristics of the materials and also minimize the amounts of the used materials which conserve these materials and lowering the total cost of operation.

Due to the world trend of replacing the fossil fuels with renewable energy, electrooxidation process in microbial fuel cell has drawn the attention recently due to the fact that this technique is efficient and clean (no waste) for wastewater management in parallel with energy production. The researchers should focus on modification of the current electrodes and searching for others to rise the anodic current densities and improve the electrodes stability, which will simultaneously reduce the onset potential for urea oxidation reaction resulting in efficient urea management and improvement the power density in the urea-based fuel cells. Electrochemical oxidation of urea wastewater in industrial scale should be economically feasible in the future.

4. Conclusions

Along the past decades, urea management in wastewater was investigated through different techniques. Some techniques are costly due to the cost of materials and maintenance. Some techniques like adsorption release additional wastes to the environment. Other techniques are complex in operation and control. But still the electrochemical oxidation method until now is the superior technique compared to others. It releases less waste, especially in the case of fuel cell where nanoparticles have been applied as anode for management of urea beside energy (hydrogen) production. Additionally, it yields low harmful environmental impacts, although the overall efficiency of the electro-oxidation processes and fuel production were improved in the last years but further research is required for industrial and transportation applications. The sustainability and success in urea management in wastewater could be achieved if the researchers work on developing all these techniques and materials.

References

[1] Rollinson A N, Jones J, Dupont V and Twigg M V 2011 *Energy Environ. Sci.* **4** 1216
[2] Urbańczyk E, Sowa M and Simka W 2016 *J. Appl. Electrochem.* **46** 1011
[3] Matijašević L, Dejanović I and Lisac H 2010 *Resour. Conserv. Recycl.* **54** 149
[4] Boggs B K, King R L and Botte G G 2009 *Chem. Commun.* **32** 4859
[5] A T Miller, B L Hassler and G G Botte 2012 *J. Appl. Electrochem.* **42** 925
[6] Bolzan A E. and Iwasita T 1988 *Electrochim. Acta* **33** 109
[7] Hernlem B J 2005 *Water Res.* **39** 224
[8] Vedharathinam V and Botte G. G. 2014 *J. Phys. Chem. C***118** 21806
[9] Ding R, Qi L, Jia M and Wang H 2014 *Nanoscale* **6** 1369
[10] Xu W, Zhang H, G Li and Wu Z 2014 *Sci. Rep.* **4** 5863
[11] Wang D, Yan W and Botte G G 2011 *Electrochem. commun.***13** 1135
[12] King R L and Botte G G 2011 *J. Power Sources* **196** 9579
[13] Yan W, Wang D and Botte G G 2012 *Appl. Catal. B Environ.* **127** 221
[14] Lohrasbi E and Asgari M 2015 *Adv Anal Chem.* 5 9
[15] Urbańczyk E, Jaroń A and Simka W 2017 *J. Appl. Electrochem.* **47** 133
[16] Ye K, Zhang D, Guo F, Cheng K, Wang G and Cao D 2015 *J. Power Source* **283** 40
[17] Szpyrkowicz L, Juzzolino C, Daniele S and De Faveri M D 2001 *Catal. today* **66** 519
[18] Guo F, K. Ye, Cheng K, Wang G and Cao D 2015 *J. Power Sources* **278** 562
[19] Yang D, Yang L, Zhong L, Yu X and Feng L 2019 *Electrochim. Acta* **295** 524
[20] Wu M-S, Ji R-Y and Zheng Y-R 2014 *Electrochim. Acta* **144** 194
[21] Barakat N A M, Yassin M A, Yasin A S and Al-Meer S 2017 *Int. J. Hydrogen Energy* **42 21741**
[22] Saleh M M, Tammam R H and Touny A H 2018 *Env. Sci.* **25** 19898
[23] Zaher A, El Rouby W M A and Barakat N A M 2020 *Int. J. Hydrogen Energy* **45** 8082
[24] Zeng M. *et al.* 2019 *ACS Sustain. Chem. Eng.* **7** 4777
[25] Li J, Li X, Huang C and Zhang J 2019 *Ionics (Kiel).* **25** 1943

[26] Alajami M, Yassin M A, Ghouri Z K, Al-Meer S and Barakat N A M 2018 *Int. J. Hydrogen Energy* **43** 5561
[27] Rahimpour M R, Mottaghi H R and Barmaki M M 2010 *Fuel Process. Technol.* **91** 600
[28] Shen S, Li M, Li B and Zhao Z 2014 *Environ. Sci. Pollut. Res.* **2** 12563
[29] Amstutz V, Katsaounis A, Kapalka A, Comninellis C and Udert K M 2012 *J. Appl. Electrochem.* **42** 787
[30] Abdel-Monem N M, Abdel-Salam O E, Nassar A F and Mahmoud M H 2013 *Int. J. Sci. Eng. Res.* **4** 1715
[31] Singla J, Sangal V K and Verma A 2019 *Process Saf. Environ. Prot.* **130** 197
[32] Simka W, Piotrowski J, Robak A and Nawrat G 2009 *J. Appl. Electrochem.* **39** 1137
[33] Mahmoud M H, Abdel-Monem N M, Abdel-Salam O E, Nassar A F and El-Halwany M A 2013 *J. Am. Sci.* **9** 96
[34] Xu W, Wu Z and Tao S 2016 *Energy Technol.* **4** 1329
[35] You J, Greenman J, Melhuish C and Ieropoulos I 2016 *J. Chem. Technol. Biotechnol.* **91** 647
[36] Barakat N A M, Alajami M, Al Haj Y, Obaid M and Al-Meer S 2017 *Catal. Commun.* **97** 32
[37] Abdelkareem M A, Al Haj Y, Alajami M, Alawadhi H and Barakat N A M 2018 *J. Environ. Chem. Eng.* **6** 332
[38] Wang L, Xie B, Gao N, Min B and Liu H 2017 *Environ. Sci. Pollut. Res.* **24** 20401
[39] Lan R and Tao S 2011 *J. Power Sources* **196** 5021
[40] Santoro C, Ieropoulos I, Greenman J, Cristiani P, Vadas T, Mackay A and Li B 2013 *Int. J. Hydrogen Energy* **38** 11543
[41] Carpenter K and Stuve E M 2019 in *Meeting Abstracts* **32** 1687.
[42] Ieropoulos I, Greenman J and Melhuish C 2012 *Phys. Chem. Chem. Phys.* **14** 94
[43] Ieropoulos I A, Greenman J and Melhuish C 2013 *Int. J. Hydrogen Energy* **38** 492
[44] Barakat N A M, Amen M T, Al-Mubaddel F S, Karim M R and Alrashed M 2019 *J. Adv. Res.* **16** 43
[45] Campos J L, Sánchez M, Mosquera-Corral A, Mendez R and Lema J M 2003 *Water Res.* **37** 3445
[46] Converti A, Scapazzoni S, Lodi A and de Carvalho J C M 2006 *J. Ind. Microbiol. Biotechnol.* **33** 8
[47] Liong Y, Ying Z and Yu L 2013 *J. Chem. Soc. Pakistan* **35** 1081
[48] Dahlan M H, Faizal M, Marsi M and Arinafril A 2015 *Int. J. Biol. Ecol. Environ. Sci.*
[49] Yavari S, Malakahmad A, and Sapari N B 2017 *Environ. Prot. Eng.* **43** 4
[50] El-Geundi M S, Ashour E A, Abobeah R M A and Shehata N 2014 *Int. J. Sci. Eng. Technol. Res.* **3** 210
[51] Egirani D E, Poyi N R and Shehata N 2020 *Int. J. Environ. Sci. Technol.* **17** 2443
[52] Naggar I M, Ahmed S A, Shehata N, Sheneshen E S and Fathy M 2018 *J Water Technol Treat Methods* **1** 104
[53] Shehata N, El-Geundi M S, Ashour E A and Abobeah R 2016 *Inter. J. Chem. Process Eng. Res.* **3** 35
[54] El-Naggar I M, Ahmed S A, Shehata N, Sheneshen E S, Fathy M and Shehata A 2019 *Appl. Water Sci.* **9** 7
[55] Parsons R, Peat R and Reeves R M 1975 *J. Electroanal. Chem. Interfacial Electrochem.* **62** 151
[56] Climent V, Rodes A, Orts J M, Aldaz A and Feliu J M 1999 *J. Electroanal. Chem.* **461** 65
[57] El-Lateef H M A, Al-Omair M A, Touny A H and Saleh M M 2019 *J. Environ. Chem. Eng.* **7** 102939
[58] Kameda T, Ito S and Yoshioka T, 2017 *J. Dispers. Sci. Technol.* **38** 1063
[59] Simha P, Ramanathan A, Thawani B, Jain P, Hussain S and Ganesapillai M 2016 *Arab. J. Chem.* **12** 5049
[60] Ooi C -H, Cheah W -K, Sim Y -L, Pung S -Y and Yeoh F -Y 2017 *J. Environ. Manage.* **197**

199
[61] Wang Y, Shen S, Li M Sun Y and Li B 2018 *Catal. Letters* **148** 1893
[62] Ooi C -H, Cheah W -K and Yeoh F-Y 2019 *Adsorption* 1
[63] Liu J, Chen X, Shao Z and Zhou P 2003 *J. Appl. Polym. Sci.* **90** 1108
[64] Abidin M N, Goh P S, Ismail A F, Said N, Othman M H, Hasbullah H, Abdullah M S, Ng B C, Kadir S H Kamal F 2018 *Carbohydr. Polym.* **201** 257
[65] Ganesapillai M and Simha P 2015 *Resour. Technol.* **1** 90
[66] Ganesapillai M, Venugopal A, Ananthkrishna V and Tapankrishna N 2015 *Asia-Pacific J. Chem. Eng.* **10** 438
[67] Safwat S M and Matta M E 2018 *J. Dispers. Sci. Technol.* **39** 1699
[68] Bergé-Lefranc D, Pizzala H, Paillaud JL, Schäf O, Vagner C, Boulet P, Kuchta B and Denoyel R 2008 *Adsorption* **14** 377
[69] Wernert V, Schäf O, Ghobarkar H and Denoyel R 2005 *Microporous Mesoporous Mater.* **83** 101
[70] Cheah W -K, Sim Y -L and Yeoh F -Y 2016 *Mater. Chem. Phys.* **175** 151
[71] Zhigang L I U, Qingliang Z, Kun W, Duujong L E E, Wei Q I U and Jianfang W 2008 *J. Environ. Sci.* **20** 1018
[72] Rahimpour M R, Mottaghi H R and Barmaki M M 2010 *Chem. Eng. J.* **160** 594
[73] Rahimpour M R, Barmaki M M and Mottaghi H R 2010 *Chem. Eng. J.* **164** 155
[74] Rahimpour M R and Mottaghi H R 2010 *Ind. Eng. Chem. Res.* **49** 9289
[75] Rychen P Scheiffelen B and Frank T 2016 Google Patents
[76] Machdar I, Depari S D, Ulfa R, Muhammad S, Hisbullah A B and Safrul W 2018 *IOP Conf. Ser. Mater. Sci. Eng.* **358** 12026
[77] Eiroa M, Kennes C and Veiga M C 2004 *Water Res.* **38** 3495
[78] Lundström A, Snelling T, Morsing P, Gabrielsson P, Senar E and Olsson L 2011 *Appl. Catal. B Environ.* **106** 273
[79] Barmaki M M, Rahimpour M R and Jahanmiri A, 2009 *Sep. Purif. Technol.* **66** 492

Catalytic thermochemical cracking of polyethylene over nanocomposite bentonite clay

A S Hamouda[a], P Acharjee[b], A Abdelrahman[c], A M Radwan[d], A H Zaki[e], A Farghali[e], H F M Mohamed[f], A Yadav[b], A Bhatnagar[b] and A Goel[b]

[a] Department of Environmental Sciences and Industrial Development, Faculty of Postgraduate Studies for Advanced Sciences, Beni-Suef University, P.O. Box 62511 Beni-Suef, Egypt
[b] The Energy and Resources Institute, TERI, Darabri Seth Block, India Habitat Center, Lodhi Road , New Delhi, India
[c] Renewable Energy Science & Engineering Department, Faculty of Postgraduate Studies for Advanced Science, Beni-Suef University, P.O. Box 62511 Beni-Suef, Egypt
[d] National Research Center, Chemical Engineering & Pilot Plant Department, Dokki, Giza, Egypt
[e] Department of Materials Science and nanotechnology, Faculty of Postgraduate Studies for Advanced Sciences, Beni-Suef University, P.O. Box 62511 Beni-Suef, Egypt
[f] Physics Department, Faculty of Science, Minia University, P.O. Box 61519 Minia, Egypt

*E-mail: asmaa_hamouda@psas.bsu.edu.eg , asmaa_hamouda@yahoo.com

Abstract. Catalytic thermochemical cracking of polyethylene has been investigated in presence of a series of bentonite clay based nanocomposite with a view toward assessing their suitability in a process for converting plastic waste to fuel. Metal oxides such as manganese oxide (MnO), titanium oxide (TiO_2) and zinc oxide (ZnO) were impregnated with sodium montomorillonite bentonite. The clay catalysts were characterized by x-ray diffraction (XRD), scan electron microscope (SEM), transmission electron microscope (TEM) and energy dispersive x-ray (EDX) spectrometers. The effect of different catalysts on the composition and the yield of gaseous was evaluated. ZnO nanocomposite bentonite clay achieved the highest methane gas formation around 93%. The superior catalytic performance of nanocomposites is probably attributed to their uniform dispersed of oxides particles on montmorillonite support.

1. Introduction

Disposal of plastic wastes are the major issues and challenges associated with the generation of large quantity of plastic waste. Conventional solutions like recycling, landfilling and plasma gasification have inherent challenges which have been listed out in various literature sources [1-8]. Two of the prominent issues are: i) release of toxic emissions such as, carbon monoxide, chlorine, hydrochloric acid, dioxin, furans, benzene, 1,3-butadiene, etc. from its combustion, and ii) leaching of additives like lead, cadmium and di(2-ethylhexyl)phthalate (DEHP) in the ground when this plastic is disposed in landfills [9]. Therefore, there is a strong need in the further Research & Development of advanced waste to energy conversion technologies for effective management of plastic waste.

Earliest references of plastic degradation using thermal conditioning date back to middle of twentieth century. Jellinek [10] performed the thermal degradation of polystyrene and polyethylene in vacuum in the temperature range of 340-400°C. An advanced plastic pyrolysis process had been developed by Western Research Institute [11]. Recently, Kalargaris *et al.* [12] published a performance analysis of various blends of plastic pyrolysis oil and diesel in a diesel engine and the result was found to be promising. Some of the factors that influence the product distribution in plastic pyrolysis include temperature, type of reactor, pressure, residence time and catalysts [13, 14]. A variety of pyrolysis studies have been conducted on industrial wastes such as tyres and plastics, and several reviews have reported on the characterization of the development of pyrolysis technologies in terms of different aspects, for example, reactor development and product characterization [15-18]; conditions for oil production, oil characteristics and upgrading [19]; the heating rate and other governing variables affecting pyrolysis process and pyrolysis products of tyres and the mechanism investigation or kinetics modeling of the pyrolysis process [19, 20]. The thermo gravimetric analysis (TGA) for polyethylene terephthalate (PET) plastic have shown that the thermal degradation starts at 400°C and no significant change is observed beyond 470°C [21]. Similarly, for high density polyethylene (HDPE) the thermal degradation is reported to start in the range of 378–404°C and it completes at 517–539°C [22]. Poly vinyl chloride (PVC) is found to have a weight loss in two different temperature ranges; i) 260-385°C wherein ~62% weight loss has been reported and ii) 385-520°C wherein ~22% of weight loss occurred [21]. Furthermore, the temperature also effects the product distribution. For maximizing the liquid production, a temperature range of 300-500 °C is recommended. The higher temperatures are found to maximize the char and gas yield [13].

Gasification of plastics was reported first in 1970's [23]. Borgianni *et al.* [24] performed gasification of wastes containing PVC using a two-stage reactor with sodium carbonate as an additive for elimination of chlorine gas from the producer gas. Air gasification for polypropylene waste has been demonstrated by Xiao *et al.* [25] in a fluidized bed reactor and inferred that the equivalence ratio had a major effect on producer gas quality and quantity. They reported a gas yield of 3.9 Nm^3/kg, with a tar content of 250 mg/Nm^3. One of the greatest challenges encountered in the plastic gasification is to control the syngas quality in terms of tar content and calorific value with changing feed [26].

Catalysts placed in contact with the feedstock inside the pyrolysis unit can not only improve conversion of plastics but also adjust the distribution of volatilized products. Catalysts can also be used downstream of the primary reactor for reformation and upgrading of pyrolysis vapors especially Lewis acid catalysts. Heterogeneous catalysts such as nanocrystalline zeolites, conventional acid solid, mesostructured catalyst, metal supported on carbon and basic oxides have been used for conversion of plastic [27]. The zeolite HZSM-5 has been found to increase the gaseous product yield to 63.5 % during the pyrolysis of HDPE [28]. Aguado *et al.* [27] had studied the effect of zeolite catalyst in a sequential two step reaction process; i) the plastic was thermally cracked in the batch reactor and ii) the vapors generated were catalytically decomposed in a fixed bed. A considerable rise in gaseous fraction up to the extent of 74.3 wt.% was observed. The effect of catalysts in different contact modes namely liquid phase and vapor phase were performed [29]. The silica-alumina catalyst in liquid phase led to higher liquid yield and the catalyst in gaseous phase assisted the decomposition of higher hydrocarbons thus increasing the gaseous yield fraction.

Clays have been used as acidic catalysts in petroleum cracking and several other organic reactions. To enhance performance, research is now focused on the synthesis of Pillared clays (PILCs), which possess larger two-dimensional pores [30]. Pillaring is achieved by the insertion of voluminous inorganic metal hydroxyl cation between layers of the clay in order to keep them apart. On thermal treatment, the metal hydroxyl cations are transformed into pillars which are responsible for the acidity [30-32]. The surface of the pillars possesses both Lewis and Bronsted acid sites [32-34]. However, the use to catalysts often makes these processes economically unviable. An important research gap can be found in the development of catalysts which are economical, available in bulk and easy to regenerate [35].

There are large deposits of bentonite clay minerals across the parts of the country. These minerals are not yet exploited for commercial purposes. The unique physicochemical properties of sodium montomorillonite clays are the result of their extremely small crystal size, variation of internal chemical composition, structural characteristics caused by chemical factors, large cation exchange capacity, large chemically active surface area, variation in types of exchangeable ions and surface charge, and interactions with inorganic and organic liquids. Clay minerals have a very strong swelling and adsorption capacity, which is particularly interesting for the impregnation of catalytically active noble metals in the inter-lamellar space of clay. Several methods were reported in the literature [36-40] for the preparation of nanoparticle/clay (especially SiO_2-based clays) composites. The impregnation of preformed nanoparticles into a clay matrix can offer a new and facile procedure for nanometal/clay composite fabrication where the size and morphology of the nanoparticles are better controlled by separate and mostly standard preparation steps.

There are several methods to prepare nano-sized materials. These materials may be used directly or in the form of supported nanoparticles on solids such as oxides, carbon or zeolites. Some usual methods for nano-catalysts preparation are impregnation, precipitation, chemical vapor deposition, and electrochemical deposition. Precipitation and impregnation methods are simple, cheap, and well-studied but it is difficult to control the size of the particles. Chemical vapor deposition is widely used in the electronics industries but it is an expensive method. Electrochemical deposition is an inexpensive method that does not need high temperature and concentration. This method would allow a good control on size and chemical properties of the deposited nanomaterials but usually forms one dimensional nanomaterials [41-43]. In most research, impregnation and precipitation methods have been used as catalysts for biomass conversion. Meanwhile, nanoparticles properties usually adjust by changing synthesis parameters such as pH of the solution, concentration, the reducing agent, and calcination temperature [44].

Various catalysts and adsorbents have been studied for upgrading the quality and debromination of the pyrolysis oil from brominated flame retarded (BFR) plastics [45, 46]. Among them, sodium montemornilite and others modified alumina and clay minerals [47, 48], fluid catalytic cracking (FCC) catalysts [47], metallic oxides [49, 50], and minerals [51] have attracted more concerns. In particular, clay metals oxides-based catalysts such as manganese oxides have shown good performances on the reduction of the amount of bromine in the pyrolysis oil [46, 48]. Meanwhile, clay composites catalysts can be used in the pyrolysis of other types of waste plastics, for example, municipal solid waste (MSW) polyalkene plastics high density polyethylene and polypropylene to improve the quality of the oils [47, 52-56]. The use of different types of clay as catalysts allows a narrower range of hydrocarbons to be produced [56].

2. Experimental

2.1. Samples preparation

Waste plastics were obtained from the cap of used disposable water bottles. These were flushed and cleaned with water, cut and grinded into small portions. High density polyethylene (HDPE B1054) was purchased from Saudi Basic Industries Corporation, SABIC, Saudi Arabia.

All chemicals which used for nano-oxide metals were purchased from Loba Cheme (India) and were used without purification. The manganese oxide (MnO) nanoparticles was prepared using co-precipitation method. Two different anions, manganese (II) sulphate and manganese oxalate slats, were used for the preparation. Same quantity of both slats with 0.2M concentration are mixed together with continuous stirring (about 1 hour) at a constant temperature of 60°C. While stirring, NaOH solution was added to the solution till the pH of the mixture becomes 12. So that, brown precipitates were formed then the mixture was filtered and washed with ethanol. Precipitates were dried for overnight at 100°C and then kept in muffle furnace at 500°C for 4 hours [57].

Nanoparticle of titanium oxide (TiO_2) powder was prepared by physical grinding. A quantity of 100 g TiO_2 powder was put in a grinding machine which having high speed rotator. It was uniformly grinded and crushed well for 15 min with utmost precaution to avoid any contamination. At the end, the finely grinded powder was separated to get nano-size powder TiO_2 which was packed in plastic pouches and stored in normal room temperature until use [58]. Zinc oxide (ZnO) nanoparticles were prepared using sol-gel method. In a typical procedure 12.6 g of zinc acetate dehydrate was added to 400 ml of double distilled water. The mixture is stirring continuous to dissolve zinc acetate completely. Then the solution was heated to 50°C and 600 ml of absolute alcohol was added slowly with stirring. After this, 6 ml of H_2O_2 (47%) was dropped to the mixture carefully and mixed it using a magnetic stirrer to get an almost clear solution. This solution was incubated for 24 hours and then was dried at 80°C for several hours to obtain white zinc oxide nanoparticles. Zinc oxide nanoparticles was washed several times with double distilled water to remove the byproducts. After washing, the ZnO nanoparticles were dried at 80 °C in hot air oven to obtain a complete conversion of zinc [59].

Bentonite powder was purchased from M-I SWACO, Egypt and its nanoparticles was prepared by physical grinding. TiO_2-bentonite, MnO-bentonite, and ZnO-bentonite nano-composites were prepared by simple impregnation method as shown in figure 1. One gram of bentonite nanoparticles was dispersed in 250 ml of water then 0.5 g of each metal oxide was added to prepare the desired composite. The TiO_2-bentonite, MnO-bentonite, and ZnO-bentonite nanocomposites were then filtered and washed several times with distilled water. All the samples were left to dry overnight at 100°C and then subjected to physical grinding method to enable size reduction. In brief, a quantity of 100g each composite was put in a grinding machine which having high speed rotator for 10 h. To prepare TiO_2-ZnO-MnO-bentonite composite, we took 1 g of each pervious prepared composite (TiO_2-Clay ,ZnO - clay, and MnO-clay) and grinded with a grinding ball machine, which having high speed rotator for 6 hrs. Then 0.5 g of the solid composite was dissolved in 100 ml of hot deionized water left overnight, then filtrate, washed several times, and dried at oven 100°C for 4 hrs.

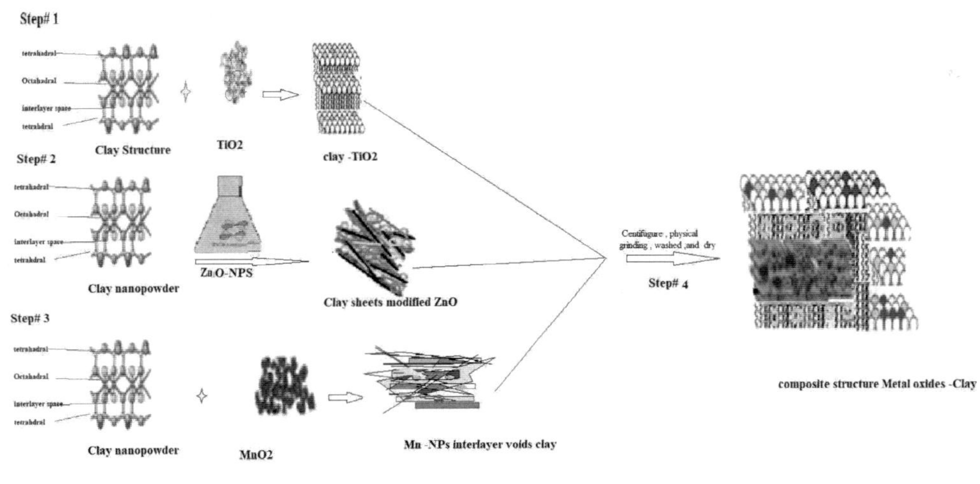

Figure 1. Schematic diagram for the impregnated process of metal oxides with Bentonite.

2.2. Characterization

The HDPE samples were studied using thermo gravimetric analyzer (TGA, NETZSCH TG 209F3 Tarsus analyzer). The alumina crucible was cleaned using distilled water and empty crucible was kept in the analyzer for calibration by creating a baseline. After completing the calibration, the grinded portion of HDPE was kept in the crucible which was placed in the sample holder for conducting analysis. The NETZSCH Proetus computer software program was used to record and analyze the sample, beginning in the range of 25 to 650°C and at 10, 20, and 50°C/min of heating rates, respectively, under a nitrogen atmosphere (flow rate 20 mL/min).

The crystalline structure of the ZnO-Bentonite, MnO-Bentonite and TiO_2- Bentonite nanocomposites was detected by X-ray diffraction (XRD). XRD patterns were recorded with a Pan Analytical, which was equipped with CuKα radiation (λ = 0.1542 nm), Ni-filter and general area detector. The diffractograms were recorded in the 2θ range of 5–80° with step size of 0.02°.

To investigate the nanostructure of the prepared samples, the microstructures were studied using transmission electron microscopy (TEM, JEOL JEM-1230). TEM has been operating at 120 kV and attached to a CCD camera. A copper grid pre-covered with a very thin amorphous carbon film. To investigate the prepared samples, small droplets of the suspended solution for each sample were placed on the carbon-coated grid. Also, scanning electron microscopy (SEM) was performed for the prepared samples using SEM (Oxford instrument INCA/Sight) at 40 kV. The measurements for TEM and SEM were carried out at the National Research Centre, Egypt. The energy dispersive X-ray analysis (EDXA) was used to investigate the pyrolytic char produced by Bentonite.

The waste plastic char samples extracted from the reactor at different conditions during gasification were analyzed by a variety of materials characterization techniques at the Central Analytical Laboratories of National Research Centre, Egypt such as elemental analysis (CHNSO elemental, Vario EL III).

The non-condensable gases from the experiment were analyzed using Gas Chromatography (VARIAN model CP-3800) which has two columns (A and B). Column A used argon as a carrier gas, which measured hydrogen and relatively heavy hydrocarbon gases. In column B, helium was used as the carrier gas to measure nitrogen, oxygen, carbon monoxide, carbon dioxide and methane. The Gas Chromatograph analysis were conducted at the Egyptian Petroleum Research Institute (EPRI), Egypt.

2.3. Pyrolysis system

Pyrolysis of polymer was carried out using a bench scale fixed-bed reactor (SS 316, 0.98 mm ID). The total height of the reactor is 38cm with total volume of 42cm^3, externally heated by an electrical furnace. The reactor consists of three main parts, to be connected together during preparation for the experiments. The upper part includes two valves and pressure gauge to monitor the pressure during the experiment as shown in figure 2. The plastic and additives are added through the lower part. The upper and lower parts are collected together through a middle fitting connection. The air in the reactor was purged for at least 5min using the selected pyrolysis agent (100% N_2). One gram of pure HDPE or waste plastic with 4% of its weight from different catalysts is loaded for one hour at 750°C. The loaded reactor was placed in the furnace at selected reaction temperature. The electrical furnace (60 x7 x 20cm^3), insulated with glass wool material, is used to supply the heat required to maintain the temperature of reaction for pyrolysis of plastic waste. Furthermore, temperature controller with a thermocouple was placed next to the outer surface of the reactor to record and control the temperature inside the reactor. After a given reaction time, the reactor was taken out of the furnace and cooled with an electrical fan for about 15min to cool down the reactor to room temperature (25°C). The non-condensable gases are flashed from the top of the reactor and measured by displacement of water in the gas sampler collector. Also, char and tar are collected and weighted after the reactor was opened.

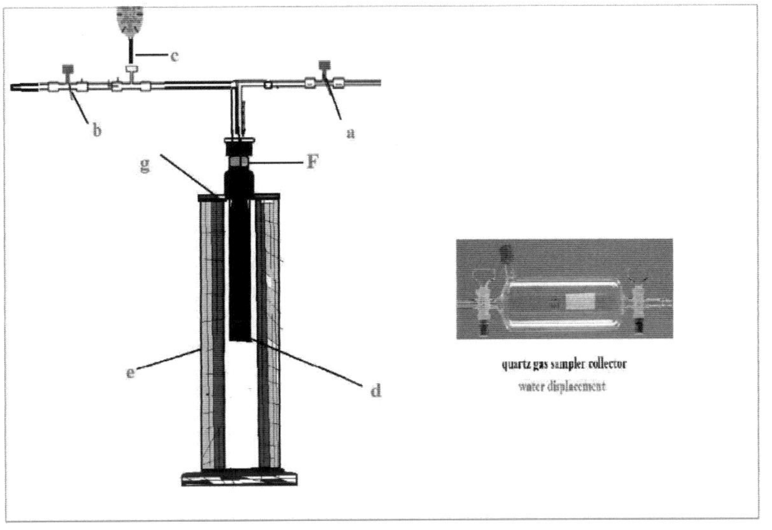

Figure 2. Schematic of pyrolysis system where (a) inlet valve, (b) outlet valve, (c) pressure gauge, (d) pyrolytic reactor, (e) furnace, (f) union connection and (g) thermocouple k-type.

3. Results and Discussion

3.1. Physical characterization of catalysts

Figure 3 shows the XRD pattern of the bentonite (clay), MnO, ZnO, TiO_2, and their composites. Table 1 lists the peaks ICDD cards for all the samples under study. By comparing the data between XRD pattern in the figure with ICDD data in the table below, which it is clear that, TiO_2 and prepared MnO and ZnOand their composites were a good agreement between them. The XRD data for all the oxides state high crystallinity. Average particle size (d) of the particles have been calculated from high intensity peak using the Debye-Scherrer equation (Eq. (1)) as [60]

$$d = K\lambda / B \cos(\theta) \quad (1)$$

Where K is the Scherrer constant (0.89), λ is the wave length of X–ray beam used, B is the full width at half maximum (FWHM) of diffraction (in radians) and θ is the Bragg's angle. As clear from figure 3(B), MnO metal nanoparticles synthesized were purely crystalline in nature. Average particle size of MnO nanoparticles was found to be 25-30 nm where the intensity of the main MnO lines sharply increased, in addition to the appearance of new lines at 2θ angles: 18°, 38°, 43° and 63°. For ZnO nanoparticles (figure 3C), X-ray diffraction shows that metal oxide is ZnO having hexagonal structure and no peak is detected due to any other material or phase indicating a high degree of purity of the as synthesized sample. The broadening of the X-ray diffraction lines reflects the nanoparticle nature of the sample. Sharpness of the peaks shows good crystal growth of the oxide particles. The calculated average crystallite is 50nm and the surface area of ZnO nanoparticles is 4.3737 m^2/g. The intensity of the main TiO_2 lines sharply increased, in addition to the appearance of new lines at 2θ angles: 21°, 25.5°, 27°, 38°, 57° and 63° as clear in figure 3D. The X-ray pattern reflected that, the TiO_2 is very pure and consists in a majority of anatase particles. The indexed peaks can be found on the powder diagrams obtained in figure 3D. The calculated average crystallite is 27 nm and the surface area of TiO_2 nanoparticles is 60m^2/g.

XRD patterns of MnO-bentonite, ZnO-bentonite and TiO_2-bentonite nanocomposites (figure 3(E, F and G) confirmed the preparation of the desired catalysts, where the patterns showed mixed peaks of Bentonite and metal oxides. It is worth mentioning that, in case of TiO_2-bentonite and ZnO-bentonite nanocomposites, the peaks of TiO_2 and ZnO are predominant in the patterns which may be attributed

IOP Conf. Series: Materials Science and Engineering 1046 (2021) 012022 doi:10.1088/1757-899X/1046/1/012022

to the good distribution of TiO_2 and ZnO on the bentonite-surface. For the ZnO-bentonite, new diffraction lines with high intensity appeared which are characteristic for the ZnO phase (ASTM 05-0664). In addition, the lamella line of the Bentonite was shifted to the left (to a higher d-spacing), which evidenced the penetration of ZnO to inside the bentonite lamella resulting in this expansion.

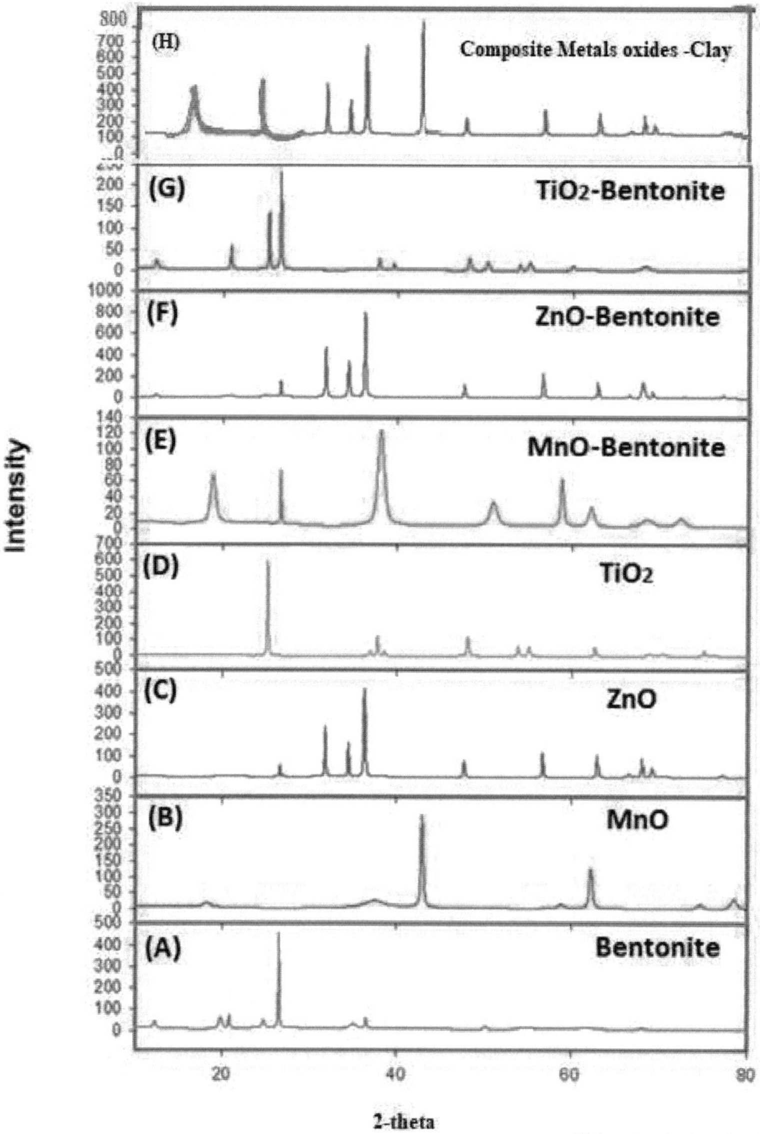

Figure 3. XRD pattern of the samples under study; (A) bentonite, (B) MnO, (C) ZnO, (D) TiO_2, (E) MnO-bentonite, (F) ZnO-bentonite, (G) TiO_2-bentonite and(H) composite metals oxide clay.

Table 1. Peak positions (2θ degree) of XRD patterns and the ICDD reference code of pure clay and different metal oxides.

Material	Peak position (2θ, degree)	ICDD Ref. code
MnO	18.18, 37.48, 42.82, 58.81, 6.21,74.58 and 78.43	04-006-0700
ZnO	20.57, 26.72, 31.84, 34.51, 36.32, 47.62, 56.67, 62.93,66.49, 68.02, 69.17 and 77.05	01-081-8838
TiO_2	25.34, 37.00, 37.85, 38.63, 48.09, 53.96, 55.13, 62.73, 68.79, 70.40 and 75.14	03-065-5714
Bentonite	6.80, 12.26, 19.80, 20.81, 24.85, 26.64, 35.11, 36.57, 50.16, 54.80, 61.72 and 68.07	Montmorillonite

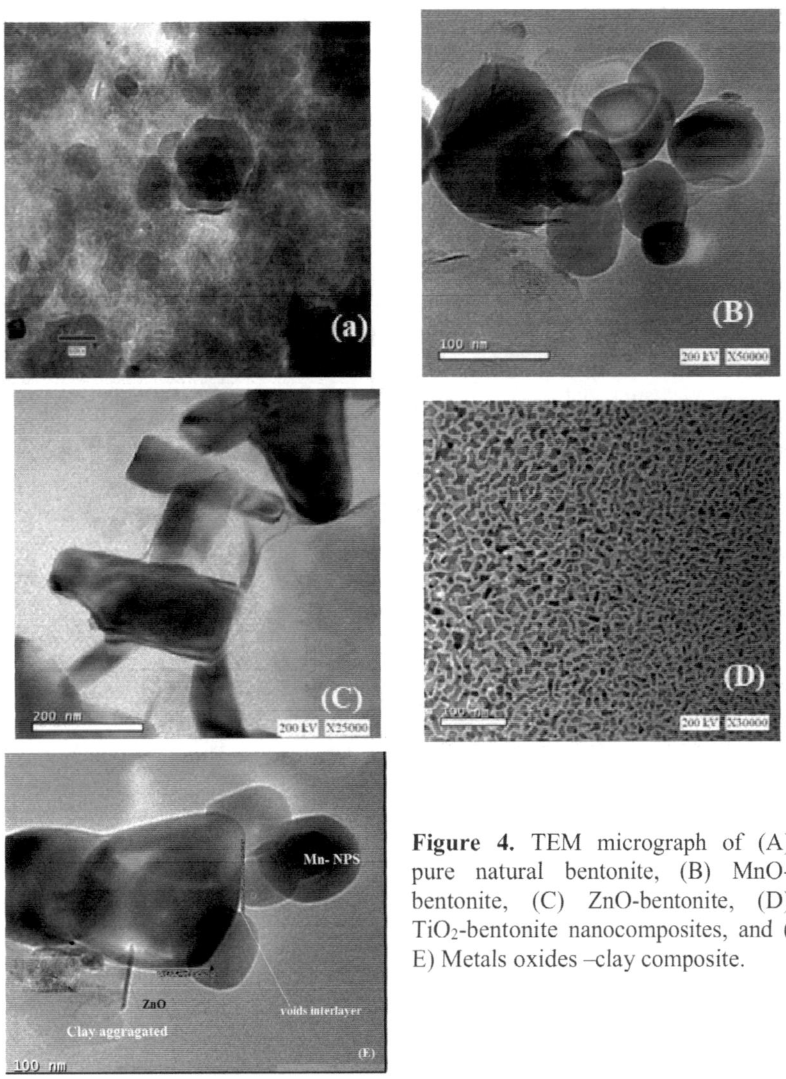

Figure 4. TEM micrograph of (A) pure natural bentonite, (B) MnO-bentonite, (C) ZnO-bentonite, (D) TiO_2-bentonite nanocomposites, and (E) Metals oxides –clay composite.

Figure 4(A-D) shows the TEM micrographs of bentonite, MnO-bentonite ZnO-bentonite and TiO_2-bentonite nanocomposites, respectively. It is clear from figure 4A that, the average particle size of natural bentonite exceeded 100 nm with no uniform shape. The MnO- and ZnO nanoparticles are deposited on the bentonite surface (figure 4(B and C)) while the TiO_2 nanoparticles uniformly covered the bentonite surface, figure 4D. Due to this uniform coverage the XRD patterns of pure TiO_2 and TiO_2-bentonite are very close to each other. Since a crystallite can be defined by studying the orientation of the lattice fringes, one can see the average crystallite size in the synthesized TiO_2 powder is about 5~10 nm in diameter. As an important functional metal oxide, MnO nanoparticles are one of the most attractive inorganic materials because of its physical and chemical properties and wide applications in catalysis, ion exchange and molecular adsorption. It is clear from figure 4B that, the average size of MnO is about 25-30 nm. There are large numbers of potential applications of MnO nanoparticles such as catalysis application in cracking of polyolefin and petroleum refinery industrial. On the other hand, the size of ZnO-bentonite nanoparticles is about 27-82 nm as shown in figure 3C. They are formed and modified by nano-bentonite which improved the catalytically activity of composite and gave good gas yield when applied on degrading of waste plastic and pure HDPE.

3.2. Thermal Characterization of HDPE

The thermal stability of HDPE sample was studied using thermos gravimetric analyzer (TGA). Figure 5 shows the mass-loss behavior as a function of temperature for HDPE sample at different heating rates. HDPE samples at all the heating rates show similar behavior with a single mass loss zone. The zone is approximately centered at (425, 450 and 460) °C for heating rate of 10, 20 and 50°C/min, respectively. For all the three heating rates, the cracking zone of HDPE is in the temperature range of (350-520) °C. Major mass loss in case of heating rate at 10 °C/min started approximately at 350 °C and continued up to 500 °C with a mass loss of around 93.66%. In case of heating rate at 20 °C/min, major mass loss started approximately at 380°C and continued up to 500°C with a mass loss of around 98.80%. Whereas for heating rate of 50°C/min, major mass started approximately at 420 °C and continued up to 520°C with mass loss of around of 96.82%, respectively. In all the heating rates, more than 93% of the mass got converted into condensable and non-condensable volatiles. Residue percentage in case of 10°C/min is high as compared to 20 and 50°C/min due to the polymer's recrystallization properties at lower heating rates.

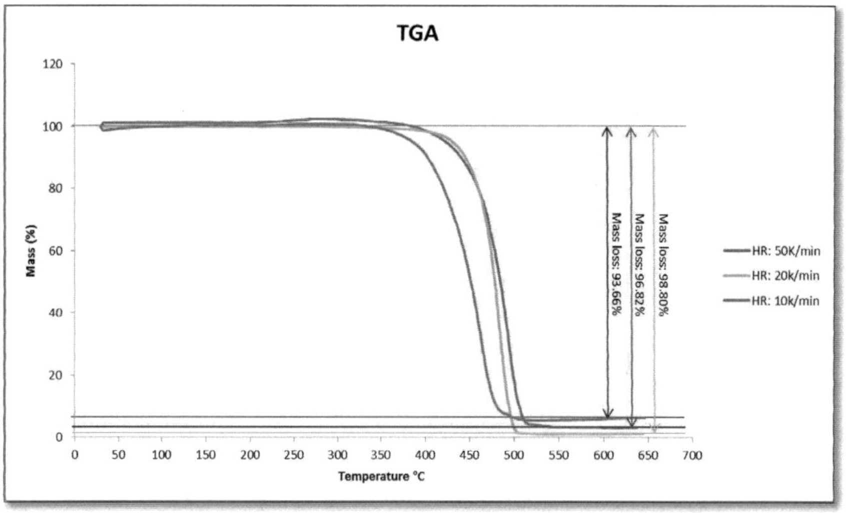

Figure 5. Mass loss curves of HDPE at different heating rates (HR).

For a closer analysis of these thermo gravimetric data, derivative curve of mass loss for all the heating rates has been shown in figure 6. The derivative curve gives a clear picture of any major or sudden loss at specific temperatures. In case of heating rate of 50°C/min, the peak mass loss is at 495°C with a mass loss rate of 124% mass loss per minute. Whereas for heating rate of 20°C/min, the peak loss is at 487°C with a mass loss rate of 57% mass loss per minute.

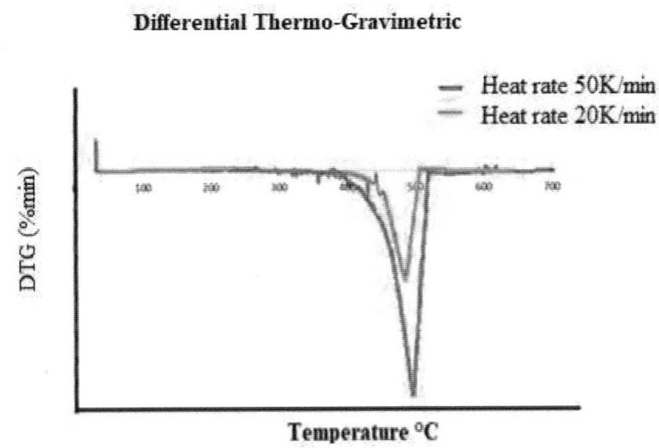

Figure 6. Derivative of mass loss for different heating rates.

3.3. Pyrolytic degradation of waste plastics and pure HDPE

Applying different prepared catalyst at optimism premieres (reaction time, temperature and loading quantity of catalyst) on both pure HDPE and its plastic waste. Table 2 shows the gas yield on the pyrolysis of pure polyethylene high density and its waste (4% of its weight) at 750 °C for 1 hour after using the prepared catalysis. The Char and condensable fractions from pyrolysis variation were also represented in the table 2. It was found that, nanocomposite catalyst gave the highest gas yield 1.02 m^3/kg comparing to other prepared catalysts. In addition, it gives a high gas ratio when it was applied on degradation waste plastic with Char and condensable fractions of 32.12%. So, it is good catalytically in cracking and deforming of polymer process. It is also clear from the table that, ZnO-bentonite is more catalytically activity for producing high gas ratio (0.936 m^3/kg) than that in MnO-bentonite (0.673 m^3/kg) and TiO_2-bentonite (0.612 m^3/kg). This can be connected to the electronic configuration in d-Orbital of ZnO.

Table 2. Effect of 4% catalytic ratio with the different catalysts on the pyrolysis of HDPE and its plastic waste at 750°C for 1 hour.

Catalytic Ratio 4%	Gas Yield (m^3/kg)	Char and condensable fractions %
HDPE	0.65	38.88
Bentonite/HDPE	0.72	37.5
MnO-bentonite /HDPE	0.673	49.9
TiO_2-bentonite /HDPE	0.612	55.58
ZnO-bentonite /HDPE	0.936	35.7
Metal oxides –Clay composites /HDPE	1.02	29.87
Metal oxides –Clay composites /PW- HDPE	0.98	32.12

3.4. Characterization of the pyrolysis gas products

3.4.1. Gas yields analysis. The gas chromatograph analysis of the gas sample obtained by the catalytic pyrolysis of both plastic waste and pure HDPE is summarized in the table 3. It has been observed that the pyrolytic sample contains many components such as aromatic, olefins, alkene, Nature gas, CO, CO_2 and Hydrogen. The greatest amount of hydrogen (15.544 %) and the smallest amount of methane (80.621%) are produced when MnO/bentonite is used, which may be due to highly oxidation sate of MnO. This is the important characterization of the transitions metal related to partially empty of d-orbital of Mn that will be easy to make cracking of polymers and release of H_2. On the other hand, TiO_2 produced lower amount of H_2 and larger amount of CH_4 which may be returned to the electronic configuration of metal ($TiO_2 >$ MnO > ZnO). Presence of free electron in d-orbital of both Ti and Mn while it is filled in Zn. So, MnO is a good catalyst deforming of polymers to get hydrogen which can be used in fuel cell application. On the other hand, composite of nano-Bentonite metal oxide gave high ratio of methane (94.82%) and low ratio of hydrogen with pyrolysis of plastic waste and pure HDPE. This may be due to impregnate the nano-bentonite metal oxide shielding of electrons or chelating with OH group of bentonite. This confirmed the improvement of the cation exchanger in composite and gave good ratio of CH_4 which can be used as a source of renewable energy.

Table 3. Comparison of gas products from catalytic pyrolysis of pure and waste HDPE with different catalysts at 750°C for 1 hour and 4% catalytic ratio.

Catalytic Pyrolysis	Components (vol. %)						
	Methane	**Hydrogen**	**Carbon Monoxide**	**Carbon Dioxide**	**Ethane**	**Benzene**	**Nitrogen**
HDPE	92.531	1.144	0.128	0.130	2.911	0.170	2.986
Bentonite/HDPE	87.722	3.811	2.899	1.537	0.751	0.159	3.121
MnO-bentonite /HDPE	80.621	15.544	0.856	0.180	0.374	0.017	2.408
TiO_2-bentonite /HDPE	85.480	4.885	6.500	1.923	0.590	0.035	0.586
ZnO-bentonite /HDPE	93.216	0.357	0.241	0.457	0.567	0.123	5.039
Metal oxides – Clay composites /HDPE	94.819	0.013	0.103	0.746	1.671	0.053	2.052
Metal oxides – Clay composites /PW- HDPE	93.785	0.001	0.068	0.162	3.702	0.228	2.053

3.5. Solid waste (char) characterize

3.5.1. Method of sampling and calculation of char samples. PEHD and PEHD waste char carbon samples extracted from the reactor at different conditions during pyrolysis were been analyzed by a variety of materials characterization techniques at the Central Analytical Laboratories of National Research Centre, Egypt. Elemental analysis was done by CHNSO elemental (Vario EL III). Elemental analysis was useful to evaluate carbon conversion efficiency. Carbon conversion was calculated by Eqs. (2-4);

$$X = (C_o - C)/C_o \quad (2)$$

$$Co = W_o \times M_o \qquad (3)$$

$$C = W \times M \qquad (4)$$

where X is the conversion of fractional carbon; C_o and C are carbon amount (gm) before and after pyrolysis, respectively; W_o and W are carbon percentage before and after pyrolysis, respectively; and M_o and M are PEHD plastic weight (gm) before and after pyrolysis, respectively.

3.5.2. Morphology analysis of composites catalyst. The surface morphologies (SEM) of the char presented nano-Bentonite catalyst is shown in figure 7. It is clear from the figure that, exists of flacks on the surface in nano-scales size average 100-120 nm, which is prove that nano-catalyst composite submersed between carbon char. Also, it was found that most of plastic polymers is converted to carbon char and other residual which appears on the surface of char. Figure 8 shows the energy dispersive X-ray analysis (EDXA) spectrum for Bentonite and table 4 listed the char composition. The char contains 65% carbon and other composition such as Silica, Al_2O_3 and Mg are 4.35, 2.6, and 0.6 of their weights, respectively. This data proved that the same chemical compositions of the papered nanocomposite were appeared which carryout under high temperature without desiccation. Also, all plastic waste or pure HDPE was converted to char.

Figure7. SEM image of the char sample pyrolysis at 750 °C for 1 hour.

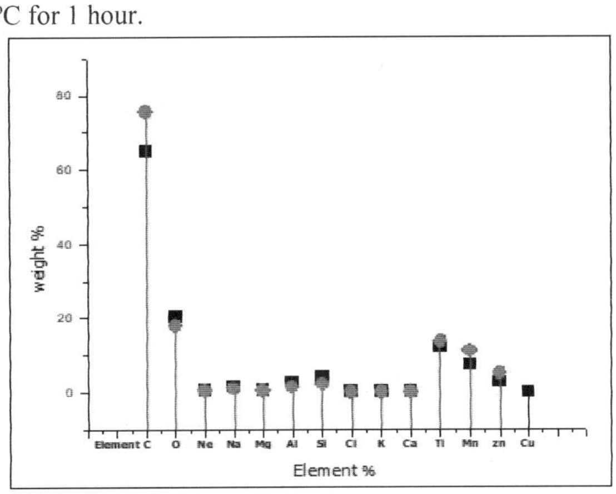

Figure 8. EDXA of catalytically pyrolysis of HDPE char with Bentonite clay.

Table 4. Residual metal element in char sample.

Element	Weight %	Atomic %	Net Int.	Error %
C	64.96	75.65	156.17	7.73
O	20.37	17.81	39.81	11.25
Ne	0.53	0.36	3.92	16.19
Na	1.61	0.98	11.86	10.55
Mg	0.61	0.35	8.06	10.93
Al	2.6	1.35	42.12	5.86
Si	4.35	2.17	78.72	4.23
Cl	0.18	0.07	2.71	20.45
K	0.29	0.1	3.82	16.09
Ca	0.29	0.1	3.29	18.94

4. Conclusion

Throughout pervious experiments and results, we conclude that, composites of MnO/Bentonite has unique specific adsorption properties with pw and PEHD. Moreover, Modification of nano-bentonite with metal oxides gives better results than using bentonite or metal oxides alone. Nano-bentonite with metal oxides increases the active sites of composite and also increases the rate of reaction. Useful gases (such as hydrogen and methane), light component paraffin and char were produced from pyrolysis process to degradation plastics. In addition, high amount of syngas was produced which can be utilized for production of ammonia (leading to different nitrogen-based fertilizers) and also, can serve as a hydrogen source. The waste plastic could be used in pyrolysis process (zero waste) better than incineration process because it is become source of green gases emission. Alternative technologies for managing end-of-life plastics exist. MnO /Bentonaite nanoparticles is a good catalyst deforming of polymers to produce hydrogen which can be used in fuel cell application.

Acknowledgment

This research was supported by the Scientific Research Developing Unit, Beni-Suef University, and the Academy of Scientific Research and Technology "ASRT" as a joint research grant under the India-Egypt agreement on science and technology cooperation. We thank our colleagues from National Research Center who provided us both of pyrolysis reactor system and experience in their laboratory. Prof. Mohamed Abdel Fatah Hamed and Prof. Ali Radwan that greatly assisted the research.

References

[1] Hillmyer and M. A 2017 *J. Sci.* **358** 868
[2] Haarman A and Gasser M 2016 *Managing Hazardous (WEEE Plastic Additives in the Indian Informal Sector)* (SRI) p1-68
[3] Vamshi A 2013 *Int. J. Eng. Res.* **2325** 0224
[4] Ruiz-Herrero J L, Nieto D V, López-Gil, A, Arranz, A, Fernández, A, Lorenzana A and Rodríguez-Pérez M Á 2016 *Constr. Build. Mater. J.* **104** 298
[5] Dave P N and Joshi A K 2010 *J. Sci. Ind. Res.* **69** 177
[6] Bhagat S, Bhardawaj A , Mittal P, Chandak P, Akhtar M and Sharma P 2016 *IIOAB J.* **7** 25
[7] Huysman S, De Schaepmeester J, Ragaert K, Dewulf J and De Meester S 2017 *Resour.Conserv. Recy. J.* **120** 46
[8] Kumar A 2013 *Int. J. Soc. Sci.* **1** 6
[9] Jellinek H and Clark J E 1963 *Can. J. Chem.* ***41*** 355

[10] Siddiqui M N and Redhwi H H 2009 *Fuel Process. Technol.* **90** 545
[11] Kalargaris I, Tian G and Gu S 2017 *Fuel Process. Technol.* **161** 125
[12] Sharuddin S D, Abnisa F, Daud W M and Aroua M K 2016 *Energy Conv. Manag.* **115** 308
[13] Panda A K , Singh R K and Mishra D K 2010. *Renew. Sust. Energ. Rev.* **14** 233
[14] Sannita E, Aliakbarian B, Casazza A A, Perego P and Busca G *2012 Renew. Sus. Energ. Rev.* **16** 6455
[15] Fivga A and Dimitriou I 2018 *Energy J.* **149** 865
[16] Paul T W 2013 *Waste Manag.j.* **33**1714
[17] Yang X , Sun L, Xiang J , Hu S and Su S 2013 *J. Waste. Manag.* **33** 462
[18] Quek A and Balasubramanian R 2013 *J. Anal. Appl. Pyrolysis* **101** 1
[19] Al-Salem S M, Lettieri P and Baeyens J 2010 *Prog. Energ. Combust. J.* **36**103
[20] Çepelioğullar Ö and Pütün A E 2013 ***J.** Nat. Appl. Sci.* **2**694
[21] Chin B L F , Yusup S, Al Shoaibi A , Kannan P, Srinivasakannan C and Sulaiman S A 2014 *J. Clean. Prod.* **70** 303
[22] Hasegawa M , Fukuda J and Kunii D 1979 *Energy: a Continuing Bibliography with Indexes* (l'Université du Michigan, United States) PP 35-36
[23] Borgianni C, De Filippis P, Pochetti F and Paolucci M 2002 *Fuel* **81**1827
[24] Xiao R, Jin B, Zhou H, Zhong Z and Zhang M 2007 *Energ. Convers. Manage. J.* **48** 778
[25] Martínez-Lera S and Ranz J P 2017 *Fuel. J.* **197** 518
[26] Aguado J, Serrano D P, San Miguel G , CastroM C, and Madrid S 2007 *J. Anal. Appl. Pyrol.* **79** 415
[27] Seo Y H, Lee K H and Shin D H 2003 *J. Anal. Appl. Pyrol.* **70** 383
[28] Sakata Y, Uddin M A and Muto A 1999 *J . Anal. Appl. Pyrol.* **51** 135
[29] Fatimah I, Narsito N and Wijaya K 2011 *J. Math. Fundam. Sci.* **43** 123
[30] Binitha N N and Sugunan S 2006 *Micropor. Mesopor. Mat.* **93** 82
[31] Vicente M A, Belver C , Sychev M, Prihod'ko R and Gil A 2009 *J. Ind. Eng. Chem.* **48** 406
[32] Okoye I P and Obi C 2011 *Inter. Arch. App. Sci.* **2** 84
[33] Salerno P, Asenjo M B and Mendioroz S 2001 *Thermochim. Acta* **379** 101
[34] Miandad R, Barakat M A, Aburiazaiza A S, Rehan M and Nizami A S 2016 *Process Saf. Environ. Protect.* **102** 822
[35] Karaborni S , Smit B , Heidug W , Urai J and Van Oort E 1996 *J. Sci.* **271** 1102
[36] Liu C , Lv P, Yuan Z , Yan F and Luo W 2010 ***Renew**. Energy* **35**1 531
[37] Szűcs A, Berger F and Dékány I 2000 *Colloid. Surface A.* **174** 387
[38] Pérez A, Centeno M A , Odriozola J A, Molina R, and Moreno S 2008 *Catal. Today* **133** 526
[39] Paek S M, Jang J U, Hwang S J and Choy J H 2006 *J. Phys. Chem. Solids.* **67** 1020
[40] Amarasekara A S, Green D and McMillan E 2008 *Catal. Commun.* **9** 286
[41] Aravind P V and de Jong W 2012 *Prog. Energ. Combust* **38**737
[42] Wilcoxon J P 2012 *Frontiers of Nanoscience* (United Kingdom , Elsevier Ltd.) **3** 43
[43] Pélisson C H, Vono L L ,Hubert C, Denicourt-Nowicki A, Rossi, L. M and Roucoux A 2012 *Catal. Today* **183** 124
[44] Bhaskar T, Matsui T, Uddin M A, Kaneko J, Muto A and Sakata Y 2003 *Appl. Catal. A-Gen.* **43** 229
[45] López A, De Marco I, Caballero B M, Laresgoiti M F and Adrados A 2011 *Fuel. Process. Technol.* **92** 253
[46] López A, De Marco I, Caballer B M, Laresgoiti M F, Adrados A and Aranzabal A 2011 *Appl. Catal.* **104** 211
[47] Hall W J and Williams P T 2008 *J. Anal. Appl. Pyrolysis.* **81** 139
[48] Terakado O, Ohhashi R and Hirasawa M 2011 *J. Anal. Appl. Pyrolysis* **91** 303
[49] Terakado O, Ohhashi R and Hirasawa M 2013 *J. Anal. Appl. Pyrolysis* **103** 216
[50] Jung S H, Kim S J and Kim J S 2012 *Fuel* **95** 514
[51] García R, Serrano D P and Otero D 2005 *J. Anal. Appl. Pyrolysis* **74** 379

IOP Conf. Series: Materials Science and Engineering 1046 (2021) 012022 doi:10.1088/1757-899X/1046/1/012022

[52] Miskolczi N, Bartha L and Deák G 2006 *Polym. Degrad. Stabil.* **91** 517
[53] Aguado J, Serrano D P , San Miguel G, Castro M C and Madrid S 2007 *J. Anal. Appl. Pyrolysis* **747** 9415
[54] Marcilla A.del Remedio, Hernández M and García Á N 2007 *J. Anal. Appl. Pyrolysis* **79** 424
[55] Olazar M, Lopez G, Amutio M, Elordi G, Aguado R and Bilbao J 2009 *J. Anal. Appl. Pyrolysis* **85** 359
[56] Cherian E, Rajan A and Baskar G 2016 *Int. J. Mod. Sci. Technol.* **1** 17
[57] Lin J T, Jong S J and Cheng S 1993 *Micropor. Mesopor. Mat.* **1** 287
[58] Alwan R M , Kadhim Q A , Sahan K M , Ali R A , Mahdi R J , Kassim N A and Jassim A N 2015 *J. Nanosci.* **5** 1
[59] Scherrer P 1912 *In Kolloidchemie Ein Lehrbuch* (Springer, Berlin, Heidelberg) pp.387
[60] Tadokoro H , H. G 1980 *Structure of Crystalline Polymers*(John Wiley and Sons, Chichester, New York, Brisbane, Toronto) pp 84

Antidiabetic effects of curcumin/zinc oxide nanocomposite in streptozotocin-induced diabetic rats

S M Abd El-Aziz[1], M Raslan[2,*], M Afify[3], M D E Abdelmaksoud[3] and K A El-Nesr[4]

[1]Faculty of Applied Medical Sciences, 6th October University, Egypt
[2]Biotechnology and Life Sciences Department, Faculty of Postgraduate Studies for Advanced Sciences, Beni-Suef University, 62511, Egypt
[3]Biochemistry Department, National Research Centre, Egypt
[4]Pathology Department, Faculty of Veterinary Medicine, Beni-Suef University, Egypt

[*]E-mail: m.raslan@psas.bsu.edu.eg

Abstract. Diabetes mellitus (DM) is a tremendously widespread endocrine disease that causes many complications risking patient's quality of life. The current study aims to evaluate the antidiabetic potential of curcumin nanoparticles (Curc-NPs), Zinc Oxide nanoparticles (ZnO-NPs), and Curcumin/Zinc oxide nanocomposite (Curc/ZnO-NC) on streptozotocin (STZ)-induced diabetic rats. Results are compared to rats treated by traditional anti-diabetic Diamicron and to normal non-diabetic rats. Adult Wistar albino rats with weight (180-200 g) were divided into 6 groups, each group contains 8 rats (4 males and 4 females). To induce type 2 DM, five groups were injected intraperitoneal with a single dose of 50 mg/kg b.w. freshly prepared STZ. Each group of diabetic rats were treated orally with a daily dose of 50 mg/kg b.w. of Curc-NPs, 10 mg/kg b.w. of both ZnO-NPs & Curc/ZnO-NC, and 5 mg/kg b.w. of Diamicron for 21 days. The antidiabetic potential of every treatment against diabetic rats was evaluated by investigating different biochemical parameters (glucose, insulin, urea, creatinine, HbA1-C, AST, ALT) and histopathological parameters as well as protein expression of Glucokinase (GK) and Glucose transporter protein 2 (GLUT-2) in the pancreas and livers of diabetic rats. All treated groups showed significant reduction in blood glucose, elevated insulin levels, regulated GLUT-2 and GK genes, however, Curc/ZnO-NC showed the most potent anti-diabetic activity compared to normal rats, the histopathological findings correlate with the achieved data.

1. Introduction

Diabetes mellitus (DM) is a metabolic disease distinguished by hyperglycaemia. It comes as a result of defects in insulin action, insulin secretion or both [1]. DM is regarded as one of the main chronic diseases all over the world. It reaches epidemic proportions in some cases. In the year 2030, it is estimated that the proportion of patients suffering from diabetes will increase by 69%. Serious diabetic problems and complications still face both the physicians and patients in spite of the increased understanding of the pathophysiological procedures involved in DM [2]. The epidemic of DM and its complications is currently considered as a major health threat on a global scale [3].

The field of Nanotechnology is one of the most rapidly growing fields. It has many applications in science and modern technology for the goal of producing new applications of different substances at the Nano scale level [4]. Nanoparticles are widely applied in many medical fields such as fluorescent biological labelling, imaging contrast agents, drug and gene delivery and in detection of bioactive molecules. The use of Nano drugs in treatment of diabetes was shown to be very effective [5]. Previous reports have indicated that selenium nanoparticles have a preserving ability towards the integrity of pancreatic β cell. It can also amplify insulin secretion (thereby reducing glucose level), and restore oxidant/antioxidant homeostasis [6].

The remarkable function of metals in metabolism of glucose and the connection of diabetes with their deficiency was reported in numerous researches. It was reported that vanadium [7], chromium, magnesium, and zinc have been included in diabetes therapy and play a role in blood sugar maintenance [8]. Zinc has a role in insulin biosynthesis, storage and secretion, also known to keep the structure of insulin. Diverse of zinc transporters do exist in the β cells of pancreas; such as zinc transporter 8. This transporter displayed a significant function in secretion of insulin. Moreover; zinc can enhance insulin signalling via various mechanisms; such as inhibition of glycogen synthase kinase-3, enhancing PI3K activity and increased insulin receptor phosphorylation [9]. Zinc is considered as an activator for more than three hundred enzymes in the body [10]; it is an essential metal in diabetes therapy. Zinc has an important role in enhancing hepatic glycogenesis by its actions on the insulin pathways; therefore it enhances glucose utilization [9].

Natural herbal products were used by people in the past for different therapeutic purposes. Thousands of secondary metabolites are produced in plants as a natural defence system [11]. Turmeric (*Curcuma longa*, family Zingiberaceae) is a medicinal rhizome well known for its major active constituent; curcumin. Chemically, curcumin is known as 7-bis (4-hydroxy-3-methoxyphenyl) - 1, 6-heptadiene-3, 5-dione, diferuloylmethane or (1E, 6E)-1, [12]. Curcumin, the major active polyphenolic constituent of turmeric rhizomes, is known for its pleiotropic effects on a wide spectrum of molecular targets including antioxidative, anti-inflammatory, chemo-preventive, chemotherapeutic, and antidiabetic activities with no considerable side effects [13]. Curcumin, when used as nano-micelle in DM treatment, caused an HbA1c lowering effect and partial decrease in serum LDL-C [14].

GK (also known as hexokinase IV) in the gut, hypothalamus, pancreas, and liver converts glucose to glucose-6-phosphate which may be used to determine the flux of glucose into these cells [15]. GK found in the β-cells serves as a ''glucostat'' by controlling glucose- stimulated insulin secretion (GSIS) [15]. GK, a strong marker for diabetic therapy, improves the glucose uptake by the liver, as well as the pancreatic insulin secretion [16].

GLUT-2 plays an important role in glucose homeostasis in living organisms and is mainly expressed in the β-cells of the pancreas, liver and the basolateral membrane of kidney proximal tubules [17]. Defective insulin secretion stimulated by glucose in diabetic lab animals was related to the reduced GLUT-2 gene expression [18].

No data were found about the effectiveness power of Curc/ZnO-NC on STZ-induced diabetic rats. Thus, a need emerged to study the biochemical, molecular and histopathological effect of Curc/ZnO-NC compared to ZnO-NPs, Curc-NPs and traditional anti-diabetic Diamicron.

2. Materials and Methods

2.1. Chemicals

Ethylene glycol pure P.A (lab – scan analytical sciences, catalogue number A00470X). Zinc nitrate (Hexahydrate) Purified (loba chemie, India), STZ was purchased from sigma – Aldrich. RQ1 RNAse-free DNAse (Invitrogen, Germany). TRIzol® Reagent (Invitrogen, Germany). RevertAid^{TM} First Strand cDNA Synthesis Kit (Fermentas, Germany). TaqTM (TaKaRa, Biotech. Co. Ltd.) Sense primer, antisense primer.

2.2. Preparation of material nanoparticles

2.2.1. Preparation of (ZnO-NPs). ZnO-NPs were prepared using the Pechini method. Briefly, the required amount of $Zn(NO_3)_2.6H2O$ was dissolved in a similar amount of ethylene glycol. The solution was then heated on a hot plate until a viscous gel-like substance was formed. With continuous heating, autoignition took place resulting in a fluffy yellowish-white powder which was then washed and grinded. Obtained powder was annealed at 500 °C for 2 hours in Lenton Furnace UAF 16/5 [19].

2.2.2. Preparation of (Curc-NPs). Commercial curcumin was suspended in acetone, and then subjected to ultrasonic waves in an ultrasonication bath for 30 minutes at 65 °C. The suspension was then stirred using a magnetic stirrer with hot plate for 1 hour until drying. After 1 hour, the powder was dried in an oven for 30 minutes at 50°C to be ready for use [20].

2.2.3. Preparation of (Curc/ZnO-NC). Firstly, the required quantities of ZnO-NPs were suspended in distilled water and acetone. The suspension was then subjected to ultrasonic waves by an ultrasonic probe for 30 minutes. Simultaneously, Curc-NPs powders were suspended in acetone and were sonicated in an ultrasonic bath for 30 minutes. Secondly, the Curc-NPs solution was added slowly drop by drop to the ZnO-NPs suspension in the ultrasonic bath by graduated pipette. The mixture was resonicated thoroughly ultrasonic probe for 31 minutes, and then transferred to a hot plate magnetic stirrer for 2 hours until drying. Powder was dried in an oven at 40°C for 15 minutes [21]. Preparation and characterizations of the nanoparticles were performed in the materials science and nanotechnology Lab. at PSAS-BSU.

2.3. Characterization of nanoparticles
The size and shape of nanoparticles was observed using high resolution transmission electron microscope (HRTEM JEOL-JEM 2100, Japan) using an accelerating voltage of 200 kV. X-ray diffraction (XRD) patterns were recorded on a PANalytical (Empyrean) X-ray diffractometer using Cu Kα1 radiation (λ=1.5406 Å) at an acceleration voltage of 40 kV, with current of 30 mA, in the angle range from 5–80°, with step scan of 0.02°. The size distribution of particles was determined by Zetasizer nano-Zs90 (Malvern, UK).

2.4. Animals
The experiment was carried out on albino Wister rats (180-200 g) that were obtained from the National Research Centre (NRC), Egypt, after the approval from NRC animal ethical committee (Approval no.16-207), in polypropylene cages (47cm × 34cm×20cm) lined with husk, renewed each day. Purified drinking water, as well as the standard laboratory pellet feed were provided, room temperature was retained at 22 ± 3°C during a cycle of 12 hours of light/12 hours of dark.

2.5. Induction of diabetes mellitus
All experimental animals were fasting overnight and diabetes was induced by an intraperitoneal injected dose of STZ (50 mg/kg b.w) freshly dissolved in 0.1 M citrate buffer (pH 4.5). After 6 hours of injecting the STZ, glucose (20%) was provided to rule the fatal STZ-stimulated hypoglycaemia. Glucose levels were measured after 2 to 7 days. Animals with high glucose levels exceeding 15 mmol (250 mg/dl) were considered diabetic, and their glucose levels were measured daily using AquaChek device [22].

2.6. Experimental design
A total of 48 albino Wistar rats were categorized into 6 groups, 8 rats per group, each of 4 males and 4 females. All treatments were dissolved in de-ionized water and administered orally once daily for 21 days. Group 1: negative control, group 2: positive control diabetic rats without treatment, group 3: diabetic rats treated with ZnO-NPs 10 mg/kg b.w [23], group 4: diabetic rats treated with Curc-NPs 50 mg/kg b.w [24] , group 5: diabetic rats treated with Curc/ZnO-NC 10 mg/kg b.w, group 6: diabetic rats treated with Diamicron 5 mg/kg b.w [25].

After one week, fasting blood samples were collected by the capillary to measure fasting glucose and insulin (initial glucose and insulin levels). Before ending the experiment, all animals were fasting during the night and the second collection of blood samples was done to measure liver and kidney function and insulin levels. Samples on sodium fluoride were collected to measure fasting glucose, and EDTA blood samples were collected to measure HBA1C. Also, different tissue samples from liver and pancreas were collected separately; for histopathological examination using 10% formalin, and for gene expression using liquid Nitrogen and kept in -80°C.

2.7. Biochemical parameters

Serum blood glucose (mg/dL) was estimated by the glucose oxidase method using the kit supplied by SPINREACT. Kit purchased from DRG international, Inc. (NJ, USA) was used to assay serum insulin concentration by an enzyme-linked immunosorbent assay (ELISA) [26]. HBA1C was estimated by immunological assays for the *invitro* quantitative determination of hemoglobin A1C in whole blood [27]. Serum AST and ALT were detected by (NADH. Kinetic UV) [28]. Serum creatinine was measured based on the reaction of creatinine with sodium picrate as described by Jaffe method [29], and urea was measured using urease method [30].

2.8. Analysis of Gene Expression

2.8.1. Isolation of RNA. TRIzol® extraction Chemical (Invitrogen) was utilized to isolate the total genomic RNA of liver and pancreas tissues of all treated animals. After completion of the isolation procedures, RNA pellet was stored in DEPC-treated water. To digest the potential DNA residues, the pellet of isolated RNA was treated with RNAse-free DNAse kit (Invitrogen, Germany). RNA aliquots were stored at -20°C or utilized immediately for reverse transcription [31].

2.8.2. Reverse transcription reaction. First Strand cDNA Synthesis Kit (RevertAidTM, MBI Fermentas) was used to synthesize the cDNA copy from liver and pancreas tissues via reverse transcription reaction (RT). A reverse transcription reaction program of 25°C for 10 min, then one hour at 42°C then 5 min at 95°C was used to obtain the cDNA copy of liver and pancreas genome. Finally, tubes of reaction containing cDNA copy were collected on ice up to use for cDNA amplification [32].

2.8.3 Quantitative Real Time-PCR. SYBR® Premix Ex TaqTM kit (TaKaRa, Biotech. Co. Ltd.) was used to perform the qRT-PCR analyses using the synthesized cDNA copies from liver tissues. Thermal cycling of the reaction was allocated to three steps; the first step was at 95.0°C for 3 min. Second step consisted of 40 cycles, each cycle was divided into 3 steps: (a) at 95.0°C for 15 sec; (b) at 55.0°C for 30 sec; and (c) at 72.0°C for 30 sec. The third step consisted of 71 cycles which started at 60.0°C and then increased about 0.5°C every 10 sec up to 95.0°C. For each reaction, a melting curve profile was conducted. The quantitative values of the target genes were normalized on the expression of the housekeeping gene (table 1). The $2^{-\Delta\Delta CT}$ method was used to determine the quantitative values of the specific genes to the β-actin gene.

2.8.4 Histopathological examinations

Different samples of the pancreas and liver tissues were obtained, fixed in buffered neutral formalin (10%), dehydrated gradually using 70%–100% ethanol, cleared using xylene, and finally inserted in paraffin. Hematoxylin and eosin (H&E) stain was used for the routinely staining of different paraffin sections (5μ in thickness) that were mounted on glass slides [36], and then examined microscopically.

2.8.5 Statistical analyses

All data were presented and analysed using IBM SPSS Statistics for Windows, Version 22.0. Armonk, NY: IBM Corp. All results are expressed as mean ± standard deviation (SD). A comparison among groups was made by students' one-way analysis of variance (ANOVA). Statistical significance was set ($p < 0.05$).

Table 1. Primer sequences used for qPCR

Gene	Primer sequence (5′–3′)[a]	References
GLUT- 2	Forward primer-TTG GCT TTC ACT GTC TTC ACT Reverse primer-CTT CCT TTT CTT TCC TCA TCT C	[33]
GK	Forward primer-AAG GAC AGG GAC CTG GGT TCC A Reverse primer-TCA CTG GCT GAC TTG GCT TGC A	[34]
β-actin (Housekeeping gene)	Forward primer-GGT ATG GAA TCC TGT GGC ATC CAT GAA A Reverse primer-GTG TAA AAC GCA GCT CAG TAA CAG TCC G	[35]

3. Results

In the present study, the therapeutic effects of Curc-NPs, ZnO–NPs and Curc/ZnO-NC were evaluated on STZ-induced diabetic rats and compared to normal rats and rats treated by traditional treatment; Diamicron.

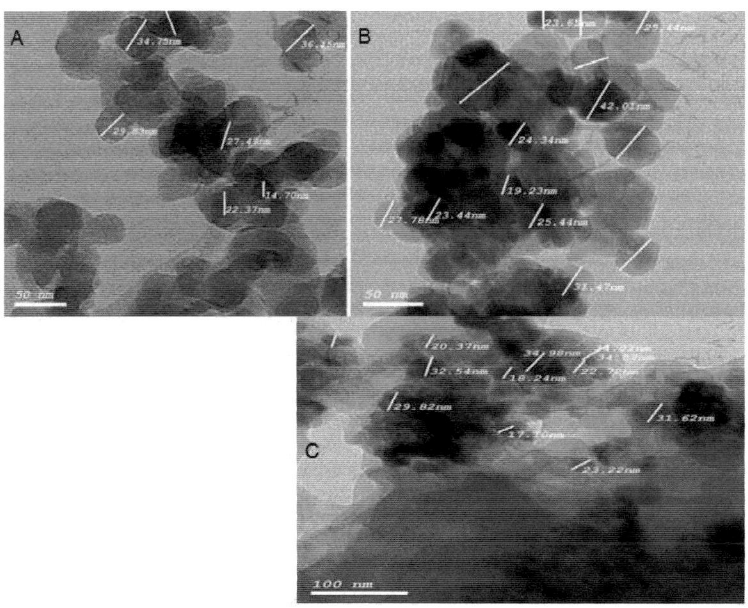

Figure 1. Characterization by HRTEM shows (A): Curc-NPs, (B): ZnO-NPs and (C): Curc/ZnO-NC.

The obtained sizes of Curc-NPs, ZnO-NPs and Curc/ZnO-NC, together with the particle shape are shown in (figure 1). Curc-NPs had an average particle size in the range of 27-30 nm. The sizes of ZnO-NPs and Curc/ZnO-NC were in the ranges of 24-30 and 20-30 nm respectively. As clearly observed, the particles have a hexagonal geometric shape with sharp boundaries in case of ZnO, and platelet shape in case of Curc-NPs. In case of the nanocomposite, hexagonal and platelet nanoparticles are stacked together. The existence of some agglomeration is obviously due to the absence of capping agent during preparation and the poor solubility of Curcumin. The hexagonal symmetry is in line with that found from the XRD analysis for the ZnO.

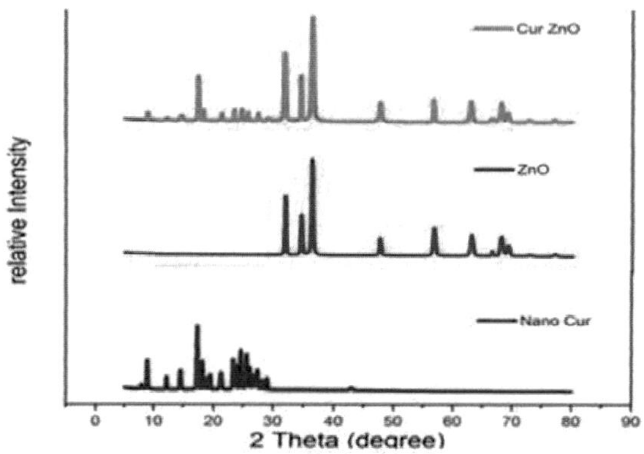

Figure 2. XRD patterns of (Curc-NPs), (ZnO –NPs) and (Curc/ZnO-NC).

XRD patterns for the nanoparticles are illustrated in (figure 2), the data revealed the crystallization of the ZnO in single phase with no additional peaks. The data of ZnO was compared and indexed with ICDD card no (01-080-6503). The reflections of ZnO were indexed as (hkl) (100), (002), (101), (102), (110), (103), (112), (201), (004).The crystal structure of the pure ZnO seen to the well-known hexagonal symmetry with space group (P63mc). The Curc-nanoparticles possessed excellent crystallinity with monolcilinic symmetry as compared with ICDD card (00-063-0943) and space group. The main reflections of Curc- nanoparticles were indexed as (hkl)(-101),(101),(110),(112),(004),(-113),(-301),(114),(-303).In the nanocomposite, the reflections of ZnO nanoparticles seems to be higher than those of curcumin despite the equivalent weight ratios in the nanocomposite (50%ZnO-50%Curc). It is also obvious owing to the difference in both densities (ZnO 5.1 and Curc 1.39 g/cm3). Moreover, the number of molecules per unit cell Z (Z=4 for Curc and =2 for ZnO). The average crystallite size was calculated using the well-known Debye Scherrer's formula [37,38,39]. It was found that the crystallite size of ZnO nanoparticles is 35.8 nm while that of Curc- is 63 nm.

Table 2. Zeta Potential for (Curc-NPs), (ZnO –NPs) and (Curc/ZnO-NC)

Name	Zeta potential (mV)	Average hydrodynamic diameter DLS (nm)
(ZnO -NPs)	-2.32	3420
(Curc-NPs)	-9.12	2620
(Curc/ZnO-NC)	-24.1	1750

The data in (table 2) reported the values of zeta potential and average hydrodynamic diameter. It is clear that the nanocomposite has the lowest value for size due to the ultrasonic waves which affect directly on size as well as distribution. The zeta potential value here is small for ZnO while it revealed large one for the nanocomposite. This result means that the latter is the most stable in its use as dispersed despite the poor solubility of curcumin.

Levels of urea, creatinine, AST and ALT (table 3) showed no significant differences among all animal groups (p values were 0.944, 0.746, 0.301 and 0.516 respectively). On the other hand, there was a significant reduction in blood glucose levels in groups treated by Diamicron, ZnO-NPs, Curc-NPs and Curc/ZnO-NC compared with the positive control group (table 4) , the percentages of reduction were 57.2, 57.8, 54.5 and 66.3% respectively (P value was 0.000).There was also a significant increase in the serum insulin levels in the diabetic groups treated with Diamicron, ZnO-NPs, Curc-NPs and Curc/ZnO-NC (P value was 0.000); increment percentages were 47.2, 45.3, 42.9 and 77.1% respectively, showing that Curc/ZnO-NC induced the highest insulin secretion(table 4).As for HBA1C, there was a significant reduction in groups treated by Diamicron, ZnO-NPs, Curc-NPs and Curc/ZnO-NC compared with the positive control group (P value was 0.000) and, noticeably, the group treated with Curc/ZnO-NC recorded the lowest value of HBA1C (table 4).

Table 3. Biochemical effect of Diamicron, Curc-NPs, ZnO-NPs and Curc/ZnO-NC on STZ-induced diabetic rats.

		Control −ve	Control +ve	Diam	Curc-NPs	ZnO-NPs	Curc/ZnO-NC	F	P value
Urea (mg/dl)	Mean	21.917	23.333	22.583	22.083	22.667	21.917	0.239	0.944
	±SD	4.575	4.048	3.040	3.302	3.197	2.325		
Creatnine (mg/dl)	Mean	0.753	0.707	0.725	0.7147	0.741	0.67	0.539	0.746
	±SD	0.124	0.148	0.146	0.146	0.118	0.114		
AST U/L	Mean	22.833	22.917	20	19.6	21	23.4	1.245	0.301
	±SD	6.4	4.5	4.8	5.9	3.1	5.3		
ALT U/L	Mean	22.08	21.5	20.5	19.9	20.6	23	0.857	0.516
	±SD	5.7	3.9	5.7	5.6	3.09	4.6		

Table 4. Effect of Diamicron, Curc-NPs, ZnO-NPs and Curc/ZnO-NC on serum Glucose, HBA1C and Insulin levels of STZ-induced diabetic rats.

		Control −ve	Control +ve	Diam	Curc-NPs	ZnO-NPs	Curc/ZnO-NC	F	P value
Initial glucose (mg/dl)	Mean	103.7	242.2	240.8	243.8	240.9	241.2	480.661	.000*
	±SD	8.05	11.48	5.84	7.48	5.85	8.68		
Final glucose (mg/dl)	Mean	101.3	347.1	148.7	158	146.4	116.8	2141.111	.000*
	±SD	4.64	10.32	4.57	4.59	5.18	5.24		
Initial insulin (μIU/ml)	Mean	12.25	8.04	8.39	8.07	8.23	12.02	583.311	.000*
	±SD	0.20	0.22	0.26	0.32	0.27	0.30		
Final insulin (μIU/ml) U/L	Mean	12.62	6.95	10.23	9.93	10.1	12.31	375.036	.000*
	±SD	0.28	0.50	0.26	0.34	0.22	0.29		
HBA1C	Mean	5.481	8.467	6.5	6.75	6.447	6.2083	106.351	.000*
	±SD	0.371	0.286	0.404	0.272	0.351	0.232		

* The mean difference is significant at the 0.001 level (P<0.001).

Results of the GLUT-2 and GK genes (tables 5, 6) and (figures 3, 4) showed that the expression levels in liver tissues were significantly lower than those in pancreatic tissues (P < 0.0001). Compared to the positive control group, all the other groups showed significant reduction in the expression levels of GLUT-2 and GK genes in liver and pancreas tissues (P < 0.0001). Furthermore, treatment with Curc/ZnO-NC reduced the expression levels of GLUT-2 and GK genes significantly compared to Curc-NPs alone, ZnO-NPs alone or with the Diamicron treatment (P < 0.0001) in both liver and pancreatic tissues.

Table 5. Comparison between the mean values of GLUT-2-mRNA in liver and pancreas with respect to the different rats groups

	Liver Mean± SD	Pancreas Mean± SD	t-statistic	P value
Control –ve	0.62 ± 0.05	0.96 ± 0.05	15.205	0.000*
Control +ve	2.35 ± 0.09	2.91 ± 0.04	17.981	0.000*
Diamicron	1.18 ± 0.02	1.39 ± 0.06	17.981	0.000*
Curc-NPs	1.42 ± 0.06	1.74 ± 0.07	10.500	0.000*
ZnO-NPs	1.23 ± 0.08	1.43 ± 0.06	10.976	0.000*
Curc/ZnO-NC	0.84 ± 0.04	1.19 ± 0.05	6.325	0.000*

*P < 0.0001

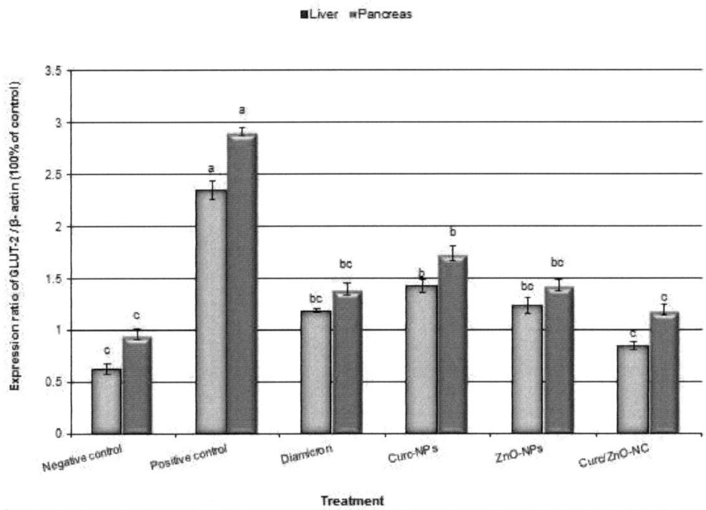

Figure 3. The alterations of GLUT-2-mRNA in liver and pancreas tissues (negative control) and positive control and DM-rats treated with Curc-NPs, ZnO-NPs, Curc/ZnO-NC and Diamicron.

Table 6. Comparison between the mean values of GK-mRNA in liver and pancreas with respect to the different rat groups.

	Liver Mean± SD	Pancreas Mean± SD	t-statistic	P value
Control –ve	0.72 ± 0.02	1.13 ± 0.04	28.991	0.000*
Control +ve	1.86 ± 0.03	2.74 ± 0.05	47.725	0.000*
Diamicron	1.21 ± 0.05	1.55 ± 0.04	16.791	0.000*
Curc-NPs	1.28 ± 0.06	1.69 ± 0.06	15.280	0.000*
ZnO-NPs	1.19 ± 0.05	1.41 ± 0.05	9.839	0.000*
Curc/ZnO-NC	0.97 ± 0.04	1.28 ± 0.03	19.606	0.000*

*P < 0.0001

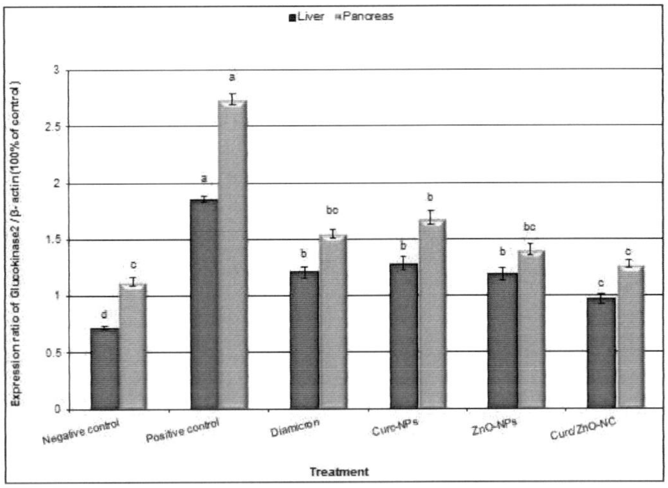

Figure 4. The alterations of GK-mRNA in liver and pancreas tissues (negative control) and positive control and DM-rats treated with Curc-NPs, ZnO-NPs, Curc/ZnO-NC and Diamicron.

Histopathological illustration (figure 5) showed that the liver of normal rats had normal histoarchitecture. The hepatic parenchyma and hepatic sinusoids appeared normal with normal distribution of Kupffer cells in contrast to the liver of diabetic rats where necrotic changes, dilatation of liver sinusoids, Kupffer cells activation and cytoplasmic vacuolization of hepatocytes were prominent. Variable degrees of improvements were noticed after treatment; liver of Diamicron-treated group, as a traditional treatment, showed moderate hepatic sinusoids (BS) dilatation with a slightly high number Kupffer cells (KC) and mild degenerative changes. In the Curc-NPs-only treated group, histoarchitecture of the liver appeared more or less normal with very mild necrobiotic changes (NT) and degeneration (FCH). In addition, mild dilatation of blood sinusoids (BS) and mild increase in number and distribution of (KC) was noticed. The Liver of ZnO-NPs-only treated group had normal tissue histoarchitecture with mild blood sinusoids (BS), widening and normal number and distribution of (KC) and congested central vein (CV).While the Curc/ZnO-Nc, treated group showed marked dilation of liver sinusoids with the presence of (KC) (LS), mild necrotic changes and congested (CV).

Histopathological illustration (figure 6) of the pancreas of the normal control rats showed normal histological architecture in the form of normal acinar structure with normal islets of Langerhans. In contrast, pancreas of positive control group showed significant decrease in islets of Langerhans (IL) with marked atrophy of acini, vacuolar degeneration and necrotosis. When diabetic groups received treatments, the following was noticed; pancreas of Diamicron-treated group showed restoration of normal size of (IL) with normal acinar cells and mild necrotic changes. Pancreas of Curc-NPs-treated group showed moderate restoration of cells of (IL) with mild necrotic changes. Pancreas of ZnO-NPs-treated group had normal-sized (IL) with mild central vacuolation and necrotic changes. Pancreas of Curc/ZnO-NC-treated group appeared more or less normal with restoration of (IL) and mild necrotic changes.

4. Discussion

Significant hyperglycemia in rats was induced by the STZ (60 mg/kg b.wt.) injection. This dose is reported to cause damage to β-cells of the (IL). Moreover, clinical diabetes emerged within 2–4 days due to autoimmune process [40]. Similar results were reported by previous studies [41], where STZ induced significant hyperglycemia which may be caused by gluconeogenesis enhancement due to the absence of insulin [42]. In this study, we evaluated the possible therapeutic effect of zinc oxide, curcumin and their nanocomposite on diabetic rats (streptozotocin-induced), compared to the diamicron treatment. There are numerous types of nanomaterials tested for their potential use in drug delivery systems [43]. Metallic nanoparticles (e.g. zinc, silver, iron and gold) and metal oxides nanoparticles have a great role in biomedical and biotechnological applications [44]. ZnO –NPs act as potent antidiabetic agents [45].

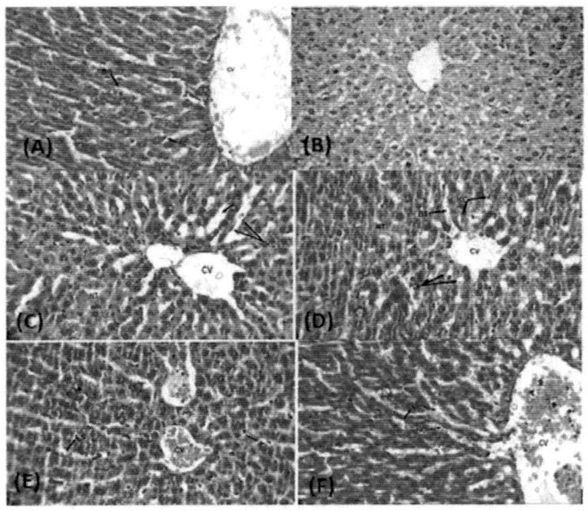

Figure 5. Histology illustration of liver for six groups (A) Negative control (B) positive control (C) Diamicron treated group (D) Curc-NPs treated group (E) ZnO-NPs treated group (F) Curc/ZnO-NC treated group.

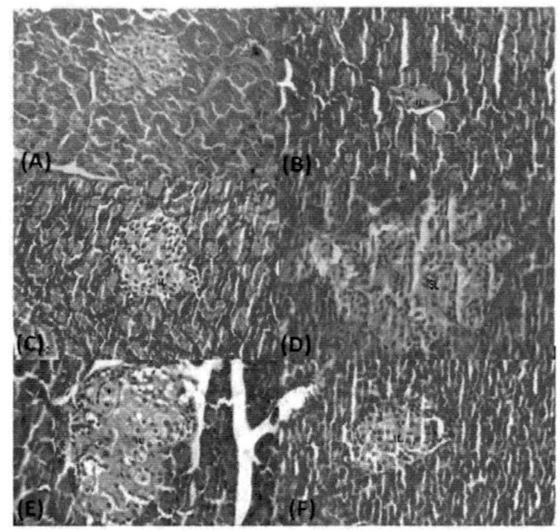

Figure 6. Histology illustration of pancreas for six groups (A) Negative control (B) positive control (C) Diamicron treated group (D) Curc-NPs treated group (E) ZnO-NPs treated group (F) Curc/ZnO-NC treated group.

Medicinal plants are recognized for their great role in replacing traditional synthetized drugs; with a similar degree of efficacy and less side effects. Curcumin was explored by many studies as a hypoglycemic agent [46], because curcumin quenches singlet oxygen effectively, leading to adduct formation. Moreover, it can be used as a cytoprotective agent for pancreatic islet cells, due to its inhibition of islet apoptosis as it inhibits the oxidative stress. However, curcumin has low bioavailability due to its sparingly solubility in water [47]; so nanocurcumin was used in this study to enhance its bioavailability, having a larger surface area in contact with the solvent leading to enhanced solubility.

Our results show the potent antidiabetic effect of the investigated nanoparticles. However, Curc/ZnO-NC induces the highest reduction in blood glucose levels and the highest elevation in insulin levels compared to Curc-NPs, ZnO–NPs and Diamicron, revealing the potent synergistic effect of Curc/ZnO-NC that improves glucose utilization and metabolism. This can be attributed to its enhancement of hepatic glycogenesis via insulin signaling pathway [9] (Table 4).

All tissues which respond to insulin can express insulin receptors (IRs) with different levels. Moreover, insulin potential requires that it binds to the receptors α–subunits leading to the phosphorylation of β–subunits to fulfill its role [45]. This supports our experimental results which demonstrated that diabetic rats liver and pancreas tissues revealed elevated GLUT-2 and GK mRNAs expression levels compared to the treated groups. Furthermore, tissues of pancreas showed elevated GLUT-2 and GK genes expression than the liver tissues of DM-rats. Also, the Curc/ZnO-NC showed the highest reduction GLUT-2 and GK gene expression compared to other treatments. These findings were asserted biochemically and histopathologicaly [48].

5. Conclusions

This research evaluates different nanoparticles; Curc-NPs, ZnO-NPs and Curc/ZnO-NC for their antidiabetic activities. All treated groups showed significant reduction in blood glucose, elevated insulin levels, regulated GLUT-2 and GK genes, however, Curc/ZnO-NC revealed the most potent anti-diabetic activity, as it showed a significant reduction in blood glucose by 66.3% and also elevated insulin levels by 77.1%. Compared to the positive control group, the histopathological findings correlate with the achieved data. Future work will be focused on the variation of both size and concnetration of nanoaprticles and its antidiabetic effect.

Acknowledgement
The authors would like to thank the National Research Centre for the financial support through Grant Project No: (11010184)

References

[1] American Diabetes Association 2014 *Diabetes Care* **37** 81
[2] Tiwari A K and Rao J M 2002 *Curr. Science* 30
[3] Zheng Y, Ley S H and Hu F B 2018 *Nat. Rev. Endocrinol.* ***14*** 88
[4] Albrecht M A, Evans C W and Raston C L 2006 *Green Chem.* **8** 417
[5] Salata O V 2004 *J. Nanobiotechnology* **2** 3
[6] Ahmed H H, El-Maksoud M D A, Moneim A E A and Aglan H A 2017 *Biol. Trace Elem. Res.* **177** 267
[7] Thompson K H, Lichter J, LeBel C, Scaife M C, McNeill J H and Orvig C 2009 *J. Inorg. Biochem.* **103** 554
[8] Chausmer A B 1998 *J. Am. Coll. Nutr.* **17** 109
[9] Jansen J, Karges W and Rink L 2009 *J. Nutr. Biochem.* **20** 399
[10] Haase H, Overbeck S and Rink L 2008 *Exp. Gerontol.* **43** 394
[11] Wink M 2015 *Medicines* **2** 251
[12] Hatcher H, Planalp R, Cho J, Torti F M and Torti S V 2008 *Cell. Mol. Life Sci.* **65** 1631
[13] Gupta S C, Patchva S and Aggarwal B B 2013 *AAPS J.* **15** 195
[14] Rahimi H R, Mohammadpour A H, Dastani M, Jaafari M R, Abnous K, Mobarhan M G and Oskuee R K 2016 *Avicenna J. Phytomed.* **6** 567
[15] Matschinsky F M 2009 *Nat. Rev. Drug Discov.* ***8*** 399
[16] Tahrani A A, Piya M K, Kennedy A and Barnett A H 2010 *Pharmacol. Ther.* **125** 328
[17] Im S S, Kang S Y, Kim S Y, Kim H I, Kim J W, Kim K S and Ahn Y H 2005 *Diabetes* **54** 1684
[18] Orci L, Unger R H, Ravazzola M, Ogawa A, Komiya I, Baetens D, Lodish H F and Thorens B 1990 *J. Clin. Invest.* **86** 1615
[19] Hozyen H F, Ibrahim E S, Khairy E A and El-Dek S I 2019 *Vet. World* **12** 1225
[20] Shirsath S R, Sable S S, Gaikwad S G, Sonawane S H, Saini D R and Gogate P R 2017 *Ultrason. Sonochem.* **38** 437
[21] Farghali A A, Khedr M H, El-Dek S I and Megahed A E 2018 *Ultrason. Sonochem.* **42** 556
[22] Murali R, Srinivasan S and Ashokkumar N 2013 *Biochimie* **95** 1848
[23] El-Gharbawy R M, Emara A M and Abu-Risha S E S 2016 *Biomed. Pharmacother.* **84** 810
[24] Zhang D W, Fu M, Gao S H and Liu J L 2013 *Evid. Based Complement. Alternat. Med.* 2013 636053
[25] Pulido N, Suarez A, Casanova B, Romero R, Rodriguez E and Rovira A 1997 *Metabolism* **46** 10
[26] Temple R, Clark P M S and Hales C N 1992 *Diabet. Med.* **9** 503
[27] Cohen M P 1992 *Eur. J. Clin. Chem. Clin. Biochem.* **30** 851
[28] Edrees H M, Elbehiry A and Elmosaad Y M 2017 *Nanotechnol* **3** 4
[29] Murray R and Kaplan A 1984 *Clinical Chemistry* (Toronto: The CV Mosby Co.)
[30] Chaney A L and Marbach E P 1962 *Clin. Chem.* **8** 130
[31] Salem N A, Wahba M A, Eisa W H, El-Shamarka M and Khalil W 2018 *Inflammopharmacology* **26** 1025
[32] Khalil W K B, Booles H F, Hafiz N A and El-Bassyouni G E 2018 *Int. J. Pharmacol.* **14** 477

[33] Amalan V, Vijayakumar N, Indumathi D and Ramakrishnan A 2016 *Biomed. Pharmacother.* **84** 230
[34] Baldini S F, Steenackers A, Olivier-Van Stichelen S, Mir A M, Mortuaire M, Lefebvre T and Guinez C 2016 *Biochem. Biophys. Res. Commun.* **478** 942
[35] Serrano R, Villar M, Martinez C, Carrascosa J M, Gallardo N and Andres A 2005 *J. Mol. Endocrinol.* **34** 153
[36] Bancroft JD, Gamble M, editors 2008 Theory and Practice of Histological Techniques. Elsevier Health Sciences, pp. 135-160
[37] Mansour S F, Abdo M A and El-Dek S I 2017 *J. Magn. Magn. Mater.* **422** 105
[38] El-Dek S. I 2010 *Philos. Mag. Lett.* ***90*** 233
[39] Khedr M H and Farghali A A 2005 *J. Mater. Sci. Technol.* **21** 675
[40] Weiss R B 1982 *Cancer Treat Rep* ***66*** 427
[41] Sellamuthu P S , Muniappan B P, Perumal S M and Kandasamy M 2009 *Glob. J. Health Sci.* **55** 206
[42] Yao X H, Chen L and Nyomba B G 2006 *J. Appl. Physiol.* **100** 642
[43] Yih T C and Al-Fandi M 2006 *J. Cell. Biochem.* **97** 1184
[44] Hirst S M, Karakoti A S, Tyler R D, Sriranganathan N, Seal S and Reilly C M 2009 *Small* **5** 2848
[45] Alkaladi A, Abdelazim A and Afifi M 2014 *Int. J. Mol. Sci.* **15** 2015
[46] Abdel Aziz M T, El-Asmar M F, El Nadi E G, Wassef M A, Ahmed H H, Rashed L A and Abdel Aziz A T 2010 *Angiology* **61** 557
[47] Modasiya M K and Patel V M 2012 *Int. J. Pharm. Life Sci.* **3** 1490
[48] Nasr M, Raslan M, Ashour S, Mahmoud S and El-Nesr K 2018 *Biosci. Res.* **15** 4520

IOP Conf. Series: Materials Science and Engineering 1046 (2021) 012024 doi:10.1088/1757-899X/1046/1/012024

Single and combined impact of silica fumes with Functionalized multi-walled carbon Nanotubes and nano silica on performance of cement mortars composites

Ibrahim M. El-Kattan[1*], M H Khedr [2], A A Farghali [2], M Elsaeidy [3] and F N Soliman[2, 3]

[1]Environmental Science and Industrial Development Department, Faculty of Postgraduate Studies for Advanced Sciences, Beni-Suef University, Beni-Suef 62511, Egypt
[2]Material Sciences and Nano Technology Department, Faculty of Postgraduate Studies for Advanced Sciences, Beni-Suef University, Beni-Suef 62511, Egypt.
[3]Wadi El Nile Cement Company, Beni-Suef, Egypt.

*E-mail: ibrahim.elkattan@psas.bsu.edu.eg

Abstract. This present work aims to study the impact of single and combined effects of silica fumes (SF) with nano silica (NS) and functionalized multi walled carbon Nanotubes (MWCNTs) on physicomechanical and dynamic properties of cement mortars (CMs). The Portland cement (PC) was partially substituted with different ratios of SF, MWCNTs and NS which, firstly was individually substituted but Secondly, were added together. PC was individually substituted with 2, 4, 6, 8, 10 wt. % of SF then combined with 0.2, 0.4, 0.6, 0.8 and 1.0 wt. % of MWCNTs and 2, 4, 6, 8, 10 wt. % of NS which mixing together. The physicomechanical and dynamic properties such as compressive strength, bending strength, water consistency; both initial and final setting times were measured. The phases of hydration products of some selected samples were additionally investigated by using XRD analyses, the morphology of hardened samples was studied by SEM techniques and TEM showed the particle size of nanoparticles. The results illustrated that, the compressive strength of cement mortars (CM) which containing SF increasing by 27% at 28days and 11 % for bending strength at early ages comparing to control sample. In combination with functionalized MWCNTS ,the compressive and bending strengths increased by 36 and 17 % in case of CMs that containing 0.4 Wt.,% functionalized MWCNTs and 8 Wt.,% SF. additionally, the combination of SF with NS showed increment in compressive strength by 28 % at 8 Wt.,% SF and 8 Wt.,% NS, respectively. SEM-photos showed that CMs containing 0.4 Wt., % functionalized MWCNTs and 8 Wt., % SF had more compacting microstructure than control sample due to bridging effect of MWCNTs and pozzolanic effect of SF.

1. Introduction

Particles with ranging lengths from 1 to 100 nm are non-materials. Cement is a fine powder acting as a hydraulic binder, i.e. a finely ground clinker and additives which, when mixed with water; forms a cement paste which hardens and sets through hydration reactions and after setting, retains its strength and stability even under water. Recently; the beneficial role of nano-materials in cement field greatly increased [1]. The cement paste matrix is in itself a multi-phase material. Basically, it is a porous material composed of calcium hydroxide (portlandite), aluminates and un-hydrated clinker embedded into an amorphous nanostructured hydration product, the so-called C–S–H gel [2]. The major most important constituent of hydrated cement (50–70%, v/v) is calcium silicates which is produced on two stages; a first then a second hydration stages resulting in low then high density hydrated phase, respectively whose mechanical properties and characteristics were reported by nano-indentation experiments [4].

Silica fume (SF), is a by-product of ferrosilicon industry known as pozzolanic material consists of small (≈100 nm) spherical shaped amorphous silica. SF was reported to increases cement strength and improves mechanical properties when mixed with optimum reported replacement of 7.5 to 10 W% to Portland cement [5-7] and predominating was investigated at 10% [8]. Studies about using SF in high strength concrete were reported along 40 years [9] showing higher compressive strengths than 100 MPa. Strengths more than 150 MPa were reported [10–12]. Advantages of SF as a replacement of cement include: Higher compressive strength with the same plant cement mix design parameters; lower cement content for a desired strength; and higher concrete durability [13–17].

Carbon nanotubes (CNTs) show two specific morphologies [18], a graphene sheet rolled singly on a cylinder known as single-walled nanotubes and multi concentric graphene sheets cylinders of nanometer and micron scales of diameters and scale lengths, respectively [19] as presented in figure 1. Those characteristics made them excellent future carbon fibers substitutes [18]; being light, strong, and can withstand bending with no torsions and breaking [20, 21]. Single walled nanotubes have Young's modulus theoretically calculated TPa range from 1 to 5 [22-24] compared to carbon fibers of 400 GPa meaning that CNTs are with very high tensile strength compared to carbon fibers. Additionally, CNTs are electrically and thermally conductive (1000 times greater conductivity than cupper and can withstand up to 2800 °C under vacuum [25]. so, they are applicable as conductors and semiconductors [18].

An oxidative treatment is performed for MWNTs synthesized in presence of a catalyst for functionalizing their surface; oxidation is performed using a mixture of concentrated nitric acid and sulfuric acid solution. Modifications of structural integrity may result due to uncontrolled conditions. The optimum surface functionalization conditions minimum nanotubes damage and maximum surface carboxylic groups (~28%) are 18 hr. treatment with 4.0 M HNO_3:10.0 M H_2SO_4 (1:3, v/v) [18], the functionalized MWCNTs were presented in (figure 1).

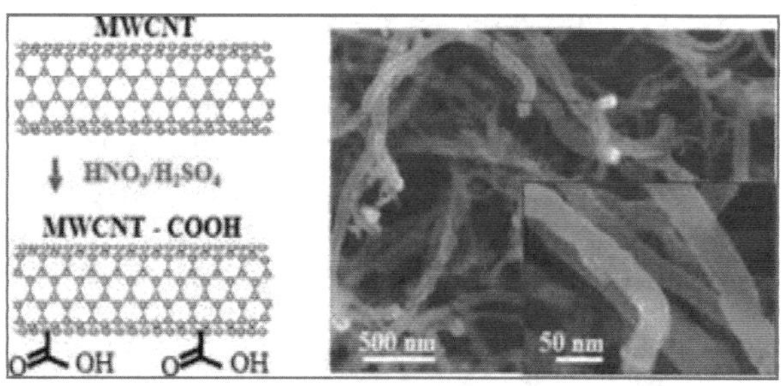

Figure 1. Illustration of MWCNTs and activated MWCNTs

The dispersion of non-functionalized CNTs in cement paste was reported to be done using ultrasonic mixing in presence of aqueous surfactant at specific time using dispersion energy. Zou et al. stated ultra-sonication energy nearly 20 J/mL per unit CNTs to cement for 84% dispersion [26]. Using high force might may dissolve CNTs or destroy CNTs, also incompatibility of surfactants affects on setting time [27]. Chemical activation is a process of creating functional groups on MWCNTs surface to enhance bonding and dispersion with matrix. Chemical activation is reported to be processed by applying ozone gas with high temperature then acid treatment lead to carboxyl groups to improve bonding with cement particles by formation of chemical reactions. Kang et al. stated the same strategy but without using surfactants [28]. Studying the effect of Nano Silica (NS) and CNTs on the properties of cement based materials revealed contradictory results, about the effect of NS on the hydration degree of cement; some papers showed an increase at the first hours [29, 30], while others showed a decrease in setting time compared to plain paste [31]. NS effect on the mechanical properties were discrepant; some showed increase [30, 32] and others reported insignificant influence [33] finally a decrease in compressive strength [34, 35] compared to blank cement samples. The latest papers was performed to synthesize the functionalized MWCNTs and studying their effect on Physico-mechanical characteristics of cement mortar [28] by using harmful greenhouse gas (ozone) and higher temperature which is consuming energy [36]. For clarification, the reported results, the surface area and type of the used NS have a different effect on physio-mechanical properties with cement composites [37]. All the up to date reported studies either using NS or MWCNTs did not follow the EN 197-1/2011 standards or any other standards [38]. Recent time, Using of nanoparticles in cement manufacturing and concrete (as substitution) leads to sustainable cement, concrete design and green environment [48].This was leading to enhancement physicomechanical properties and durability such as aggressive Medias and firing resistivity [49, 50].

The main objective of the present work is studying, the single and dual effect of SF with functionalized MWCNTs and NS on dynamic and physico-mechanical properties of cement mortars (CMs). The MWCNTs were functionalized without using of greenhouse gases or energy consuming temperature [36]. The NS was performed by using commercial precipitation method. Different experimental and instrumental techniques were conducted to evaluate the behaviour of CMs. Physico-mechanical and dynamic properties such as compressive strength, bending moment, water consistency, initial setting time and final setting time were conducted to evaluate the performance of CMs. The results of this paper confirmed that, the modification of PC- properties with nanoparticles decreased the global environmentally harmful carbon dioxide emissions combined with cement industry [39].

2. Experimental

2.1. Synthesis, purification, characterization, and functionalization of multi-walled carbon nanotubes (MWCNTs) by thermal chemical vapor deposition (CVD)

Synthesis of MWCNTs in presence of Co-Fe/$CaCO_3$ catalyst, characterized by TEM (figure 2) [40], was performed at 600 °C for 30 minutes using commercial acetylene gas cylinder, a procedure reported to be of a high MWCNTs yield [40]. By energy, acetylene molecules were cracked into reactive atomic carbon which diffused into the substrate, which was then heated and coated with the catalyst. CVD used in the synthesis was presented as diagram in (figure 3). After purification, functionalized using Conc. H_2SO_4 and HNO_3 was performed [40]. The produced MWCNTs were investigated by TEM before (figure 4) and after (figure 5) purification and by XRD (figure 6). TEM for functionalized MWCNT shows the dispersion and the absence of impurities as catalyst particles. The crystal structure and phases of the prepared Functionalized MWCNT were scanned by X-ray diffraction (XRD, PANalytical Empyrean) with scanning range from 5° to 80°, the detector step size was 0.03°, and the step time was 0.3 s, the diffract meter applied is Cu Kα radiation with the wavelength of 1.54060 Å and the Generator Settings operated at 40 KV and 30 mA with scanning type, continuous. XRD for Functionalized MWCNT showed that the crystal size is 7.8 nm which had been calculated automatically by the XRD instrument using Debye sherrer formula.

2.2. Synthesis and characterization of precipitated nanosilica (NS)

Conc. HCl was added drop wisely added to sodium silicate solution with constant stirring to at room temperature (21-24 °C). When it's 8-9 pH; silica gel are formed which are aged for 24 h. at room temperature, filtered, washed several times with water till absence of chloride tested by silver nitrate, then dried at 90 °C for 18 hrs., finally ground in a ball mill for 2 hrs., [41] then characterization by TEM (figure 7) forms NS was amorphous as characterized by XRD (figure 8), the particle size ranged from 41 - 102 nm as characterized in SEM (figure 9). XRF (table. 1) showed that the silica content is nearly 91%. TEM of NS (figure 7) confirmed that a spherical morphology of almost particles with low aggregation. The TEM proved that, NS particles have an average diameter from 41 nm to 102 nm. The TEM photos reported that the crystals are mostly angular with slightly elongated shape. The light pattern showing bright points and rings supported the amorphous structure of NS particles which was previously

IOP Conf. Series: Materials Science and Engineering 1046 (2021) 012024 doi:10.1088/1757-899X/1046/1/012024

confirmed by the XRD pattern for amorphous NS (figure 8) which showed a typical broad peak which was corresponding to the amorphous phase of NS particles. Peak broadness may be due to the small size and incomplete inner structure of the prepared particles. This proved that a high percentage of amorphous particles. SEM micrograph revealed that the average particle size of powdered-NS ranged from 41 nm to 102 nm and the particles are spherical, non-agglomerated and possess a smooth surface morphology (figure 9). XRF (table 1) showed that the silica content is nearly 91%.

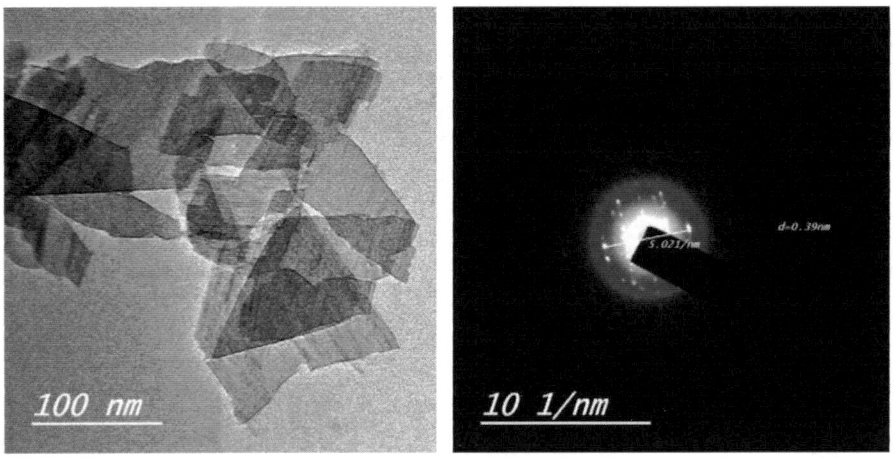

Figure 2. TEM-photos of the catalyst (Co-Fe/$CaCO_3$) used for MWCNTs preparation.

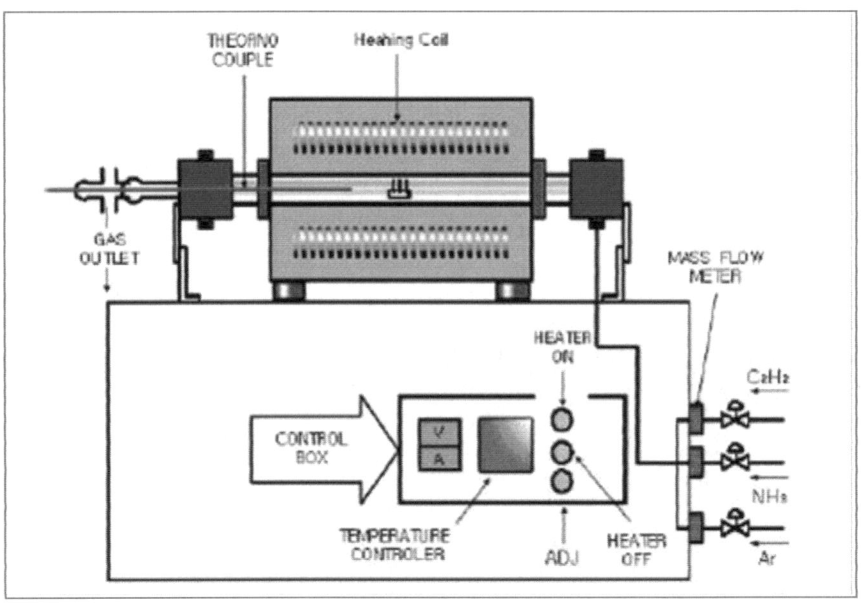

Figure 3. Schematic diagram of thermal CVD apparatus.

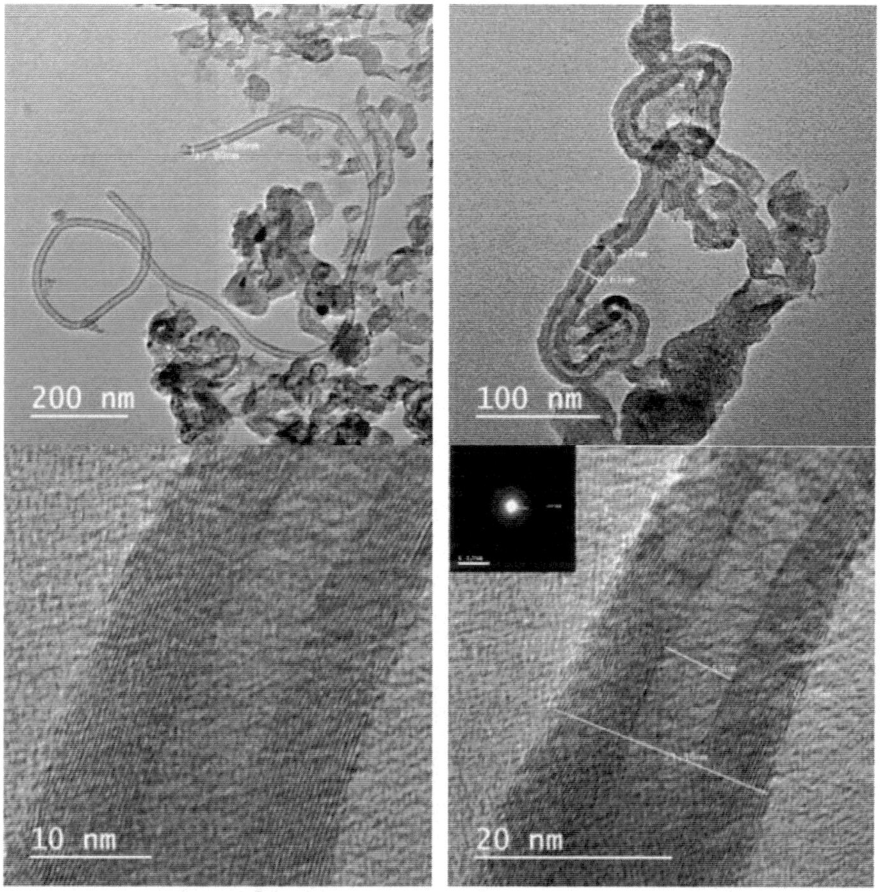

Figure 4. TEM-photos of the MWCNTs prepared by CVD before purification and functionalization.

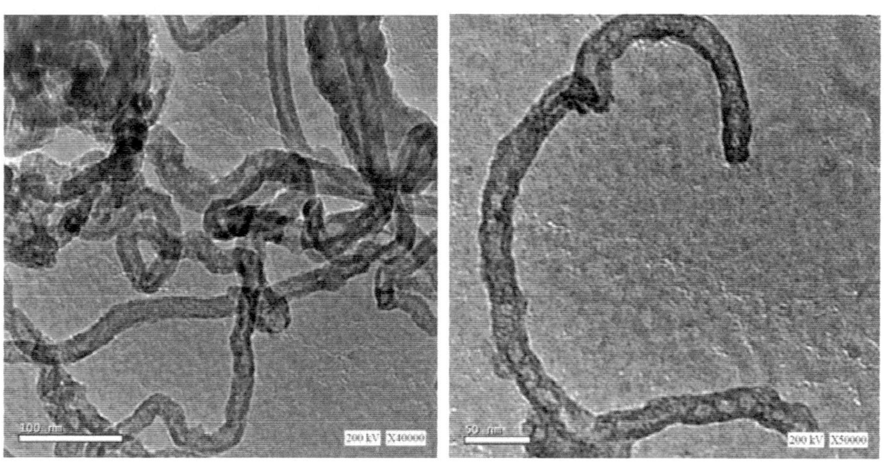

Figure 5. TEM-photos of the purified and functionalized MWCNTs.

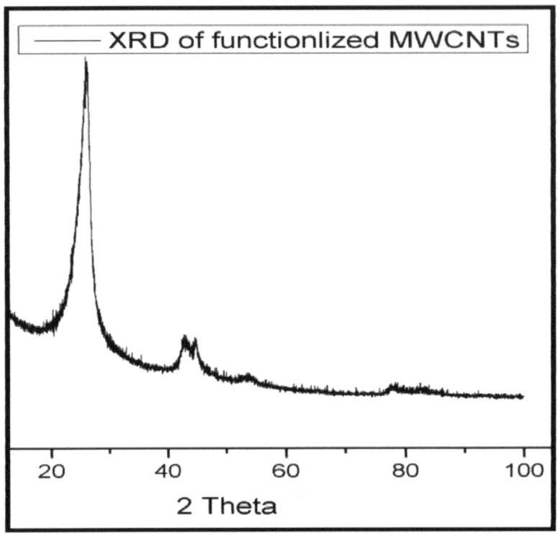

Figure 6. XRD-pattern of the prepared Functionalized MWCNTs.

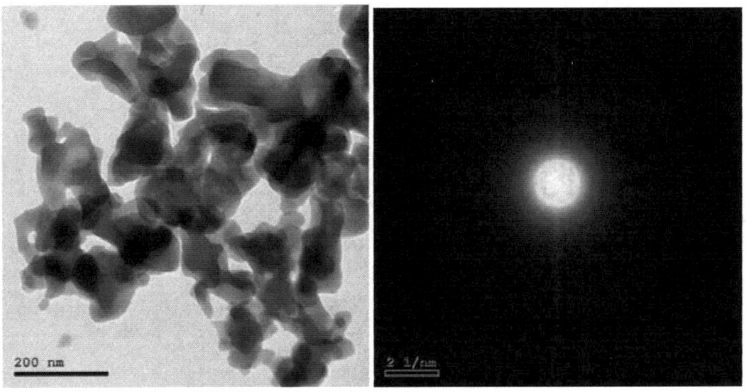

Figure 7. TEM-photos and particle light diffraction for the precipitated amorphous nano silica (NS).

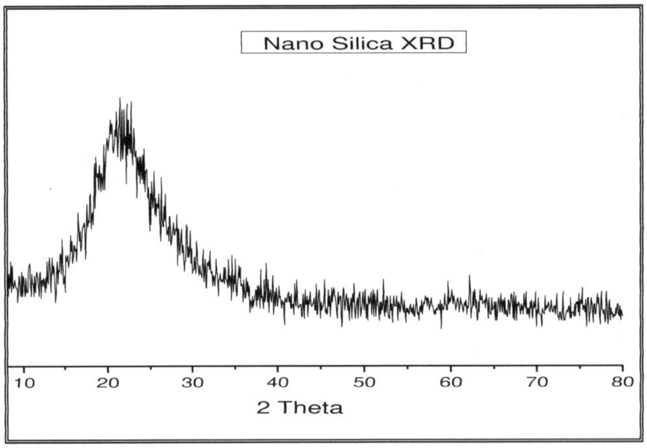

Figure 8. XRD pattern for the precipitated amorphous Nano Silica (NS).

Figure 9. SEM-photos of the precipitated amorphous Nano Silica (NS).

Table 1. XRF analysis of the precipitated Nano Silica (NS).

SiO_2%	Al_2O_3%	Fe_2O_3%	CaO%	MgO%	SO_3%	K_2O%	Na_2O%	Chloride %
90.82	0.07	0.00	0.48	0.10	0.20	0.400	0.020	0.10

2.3. Characterization of SF

The purchased SF from Sika Egypt for Construction Chemicals Co., was characterized by XRF showed in table 2 with SiO_2% content 88.48%, XRD showed amorphous SF (figure 10) TEM as shown in figure 11, shows particles diameter ranges from 50 to 150 nm.

2.4. Samples preparation

Portland Cement (PC) CEM I (42.5 N) used as a blank cement type in the suggested study is complying with BS EN 197-1/2011 [38] produced by Wadi El Nile Cement Company, Egypt, the physical and chemical analysis of the used plain CEM I (42.5 N) was represented in table 2 and XRD characterization was shown in figure 12. Cement was mixed in different ratios with, SF at first, and then both SF and functionalized MWCNTs, and finally both SF with NS for preparing samples to be tested, the summarized nominations and mix design of prepared samples were represented in table 3.

Table 2. XRF analysis of the silica fumes (SF).

SiO_2%	Al_2O_3%	Fe_2O_3%	CaO%	MgO%	SO_3%	K_2O%	Na_2O%	Chloride %
88.48	1.44	1.68	0.07	0.38	0.10	0.47	0.54	0.003

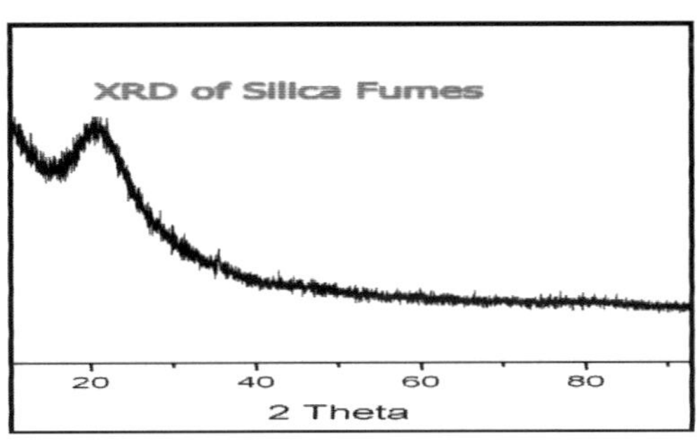

Figure 10. XRD-pattern for the used silica fumes (SF).

IOP Conf. Series: Materials Science and Engineering 1046 (2021) 012024 doi:10.1088/1757-899X/1046/1/012024

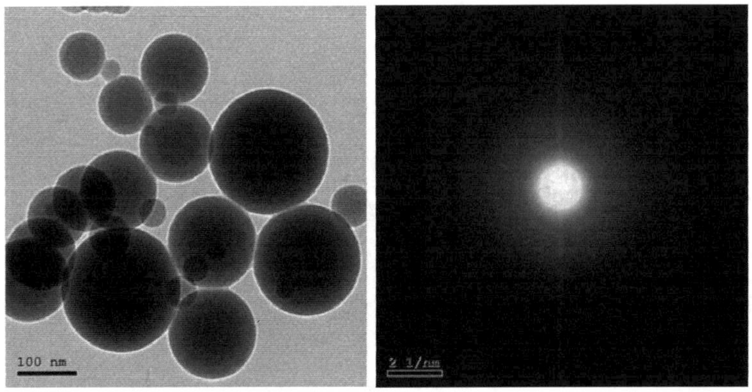

Figure 11. TEM-photos for Silica Fumes (SF).

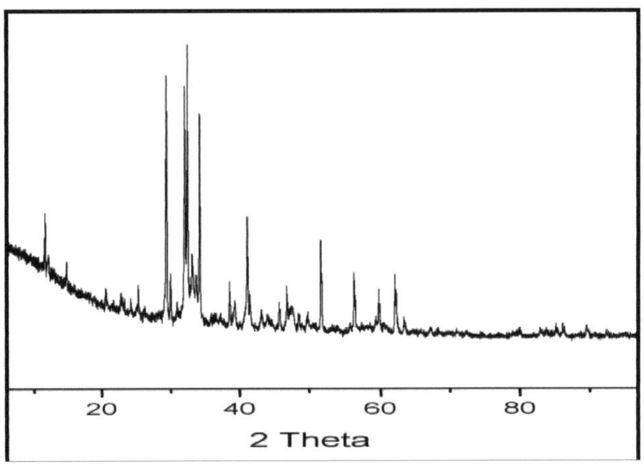

Figure 12. XRD-pattern of the used plain CEM I (42.5 N).

Table 3. Chemical and physical analysis of CEM I (42.5 N) according to European Standard: Cement Composition 2011).

SiO_2%	19.67	Na equivalent	0.77	Wet Chloride (%)	0.05
Al_2O_3%	4.96	LSF	99.33	Wet Free CaO (%)	1.46
Fe_2O_3%	3.64	SIM	2.29	Insoluble Residue	0.60
CaO%	62.85	ALM	1.36	45μ (%)	18.70
MgO%	1.16	C3S %	52.44	90μ(%)	1.80
SO_3%	2.82	C2S %	16.84	Blaine (cm^2/m)	3018
K_2O%	0.288	C3A %	7.36	Soundness (mm)	0.00
Na_2O%	0.58	C4AF %	11.06	Color degree	26
Chloride%	0.06	Moisture %	0.40	Cr^{+6} wet (PPM)	1.6
Free CaO%	1.83	L.O.I %	3.48		

According to BS EN 197-1/2011 guidelines [38]; the effect of the cement replacement with firstly SF alone, then both SF with functionalized MWCNTs, and finally both SF with NS in different weight ratios illustrated in table 4 was tested. For studying compressive strengths (2 days, 7 days, and 28 days), bending strengths (2 days, 7 days, and 28 days), the test was applied on 3 prismatic specimens 40 × 40 × 160 mm diameters, with water cement ratio of 0.45, the samples were mechanically mixed then compacted in a mould using a jolting apparatus, then stored for 24 h. in moist atmosphere at 20 ± 1.0 °C and 90% relative humidity. After demolding, samples were stored under water until strength testing, at the required age, samples were then broken in flexure, for determining bending strength where required, and each half was tested for strength in compression, the physical laboratory room shall be kept at 20 ± 2 ˚C and 50% relative humidity [42,51].

Table 4. Mix composition of cement mortars with or without SF, SF and functionalized MWCNTs or SF with NS.

Sample Notation	Cement %	SF Wt. %	Functionalized MWCNTs Wt. %	NS wt.%	Water/Cement Ratio
SF_0	100.0	0.0	0.0	0.0	0.45
SF_2	98.0	2.0	0.0	0.0	0.45
SF_4	96.0	4.0	0.0	0.0	0.45
SF_6	94.0	6.0	0.0	0.0	0.45
SF_8	92.0	8.0	0.0	0.0	0.45
SF_{10}	90.0	10.0	0.0	0.0	0.45
$C_{0.2}SF_2$	97.8	2.0	0.2	0.0	0.45
$C_{0.4}SF_4$	95.6	4.0	0.4	0.0	0.45
$C_{0.6}SF_6$	93.4	6.0	0.6	0.0	0.45
$C_{0.8}SF_8$	91.2	8.0	0.8	0.0	0.45
$C_{01}SF_{10}$	89.0	10	1	0.0	0.45
$C_{0.2}SF_{10}$	89.8	10	0.2	0.0	0.45
$C_{0.4}SF_8$	91.6	8	0.4	0.0	0.45
$C_{0.8}SF_4$	95.2	4	0.8	0.0	0.45
$C_{01}SF_2$	97.0	2	1	0.0	0.45
NS_2SF_{10}	88.0	10	0.0	2	0.45
NS_4SF_8	88.0	8	0.0	4	0.45
NS_6SF_6	88.0	6	0.0	6	0.45
NS_8SF_4	88.0	4	0.0	8	0.45
$NS_{10}SF_2$	88.0	2	0.0	10	0.45
$NS_{10}SF_{10}$	80.0	10	0.0	10	0.45
NS_8SF_8	84.0	8	0.0	8	0.45
NS_4SF_4	92.0	4	0.0	4	0.45
NS_2SF_2	96.0	2	0.0	2	0.45

Standard water required for optimal consistency and setting times were tested as per EN 196-3:2005 [43], mixed cement samples total weight was 500 gm ± 1 gm, the required water was measured, then each sample was mechanically mixed with standard on and off times then moulds were filled determining water required for optimal consistency, and setting times according to the standard procedures using Vicate apparatus [43].

3. Results and discussion

3.1. Single effect of SF on physico-mechanical properties of CMs

3.1.1. Compressive strength. The compressive strength values of CMs containing 2, 4, 6, 8, 10 wt., % of SF cured at 2, 7, and 28 days are reported in figure 13. The compressive strength of CMs increased with increasing the levels of replacement of SF. Generally, the compressive strengths at 2, 7, and 28 days increased gradually by increasing SF ratio from 2 to 6 Wt.,%. A maximum increase up to 9, 2, and 27% was observed in the 2, 7, and 28 days, respectively in the levels containing 6 Wt., % of SF. The highest levels of replacement at 8 and 10 Wt.,% showed a gradual decrement in compressive strengths at all ages of hydration. This is due to positive impact of amorphous silica in NS which consuming $Ca(OH)_2$ and increasing CSH which responsible for binding capacity. The additional formation of hydrated calcium silicate (C-S-H) led to more densification in microstructure [44].

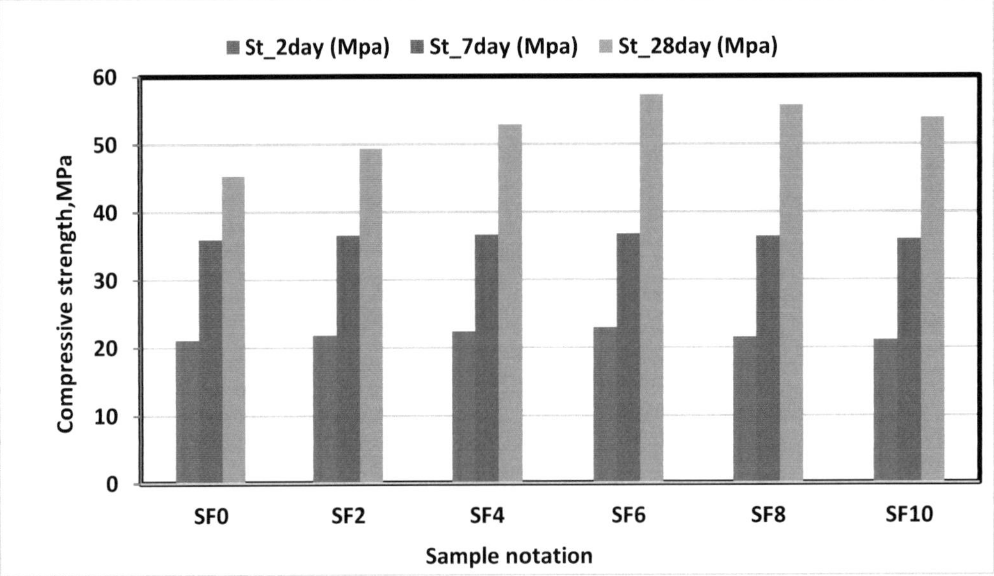

Figure 13. Compressive strength of CMs containing 2, 4, 6, 8, 10 wt. % of SF cured at 2, 7, and 28 days.

3.1.2. Bending strength. The bending moment results of CMs containing 2, 4, 6, 8, 10 wt., % of SF cured at 2, 7, and 28 days are showed in figure14. Clearly, the bending strength values increased with increasing SF up 6 Wt., % then decreasing up to 10 Wt., % of SF replacement. The bending strengths at 2, 7, and 28 days of curing gradually increased by increasing SF from 2 to 6 Wt., %. A maximum increment up to 4, 7, and 11 % was observed in the 2, 7, and 28 days, respectively in the samples containing 6 Wt., % of SF.

In general, this increment in bending strength may be due to that, pozzolanic activity of SF with the smallest particle size relative to the size of cement grain. The smallest particle size of silica fume (in range from 50 nm to 150 nm) was found to act as a physical barrier [45].These results of bending strengths confirmed the pervious results of compressive strength.

3.1.3. Water for standard consistence for complete hydration. Water absorption of CMs with or without SF at 2, 7, and 28 days of curing showed in figure 15. The replacement of CMs with SF (2, 4, 6, 8, 10 Wt., %) showed an increment in water consumption for standard consistence (WC) for complete hydration. The water consistency of CMs which containing 10 Wt., % of SF illustrated the highest values. On the other side CMs which containing 6 Wt., % of SF showed decrement in WC comparing to other mixes containing SF. This is attributed to the pozzolanic reactivity of SF and particle size of silica fume (in range from 50 nm to 150 nm)[45].

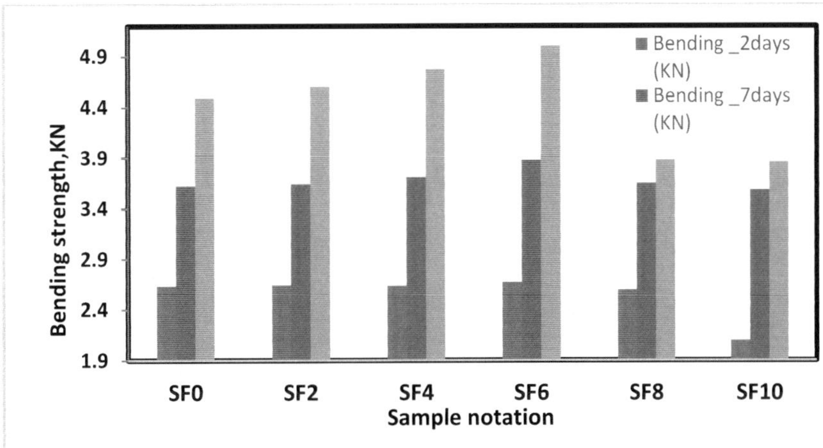

Figure 14. Bending moment of cement mortar contacting 2, 4, 6, 8, 10 wt. % of SF at 2, 7, and 28 days.

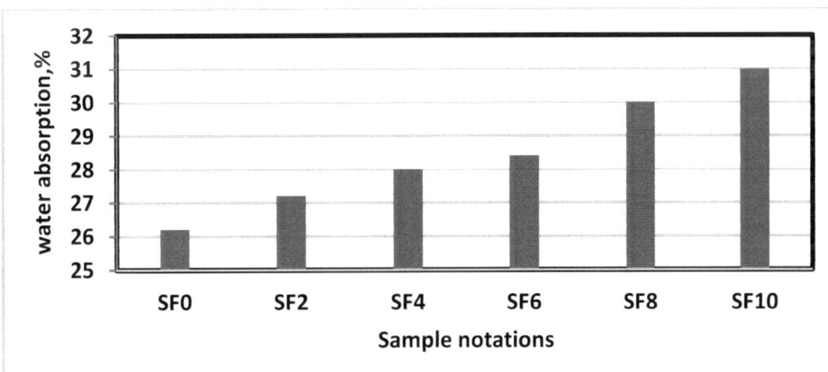

Figure 15. Water for standard consistence of cement mortar containing 2, 4, 6, 8, 10 wt. % of SF.

3.1.4. Initial and final setting times. Both Initial setting time (IST) and final setting time (FST) were reported in figure 16. The initial setting time decreased with increasing the replacement levels of SF. On the other side, the final setting time increased with decrement of SF ratios. This is mainly due to impact of amorphous SF which accelerated the hydration reactions in cement matrix. The acceleration of cement provided nucleation sites for $Ca(OH)_2$ within minutes after reaction commences, and also by reacting with Ca^{++} ions. The Reduction of Ca^{++} ion affected the nature of hydration products [46].

3.2. Combined effect of Silica fumes and functionalized MWCNTs on physico-mechanical properties of CMs

3.2. 1. Compressive strength. The compressive strength results of CMs containing 2, 4, 6, 8, 10 Wt., % of SF and 0.2, 0.4, 0.6, 0.8, 1.0 Wt., of functionalized MWCNTs % cured at 2, 7, and 28 days of hydration are plotted in figure 17. A gradually increment in 2, 7, and 28 days compressive strengths were observed by increasing replacement levels of both SF and functionalized MWCNTs. maximum increasing was observed at (C0.8SF8) which containing 0.8 Wt.,% functionalized MWCNTs and 8 Wt., % SF at early ages of hydration. The compressive strength increased by 15 and 12 % of control sample strength. Notably, The compressive strength of (C0.4SF8) which containing 0.4 W% functionalized MWCNTs and 8 W% SF at 28 days of curing showed increment by 36 %. This is due to duel effect of SF with pozzolanic activity and bridging effect of MWCNTs.

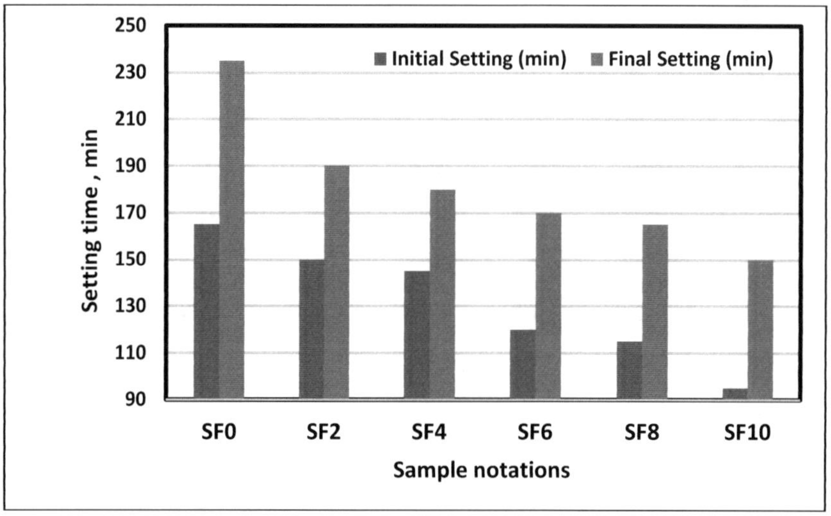

Figure 16. Initial and final setting times (min) of CMs with or without SF.

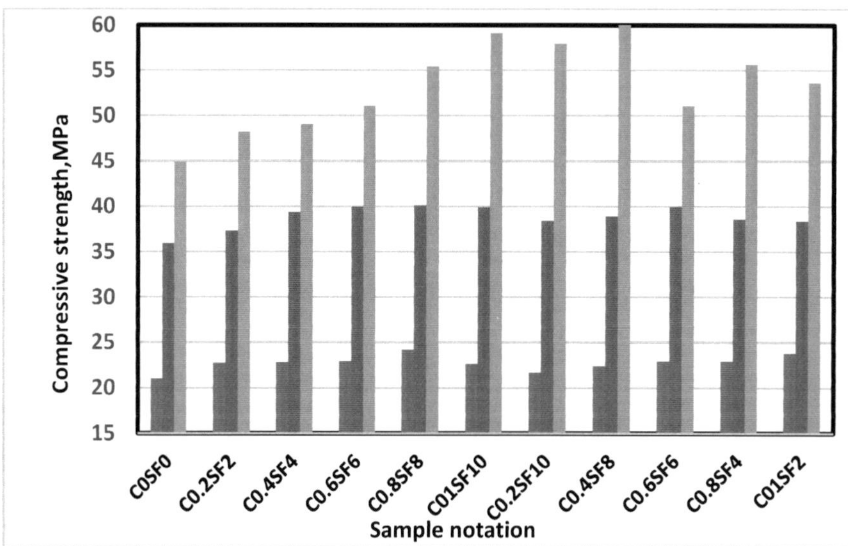

Figure 17. Compressive strength of cement mortar containing Functionalized MWCNTs and SF at 2, 7, and 28 days.

3.2.2. Bending strength. The bending strength results of CMs containing 2, 4, 6, 8, 10 Wt., % of SF and 0.2, 0.4, 0.6, 0.8, 1.0 Wt., of functionalized MWCNTs % cured at 2, 7, and 28 days of hydration are showed in figure 18. A gradually increment in the bending strength was observed upon increasing the replacement levels of both of them. The maximum value in bending strength was observed at the samples which containing 0.4 Wt.,% of MWCNTs and 4 Wt.,% of SF at early ages of hydration. The bending strength increased by 8 and 11 % at 2 and 7 days, respectively. On other side, at later ages of hydration, the bending strength recorded increment by 17 % at the mix which containing optimum ratio (0.6 Wt., % functionalized MWCNTs and 6 Wt.,% SF.

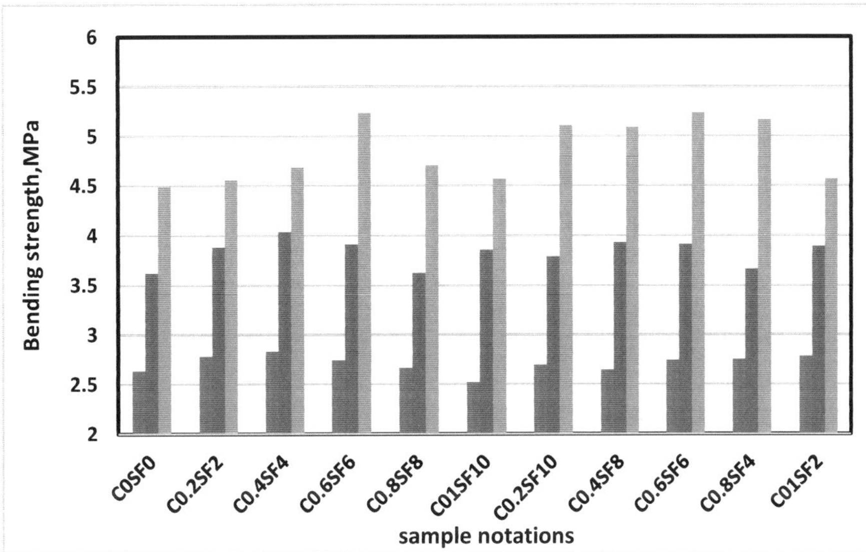

Figure 18. The bending moment of CMs contains Functionalized MWCNTs and SF with different ratios at 2, 7, and 28 days.

3.2.3. Water for standard consistence for complete hydration. The water consistence of CMs with or without both SF and Functionalized MWCNTs together at 2, 7, and 28 days of hydration showed in figure 19. Obviously, gradually increment in water for standard consistence was observed with high volume SF (10 Wt.,%) and low content Functionalized MWCNTs. The (C0.1SF10-mix) showed the highest value in water absorption. This is due to the fact that; the finer spherical particles surface area demands more hydration to attain the required cement properties [47].

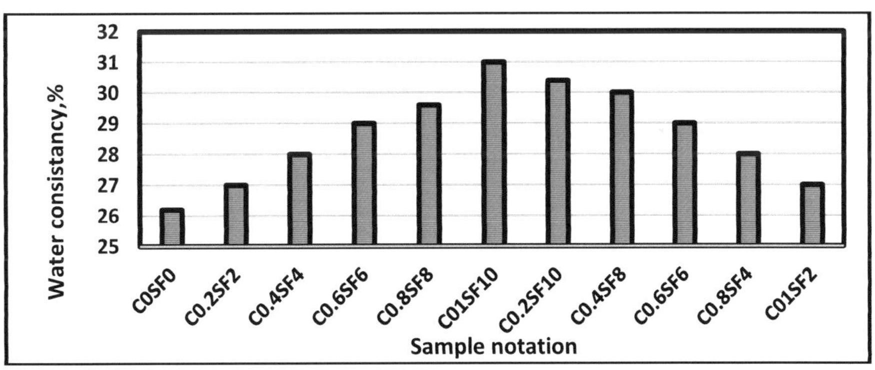

Figure 19. Water for standard consistence of cement paste containing Functionalized MWCNTs and SF at 28days of curing.

3.2.4. Initial and final setting times. Both Initial setting time (IST) and final setting time (FST) were showed graphically in figure 20. IST and FST decreased followed by increasing the content of Functionalized MWCNTs and SF together. All CMs which contained both SF and Functionalized MWCNTs showed the shortest IST and FST comparing to CMs without of them. The mix (C0.2SF10) showed shorter setting times than that the same mixes containing Functionalized MWCNTs and SF. This is due to effect of finer particles of SF and high surface area.

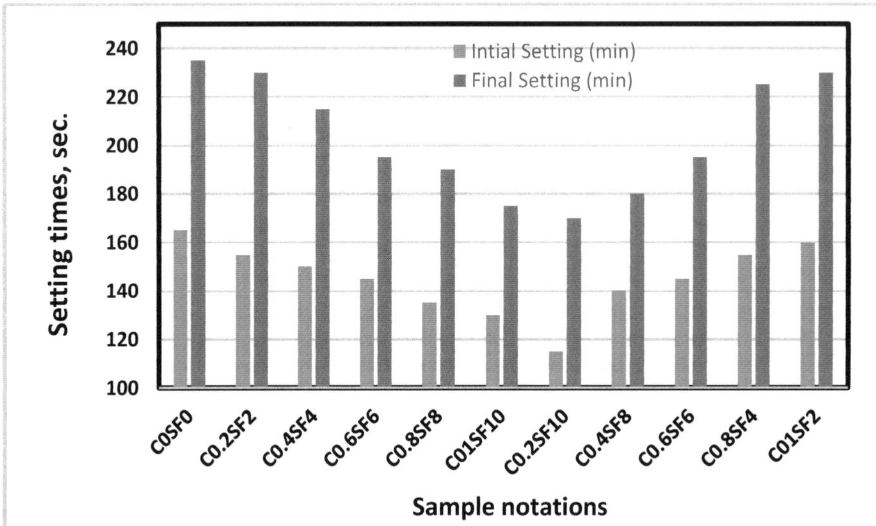

Figure 20. Initial and final setting times (min) of cement and cement paste containing Functionalized MWCNTs and SF.

3.3. Duel effect of Silica fumes and Nano silica on physico-mechanical properties of CMs

3.3.1. Compressive strength. The compressive strength results of CMs containing 2, 4, 6, 8, 10 Wt., % of SF and the same values of NS together cured at 2, 7, and 28 days of hydration are reported in figure 21. Notably, at early ages of hydration, the -NS8SF4 sample which containing (8 Wt., % NS and 4 Wt.,% SF) showed marginally increasing in compressive strength than other mixes by 8%. The highest increment in 7 and 28 days compressive strengths was observed in case of (NS8SF8) that contained 8 Wt., % NS and 8 Wt., % SF, which increased by 9 and 28 %, respectively. This is due to combined effect of NS and SF which improved physical and mechanical properties of cementations composites. NS consumed Ca(OH)2 and formed additional amount of CSH with high binding capacity.

3.3.2. Bending strength. The bending moment results of CMs containing 2, 4, 6, 8, 10 Wt., % of SF and the same values of NS together cured at 2, 7, and 28 days of hydration were illustrated in figure 22. Minor effect was observed for bending strength (2, 7, and 28 days) of CMs containing both NS and SF with different ratios. At early ages (2 and 7 days) of hydration, the NS4SF8-mix showed the highest bending strength but other mixes decreased by 17 % and 12% in case of NS10SF2- mix. Finally, the bending strength of NS8SF8 –mix showed the highest bending strength at 28days.

3.3.3. Initial and final setting times. Both Initial setting time (IST) and final setting time (FST) of CMs with or without SF and NS together were showed graphically in figure 24. Generally, the decrement in IST and FST was observed by increasing the replacement levels of SF and NS together. The (NS10SF10-Mix) showed the shortest setting times comparing to other mixes. This is due to the high surface area for both of them. Amorphous SF and NS accelerated the hydration reactions in cement matrix especially with high levels of replacement.

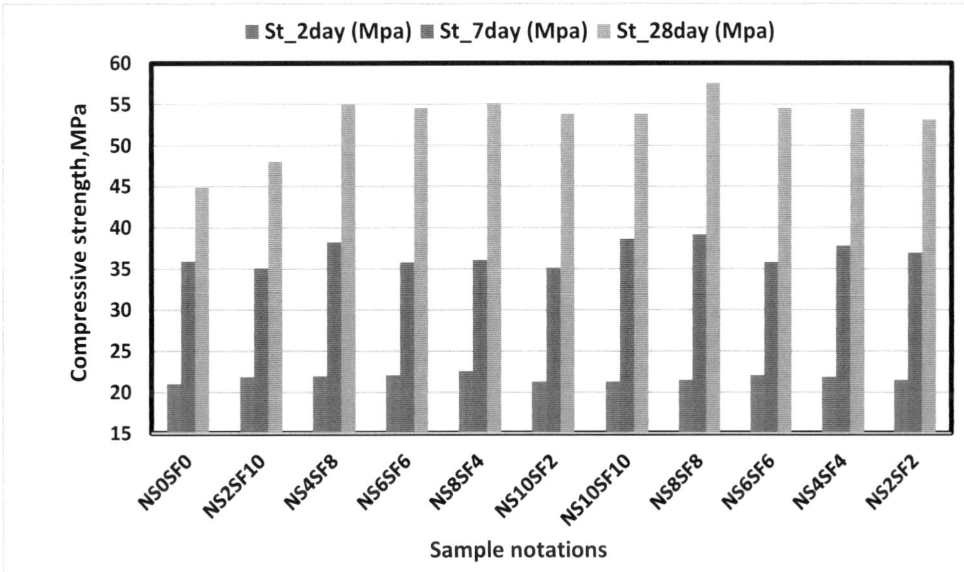

Figure 21. Compressive strength of cement mortars containing NS and SF with different ratios at 2, 7, and 28 days.

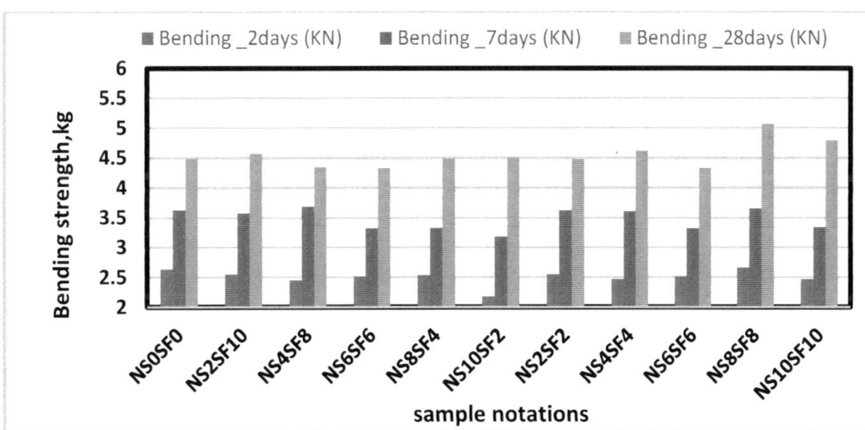

Figure 22. The bending moment of cement mortar containing NS and SF with different ratios at 2, 7, and 28 days.

3.3.4. Water for standard consistence for complete hydration. Water consistency of CMs with or without SF and NS at 2, 7, and 28 days of curing was presented in figure 23. Clearly, the SF and NS content replaced together with high levels, the Water for standard consistence for complete hydration increased. The sample which didn't contain SF and NS showed the lowest water absorption. NS10SF10- Mix showed the highest value of water absorption. This is due to high surface area and fine particles of NS and SF which required to absorbed additional water.

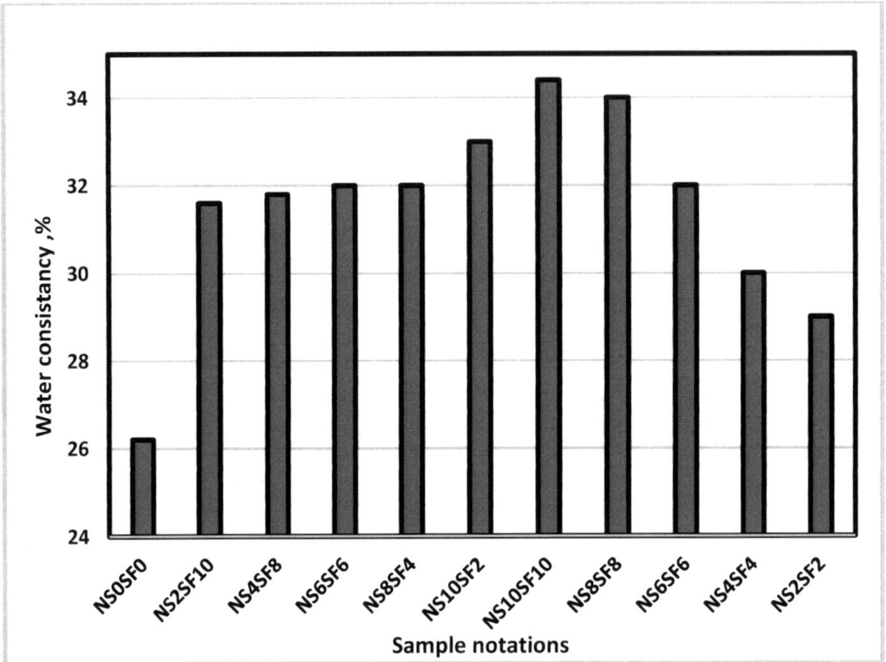

Figure 23. Water for standard consistence of plain CEM I (42.5N) and mixtures containing both NS and SF in different ratios.

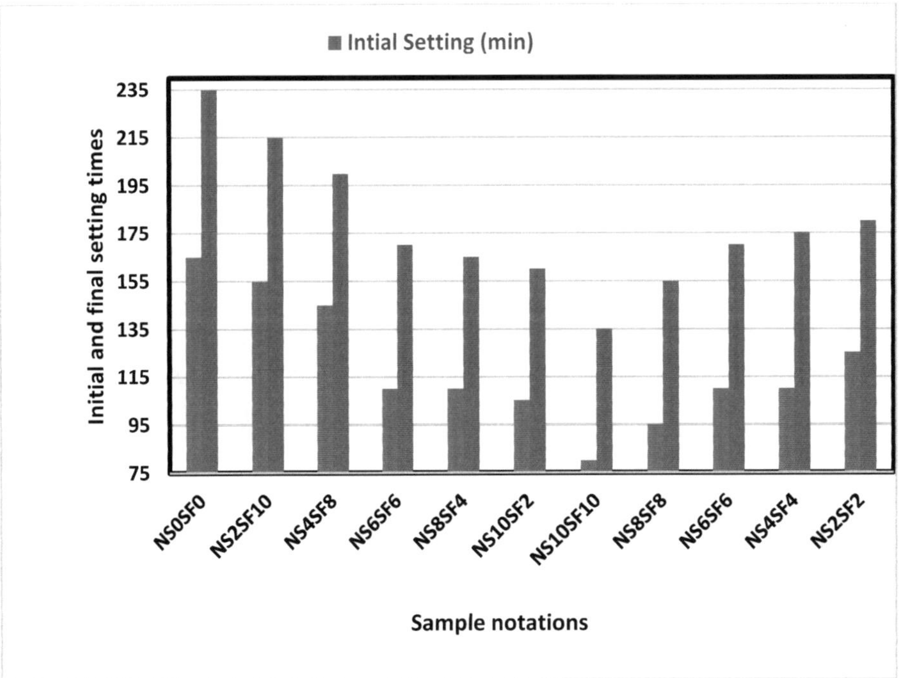

Figure 24. Initial and final setting times (min) of plain CEM I (42.5N) and mixtures containing both NS and SF in different ratios.

3.4. Microstructure of CMs

The SEM-morphology of cement pastes with or without SF singly, cement paste with mixture of SF and Functionalized MWCNTs and cement paste with mixture of SF and NS illustrated in figure 25. The SEM-morphology of cement paste containing SF (MCSF) showed the rods of CSH which formed due to amorphous silica of SF comparing to control sample (MC0) due to normal hydration of cement. The-MCSF- sample showed high microstructure compared to the blank sample due to filling effect of SF. On the other side, the cement paste with mixture of SF and Functionalized MWCNTs (MCSF-MWCNT-mix) showed the bridging effect of functionalized MWCNTs and effect of spherical amorphous silica in SF. This MCSF- MWCNT-mix showed the dual effect of crack bridging with MWCNTs or inters cracks filling with SF. This combined effect showed the densification in microstructure and reflect the good results of compressive strength and bending moment. This sample reported the highest compact comparing to control sample. Clearly, the cement paste with SF and NS together (MCSF- NS-mix) showed the additional CSH with more compact in microstructure. This reported the duel effect of nano-spheres (NS) and amorphous silica (SF) which increased the densification of microstructure. By comparing with control sample, MCSF- NS-mix showed denser in micro-porous structure.

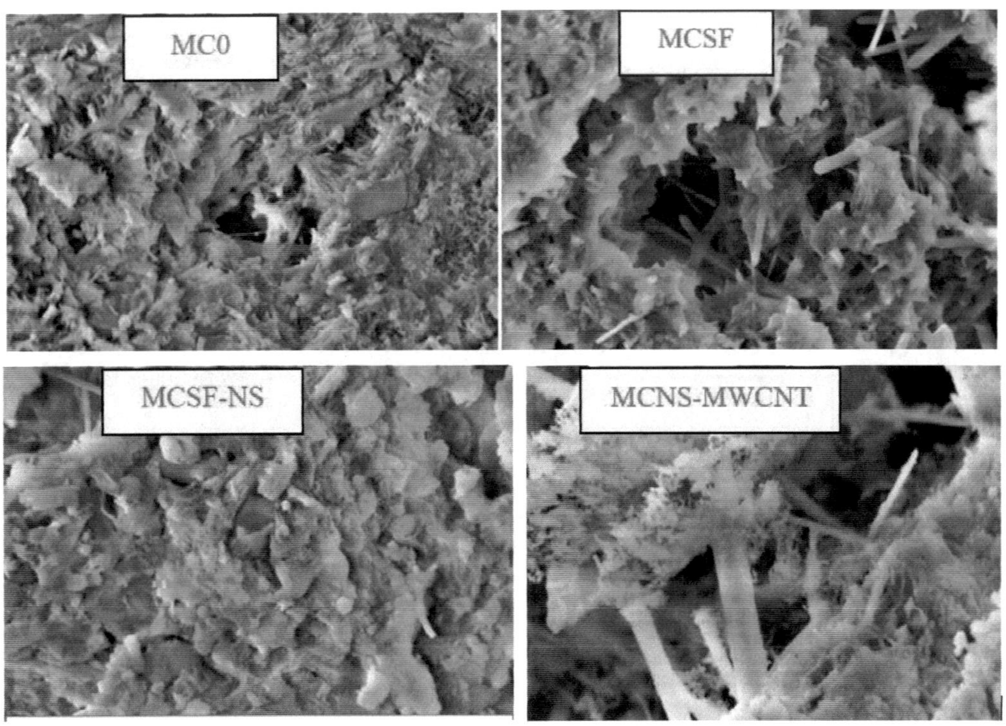

Figure 25. SEM-photos of cement pastes with SF singly, cement paste with mixture of SF and Functionalized MWCNTs and cement paste with mixture of SF and NS.

4. Conclusion

Performance of CMs incorporated of silica fumes, purified functionalized MWCNTs and nanosilica was studied. Generally, increment of compressive and bending strengths (due to the fact that MWCNTs bridge the cracks) was observed. No significant effect on water consumed for optimum consistence was observed, while initial and final setting times were generally decreased. This effect may be due to the fact that MWCNTs are highly dispersive in water so increase mixing homogeneity. Meanwhile the replacement of cement composites with NS and SF showed increasing in physico-mechanical properties such as compressive, bending strengths, and water consistency due to the finer spherical particles of NS and surface area that demanded more hydration to attain the required cement properties but setting times were significantly decreased. The combination between MWCNT and SF showed the good duel effect on mechanical properties more than the impact of NS and SF. This combination between MWCNT and SF led to enhance the physico-mechanical properties of CMs. The results of compressive, bending strengths, water consistency and setting times were confirmed by SEM-morphology which showed more compact in microstructure. This investigation, presented eco-friendly technique to prepare the nano-particles to enhance the mechanical properties of CMs.

References

[1] European Standard EN 197-1/2011 (2011); Cement -Part 1:6 August; 9-11
[2] Taylor H F W 1997 *Cement Chemistry* (London: Thomas Telford Services Ltd) P 3
[3] Jennings H M 2000 *Cem. Concr. Res.* **30** 101
[4] Constantinides G and Ulm F J 2004 *Cem. Concr. Res.* **34** 67
[5] Valipour M, Pargar F, Shekarchi M, Khani S and Moradian M 2013 *Constr. Build. Mater.* **46** 63
[6] Afroughsabet V and Ozbakkaloglu T 2004 *Constr. Build. Mater.* **18** 669
[7] Medina N F, Barluenga G and Hernández-Olivares F 2015 *Constr. Build. Mater.* **96** 556
[8] Wongkeo W, Thongsanitgarn P, Ngamjarurojana A and Chaipanich A 2014 *Mater. Des.* **64** 261
[9] Ahmed M A, Hassaneen Y A, Assaf K A and Shawkey M A 2015 *Open J. Civ. Eng.* **5** 131
[10] Ahmed M A, Hassaneen Y A, Assaf K A and Shawkey M A 2015 *Int. J. Civ. Struct. Eng. Res.* **4** 159
[11] Ahmed M A, Hassaneen Y A, Assaf K A, El-dek S E and Shawkey M A 2015 *Microelectron. Eng.* **146** 53
[12] Suzuki T 1987 *First International Symposium on Utilization of High Strength Concrete* (Stavanger, Norway, Norwegian Concrete Association) P 51
[13] Gautefall O 1986 *J. Am. Concr. Inst.* **2** 991
[14] Radjy F F 1986 *J. Am. Concr. Inst.* **2** 1135
[15] Kohno K 1989 *J. Am. Concr. Inst.* **2** 815
[16] Yamato T 1989 *J. Am. Concr. Inst.* **2** 897
[17] Marusin S L 1989 *J. Am. Concr. Inst.* **2** 924
[18] Ebbesen T W 1996 *Carbon Nanotubes* (Florida: CRC Press) p 111
[19] Breuer O and Sundararaj U 2004 *Polym. Compos.* **25** 630
[20] Endo M, Iijima S and Dresselhaus M S 1996 *Carbon Nanotubes 1st Edition Pergamon* (New York: Elsevier Science Inc.) P 198
[21] Falvo M R, Taylor R M, Helser A, Chi V, Brooks Jr F P, Washburn S and Superfine R 1999 *Nature* **397** 236
[22] Robertson D H, Brenner D W and Mintmire J W 1992 *Phys. Rev. B Condens. Matter* **21** 12592
[23] Yacobson B I, Brabec C J and Bernholc J 1996 *Phys. Rev. Lett.* **76** 2511
[24] Hernández E, Goze C, Bernier P and Rubio A 1998 *Phys. Rev. Lett.* **80** 4502
[25] Collins P G and Avouris P 2000 *Sci. Am.* **283** 62
[26] Zou B, Chen S J, Korayem A H, Collins F, Wang C M and Duan W H 2015 *Carbon* **85** 212
[27] Abu Al-Rub R K, Tyson B M, Yazdanbakhsh A and Grasley Z 2012 *J. Nanomech. Micromech.* **2** 1
[28] Kang S T, Seo J Y, and Park S H 2015 *Adv. Mater. Sci. Eng.* **20** 1
[29] Jo B W 2007 *Constr. Build. Mater.* **21** 1351
[30] Shih Y J, Chang T P and Hsiao T C 2006 *Mater. Sci. Eng. A.* **424** 266
[31] Gaitero J J, Campillo I and Guerrero A 2008 *Cem. Concr. Res. Res.* **38** 1112
[32] Chandra S and Maiti S C 1998 *Sixth NCB international Seminar on Cement and Building Materials*, The Council, New Delhi, **5** 24
[33] Seff L N, Hortza D, Repette W L, Ferreria V M and Labrincha J A 2012 *Mater. Sci. Eng.* A. **532** 354
[34] Ji H 2005 *Cem. Concr. Res.* **35** 1943

[35] Ltifi M, Guefrech A, Mounanga P and Khelidj A 2011 *Procedia. Eng.* **10** 900
[36] Karl T R and Trenberth K E 2003 Science **302** 1719
[37] Bagheri A, Parhizkar T, Madani H and Raisghasemi A M 2012 *Mater. Struct* **46** 34
[38] BS EN 197-1:2011, European Standard: *Cement Composition, specifications and Conformity Criteria for Common Cements* ICS 91. 100. 10
[39] Tantawy M A, El-Roudi A M, Abdalla E M and Abdelzaher M A 2013 *J. Eng.* **2013** 1
[40] Bahgat M, Farghali A A, El Rouby W M A and Khedr M H 2011 *J. Anal. Appl. Pyrol.* **92** 307
[41] El-Didamony H 2019 *Bol. Soc. Esp. Ceram.* V. **59** 31
[42] EN 196-1:2005 (E), Methods of testing cement – Part 1: *Determination of Strength*, page. 6, ICS 91.100.10.
[43] EN 196-3:2005 (E), Methods of testing cement – Part 3: *Determination of Setting Time and Soundness*, pp. 6, ICS 91.100.10.
[44] Bjornstrom J, Martinelli A, Matic A, Borjesson L and Panas I 2004 *Chem. Phys. Lett.* **392** 242
[45] Abdelzaher M A, Hamouda A S, Ismail I. M and El-Sheikh M A 2018 *Key Eng. Mater.* **786** 248
[46] Balboul B A A, Abdelzaher M, Hamouda A S and Zaki A H 2019 *Egypt. J. Chem.* **62** 1105
[47] Aggarwal P, Singh R P and Aggarwal Y 2015 Cogent Eng. **2** 1078018
[48] Abo-EL-ENein S A, Abdel-Gawwad H A, Hodhod O A and El-Kattan I M H 2016 *Am. J. Sci.* **12** 3
[49] Abdel-Gawwad H A, Hassan H S, Vásquez-García S R, Israde-Alcántara I, Ding Y C, Martinez-Cinco M A and El-Kattan I M 2020 *J. Clean. Prod* **252** 119875
[50] Abdel-Gawwad H A, El-Enein S A, Heikal M, Abd El-Aleem S, Amer A A and El-Kattan I M 2018 *Constr. Build. Mater.* **176** 676
[51] El-Kattan I M, Abdelzaher M A and Farghali A A 2020 *J. Mater. Res. Technol.* https://doi.org/10.1016/j.jmrt.2020.05.087

Effect of hydrothermal time and acid-washing on the antibacterial activity of Sodium titanate nanotubes

H Mohamed[a], A H Zaki[a,*], Fatma I Abo El-Ela[b] and S I El-dek[a*]

[a]Materials Science and Nanotechnology Department, Faculty of Postgraduate Studies for Advanced Sciences, Beni-Suef University, Egypt
[b]Department of Pharmacology, Faculty of Veterinary Medicine, Beni-Suef University, Beni-Suef, Egypt

*E-mail: ayman.zaki@psas.bu.edu.eg (AHZ); samaa@psas.bu.edu.eg (SIE)

Abstract. Bacterial resistance to antibiotics has been a major concern globally. In this work, sodium and hydrogen titanate nanotubes were successfully synthesised using a facile hydrothermal technique. The prepared samples were characterized by XRD, FTIR , HRTEM surface area analyser, hydrodynamic size analyser and zeta potential. Minimum inhibitory concentration (MIC) and the minimum bactericidal concentration (MBC) measurements besides agar well diffusion method showed good antimicrobial activity of both nanotubes. The antimicrobial activity of sodium titanate nanotubes prepared at 6 and 23 hr., and their hydrogen exchanged forms; (HTNT) and (HTNS) explored good antimicrobial activity against both Gram positive and Gram negative bacteria besides their antifungal activity which reflects their importance in treatment of bacterial infections causing serious diseases in both animal and human.

1. Introduction

Nanotechnology has emerged as one of the most useful tools for facing numerous biological and biomedical challenges [1]. Nanoparticles can offer enhanced properties and better results due to their higher surface to volume ratio compared to bulk materials. Nanoparticles have been shown to possess improved chemical, mechanical, electrical, optical and magnetic properties [2]. One of most promising biomedical application of nanomaterials is their use as antibacterial and antimicrobial agents for drug resistive pathogens (DRPs). Resistance to antibiotics is considered a big challenge in pharmaceutical and biomedical fields of study [3]. DRPs represents a serious threat for containing their spread and for their serious impact on human health [4,5]. Patients infected with DRPs are susceptible to increased treatment costs, longer treatment periods in hospitals and require the use of 2-3 different antibiotics which is a less effective and more toxic technique [3,4]. Studies have shown that multi-drug resistive bacteria has caused thousands of deaths and novel methods and strategies are of urgent need [5].

Several types of nanoparticles have shown great promise as antibacterial agents compared to standard antibiotics [6–9]. Silver [10], ZnO [11], TiO_2[12,13]and chitosan [14]nanoparticles are among the most heavily studies nanomaterials as antibacterial agents. Titanium dioxide and titanate nanostructures offer several advantages such as being cheap, available, ion exchangeable and easy to synthesize in different morphologies[15]. Several studies available in the open literature reported promising results for the use of titanate and modified titanate nanomaterials for antimicrobial applications. Yada et al modified the surface of a sodium titanate nanofiber with silver and utilized it as antibacterial agent against *Staphylococcus aureus*[16]. Results showed that silver exchanged titanate thin films showed promising antibacterial effect against the tested microorganism. Lee et al reported that silver exchanged titanate structures are very favourable materials for antibacterial applications [17]. Matsushita et al. have showed that calcium modified titanate films on top of Ti metal structures show strong antibacterial properties against the tested *Staphylococcus aureus* microorganism [19]. Stoyanova et al. studied the antimicrobial effect of Zinc titanate nanostructres prepared by sol-gel technique and reported promising results against *Esherihia coli* [18]. Moreover, Shah et al reported significant results against *P. aeruginosa* and *S. aureus*when using bariumtitanate nanoparticles as novel antibacterial powder [19]. To this extent, no study has reported the effect of hydrothermal time and acid washing on their antibacterial and antimicrobial activities against pathogens. The aim of the current work is synthesized and characterize sodium titanate nanotubes at different hydrothermal times and also their hydrogen exchanged nanotubes. The prepared samples are to be tested as antimicrobial agents against several pathogens including *P. Aeruginosa*, *E. Coli, Staphylococcus and Streptococcus.*

2. Materials and methods

2.1. Materials
Titanium dioxide, sodium hydroxide and Hydrochloric acid were suppliedby Loba chemie, Piochem for laboratory chemicals and Adwic respectively. All chemicals were used as received without further purification.

2.2. Synthesis of sodium and hydrogen titanate nanotubes
Sodium titanate nanotubes(STNS) were prepared following a facile hydrothermal technique as previously reported in our previous work [20–22]. Briefly, 10 g of TiO_2 are suspended in 500 mL of 10 M NaOH solution. The solution is kept under vigorous stirring for 30 min. Then the solution is transferred to a 1L stainless steel autoclave which was heated at 160 °C for 6 and 23 hours to form STNS and STNT respectively. The autoclave was then left to cool naturally and the formed powder was then separated and washed extensively with distilled water then dried at 80 °C overnight. The Sodium ions replacement process was conducted by washing STNT or STNS with 0.1M HCl solution, where the mixture was subjected to sonication for 30 min. using ultrasonic with probe (20 KHz).

2.3. Characterization
X-ray diffraction experiments were conducted on a PANalytical (Empyrean) X-ray diffraction using Cu Kα radiation (wave length 0.154 cm^{-1}) at an accelerating voltage 40 KV, current of 35 mA, scan angle range from 20 to 70° and scan step 0:02°. The BET specific surface area, specific pore volume, and pore sizes of the adsorbent materials were determined by N_2 adsorption isotherms using an automatic surface analyzer (TriStar II 3020, Micromeritics, USA). Hydrodynamic particle sizes and zeta potentials were determined on a Malvern zetasizer instrument (Malvern Instruments Ltd).

2.4. Isolation and identification of bacterial pathogens

2.4.1. Isolation of bacterial pathogens. E. coli strain was grown on its selective media MacConkey agar media (Oxoid; CM 0115) and Eosin Methylene Blue agar (EMB; Oxoid; CM 69) plates whereas, the other strains (*Staphylococcus* and *Streptococcus*) were determined on agar plates (Gram stain and colony morphology). The plates were incubated at 37°C for 48 hours and the bacterial growth was observed for all of the isolates[23].

2.4.2. Biochemical and serological identification of bacterial pathogens. Standard Kits (Biomerieux, Marcy L'etoil, France) API had been used for both physiological and biochemical identification of each bacterial isolates of both gram positive and negative bacteria.

2.5. Bacterial strains and culture conditions
For antibacterial studies, Gram-positive *St.coccus*, *S. aureus* and Gram-negative *E. coli* and *P. Aeruginosa* were four microbial agents of multiple infective diseases in humans and animals, chosen as the target organism for this investigation. *E. coli* (ATCC 25922), *S. aureus* (ATCC 25913), *St.coccus* (ATCC *49619*), and *P. aeruginosa* (ATCC *27853*) were purchased from Cairo Microbiology Research Centre besides two different strains of fungi species (*Aspegillus Flocculosus* and *Aspegillus nigricans*). Sabouraud 2% Glucose Agar with Chloramphenicol media for microbiology was used for fungal culturing before start working. All tubes were sterilized in an autoclave before the experiments. Muller Hinton broth was used to culture *E. coli*, *S. aureus*, *St.coccus* and *P. aeruginosa* at 37 °C for 24 hour in an incubator. The liquid cultures were finally diluted to obtain bacterial cell concentration of approximately 10^7 colony forming units (CFU)/ml for following antibacterial test.

2.6. Determination of minimum inhibition concentration (MIC)
Using broth dilution method according to Clinical Laboratory Standards Institute (CLSI)[24]:The MIC was defined as the lowest concentration that completely inhibited the growth of microorganisms for 24 hours. Bacterial strains were grown overnight on MHA plates at 37°C before being used. The antimicrobial activity of the-obtained nanotubes was examined using the standard broth dilution method [24].

For each sample, different concentrations were diluted with Muller Hinton broth to give a final concentration ranging from (1000 µg/ml-6.25 µg/ml) in the presence of both positive and negative control tube. The bacterial isolate was sub cultured on Muller Hinton Agar (M.H.A) and incubated at 37°C for 24 h [25]**.** Colony from the tested microorganism was suspended in 5ml saline, and the suspension was adjusted to 0.5 McFarland standard to give organism suspension of (1×10^8 CFU/ml). All tubes were incubated at 37°C for 24 hrs. Results were recorded in terms of MIC, which is the lowest concentration of antimicrobial agent causing almost complete inhibition of growth or giving no visible growth.

2.7. Minimal bactericidal concentration (MBC)
After MIC determination of the nanoparticles tested, aliquots of 100 µl from all tubes in which no visible bacterial growth was observed were seeded in MHA plates and were incubated for 24h at 37°C. The MBC endpoint is defined as the lowest concentration of antimicrobial agent that kills 100% of the initial bacterial population. The number of plates without colonies was noted [24].

The MIC of STNT, STNS, HTNT and HTNS was measured by broth dilution test in dark condition. The tubes were incubated at 37°C for 48 h on a rotary platform. The visual turbidity of the tubes was noted before and after incubation. The MIC was defined as the endpoint where no visible turbidity could be detected.

2.8. Antibacterial Sensitivity of the Synthesized Nanomaterials
Each tested Nanomaterials was prepared to give a final stock concentration of 10 mg/ml. This stock solution was used for preparing different diluted concentrations of each material in distilled water. The antibacterial activity of the synthesized nanomaterials was measured using agar well diffusion method in presence of Difloxacin 92.9% as a standard reference drug to Gram negative bacteria *as E-coli* and Tylvalosin to Gram positive bacteria as *S. aureus*.

Agar diffusion method is one of the formal techniques, in this test; the tested samples were spread on agar layer in a Petri-dish. The growth of the tested microorganisms was limited to a circular area or zone around the pores which contained tested material. The antimicrobial effects were determined and measured as a diameter zone in millimetres by using calibre.

The experimental procedures were as follow: The bacterial strains were suspended in saline solution with an intensity of 1.5x10^8 viable organism/mL, the population of the investigated pathogen was matched to McFarland tube No.0.5 (1.5x10^8) colony forming units CFU/ml. Standard sterilized discs were impregnated overnight with the three different tested concentrations of the tested nanomaterials using standard size (50 mm diameter) of Whatman filter paper. The impregnated discs were placed aseptically with some sterile forceps on Mueller-Hinton agar plates. All plates were inculcated at 37°C for 24 hr. then inhibition zone diameter of all tested bacteria was determined via two-fold serial dilution. All the readings were taken in triplicate.

3. Results and discussion

3.1. Samples characterization

The powder XRD diffractograms for all prepared samples are shown in Figure 1. The diffractograms of both samples HTNS and HTNT were matched with the standard JCPDS 47-0561 file. The diffraction peaks at 24.6, 25.6, 28.4, 38.1 and 48.7 were indexed to the (110), (202), (111), (60-2) and (114) planes. The STNT and STNS samples were indexed in reference to the ICDD-04-0157486 standard data. The peaks at 10.2, 24.1, 25.5, 28.7 and 48.2 could be indexed to the (001), (201), (011), (111) and (303) planes. The average crystallite sizes of STNT, HTNT, STNS and HTNS were calculated to be 31.86, 31.87, 63.82 and 63.84 nm respectively. As expected, no significant change in the average crystallite size was caused by hydrogen replacement. FTIR spectra of all prepared samples is shown in figure 2. The peaks at 3400 cm^{-1} and 1620 cm^{-1}can be attributed to the stretching vibrations and deformation of hydroxyl groups of physically adsorbed water molecules [28]. The low intensity peaks are typical of the Ti-O and Ti-O-Ti bonds [29].

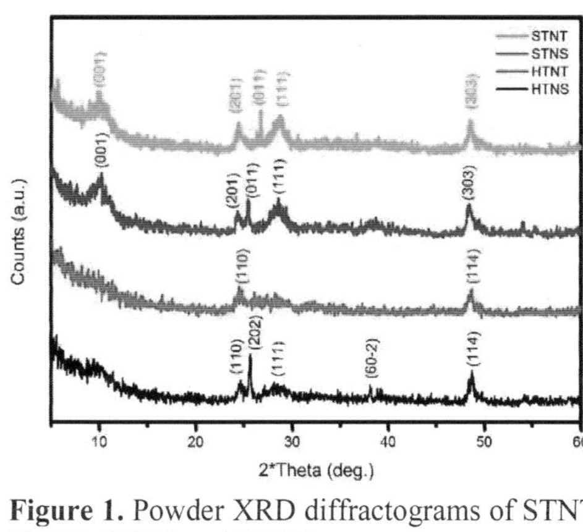

Figure 1. Powder XRD diffractograms of STNT, HTNT, STNS and HTNS.

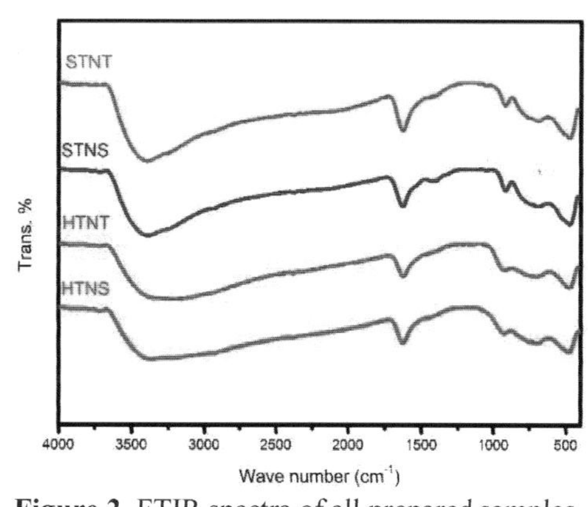

Figure 2. FTIR spectra of all prepared samples.

Figure 3 shows the HRTEM images of the prepared titanates, it is clear from that the sodium titanate nanotubes were successfully prepared at 6 and 23 hr. The HRTEM images reveal that increasing the hydrothermal time from 6 to 23 hr enhanced the quality of the prepared nanotubes, where the sample obtained at 6 hr contains some unrolled nanosheets. After acid-washing no significant change occurred in the morphology of the samples. It is clear that the average diameter of the prepared nanotubes is lower than 20 nm, while the average length is lower than 1000 nm in all samples.

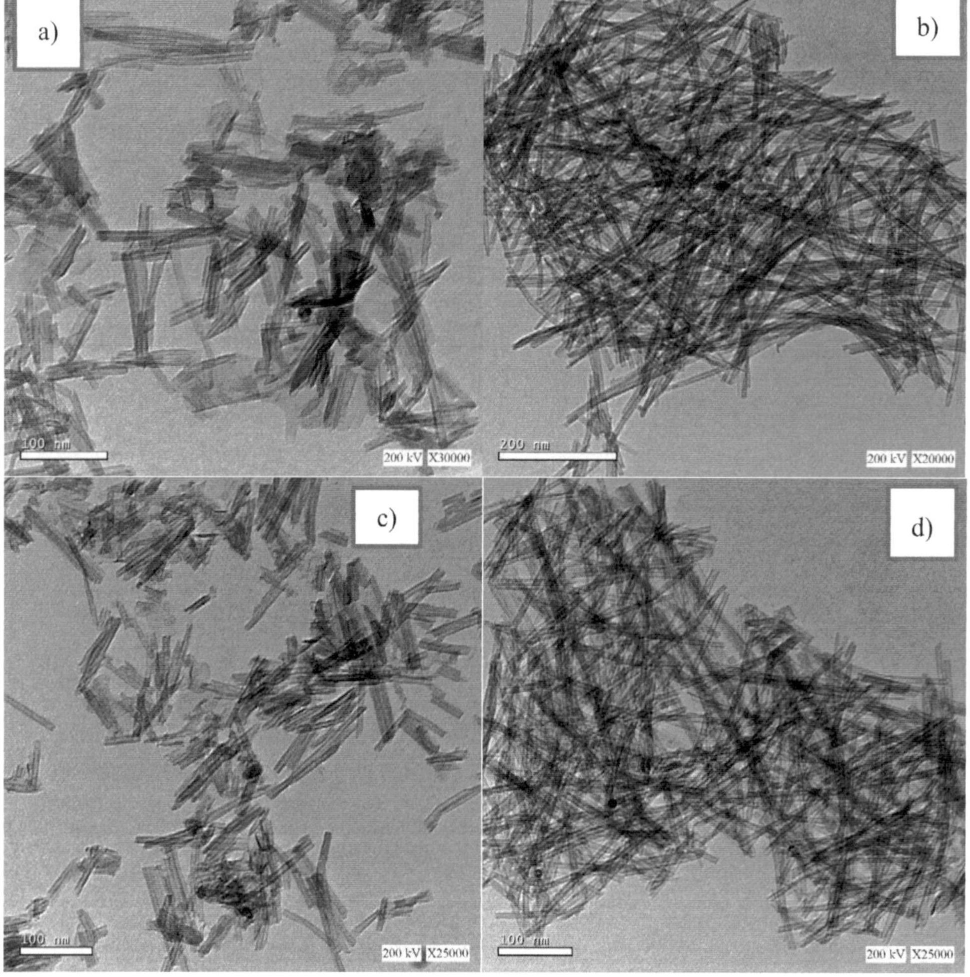

Figure 3. HRTEM image of (a) STNS, (b) STNT, (c) HTNS, and (d) HTNT

The BET surface area for all prepared samples was estimated using Nitrogen adsorption isotherms as shown in figure 4 (a). The BET surface area of STNT, HTNT, STNS and HTNS were calculated to be 114.57, 141.84, 96.48 and 135.9 m^2/g respectively. For all samples the isotherm follows a type IV behaviour indicating the presence of mesoporous pores within all the samples [24]. The pore size distribution of the samples is shown in figure 4 (b). The pores present are probably interparticle pores formed due to the aggregation of the nanomaterials as no surfactant is used during preparation.

Zeta potential for all samples is tabulated in Table 1. STNT and STNS showed a zeta potential of -11.2 and -11.7 mV respectively. Sodium titanate based structures had close zeta potential values irrespective of the morphology indicating the role of sodium ion on the overall surface charge. When sodium was replaced with hydrogen, HTNT and HTNS showed a zeta potential of 2.5 and 0.13 mV respectively. The hydrodynamic sizes for all samples are shown in Table 1. The large values reflect the extensive agglomeration of the particles in solution owing to their low zeta potential values. Another plausible reason for the large values is the particle shape (tubes and/or sheets)

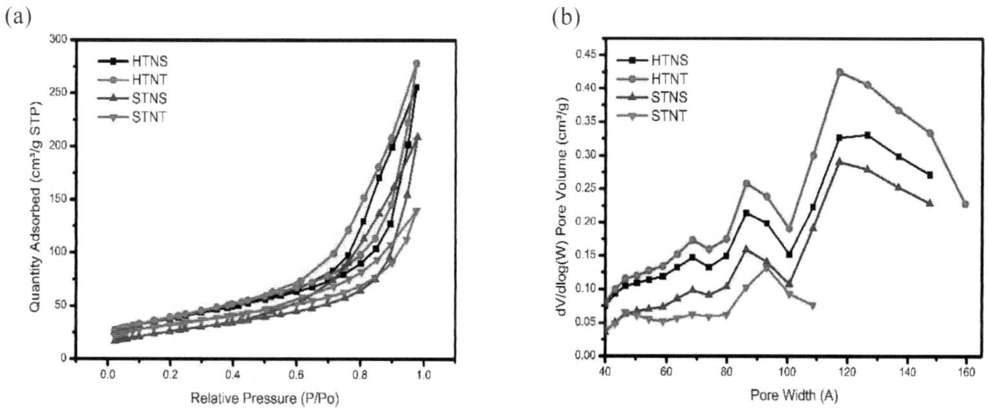

Figure 4. Nitrogen adsorption isotherm (a) and pore size distribution (b) for all prepared samples.

Table 1. Zeta potential and hydrodynamic size of STNT, HTNT, STNS and HTNS.

	Zeta potential (mV)	Hydrodynamic size (nm)
STNT	-11.2	1704
HTNT	2.5	1822
STNS	-11.7	1758
HTNS	0.13	1686

3.2. Antimicrobial study

In this work, the antimicrobial effect of STNT, HTNT, STNS and HTNS was investigated. Titanate nanostructures are known to have promising antimicrobial potential under photocatalytic conditions [26]. In this work all samples were tested under dark conditions. As shown in figures 6 and 7, all samples showed antimicrobial activity against the tested pathogens. Sodium titanate nanotubes (STNT) in this study showed the best antimicrobial activity than other prepared shapes, even though slightly higher MIC than other shapes except in pseudomonas infection; showed also excellent antimicrobial activity (MIC) but that's due to the visual accuracy for the MIC evaluation while the actual antimicrobial activity achieved through the disc diffusion method.. Only few studies in the literature reported antimicrobial activity for titanate nanostructures under dark conditions. Hydrogen titanate nanofibers and nanotubes possess significant antibacterial activity against Gram-positive bacteria and also against Gram negative bacteria (*E. Coli*) but in lower activity than the positive one according to the measured zone of inhibition when compared to the Gram positive one as also mentioned by Kundu *et al.* [27]. Neagu *et al.* [28] also showed that hydrogen titanate nanotubes has a significant antibacterial effect against *Bacillus subtilis* and *Virgibacillus halodenitrificans* which may be due to the interaction of the nanotubes (synthesized at different conditions) with the cell wall composition of the microorganism. The results reported in this study shows that the activity of hydrogen titanate nanostructure (HTNT and HTNS) against Gram-positive bacteria is not always better than those for Gram-negative ones as opposed to results by Kundu *et al.* [27]. It can be observed from figure 5 that the MIC of both HTNT and HTNS is much lower in case of *Escherichia coli* compared to *Streptococcus Pneumonaie* and *Staphylococcus aureus*. It can be also shown that the MIC for *Escherichia coli* is significantly lower than that measured for *Pseudomonas aeruginosa* which suggests that the antibacterial effect depends on the interaction between the nanostructure and the cell wall of the microorganism as reported by Neagu *et al.* [28] even for Gram-negative strains.

Sodium titanate nanotubes possess an antibacterial effect against methicillin-resistantstaphylococcus aureus (MRSA) Yada *et al.* [16,29]. The authors showed that the tested nanotubes have similar antibacterial effect in dark as that under UV illumination. The results in our study confirm the antibacterial effect of sodium titanate nanostructures (STNT and STNS) against both Gram-negative and Gram-positive bacteria. As illustrated in Figure 5, both structures showed MIC values comparable to both HTNT and HTNS for the Gram-negative and Gram-positive bacteria strains tested. This shows that one of the main factors defining the difference between sodium and hydrogen based titanate nanostructures is their zeta potential. For the hydrogen based titanate nanostructures the zeta potential and hence the surface charge was insignificantly different as previously discussed. Similarly, sodium and based titanate nanostructures showed close zeta potential values. This illustrates that zeta potential is the main driving factor behind the interaction between the nanostructures and the bacteria cell wall. Irrespective of the surface area and the crystallite size, the surface charge will affect the electrostatic attraction or repulsion between the surface of the nanostructures and the bacteria cell wall. This will control the nature of interaction (bacteria – nanoparticles) and hence the MIC for each bacteria strain. Finally, STNT is the best and the more efficient one in the treatment of both Gram positive and Gram negative bacteria.

(a) (b)

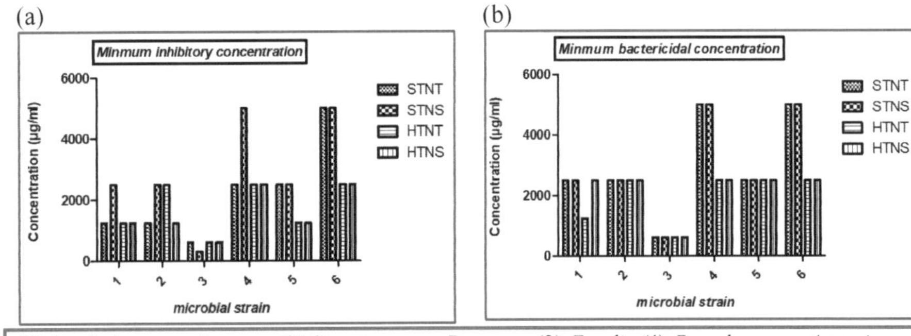

(1) Staphylococcus Aureus; (2) *Streptococcus Pyogens;* (3) *E.coli;* (4) *Pseudomonas Aeruginosa;* (5) *Aspergillus Flaculous;* (6) *Aspergillus nigricans*

Figure 5. MIC (a) and MBC (b) of STNT, STNS, HTNT and HTNS. MIC was determined by micro broth dilution technique and values reported are the values (Mean) obtained in triplicate.

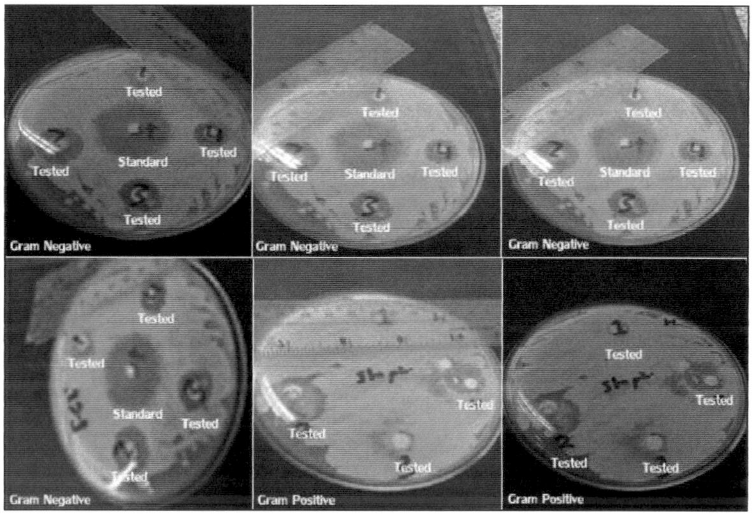

Figure 6. Zone of inhibition of *STNT (1), STNS (2), HTNT (3) and HTNS (4)* against both Gram positive and Gram negative bacteria.

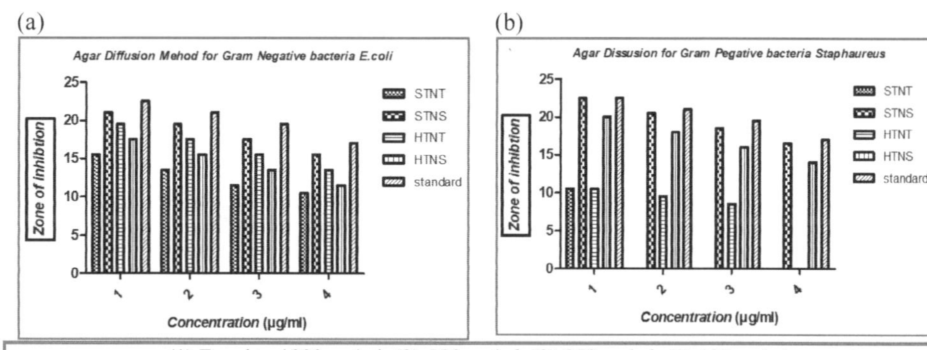

(1) Equal to 1000 µg/ml; (2) *500 µg/ml;* (3) *250 µg/ml;* (4) *125 µg/ml*

Figure 7. The actual measured zone of inhibition of STNT, STNS, HTNT and HTNS against both Gram positive and Gram negative bacteria.

4. Conclusion

A facile hydrothermal technique was used to prepare sodium and hydrogen based titanate nanotubes . The crystallene structure of the prepared samples was verified by powder XRD diffraction analysis. Samples were successfully characterized by surface area analyser, hydrodynamic size analyser and zeta potential. The antibacterial activity of STNT, STNS, HTNT and HTNSon *E Coli*, *S.aures, St.coccus*, and *P. aeruginosa* besides, two different strains of fungi (*Aspegillus Flocculosus* and *Aspegillus nigricans*) was investigated. The experimental results indicated that STNT, STNS, HTNT and HTNS have an important antibacterial effect against different microorganisms causing serious infectious diseases in human and animals. To summarize, the resistant bacterial strains to antibiotics have become a serious public health problem, so increasing the demand for discovering or development new antibacterial materials is a strong demand. Nanotechnology has an important role nowadays in creating new antimicrobial agents through increasing the penetration power, shape and increasing the surface area. In the present study, STNT, STNS, HTNT and HTNS showed good antibacterial activity at different concentrations against Gram Positive and Gram-negative bacteria besides antifungal activity with increasing antimicrobial activity towards the STNT shape than other shape. The synthesized nanomaterials possessed a well-developed surface chemistry; chemical stability and the shape of the nanomaterials are the main factors responsible for the antimicrobial activity.

References

[1] Naqvi S Z H, Kiran U, Ali M I, Jamal A, Hameed A, Ahmed S and Ali N 2013 *Int. J. Nanomedicine* **8** 3187
[2] Hajipour M J, Fromm K M, Akbar Ashkarran A, de Aberasturi D, eLarramendi I R D, Rojo T, Serpooshan V, Parak W J and Mahmoudi G 2012 *Trends Biotechnol.* **30** 499
[3] Rai M K, Deshmukh S D, Ingle A P and Gade A K 2012 *J. Appl. Microbiol.* **112** 841
[4] Lara H H, Ayala-Núñez N V, Turrent L del C I and Padilla C R 2010 *J. Microbiol. Biotechnol.* **26** 615
[5] Wang X,Yang F,Zhao J ,Xu Y, Mao D, Zhu X ,Luo Y and Alvarez P J J 2018 *NanoImpact.* **10** 61
[6] Allahverdiyev A M, Kon K V, Abamor E S, Bagirova M and Rafailovich M 2011 *Expert Rev. Anti. Infect. Ther. 9 1035*
[7] Sirelkhatim A, Mahmud S, Seeni A, Kaus N H M, Ann L C, Bakhori S K M, Hasan H and Mohamad D 2015 *Nano-Micro Lett.* **7** 219
[8] Marambio-Jones C and Hoek E M V 2010 *J. Nanoparticle Res.* **12** 1531
[9] Hafshejani T M, Zamanian A, Venugopal J R, Rezvani Z, Sefat F, Saeb M R, Vahabi H, Zarrintaj P and Mozafari M 2017 *J. Control. Release* **262** 317
[10] Al-Jumaili A, Alancherry S, Bazaka K and Jacob M 2017 *Materials (Basel).* **10** 1066
[11] Panáček A, Kvítek L, Smékalová M, Večeřová R, Kolář M, Röderová M, Dyčka F, Šebela M, Prucek R, Tomanec O and Zbořil R 2018 *Nat. Nanotechnol.* **13** 65
[12] Qi K, Cheng B, Yu J and Ho W 2017 *J. Alloys Compd.* **727** 792
[13] Bui V, Park D and Lee Y-C 2017 *Polymers (Basel).* **9** 21
[14] Yadav H M, Kim J-S and Pawar S H 2016 *Korean J. Chem. Eng.* **33** 1989
[15] Ma Z, Garrido-Maestu A and Jeong K C 2017 *Carbohydr. Polym.* **176** 257
[16] Chouirfa H, Bouloussa H, Migonney V and Falentin-Daudr C 2019 *Acta Biomater.* **83** 37
[17] Yada M, Inoue Y, Noda I, Morita T, Torikai T, Watari T and Hotokebuchi T 2013 *J. Nanomater.* **2013** 1
[18] Lee S-B, Otgonbayar U, Lee J-H, Kim K-M and Kim K-N 2010 *Surf. Coatings Technol.* **205** 172
[19] Kizuki T, Matsushita T and Kokubo T 2014 *J. Mater. Sci. Mater. Med.* **25** 1737.
[20] Stoyanova A, Hitkova H, Bachvarova C N, Irodanova R, Ivanova N and Sredkova M 2012 *Dig. J. Nanomater. Biostructures* **7** 777
[21] Shah A A, Khan A, Dwivedi S, Musarrat J and Azam A 2018 *Mater. Lett.* **229** 130
[22] Farghali A A, Zaki A H and Khedr M H 2016 *Nanomater. Nanotechnol.* **6** 12
[23] Zaki A M, Zaki A H, Farghali A A and Abdel-Rahim E F 2017 *J. Pure Appl. Microbiol.* **11** 725
[24] Rashad S, Zaki A H and Farghali A A 2019 *Nanomater. Nanotechnol.* **9** 1
[25] Collee J G, Marmion B P, Fraser A G and Simmons A Eds. 1996 Mackie & McCartney Practical Medical Microbiology, 14th Edition, Churchill Livingstone, New York
[26] Wikler M A 2006 Methods for dilution antimicrobial susceptibility tests for bacteria that grow aerobically: approved standard, CLSI. 26 (2006) M7-A7.
[27] Zhang H, Cheng K, Yanling J I, XiaoLan L I U, LinSong L I, Xing T Z and Zuliang D U 2008 *Sci. China Ser. B-Chem.* **51** 976

[28] Muniyappan S, Solaiyammal T, Sudhakar K ,Karthigeyan A and Murugakoothan P 2017 *Modern. Elec. Mat.* **3** 174
[29] Tarek N, Hassan H M, AbdelGhani S M M, Radwan I A, Hammouda O and El-Gendy A O 2014 *Beni-Suef Univ. J. Basic Appl. Sci.* **3** 149.
[30] Jose M, Sienkiewicz P, Szyma S K, Darowna D, Moszy S D, Lendzion-Bielu Z, Szyma S K and Mozia S 2019 *Nanomaterials* **9** 795.
[31] Kundu S, Sain S, Choudhury P, Sarkar S, Das P K and Pradhan S K 2019 *Mater. Sci. Eng. C* **99** 374.
[32] Neagu S, Preda S, Zaharescu M, Kamekura M, Cojoc R and Enache M D 2018 *Rom. Biotech. Lett.* 23 13814
[33] Rahman M M and Asiri A M 2016 *Nanofiber Research: Reaching New Heights* (Croatia: Intech) pp 191

IOP Conf. Series: Materials Science and Engineering 1046 (2021) 012026 doi:10.1088/1757-899X/1046/1/012026

Effect of La doping on the structural and gas sensing properties of PbS thin films deposited by spray pyrolysis

A Hamdedein[a], A A Aboud[b], W M A El Rouby[a], M D Khan[c], A A Farghali[a], M H Khedr[a,*] and N Revaprasadu[c,*]

[a] Materials Science and Nanotechnology Department, Faculty of Postgraduate Studies for Advanced Sciences (PSAS), Beni-Suef University, 62511 Beni-Suef, Egypt
[b] Physics Department, Faculty of Science, Beni-Suef University, Egypt
[c]Department of Chemistry, University of Zululand, Private Bag X1001, KwaDlangezwa 3886, South Africa

*E-mail: RevaprasaduN@unizulu.ac.za (NR); dkheddr@yahoo.com (MHK)

Abstract. Phase pure PbS and lanthanum doped PbS thin films have been deposited by the spray pyrolysis technique. All the deposited films were gray to black, pin hole free and strongly adherent to the ITO glass substrates. The substrate temperature was varied between 450 and 500 °C, with doping concentration ratios of 1, 5 and 10% for La. The doping performed at 450 °C, showed phase segregation whereas successful incorporation of La in the PbS lattice was achieved at 500 °C. The films were characterized by x-ray diffraction (p-XRD), scanning electron microscope (SEM), transmission electron microscope (TEM), Energy-dispersive X-ray spectroscopy (EDX), and ultraviolet visible spectroscopy (UV). The sensitivity of the as deposited thin films was also tested against hydrogen gas. The sensor material was exposed to a low concentration of 200 ppm of the target gas to determine the sensitivity performance of the sensor, at different temperatures while measuring the variation in the resistance. The doping of lanthanum significantly enhances the sensitivity of the PbS thin films against the target gas. The gas sensor exhibited n-type conductivity as confirmed by the decrease in resistance on exposure to hydrogen. The as prepared catalysts show good sensing behavior near room temperature and at low concentration of hydrogen at 200 ppm without the need to apply an external voltage.

1. Introduction

Hydrogen detection has gained considerable interest as it is the main product in fuel cells, monitoring faults and combustion engines. Hubert *et al.* indicated that more than 1000 articles have been published in gas sensing between the period of 1975 and 2010 [1]. The physical and chemical properties of the hydrogen gas, *i.e.* colorless, odorless, tasteless, and flammable gas, makes its detection by human senses almost impossible and highly dangerous [2]. It is well known that a mixture of oxygen and hydrogen causes explosions and degradation of steels [3]. At a higher concentration of more than 4%, hydrogen becomes flammable and explosive. Extensive research has been carried out to use hydrogen as an alternate fuel, to meet the upsurge in energy demand. Hence, there is an urgent need to fabricate high performance hydrogen sensors, for safe use of hydrogen as a fuel [4].

Different techniques have been employed for the detection of hydrogen, these include, use of catalyst, mechanical methods, thermal conductivity, electrochemical, resistance, and optical processes [2]. The electrochemical and resistance-based techniques are most distinctive because of their selectivity and capability for detection even at much lower concentrations [2].

It is well established that the surface absorption or desorption of a gas by metal oxides or sulfides, results in change in the conductivity of these materials. The phenomenon was first investigated on ZnO thin films [5,6]. The sensitivity of a catalyst surface to the gas can be as low as parts per billion (ppb). Moreover, it was noticed that some of the catalysts showed higher selectivity towards a certain gas even in a mixture of different gases. The selectivity of a catalyst is at a maximum at a definite temperature, also known as the optimum working temperature, where the gas adsorption to the catalyst surface is at a maximum. However, the optimum temperature may vary considerably from one catalyst to the other and from gas to gas. Hence, a catalyst can show selective behavior towards different gases at different optimum temperatures. It is also known that the higher surface area of the catalyst, can lead to enhanced adsorption of a target analyte, resulting in a stronger and prominent response (specifically at the ppm concentration level). The interaction of the target gas with the surface of the metal sulfides nano-particulate film (generally through surface adsorbed oxygen ions), results in a sudden and quick alteration in the charge carrier concentration of the catalyst. As a consequence, the conductivity (or resistivity) of the material also changes significantly [5].

Palladium based materials and metal-oxide semiconductors, such as, SnO_2, TiO_2, and NiO are considered good sensing materials because of their capability to detect hydrogen at lower concentrations and the ability to work near room temperature. Although these sensors display good sensitivity towards hydrogen, the high costs and inherent defects in some are limitations, *e.g.* Pd shows hysteresis behavior in electrical resistance as a result of the adsorption of hydrogen on its surface [7].

Metal sulfides are versatile materials with vital applications in diverse fields such as, solar cells, gas sensing, remote sensing and photo resistors [8–13]. Lead sulfide (PbS) is an important member of the IV–VI class of the semiconductors with a band gap of 0.38 eV, and a relatively large excitonic Bohr radius (20 nm) [14]. Narrow band gap, large Bohr radius and quantum size effects are the prime features of PbS making it suitable for optoelectronics applications, such as, infrared detectors [15]. Different methods have been used successfully for the fabrication of PbS nanostructures, such include colloidal routes, spray pyrolysis, photo chemical deposition, chemical bath deposition, aerosol assisted chemical vapor deposition (AACVD), electro deposition and successive ionic layer adsorption and reaction (SILAR) methods [16–19]. Although, the synthesis of shape and size controlled PbS has been achieved by different routes, the drawback in some cases are the synthetic protocols which are tedious requiring high reaction temperatures, and toxic or expensive capping agents.

As compared to the other deposition techniques, spray pyrolysis is relatively facile, quick, cost effective and potentially suitable for the deposition of a relatively larger area. The films quality can be precisely controlled by altering the deposition parameters, such as, precursor concentration, substrate temperature, gas flow rate, irradiation time, volume of the solution sprayed and doping (type of metal and its ratio) [20].

The fabrication of reliable and sensitivity enhanced sensors, at relatively low cost, is highly desirable. The ability of the metal sulfides to be produced easily on a bulk scale with simple controllable process makes them potentially suitable materials for device fabrication. There is only limited literature available on the contribution of lead sulfide thin films in hydrogen sensing devices. In the current work, phase pure PbS and La doped PbS nano-particulate thin films have been deposited using the spray pyrolysis technique. The effect of doping on the structure has been investigated. Finally, the performance of the deposited thin films for 200 ppm hydrogen gas sensing has been recorded at different temperatures in the range of 50 to 150 °C.

2. Materials and methods

2.1. Materials

Lead acetate ($Pb(CH_3COO)_2$ (99%)), thiourea ($CS(NH_2)_2$, (99%)), lanthanium (III) acetate ($La(Ac)_3.xH_2O$), (99%)), and methanol (CH_3OH, (98%)) were obtained from (Sd fine-Chem limited – India) and used as such.

2.2 Samples preparation

The spray pyrolysis setup consists of a substrate heater with an iron–constant thermocouple, a solution reservoir, air compressor, spray gun, and a gas exhaust unit. The optically cleaned Indium doped tin oxide (ITO) substrates were heated to the desired temperature by placing them over the hot plate with a temperature probe. The aqueous solution containing a mixture of desired metal salt(s) and thiourea, was then sprayed on the pre-heated substrates, at different temperatures (450 °C or 500 °C) to obtain homogeneous thin films.

To prepare pure PbS thin films, lead acetate (0.05 M) and thiourea (0.051 M) were added in 100 mL of water and stirred for half an hour. A slightly higher molarity of thiourea over lead acetate ensures the completion of the reaction. After half an hour, the solution was ready to be sprayed by addition of an equal volume of methanol in the reaction mixture. The distance between the substrate and the spray gun nozzle was fixed at 40 cm and compressed air was used as a carrier gas. The flow rate of carrier gas was kept constant during the whole experiments. The solution containing the reaction mixture (5.0 mL) was sprayed as fine droplets on the substrate and allowed to be decomposed thermally over pre-heated ITO glass substrates. To avoid excess cooling of the substrate and to maintain the desired deposition temperature (450 °C or 500 °C), the solution was sprayed slowly (one mL of the sprayed solution pulsed every 5 mins.). The spraying processes continued for 30 min. As a result, continuous lead sulfide films were obtained with the substrates allowed to cool to room temperature naturally.

The same experimental conditions were used to prepare La doped lead sulfide thin films by adding the calculated amount of lanthanum salt, for 1, 5 and 10% doping ratio, the films were deposited at temperatures of 450 °C and 500 °C. The reactions involved in the formation of lead sulfide films are as follows:

$$Pb(CH_3COO)_2 \rightarrow Pb^{+2} + 2CH_3COO^- \quad (1)$$

$$CS(NH_2)_2 + OH^- \rightarrow CH_2N_2 + H_2O + SH^- \quad (2)$$

$$SH^- + OH^- \rightarrow S^{-2} + H_2O \quad (3)$$

$$Pb^{+2} + S^{-2} \rightarrow PbS \quad (4)$$

2.3. Characterization of the prepared films

X-ray diffraction analysis of the films was performed on a PANalytical (Empyrean) X-ray diffractometer, with Cu Kα radiation (λ=1.54045 Å) and accelerating voltage of 40 kV, current of 35 mA, scan angle 20 - 70° range and scan step of 0.02°. The average crystallite size of the prepared nano-crystallites was determined by Scherrer's equation (Eq. 5) [21–23].

$$D = 0.94\lambda / \beta \cos\theta \quad (5)$$

Where D represents average crystallite size, λ is the radiation wavelength, β shows the corrected full width at half maximum (FWHM) and θ is the diffraction angle. The morphology of the films was analyzed by FE-SEM, Quanta FEG 250. TEM micrographs were captured using the JEOL JEM 1010 transmission electron microscope, which operates at 100 kV. The diffused reflectance spectra of all the deposited films were obtained from the double beam UV-Vis-NIR Shimadzu spectrophotometer model type UV-3600.

2.4. Sensing measurements

Ingenious gas sensing unit has been used to measure the change in thin film resistance upon exposure to the target gas, with variable concentrations mixed with a certain amount of dried air, in order to obtain the maximum sensitivity at 9% of hydrogen. Thin film was placed between two metallic crocodile forceps connected by Teflon bars to estimate the film resistivity. A coating of silver paste was introduced on the film surface to provide electrical contact with the external electrodes. The system was placed inside a digital horizontal furnace with a temperature controller unit to adjust the temperature to the desired values. The films have been tested at temperatures of 50, 80, 120 and 150 °C and the variation in the resistance recorded using a computerized digital multimeter (Avometer) of type Protex506.

Firstly, a continuous flow of dried air was allowed to pass over the as prepared thin films, and the resistance of the film was recorded for few minutes before the introduction of the target gas (hydrogen) flow, to obtain a baseline value. A known amount of hydrogen gas was introduced in the air stream and the change in the resistance recorded. The target gas could pass for 5 mins, after which the gas flow was stopped, the resistance was continuously recorded until it reached its original value. This process was repeated several times to confirm the reproducibility and stability.

3. Results and discussion

The phase determination and purity of deposited thin films was analyzed by p-XRD analysis. The X-ray diffraction patterns of the pure and the La-doped PbS films deposited at a substrate temperature of 450 °C are shown in figure 1. All films were found to be black in color, pin hole free and in good adhesion with the substrate. The diffraction planes recorded are (111), (200), (220), (311),(222), (400), (331) and (420) which correspond to pure PbS and matches well with the standard pattern (ICDD # 00-005-0592) [24]. In pure PbS films the (200) plane has the maximum intensity, indicating the preferred orientation of crystallites along this plane. However, upon doping of lanthanum, the intensity of the (111) reflection starts to change as a function of concentration. It may indicate that the La-doping changes the preferred orientation of the deposited lead sulfide films. At La doping concentration of 1 and 10 %, two new peaks, marked as (*), were observed at 2-θ value = 22.99 and 24.96°, which matches with the (200) and (111) diffraction peaks of the α-La_2S_3 phase (ICDD# 22-645) [25]. The intensities of diffraction peaks of the PbS phase were also found to increase at La concentration of 10 %. This is attributed to the formation of the La_2S_3 phase which consumes the lanthanum in the separated phase instead of getting it incorporated into the PbS lattice. In all deposited films a slight shift in the 2-θ value was observed as the La-content increases, as shown in figure 1 and table 1. No impurity peaks were observed for La concentration of 5%, which may indicate a good incorporation of La into the PbS lattice.

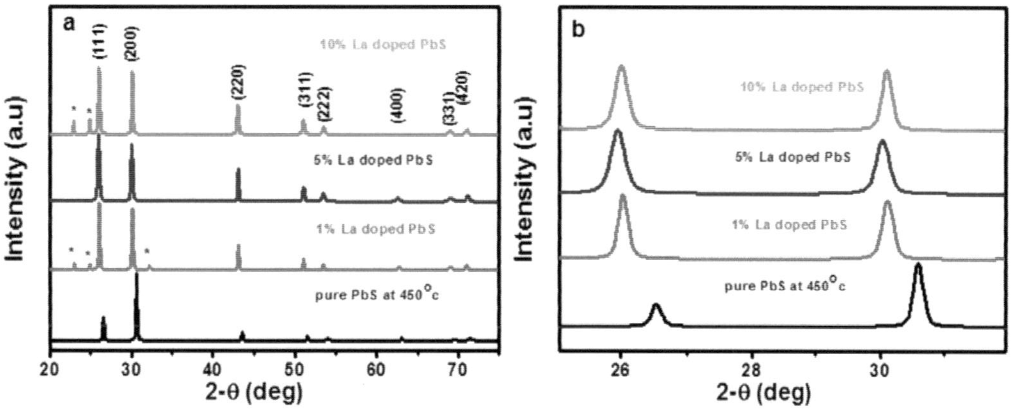

Figure 1. (a) p-XRD of the pure PbS and La-doped PbS films deposited at a substrate temperature of 450oC and (b) an extended part of the diffraction pattern.

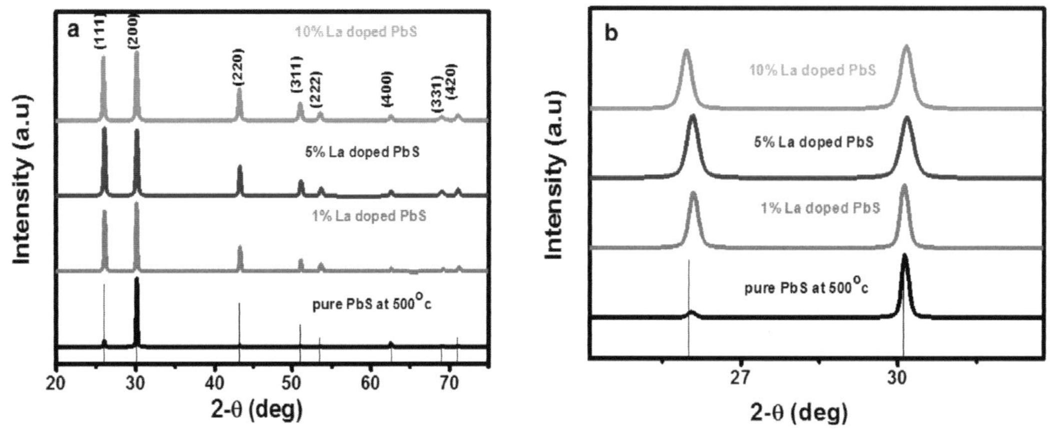

Figure 2. (a) p-XRD of pure PbS and La-doped PbS films deposited at a substrate temperature of 500 °C, and (b) extended part indicating a slight shift due to doping.

Figure 2(a) shows the diffraction patterns for pure and La-doped PbS films deposited at 500 °C. The diffraction planes are indexed as (111), (200), (220) and (400), which correspond to the cubic PbS phase [23]. The PbS films seem to be highly textural as the (200) peak appears as the only main high intensity peak. By incorporation of La, the intensity of the (111) peak and other planes *i.e.* (311), (222), (311) and (420) also started to appear, with respect to the (200) plane [24]. Unlike films prepared at 450 °C, no phase segregation was observed as peaks belonging to La_2S_3 or another impurity were absent. It indicates the successful incorporation of La in the PbS lattice by increasing the temperature from 450 to 500 °C. The intensity of the diffraction peaks changes as the La- content increases in all the films and a slight shift in the 2-θ value was also observed as the La-content increased (figure 2(b)).

The grain size of all the films was estimated using both Scherrer's equation and the Willamson-Hall equation [23,26]. The average crystallite size was obtained by the Scherrer's equation (Eq. 5) as mentioned in the experimental section.

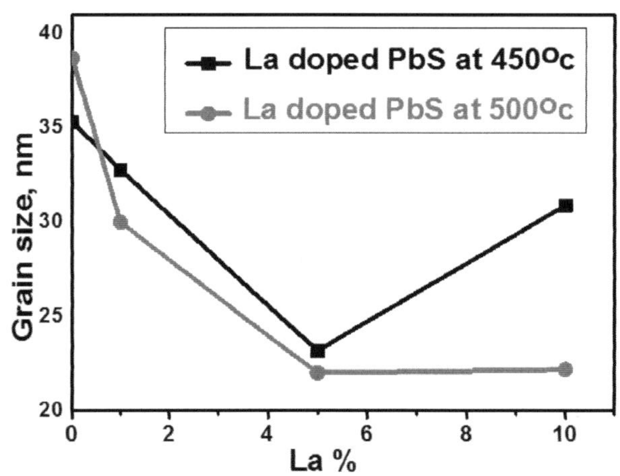

Figure 3. Change in grain size with respect to doping ratio and reaction temperature.

Figure 3 presents the variation of the grain size values with respect to the dopant concentration at the deposition temperatures of 450 and 500 °C. As observed for pure PbS, the size changes from 35.3 to 37.8 nm by changing the deposition temperature from 450 to 500 °C. Such results could be attributed to the increase in the deposition temperature which indicates agglomeration of the particles and improved crystallinity. Introducing dopants such as La has a significant effect on the grain size. Although not regular, but generally a reduction in the grain size is observed at both temperatures and different dopant concentrations. The irregular behavior of size reduction was comparatively prominent in films deposited at 450 °C [27].

The overall reduction in the crystallite size may be attributed to the substitution of the divalent Pb^{2+} ions with ionic radii equal to 119 (pm), by a smaller trivalent metal ion (for *e.g.* La^{3+} with ionic radii 103 pm), respectively. As a result, the sulfur vacancy formed, and the crystal lattice reduced. The reduction in lattice parameters was also observed by a slight shift of diffraction peaks towards the higher 2-theta values, in the p-XRD pattern [28]. The lattice reduction reduces the particle size and could be explained by an increase in the point defects concentrations [29,30].

The strain induced in the crystal lattice, due to the presence of the defects can also contribute to peak broadness. As substitution of a divalent atom by trivalent atoms introduces defects in the lattice, hence the Williamson-Hall method was used to determine the effect of strain on the peak broadening and determination of the crystallite size by excluding the strain factor. The difference between crystallite size induced broadening and the strain induced broadening can be calculated by the Williamson–Hall plot using Eq. (6) as shown:

$$\beta \cos\theta = \frac{m\lambda}{D} + 2\varepsilon \sin\theta \qquad (6)$$

where β is FWHM in radians, λ is CuKα wavelength (0.154186 nm), θ is the Braggs diffraction angle, D is the crystallite size in nm, ε is the strain and m is a correction factor which is considered to be 1. Figure 4, shows a linear fit plot, obtained by plotting $\beta \cos\theta$ vs $\sin\theta$ at which the crystallite size and strain could be determined plotting the intercept with y-axis and the slope, respectively. The plot indicates that positive strain values were obtained for all prepared samples [31]. The strain values and a comparison of the size calculated from both the Scherrer and Williamson-Hall equation is shown in table 1.

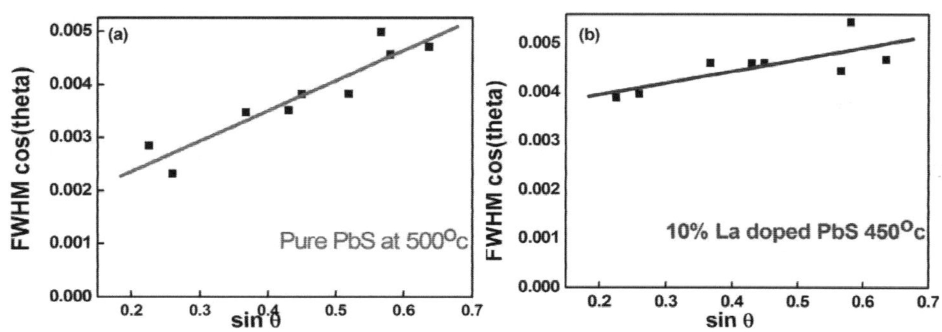

Figure 4. Williamson-Hall effect on the grain size of PbS thin films.

Table1. Grain size and strain values calculated for all prepared samples.

Material	Grain size by Scherrer's equation(nm)	Grain size by Williamson-hall method(nm)	Strain
Pure PbS at 450 °C	35.3	105.4	0.00307
1% La at 450 °C	32.8	85.7	0.00316
5% La at 450 °C	23.2	32	0.00135
10% La at 450 °C	30.9	44.9	0.00019
Pure PbS at 500 °C	38.7	127.8	0.00285
1% La at 500 °C	30	126.8	0.00448
5% La at 500 °C	22	30.3	0.00134
10% La at 500 °C	22.2	43.2	0.00362

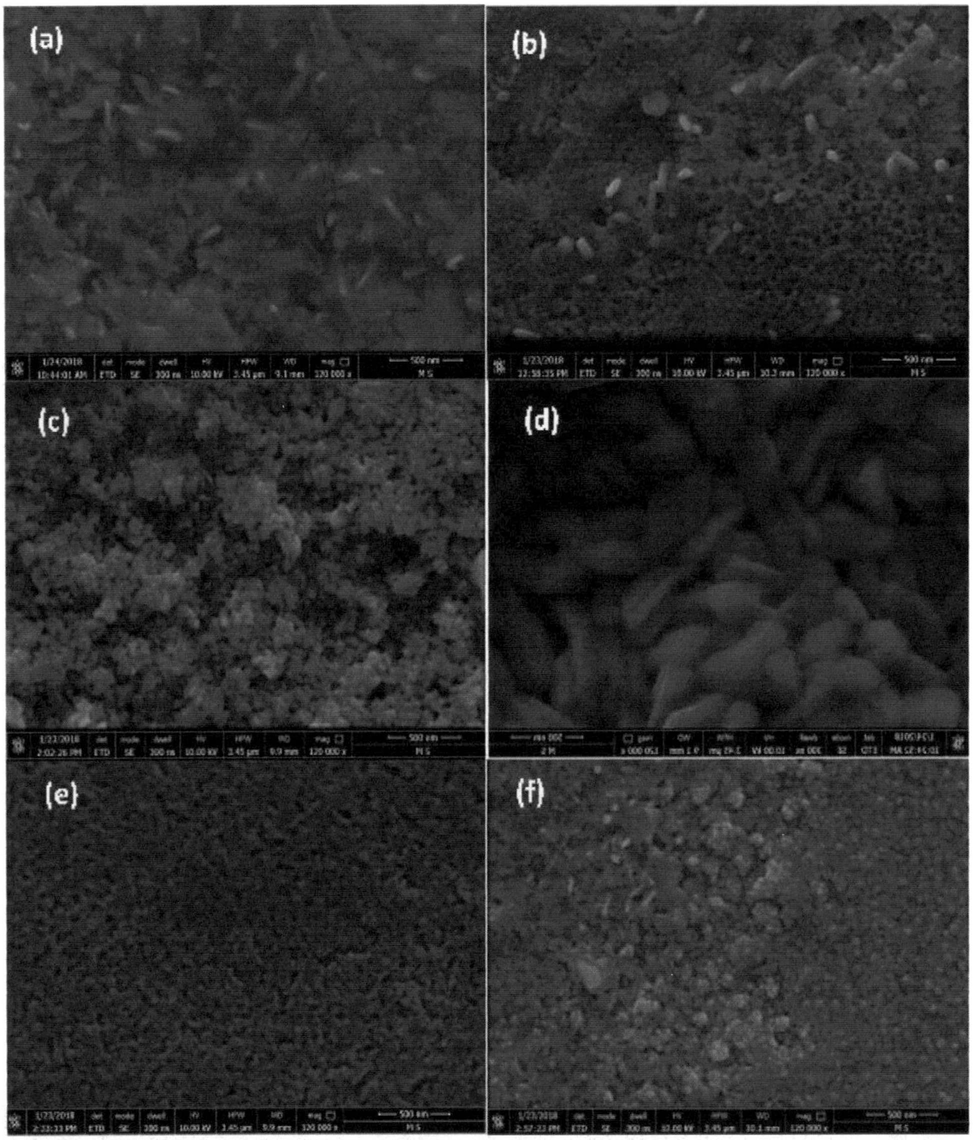

Figure 5. SEM images of (a) pure PbS, (b) 1% La doped PbS and (c) 5% La doped PbS thin films prepared at 400 °C, whereas, (d) pure PbS, (e) 1% La doped PbS and (f) 5% La doped PbS thin films prepared at 500 °C.

3.1. Scanning electron microscopy

The morphology of the samples was observed using SEM analysis. Figure 5 (a-c) shows SEM images of pure lead sulfide, 1 and 5% La doped lead sulfide films deposited at 450 °C, whereas figure 4 (d-f) shows pure lead sulfide, 1 and 5% La doped PbS films deposited at 500 °C, respectively. The SEM images for pure PbS (figure 5(a)) prepared at 450 °C reveals that, highly condensed and agglomerated particles with stripes and rod- like structures were formed. While the SEM images of the La doped lead sulfide shows the formation of small dense particles which aggregate to form a porous surface, however, there are still some rod-like structures present in the sample with a small amount of La. By increasing the doping ratio from 1 to 5% the elongated particles began to evolve into particles that are smaller in size (figure 5(b, c)). Figure 4(d) shows the SEM image of a nanostructured PbS film prepared at a higher temperature of 500 °C, it is clearly observed that the morphology is significantly affected because of raising deposition temperature. Slightly elongated, irregularly shaped, and agglomerated particles form a highly compact film. The higher reaction temperature may enable early formation of a metal sulfide with larger crystallite size before substrate adhesion and higher rate of diffusion on the substrate can lead to the bulky structures. The SEM images of La doped lead sulfide showed that small dense particles with porous surfaces were formed. By increasing the doping ratio from 1 to 5% the elongated structured particles disappeared but the particles were still agglomerated. The higher temperature may result in the formation of the product just before the interaction of the solution mixture with the substrate, which may cause increased agglomeration.

3.2. Transmission electron microscopy

The transmission electron microscopy (TEM) image of pure lead sulfide and the particle-size distribution histogram after scratching the powder over the film, which shows the size range for the width of particles for a count of over 100 particles, have been obtained and shown in figure 6. The morphology of the nanoparticles showed some anisiotropic morphology with slight agglomeration. These obtained morphologies are comparable to those reported [16]. The estimated average particle diameters were found to be 40.32 nm.

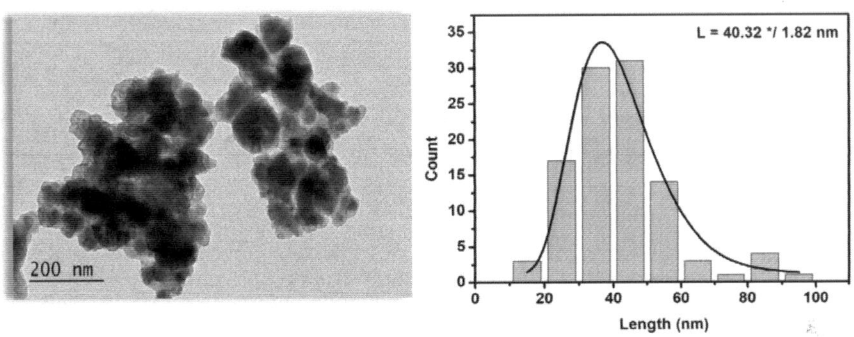

Figure 6. TEM image of pure PbS with its respective particle size distribution histogram deposited at 500 °C.

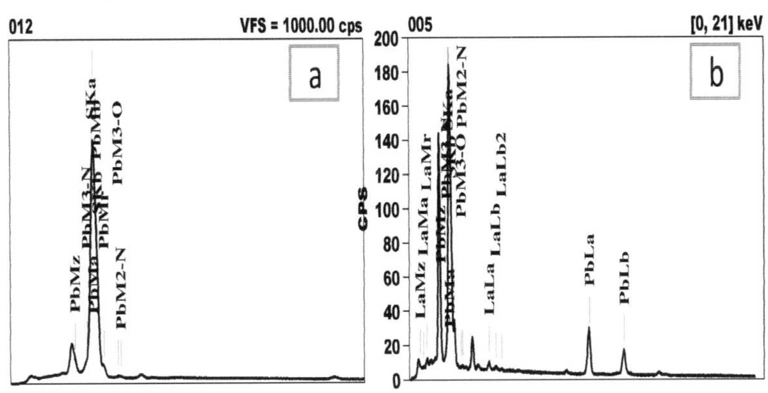

Figure 7. EDX images of pure PbS sprayed at 400°C and 10% La doped PbS at 500°C.

Energy dispersive X-ray (EDX) analysis was performed on pure lead sulfide film and films doped with 1 and 10 % of La at both temperatures and the representative spectra are shown in figure 7. The obtained results confirm the presence of La in the doped samples. The spectrum exhibits signals corresponding to lead, sulfur, and lanthanum. All the films were slightly sulfur enriched. The atomic % of La was found nearly like what was expected, as calculated for feed ratio and is shown in table 2.

Table 2. Mass and atomic percent of the as prepared samples.

Material	Elements	Atom % of expected feed	Atom %
Pure PbS at 450 °C	Pb	39.044	39.62
	S	59.504	60.38
1% La at 450 °C	Pb	39.93	43.00
	S	52,64	56.69
	La	0.29	0.31
10% La at 450 °C	Pb	37.9	41.34
	S	49.7	54.29
	La	3.12	4.37
Pure PbS at 500 °C	Pb	38.8	39.65
	S	59.05	60.35
1% La at 500 °C	Pb	39-4	40.79
	S	56.97	59.06
	La	0.136	0.14
10% La at 500 °C	Pb	38.5	44.08
	S	44.6	51.05
	La	4.22	4.87

3.3. Optical properties

The incorporation of La in PbS films was better in films deposited at 500 °C, no phase segregation was observed, hence, the optical properties of these films were studied by using double beam UV-Vis-NIR spectroscopy. The optical absorbance spectra were studied in the range of 200-1200 nm wavelength for pure and La doped lead sulfide deposited on ITO glass substrates at 500 °C (figure 8(a)). It is clearly seen that the strong peak belongs to pure lead sulfide film at 380 nm absorbed in the UV region and a small broad peak above 700 nm in the region of near-IR. For La-doped films, a strong peak for all doping ratio also appears in the UV region but shifted to the left, indicating the reduction of crystallite size as observed by p-XRD analysis.

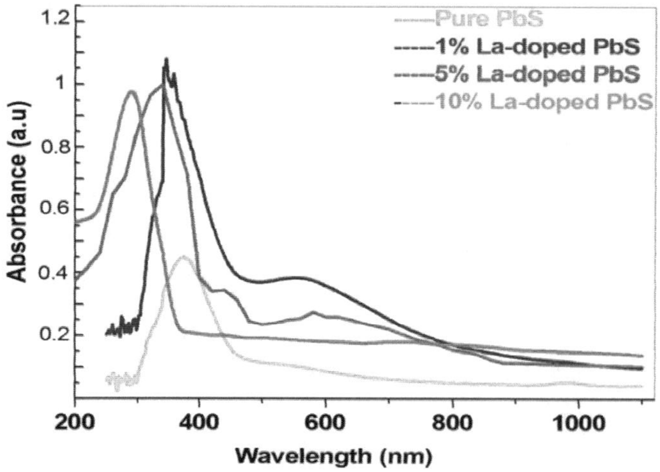

Figure 8. UV-Vis-NIR spectra of pure PbS and La-doped PbS films.

Also, a broad peak appears in all the doped samples at 580 nm in the visible range. This absorbance increases with the increase of the doping ratio of lanthanum in the pure phase from 1 to 10 %. The new absorption in the visible region played a good role to enhance the activity of the prepared catalysts. Based up on the type of direct allowed transition, the optical band gap values were estimated using Tauc's equation (Eq.7), for all thin films prepared at 500 °C [32].

$$\alpha = A\frac{(h\nu-E_g)^{1/2}}{h\nu} \quad (7)$$

where α is the absorption coefficient, A is the samples absorbance, ν is reciprocal of the wavelength, E_g is the optical band gap and h is the Planck constant. As the calculations did not consider the reflected light so, for better accuracy, it is important to calculate reflected light first. As the light is either absorbed or transmitted or reflected, the absorption coefficient α can be estimated by calculating first the reflected radiations according to the following equations (8-10).

$$A = 1 - R - T \quad (8)$$

$$R + T = e^{-\alpha d} \quad (9)$$

$$\alpha = \frac{-1}{d} \ln\left(\frac{T}{(1-R)^2}\right) \quad (10)$$

where d is the thickness of the sample. The band gap values were measured by extrapolating $(\alpha h\upsilon)^2$ versus hυ and the band gap energy can be obtained at the intersection between the photon energy axis and the steep part of the curve (figure 9). Pure lead sulfide, 1, 5 and 10% lanthanum doped lead sulfide showed band gap values of 1.62, 1.52, 1.5 and 1.48, eV, respectively. The values are strongly blue shifted as compared to the bulk value of 0.38 eV, for PbS [14]. This type of blue shift in absorption spectra has been reported due to the well-known quantum confinement effect for the lead sulfide nanoparticles as compared to the bulk. The reduction in the band gap can be related to the small doping ratios, which might create ground and excited f-energy states in the band gap of lead sulfide, these energy states of lanthanum take up many excited electrons coming from O 2P level [1], in addition, the concentration of Pb^{2+} in the lattice boundary was limited by doping. For these two reasons the band gap decreased.

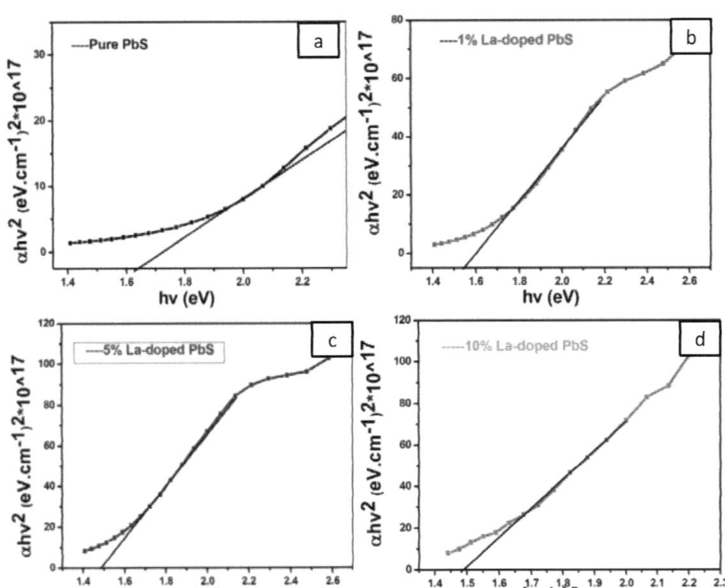

Figure 9. estimated band gap of (a) pure PbS, (b) 1% La doped PbS and (c) 5% La doped PbS thin films prepared at 500 °C.

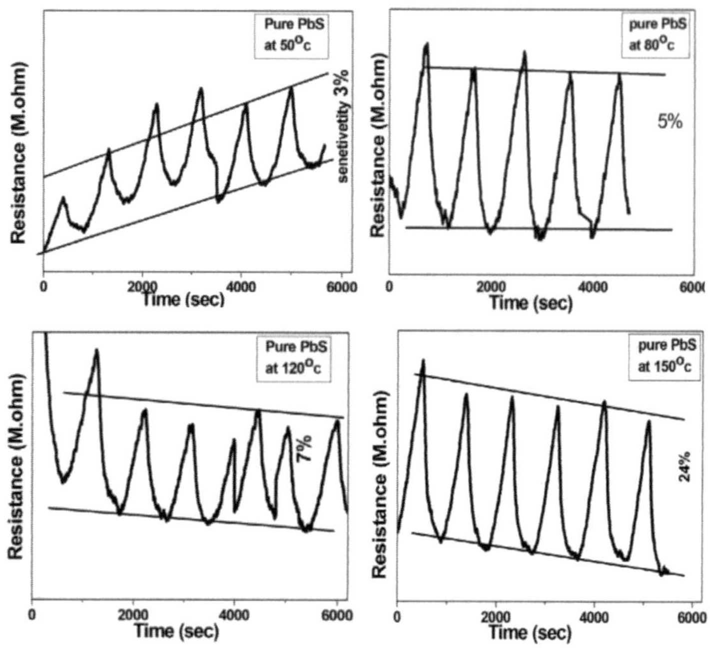

Figure 10. Gas sensing response of pure PbS thin films at different temperatures.

3.4. Gas sensing properties

The sensing performance of all the deposited films was measured based on their resistance variation upon absorption and desorption of H_2 gas. The baseline value was obtained by measuring resistance of the samples in the absence of hydrogen. Figure 10 shows the sensitivity of pure lead sulfide thin films on exposure to hydrogen gas (H_2) at different reaction temperatures. With increase in the reaction temperature, up to 150 °C, the sensitivity of the pure lead sulfide films increased, which means that the optimum operating temperature is probably at 150 °C or above. The temperature dependent sensitivity can be explained based on the kinetics and adsorption/desorption mechanism of the gas to the materials surface. The sensitivity is affected by the speed of the chemical reaction and the diffusion rate of the gas molecules to the surface of material. At low temperatures, the sluggish rate of the chemical reaction and reduced surface adsorption of the gas is responsible for a low response, whereas the rate of reaction and adsorption at surface is much efficient at higher temperature, therefore increasing the sensitivity at elevated temperatures.

Similarly, figure 11 shows sensitivity of La doped PbS thin films, with different concentrations of La at different temperatures, the figures are arranged in increasing order of concentration from left to right at a temperature. Increase in temperature for a doping concentration is depicted from top to bottom. As clearly observed the resistance of all samples decreased by the introduction of the gas to the reaction chamber. Such reduction in resistance, as indicated by the literature, ensures the n-type behavior of lead sulfide by exposure to a reducing gas [33]. In n-type semiconducting materials, electrons are the majority charge carriers, and such materials upon interaction with the reducing gas, results in an increase in the conductivity (decrease in the resistivity), whereas, the oxidizing gas decreases the overall conductivity (increase in the resistivity) by diminishing the sensing layer of the charge carrying electrons from the surface. Likewise, a p-type semiconductor is a material that conducts through the movement of positive holes, as a result reverse effects are observed with the material. Therefore, the conductivity of p-type semiconductor will be increased in the presence of an oxidizing gas (whereby the gas increases the number of positive holes) [34], and the reducing gas will result in decrease in conductivity (high

IOP Conf. Series: Materials Science and Engineering 1046 (2021) 012026 doi:10.1088/1757-899X/1046/1/012026

resistivity), by providing the negatively charged electrons into the material, thereby decreasing the overall positive (hole) charge carrier concentration.

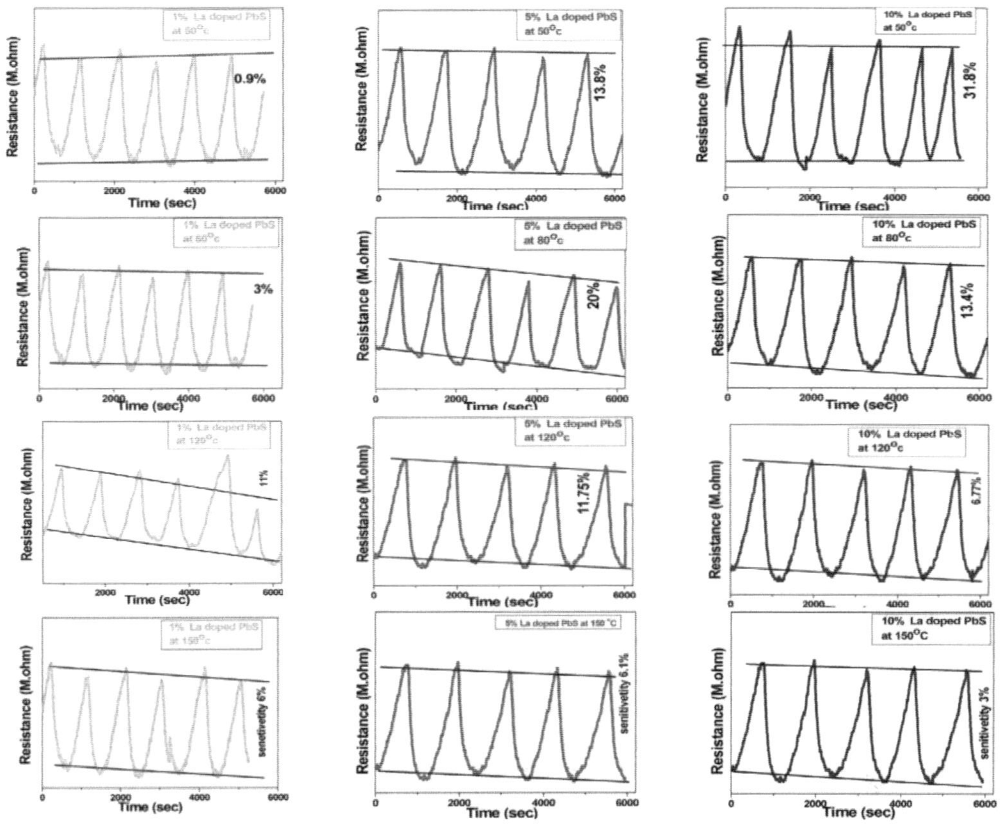

Figure 11. Gas sensing response of La-doped PbS thin films at different temperatures.

Figure 12 shows the sensitivity performance of the samples at different temperatures (50-150 °C). The sensitivity was measured by using the following eq. 11 [35];

$$S = \frac{R_g - R_o}{R_o} x100 \tag{11}$$

Where R_g is the resistance of the target gas and R_o is the resistance under N_2 atmosphere. It is clearly observed that doping increases the sensitivity of the PbS thin film. By increasing the doping ratio of lanthanum from 1 to 10%, the optimum working temperature is shifted from 150 °C to 50 °C. For gas sensing phenomena the surface interaction must not be neglected, considering what happens when H_2 gas interacts. Firstly, the presence of lanthanum plays an important role for hydrogen sensing, as it helps in the ionization of (H_2 to $2H + e^-$) [36]. The lanthanum on the surface of the lead sulfide thin film forms Schottky junctions and take electrons from lead sulfide, resulting in an increase in widths of the interfacial electron depletion regions. Hence, the lead sulfide thin film become more sensitive to adsorb the gas on the surface on picking electrons. Therefore, it can be suggested that lanthanum can help in the dissociation-ionization mechanism of H_2 gas. Also, when lead sulfide thin films interact with the target gas, it directly diffuses to the interior part filling the lattice vacancies. This removes the holes formed and increases the charge carrier concentration. These effects accompanied with the reduction in the particle size causes an increase in the sensitivity of doped thin films.

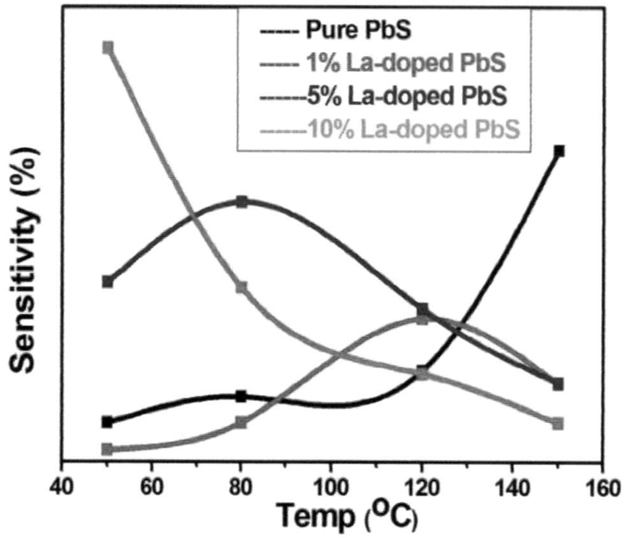

Figure 12. Sensors sensitivity behavior of pure and La-doped PbS thin films at different temperatures.

4. Conclusion

Pure PbS and lanthanum doped PbS thin films have been fabricated by the spray pyrolysis technique at temperatures of 450 and 500 °C, with a doping concentration ratio of 1, 5 and 10% for La. The sensor materials have been exposed to the target gases with concentration of (200 ppm) mixed with certain amount of dried air The sensing performance of the films have been tested at temperatures of 50, 80, 120 and 150 °C where it was found that doping increases the sensitivity of the PbS thin film. The synthesized materials showed comparable activity to other sensing materials prepared by different synthetic ways. All films have large surface area and good sensing behavior near room temperature without the need to apply external voltage to enhance the electrons movement.

Acknowledgments

Authors are thankful to a co-financed project by the Egyptian and South Africa governments, granted by the Ministry of Scientific Research (MSR) in Egypt and National Research Foundation (NRF) in South Africa, ID no. 18-3-14.

References

[1] Hübert T, Boon-Brett L, Palmisano V and Bader M A 2014 *Int. J. Hydrogen Energy* **39** 20474
[2] Hübert T, Boon-Brett L, *Sensors Actuators, B Chem.* **157** 329
[3] Hirth J P 1980 steel *Metall. Trans. A* **11** 861
[4] Jones T A and Walsh P T 1988 *Met. Rev.* **32** 50
[5] Aboud A A, Al-Kelesh H, Rouby W M A E, Farghali A A, Hamdedein A and Khedr M H 2018 *J. Mater. Res. Technol.* **7** 14
[6] Nisha R, Madhusoodanan K N, Vimalkumar T V. and Vijayakumar K P 2015 *Bull. Mater. Sci.* **38** 583
[7] Santandrea R P and Behrens R G 1986 *High Temp. Mater. Process.* **7** 149
[8] Sayed S G, El Rouby W M A and Farghali A A 2019 *Rare Met.* **39** 218
[9] Liu Y, Li Y, Kang H, Jin T and Jiao L 2016 *Mater. Horizons* **3** 402
[10] Tian Q, Wang G, Zhao W, Chen Y, Yang Y, Huang L and Pan D 2014 *Chem. Mater.* **26** 3098
[11] Wan H, Ji X, Jiang J, Yu J, Miao L, Zhang L, Bie S, Chen H and Ruan Y 2013 *J. Power Sources* **243** 396

[12] Tada H, Fujishima M and Kobayashi H 2011 *Chem. Soc. Rev.* **40** 4232
[13] Chianelli R R, Berhault G and Torres B 2009 *Catal. Today* **147** 275
[14] Zhou H S, Sasahara H, Honma I, Komiyama H and Haus J W 1994 *Chem. Mater.* **6** 1534
[15] Hwang D K, Lee Y T, Lee H S, Lee Y J, Shokouh S H, Kyhm J H, Lee J, Kim H H, Yoo T H, Nam S H, Son D I, Ju B K, Park M C, Song J D, Choi W K and Im S 2016 *NPG Asia Mater.* **8** e233
[16] Tshemese Z, Khan M D, Mlowe S and Revaprasadu N 2018 *Mater. Sci. Eng. B Solid-State Mater. Adv. Technol.* **227** 116
[17] Khan M D, Hameed S, Haider N, Afzal A, Sportelli M C, Cioffi N, Malik M A and Akhtar J 2016 *Mater. Sci. Semicond. Process.* **46** 39
[18] Akhtar J, Sher M, Dilshad M, Khalid W, Revaprasadu N and Azad Malik M 2015 *Mater. Sci. Semicond. Process.* **36** 20
[19] Zhang J, Gao J, Miller E M, Luther J M and Beard M C 2014 *ACS Nano* **8** 614
[20] Sivaraman T, Nagarethinam V S and Balu A R 2014 *Res. J. Mater. Sci. CdS* **2** 6
[21] Esmat M, Farghali A A, Khedr M H and El-Sherbiny I M 2017 *Int. J. Biol. Macromol.* **102** 272
[22] El Rouby W M A, Farghali A A and Hamdedein A 2016 *Water Sci. Technol.* **74** 2325
[23] Khorsand Zak A, Majid W H A, Ebrahimizadeh Abrishami M, Yousefi R and Parvizi R 2012 *Solid State Sci.* **14** 488
[24] Obaid A S, Mahdi M A, Yusof Y, Bououdina M and Hassan Z 2013 *Mater. Sci. Semicond. Process.* **16** 971
[25] Bagde G D, Pathan H M, Lokhande C D, Patil S A and Muller M 2005 *Appl. Surf. Sci.* **252** 1502
[26] Khorsand Zak A, Abd. Majid W H, Abrishami M E and Yousefi R 2011 *Solid State Sci.* **13** 251
[27] Vernè E, Di Nunzio S, Bosetti M, Appendino P, Vitale Brovarone C, Maina G and Cannas M 2005 *Biomaterials* **26** 5111
[28] Tawfik W Z, Esmat M and El-Dek S I 2017 *Appl. Nanosci.* **7** 863
[29] Liu C, Bai R and San Ly Q 2008 *Water Res.* **42** 1511
[30] Jia Y Q 1991 *J. Solid State Chem.* **95** 1847
[31] Goodwin A L, Calleja M, Conterio M J, Dove M T, Evans J S O, Keen D A, Peters L and Tucker M G 2008 *Science* **319** 794
[32] Dolgonos A, Mason T O and Poeppelmeier K R 2016 *J. Solid State Chem.* **240** 43
[33] Moreno-García H, Nair M T S and Nair P K 2011 *Thin Solid Films* **519** 2287
[34] Varghese O K, Kichambre P D, Gong D, Ong K G, Dickey E C and Grimes C A 2001 *Sensors Actuators B Chem.* **81** 32
[35] Gu H, Wang Z and Hu Y 2012 *Sensors* **12** 5517
[36] Imanaka N and Adachi G Y 1997 *J. Alloys Compd.* **250** 492

Synthesis and characterization of Ni/NiCo$_2$O$_4$ modified electrode for methanol electro-catalytic oxidation

A Hamdedein[a], W M A El Rouby[a], M D Khan[c], M M EL-Deeb[b], A A Farghali[a], M H Khedr[a*] and N Revaprasadu[c*]

[a] Materials Science and Nanotechnology Department, Faculty of Postgraduate Studies for Advanced Sciences (PSAS), Beni-Suef University, 62511 Beni-Suef, Egypt.
[b] Chemistry Department, Faculty of Science, Beni-Suef University, Egypt.
[c] Department of Chemistry, University of Zululand, Private Bag X1001, KwaDlangezwa 3886, South Africa.

*E-mail: RevaprasaduN@unizulu.ac.za (NR); dkheddr@yahoo.com (MHK)

Abstract. The present work reports the study of hydrothermally synthesized Ni/NiCo$_2$O$_4$ with different ratios of nickel (10, 20 and 30) %. The crystallite size was decreased from pure nickel cobaltite (NiCo$_2$O$_4$) to the 10, 20 and 30% Ni/NiCo$_2$O$_4$ samples. In a basic media by utilizing the chrono-amperometry method, the electro-chemical behavior and persistence on glassy carbon adjusted electrode was followed across methanol electro-oxidation. It is clearly seen with growing the methanol concentration up to to 2 M, a clear increase of the current density was obtained for the pure nickel cobaltite catalyst which may be related to the increased oxidation at the surface. Scanning electron microscopy (SEM) images of the pure catalyst shows needle-like structures without agglomeration, recording a highly current density of 80 mA/cm^2 at 2 M of methanol and scan rate of 60 mV/s. All the tested electrodes show a higher stability hence, the current decayed more slowly with less corrosion impact. Surface modification, barrier and block action effect by Ni on the surface of the catalyst, nearly 80% shortage in the efficiency was observed by using 10 % of Ni. By increasing the ratio to 20 and 30 % the efficiency dropped sharply. Hence, the following order of reactivity was observed: pure nickel cobaltite > 10% Ni-nickel cobaltite > 20% nickel cobaltite> 30% nickel cobaltite, according to their current density values.

1. Introduction

Two major technological challenges that scientists face in the current century are climate change and the limited amount of fossil fuels. To solve these critical problems, there is a need to develop clean alternative energy sources with high energy capacity, such as alcohol oxidation fuel cells, batteries and electrochemical capacitors (EC) or supercapacitors [1–4]. Unfortunately, commercially available supercapacitors are hindered by their heavy cost materials [5]. Although, transition bimetallic oxides widely applied for many energy storage and electronic systems as their synergistic enhancement effect on the catalytic reactions, they have been found with certain limitations. Amongst them, RuO$_2$ exhibits an ideal superior performance behavior but its applicability is limited due to its highly cost and toxicity levels [6–9]. The electro-catalytic oxidation of alcohol plays an interesting role in many reactions such as dyes, fine chemicals, drugs and fuel cells. Direct methanol fuel cells (DMFCs) attracted much attention due to its friendly impact on electronic and portable equipment. Amongst its two types, i.e., alkaline, and acidic DMFCs, the alkaline has more priority and is widely used because of its less corrosive alkaline action which makes the electrodes last longer. Many attempts have been made to synthesize inexpensive substitution electrode materials and to develop single-source oxides materials. However, binary oxides such as metal cobaltite allow high redox reactions using the dual action of both cobalt and the metal ions. Nickel cobaltite is considered one of the most promising material in the cobaltite family with a spinel structure AB_2O_4 (A=Ni and B=Co) [11-13]. The higher redox chemistry of the binary system over the single metal oxides comes from the synergistic effect of the two metals which arises from the co-existence of both the Ni and Co species [10]. Other advantages include their higher conductivity, catalytic behavior, low cost, stability, low electrically resistance and environmentally friendly nature. These features enabled them to be widely applied as electro-catalysts, magnetic materials, optical limiters, chemical sensors and in lithium ion batteries [11,12]. Hence, the fabrication of nano-structured nickel cobaltite has recently attracted much attention. In previous reports, nickel cobaltite has been synthesized by various ways such as spray pyrolysis, co-precipitation hydroxide decomposition route, pulsed laser, combustion and reverse micellar pathway. Other methods used alternative growth control such as poly vinyl chloride to obtain uniformity in morphology [13–15]. Simple techniques have been used to prepare highly modified nanorods of catalysts used to detect glucose levels [16,17]. Mesoporous materials have been synthesized by the microwave-assisted method showing promise for further applications [18]. Of all these methods, traditional and facile chemistry methods have been widely used to synthesize nanosized materials with controlled morphology. In particular, the hydrothermal treatment technology is an easy technique based on solutions, without the use of any mold or surfactant. Until now, various nickel cobaltite nanostructures have been extensively synthesized showing markedly improved properties. Recently, nickel cobaltite nano-structures materials based on hollow, meso-porous, nanowire and nano-flower morphologies have demonstrated that rational design of morphology and microstructure is a practical way to improve the electro-chemical properties [19,20].

Poor capacity rate is still the main drawback of nickel cobaltite as pseudo capacitive materials. Chemical doping and surface modification are the most important ways to modify the electrical properties of these materials. The development of doping methods for this two-dimensional material will be a key factor in its future applications and will require completely different methods from conventional methods of bulk materials [21,22]. It is well known that electro-chemical performance depends on electrode properties such as size, porosity, surface area, crystallinity, structure, etc. Therefore, many researchers are exploring new ways to prepare nano-cobaltite with multiple morphologies [23].

The main objective of this study is to prepare a pure nickel cobaltite electrode by a facile method while trying to modify the surface using nickel metal at three variable ratios 10, 20 and 30 %. The particle size and the morphology change of the prepared catalysts have been studied. The characterized catalysts were applied in a fuel cell as electrodes with an aim to enhance catalytic performance and computable current density output.

2. Experimental procedure

2.1. Chemicals

All high analytical grade chemicals such as cobalt acetate ($Co(Ac)_2.6H_2O$), nickel nitrate ($Ni(NO_3)_26H_2O$), urea, methanol, Nafion soln. (5 wt%), sodium borohydride ($NaBH_4$), isopropanol, potassium hydroxide were used as received.

2.2. *Sample preparation*

2.2.1. Synthesis of $NiCo_2O_4$. About 0.25 g of Ni $(NO_3)_2.6H_2O$ (1 mmol), 0.5 g of Co $(AC)_2.4H_2O$ (2 mmol), and 0.72 g urea (12 mmol) solutions are allowed to dissolve in a definite volume of distilled water. The solution is then sonicated (SONICS, probe diameter 5.8 mm, on / off pulses 9.9 and 3 s, respectively) for 20 minutes. After which the solution is placed in a stainless steel autoclave lined by Teflon at 150 °C for 16 h. The precipitate is separated from the supernatant, collected, filtered, washed many times, then dried at 60 °C for 24 h. In order to remove the unreacted materials, the synthesized samples are roasted at 350 °C for 3 h in the presence of dried air using a muffle furnace to produce a single phase nickel cobalt powder.

2.2.2. Preparation of χ%Ni/$NiCo_2O_4$ (χ = 10, 20, and 30)
0.25 g of $NiCo_2O_4$ is dispersed in 100 mL of distilled water under mechanical stirring for 10 min. An aqueous solution of $Ni(NO_3)_2{\cdot}6H_2O$ (0.248, 0.469 and 0.744 g in 20 mL water) was added to the first solution drop by drop with continuous stirring for 18 h at 25 °C. Then, 0.5 g of sodium borohydride is mixed drop by drop under stirring for 16 h and temperature of 50 °C. The catalyst is allowed at room temperature and collected, washed with methanol, and dried under reduced pressure at 80 °C.

2.3. *Characterization*
X-ray experiments are performed on the PANalytical X-ray diffraction meter (Empyrean) using Fe Kα radiation (wavelength = 1.54045 Å) at an accelerating voltage of 40 kV, current of 35 mA, a scan range of 20 - 70° and a step scan of 0.02°. The Scherrer equation (eq. 1) was used to determine the average crystallite size of the prepared nanoparticles [24–27].

$$D = 0.94\lambda/\beta\cos\theta \quad (1)$$

where D is the average crystalline size, λ is the radiation wavelength, β is the full width at half maximum and θ is the diffraction angle. Micrographs of a scanning electron emission were recorded by Quanta FEG 250 (Switzerland)(SEM). TEM micrographs were captured using the JEOL JEM 1010 transmission electron microscope, which operates at 100 kV. Elemental mapping by EDX analysis are carried out with the JEOL JEM 2010F scanning electron microscope working at 200 kV acceleration voltage. XPS data are taken with a triple Channeltron CLAM2 analyzer under a vacuum better than 10-8 Torr by using AlKa radiation and a constant analyzer transmission energy of 20 eV. A computer fitted using a program is used to record the spectra [28]. The atomic concentrations were measured by using atomic sensitive factors [29].

2.4. *Glassy carbon electrode adjustment*
The glassy carbon electrode was adjusted by adding 0.002 g of $NiCo_2O_4$ or χ% Ni/ $NiCo_2O_4$ (χ = 10, 20 &30) to 25 mL of Nafion solution (5 wt%) and 300 mL of isopropanol. The obtained suspension undergoes 30 minutes of continuous sonication at 25°C, then 15 microns of the well suspended solution was put over the active area of the glassy carbon electrode (3 mm diameter and 0.0706 cm^2 surface area). The electrode is allowed for 25 minutes at 80 °C and cooled down to room temperature.

2.5. *Electrochemical analysis*
Electrochemical data were analyzed by using Potentiostat/Galvanostat (AUTOLAB PGSTAT 302N) with the well-known electrodes glassy carbon (GC) as working, Pt as counter and Ag/AgCl as a reference. All experiments performed at room temperature. The working electrode was polished by diamond and alumina, then washed with distilled water then finally dried. NOVA 1.11 software was used to record and fit the obtained results.

3. Results and discussion

3.1. Characterization of the prepared samples by XRD

Figure 1 illustrates the X-ray diffraction patterns of the synthesized $NiCo_2O_4$ and χ % Ni/$NiCo_2O_4$ compounds. In case of all prepared samples well-defined and matched peaks are clearly seen located at 2θ = (18.9, 31.1, 36.6, 38.4, 44.6, 55.4, 59.1 and 64.4) which are assigned to (111), (220), (311), (222), (400), (422), (511) and (440) planes respectively (card no. 00-020-0781) [30]. All the peaks of the synthesized materials can be indexed to the centered cubic phase of nickel cobaltite. Pure and crystalline phases were formed without secondary phase peaks with no peak related to any impurities. These results were matched with the data obtained from XPS analysis at which wide scan spectra recorded for the 20% Ni-nickel cobaltite catalyst confirms no peaks could belong to any contamination. We noticed a crystallite size decrease from 9.8 to 9.6 to 8.5 and 8 nm in case of pure nickel cobaltite, 10, 20 and 30% Ni- nickel cobaltite, respectively. In case of pure nickel cobaltite and 10 % Ni- nickel cobaltite phases, the (311) plane has the maximum intensity that indicates the preferred orientation of crystallites along this plane. However, upon increasing the ratio to 20 and 30% hence, the intensity of the (311) reflection starts to change with the concentration and the plane of the highest intensity changed to (400). The Ni-concentration changes the preferred orientation of the modified nickel cobaltite. The grain size was calculated using both Scherrer's and Willamson-Hall equations [31,32]. As mentioned Scherrer's equation (Eq. (1)) was used to calculate the average crystallite sizes.

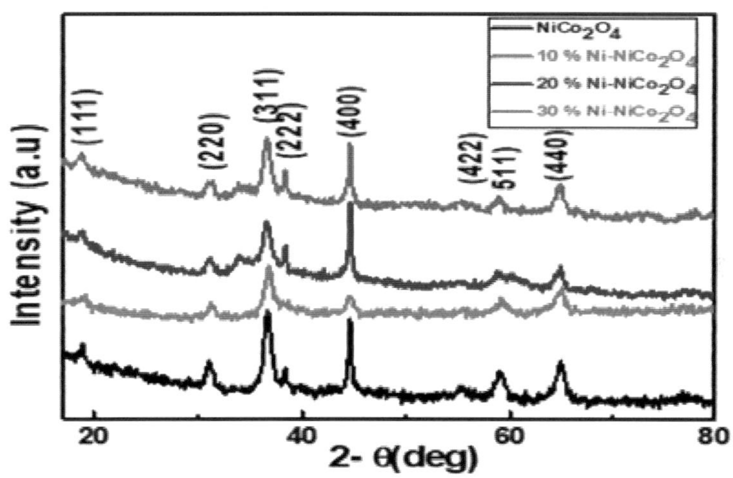

Figure 1. XRD of $NiCo_2O_4$, 10, 20 and 30 % Ni/$NiCo_2O_4$,

Figure 2 shows the variation of the grain size values with respect to nickel ratio %. Introducing metal like Ni has an effect on the grain size. The reduction in the grain size may be attributed to the creation of the oxygen vacancy which is accompanied by a reduction of the crystal lattice. The lattice parameter reduction was also observed by a slight shift of diffraction peaks towards the higher 2-theta values, in p-XRD as shown with 10 % nickel. The lattice reduction reduces the particle size and maybe attributed to the concentration increase of the point defects [33,34].

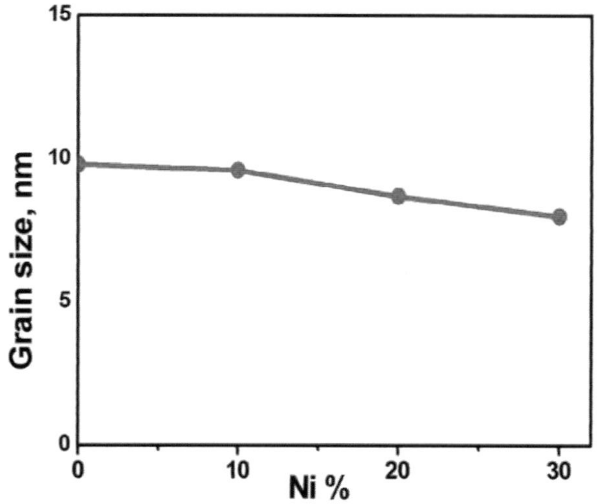

Figure 2. Change in the grain size of $NiCo_2O_4$ with respect to doping ratio%.

The strain caused by the crystalline lattice due to the defects can contribute to peak broadness. Since the production of oxygen vacancies leads to lattice defects, the Williamson Hall method was used to determine the effect of strain on the peak broadening and determine the size of the crystal by excluding the strain factor. The effect of crystal-induced expansion in size can be distinguished from strain-induced expansion on the full width at high maximum (FWHM), by the Williamson–Hall plot according to Eq. (2) [35].

$$\beta \cos \theta = \frac{m\lambda}{D} + 2\varepsilon \sin \theta \text{ s} \quad (2)$$

where β is FWHM in radians, D is the crystallite size in nm, λ is X-ray wavelength (CuKα=0.154186 nm), θ is the Bragg diffraction angle, ε is the strain and m=1 and called the correction factor. Figure 3 shows a linear fit plot, obtained by a plot between β cos θ and sin θ at which both crystallite size and strain could be determined by plotting the intercept with y-axis and the slope, respectively. Positive strain values for all prepared samples were obtained [36]. A comparison of size calculation from both Scherrer and Williamson-Hall equation is shown in table 1.

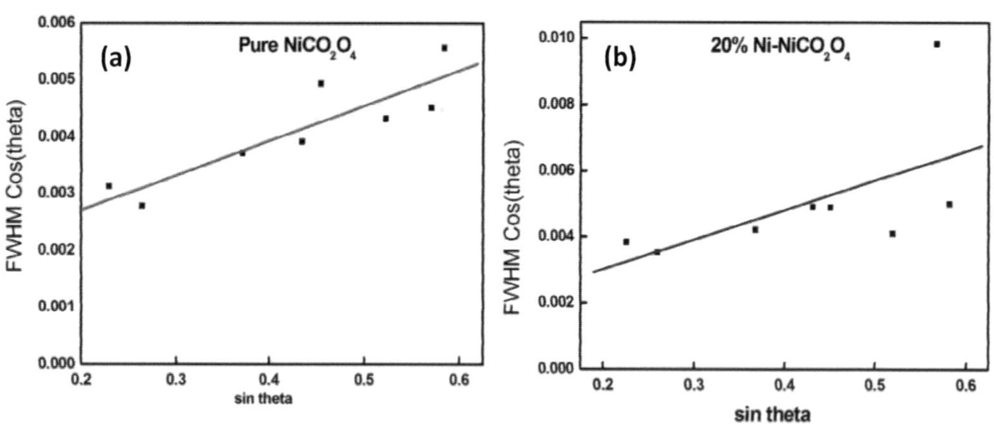

Figure 3. Williamson-Hall way and its effect on the grain size of $NiCO_2O_4$.

IOP Conf. Series: Materials Science and Engineering 1046 (2021) 012027 doi:10.1088/1757-899X/1046/1/012027

Table 1. Grain size and strain values calculated for all prepared samples

Material	Scherrer's equation (nm)	Williamson-hall method (nm)	Strain
Pure $NiCO_2O_4$	9.8	89	0.00389
10% Ni- $NiCO_2O_4$	9.6	80	0.00432
20% Ni- $NiCO_2O_4$	8.5	75	0.00483
30% Ni- $NiCO_2O_4$	8	60	0.00592

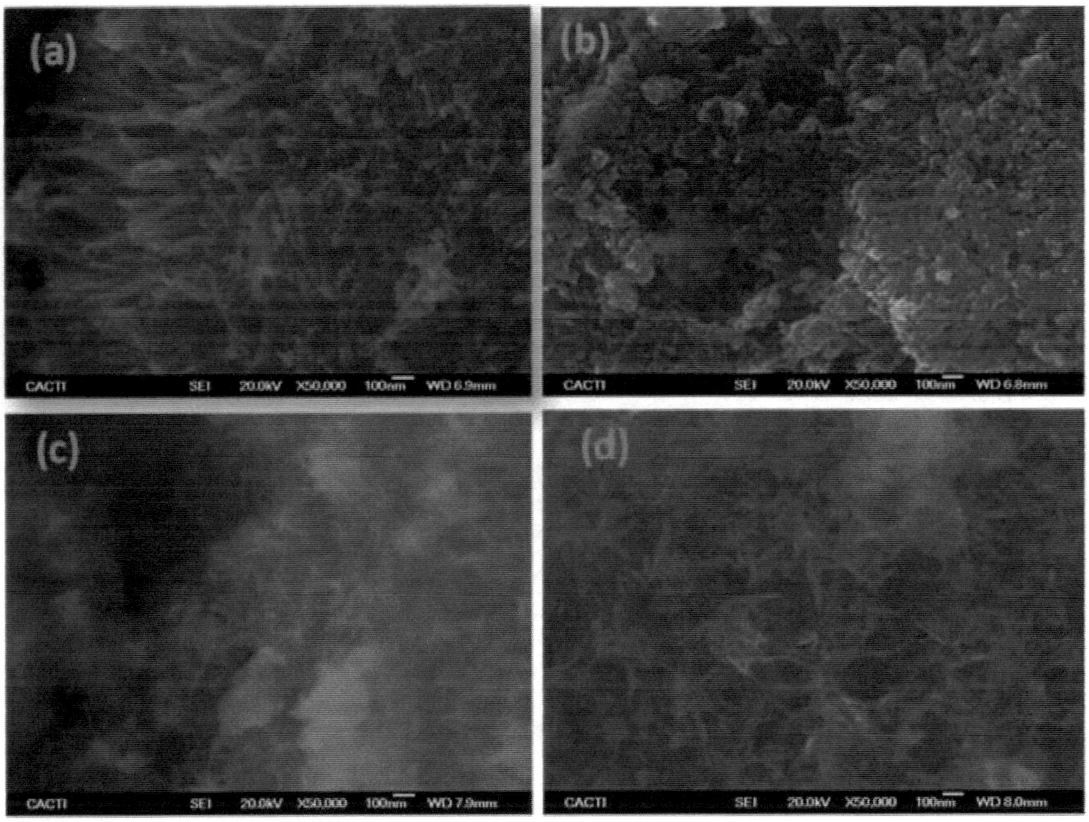

Figure 4. SEM images of $NiCo_2O_4$ and 10, 20 and 30% Ni/$NiCo_2O_4$ nanoparticles.

It clearly seen that the nickel cobaltite surface looks like a needle structure as indicated in figure 4(a). In order to modify the nickel cobaltite surface for that purpose, nickel nitrate was used as a source of nickel. By using 10% of Ni the produced nanoparticles show a porous structure, the needle-like morphology seemed to have disappeared. This may be attributed to the formed nickel atoms that prevent the surface needle growth of nickel cobaltite particles (see figure 4(b)). By increasing the doping ratio to 20 % a strip like structure started to form with agglomeration while by further increasing the doping ratio to 30% the same strip like behavior continues to grow without any agglomerated particles (figure 4 (c,d).

TEM images for $NiCo_2O_4$ and 30% Ni/$NiCo_2O_4$ nanoparticles hydrothermally prepared are illustrated in figure 5(a,b). The $NiCo_2O_4$ nanoparticles are closely strip- like particles with agglomeration (figure 5(a)). With further increase of the Ni ratio to 30%, the resulting nanoparticles seem to be more dispersed and less agglomeration was observed (figure 5(b)).

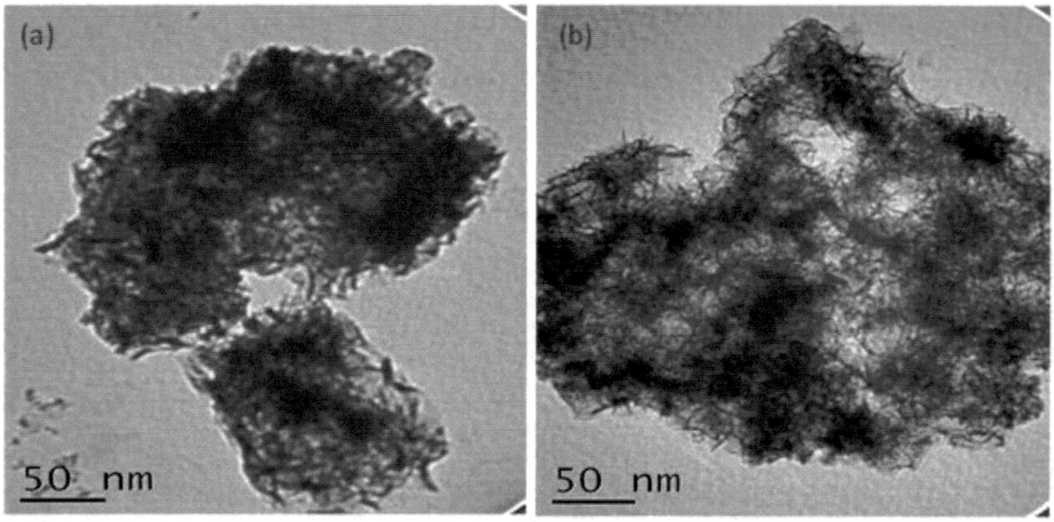

Figure 5. TEM images of (a) pure $NiCo_2O_4$, (b) 30% Ni-$NiCo_2O_4$.

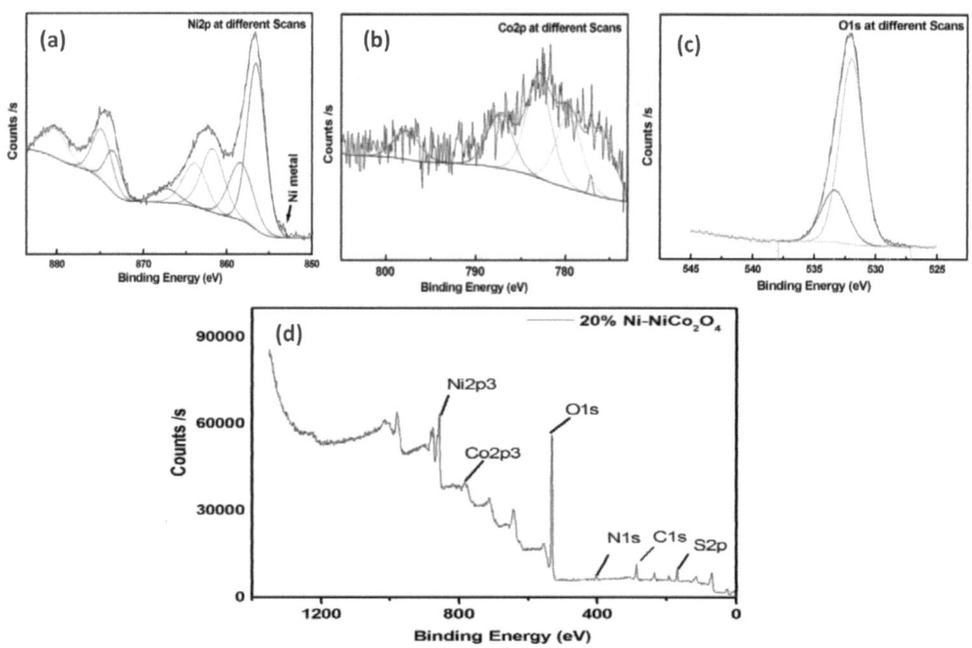

Figure 6. XPS pattern spectra of 20% Ni -$NiCo_2O_4$.

The wide scan spectra recorded from the 20% Ni-nickel cobaltite catalyst with the expected elemental contribution Co, Ni, O, and C is shown figure 6. No peaks could belonging to any contamination are found. The photoelectron spectra role of the Ni2*p* X-ray in compounds enriched with oxygen is still a dialectical topic as mentioned previously [37]. It is well known that the fitting approach of Ni2*p* photoelectron spectra pattern recorded from the target material (figure 6(a)), not considered unrivaled at which for the same spectrum different models give high sensible results of x-axis. It is clearly seen that for the existing case, the spectra of the 20% Ni-nickel cobaltite sample were matched with four cases. Firstly, the line at 854.483 eV and FWHM equal to 4.47 eV, related to Ni^{+2} which centered in the octahedral sites of the spinel structure. Secondly, the line located at binding energy equal to 856.6 eV, is related to Ni^{+3} presences [38–41]. Moreover, the peaks located at 862 eV (FWHM"6.0 eV) related to both Ni^{+2} and Ni^{+3} was fitted as a one broad peak. While the nickel metal spectrum which is well known by its two satellite intensities fitted at bending energy nears to 6 eV (major) and 3.7 eV (minor) started from the main line located at 852.6 eV [42]. Co2*p* photoelectron spectra obtained from the 20% Ni-nickel cobaltite catalyst is shown in figure 6(b). It was found the peaks are similar and matched with that peaks recorded for Co_3O_4. The different components were found in the spectra were due to the Co^{+2} and Co^{+3} [43,44]. The O 1*s* spectra recorded contains contributions of four different types of O_2 which fingered out as O (1-4) (figure 6(c)). Firstly, O1 peak located at 529.2 eV, refers to the M-O bonds (where M refer to metal). The component O2 at 530.6 eV is related to the OH groups as soon as they found at the surface directly hydroxylated and this maybe reviled to the surface oxygen was substituted by OH group or oxyhydroxide happened at the surface. While O3 peak is located at binding energy of 531.5 eV and this peak associated with low coordination O ions located at the surface as mentioned in the literature. Finally, O4 peak related to the surface chemical and physic-sorbed H_2O molecules interaction [45,46]. In conclusion XPS data were summarized in figure 6(d) with the expected elemental contribution Co, Ni, O, and C

Energy dispersive X-ray (EDX) examination for the pure nickel cobaltite and 10 and 20 % Ni- nickel cobaltite (Ni-$NiCo_2O_4$) nanoparticles are illustrated in figure 7. All obtained results confirmed the existence of nickel metal in the modified samples. The spectrum exhibits signals corresponding to nickel, cobalt and oxygen. All the materials were slightly oxygen enriched. The atomic % of Ni signal increased from pure to 10 and 20 % Ni- nickel cobaltite. The carbon signal is attributed to the carbon matrix used to prepare the catalyst before measurement. The mass and atomic percent of the as prepared samples are shown in table 2.

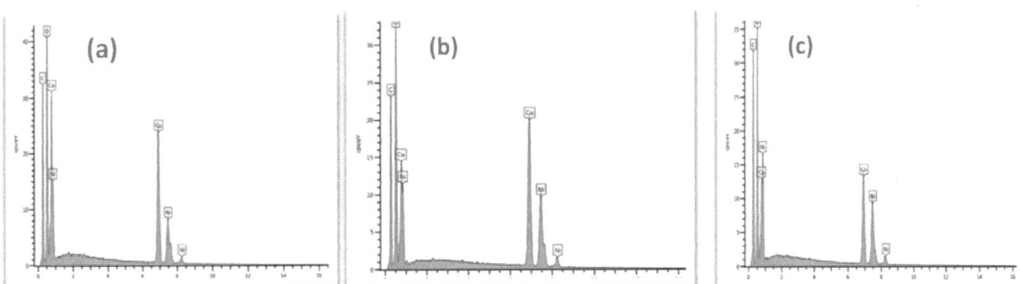

Figure 7. EDX images of (a) pure $NiCo_2O_4$, (b) 10% Ni doped $NiCo_2O_4$ and (c) 20% Ni doped $NiCo_2O_4$.

Table 2. Mass and atomic percent of the as prepared samples.

Material	Elements	Weight%	Atom %
Pure $NiCo_2O_4$	Ni	9.27	3.25
	Co	24.41	8.35
	O	20.88	31.51
	C	45.42	57.02
10% Ni- $NiCo_2O_4$	Ni	17.97	5.64
	Co	23.02	7.51
	O	20.64	24.78
	C	38.54	62.07
20% Ni- $NiCo_2O_4$	Ni	19.12	5.68
	Co	14.28	4.37
	O	27.89	31.43
	C	38.72	58.52

3.2. Electrochemical activity

In order to follow up the reduction-oxidation process and the performance rate, for this reason cyclic voltammetry analysis were done. Hence, the electro-chemical activity of $NiCo_2O_4$ and 10, 20 and 30 % Ni/$NiCo_2O_4$ the glassy carbon (GC) modified electrodes are checked. The experiments were performed in the same experimental conditions of 1 M potassium hydroxide solution in the potential range between - 0.2 and 0.6 V (vs. Ag/AgCl) at 25 °C and at different scan rate ranges between (10 -100) mv/s. All the data are graphically represented in figure 8(a-d). Before running the experiments, the glassy carbon electrode (GC) was tested alone at different scan rate range from10 to100 mV/s in order to clarify the absence of any interference peaks within the applied potential. The cyclic voltammetry format shape at all the recorded scan rates clearly confirms the pseudo-capacitive characteristics of Faradaic reactions. Nickel cobaltite display highly broadened strong characteristic oxidation and reduction peaks due to the combination of Ni^{+2}/Ni^{+3} and Co^{+2}/Co^{+3} ions which appear in the main oxidation state for pure nickel cobaltite ($NiCo_2O_4$) and 10% Ni- nickel cobaltite electrodes (figure 8 (a, b)) [36]. Those curves shapes reveal that the capacitive properties are distinct for those of the electric double-layer derived from the OH^- ions, where the shape is usually near an ideal rectangular shape. It is clearly observed that positive and negative shifts in the anodic and cathodic peaks by increasing the scan rate. Such a continuous flow contribution of both reduction-oxidation reactions of Ni^{+2}/Ni^{+3} and Co^{+2}/Co^{+3} ions produce electron rich reactions with two electro-active center which enable these electrodes to be applied for many applications [47,48]. In case of 20 and 30% Ni-nickel cobaltite modified glassy carbon electrodes (GC) undergo the same experimental conditions and there are no characteristic oxidation-reduction peaks (figure 8(c,d)). The redox reactions in alkaline solution (Eq. (3)) for the electro-active centers on the modified electrodes are illustrated by many researchers [49].

$$NiCo_2O_4 + OH^{-1} + H_2O \rightarrow NiOOH + 2CoOH + e^{-1} \quad (3)$$

From figure 8, the highly electro-active signal of pure nickel cobaltite over the other electrodes maybe due to the different morphology of their surfaces as mentioned previously. SEM images of pure nickel cobaltite illustrated that all synthesized materials have dispersal needles like structure as shown in figure 5. It is well known that the crystallinity and the morphology specially the needle like structure have been chosen to achieve more favorable diffusion and separation effects for the electro-excited charge carriers affecting the electro-chemical reaction rate and concluded from its high anodic and cathodic current densities [50–52]. The higher electro-catalytic activity of pure nickel cobaltite catalyst is related to its high tendency for porosity formation as a result of the presence of the dual action of both cobalt and metal ions which has a huge influence on the electrochemical activity [53]. Here, $NiCo_2O_4||OH^-$ represents the product formed because of the cathodic reactions and the electric double layer and all including the contribution of OH^- ion [54]. Cyclic voltammetry studies of $NiCo_2O_4$ nano-structures (figure 8) shows that the area of the CV curve for the $NiCo_2O_4$ is higher in case of nano-needles than both nano-spheres and nano-strips of Ni-$NiCo_2O_4$. The peak positions difference in the redox reactions comes from the individual electronic differences in the transition of Ni^{+2}/Ni^{+3} and Co^{+2}/Co^{+3} or the combined effect of transition states of Ni^{+2}/Ni^{+3} and Co^{+2}/Co^{+3}, which are greatly affected by both difference between the morphology and the crystallinity. In contrast, the redox peaks were found to be more ill-defined in case of Ni^{+2}/Ni^{+3}. Also, the peak currents were found to be low at the $Ni(OH)_2$/GC electrode than the current found at the LDHf/GC electrode [54,55].

It is clearly seen that the linear dependence of both anodic and cathodic current density of the oxidation-reduction peaks associated with Ni^{+2}/Ni^{+3} and Co^{+2}/Co^{+3} with the square root of the scan rate (10-100) $mV.s^{-1}$of pure nickel cobaltite and 10% nickel-nickel cobaltite shown in figure 8 (e and f), respectively. There is an augmentation in oxidation-reduction reactions with altering the scanning rate and the electrochemical reaction is governed by the propagation of hydroxyl ion in the meso-pore sites. Moreover, the spread rate of the hydroxyl ions through the pore geometry of the needles structure of nickel cobaltite to the inner wall active sites pores is higher than the narrow pore geometry of the nano-spheres and nano-strip like structures of other electrodes which reflected to the current density amounts in all catalysts [53].

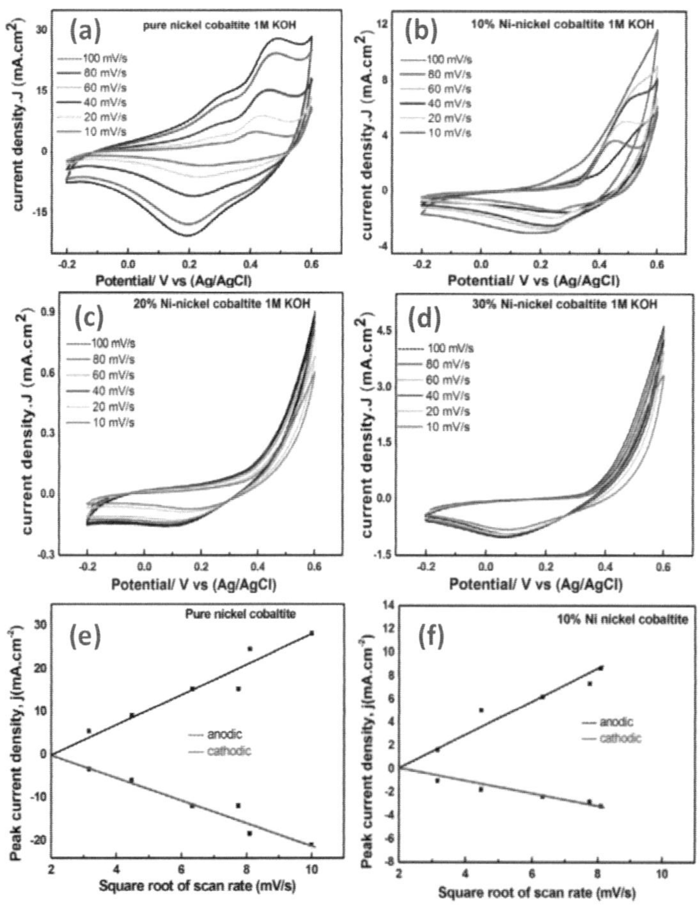

Figure 8. Cyclic voltammograms of (a) pure $NiCo_2O_4$, (b) 10% Ni-$NiCo_2O_4$, (c) 20% Ni-$NiCo_2O_4$ and (d) 30% Ni-$NiCo_2O_4$ in 1.0 M KOH solution at different scan rate from 10 to 100 mV/s at room temperature, the relationship between anodic and cathodic current densities with the square root of the scan rate for (e) pure $NiCo_2O_4$ and (f) 10% Ni-$NiCo_2O_4$.

3.3. Electro-catalytic behavior in MOH

The electro-catalytic activity of the prepared $NiCo_2O_4$, 10, 20 and 30% Ni/$NiCo_2O_4$ modified glassy carbon electrodes towards methanol electro-oxidation is investigated using the CV method in a potential range of (-200 and 600) mV as indicated in figures (9 and10). Hence, the pure electrode has priority and higher current density over the other electrodes. Also, the redox peaks for Ni^{+2}/Ni^{+3} and Co^{+2}/Co^{+3} are disappeared and only one broad peak belonging to methanol oxidation started to appear around 0.35 V (vs.Ag/AgCl). There is a shift in the negative direction for the MOH electro-oxidation potential at pure nickel cobaltite comparing with 10, 20 and 30 % Ni- nickel cobaltite electrodes uncovers its higher electro-catalytic activity for MOH electro-oxidation. Catalytic current density is also used to check the electro-catalytic efficiency behavior for the electro-oxidation reactions, a high anode current density of 80 mA/cm^2 at 60 mV/s and 2 M of MOH with a high stability of 90.6% current retention after many repeated cycles for pure electrode compared to 37.8 mA/cm^2 at 0.38 V of 10 % Ni-nickel cobaltite. These results are in a line with the others obtained in case of the experiments done in the absence of MOH. To clarify the methanol concentration effect on the electro-catalytic current intensity, it is well known that the anodic current oxidation peak of MOH is directly proportional to the square root of the

IOP Conf. Series: Materials Science and Engineering 1046 (2021) 012027 doi:10.1088/1757-899X/1046/1/012027

scan rate (r2 = 0.9962) [56]. It can be observed that the current response of 20 and 30 % Ni-nickel cobaltite electrodes for the addition of methanol is small and negligible. This inability is indicative of the bad electro-catalytic activity of these modified electrodes toward methanol. Also, this may be reviled to the low efficiency of the electrodeposited, morphology difference and the higher amount of nickel used. On the other hand, in the LDH layer in order to form a 3D structure, all of H_2O molecules, interlayer anions and the positively charged layers, are attached together via electrostatic interactions and hydrogen bonds. Hence, few amounts of Ni^{+3} ions leads to an unstable LDH structure and hence low electro-catalytic current intensity [55,57].

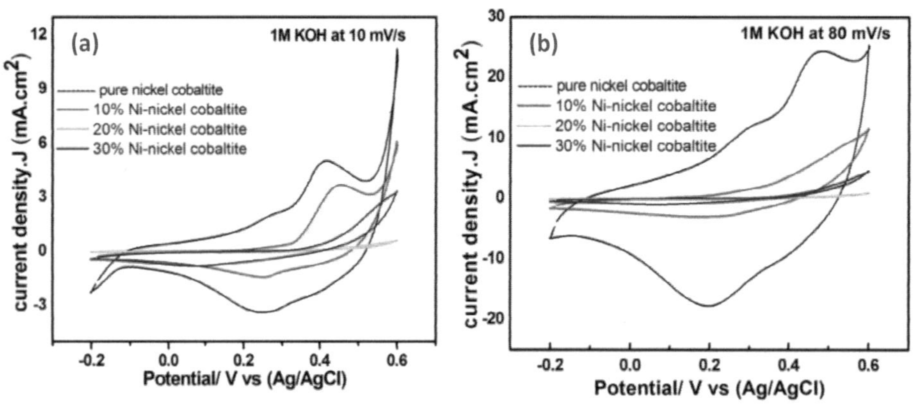

Figure 9. Cyclic voltammograms of pure $NiCo_2O_4$, (10, 20 and 30) % Ni -$NiCo_2O_4$ in absence of methanol and scan rate of (a) 10 mV/s and (b) 80 mV/s.

3.4. Mechanism of methanol electro-oxidation

The electro-catalytic mechanism for MOH is indicated according to the following reactions [58]:

$$\mathrm{Ni}Co_2O_4 + OH^{-1} + H_2O \rightarrow NiOOH + 2CoOH + \mathrm{e}^{-1} \tag{4}$$

$$\mathrm{NiOO}H + \mathrm{CH_3OH} + 1.25O_2 \rightarrow Ni(OH)_2 + CO_2 + H_2O \tag{5}$$

$$\mathrm{NiOO}H + \mathrm{CH_3OH} + 2.5O_2 \rightarrow Ni(OH)_2 + 2Co(OH)_2 + 2CO_2 + 3H_2O \tag{6}$$

The methanol oxidation mechanism was discussed on different materials containing nickel as modified nickel electrodes. Fleischmann [59] assumed that NiOOH was produced electrochemically and its main role is to oxidize the organic species through a chemical interaction rather than the direct electron transfer between the organic compound and the anode. This explanation was built up on the fact that the potential at which a high valent NiO was produced the organic compounds were oxidized. But, a new anodic peak was found after the oxidation of $Ni(OH)_2$ to NiOOH which cannot be explained using the Fleischmann theory. Hence, Taraszewska and Goslonek [54] explained this new peak according to the trapped hydroxyl ions in the nickel hydroxide film which plays an interesting role in the methanol oxidation. Despite the new theories used to explain methanol oxidation, the Fleishmann mechanism was still supported. El-Shafei fingered out that the presence of two peaks in the positive sweep instead of one anodic peak, maybe due to the catalytic reaction between methanol and NiOOH which is so slow when compared with the surface adsorbed NiOOH/$Ni(OH)_2$ reaction [60]. Fleischmann's mechanism can be used to illustrate what happened during methanol addition whereby an anodic increased and cathodic decreased of the peak current of Ni(II) oxidation.

The synthesized electrodes stability was studied in 2 M of methanol, 1 M of KOH and 0.6 V lasted for 36000 seconds cycle. The result showed that the methanol oxidation current slowly decayed at the electrodes, ensuring the higher stability of the prepared electrodes as shown in figure 11. Also, it is clearly seen that the higher electro-catalytic activity and stability of the pure nickel cobaltite over the others for MOH oxidation which is in accordance with the results shown in figures (9 and 10).

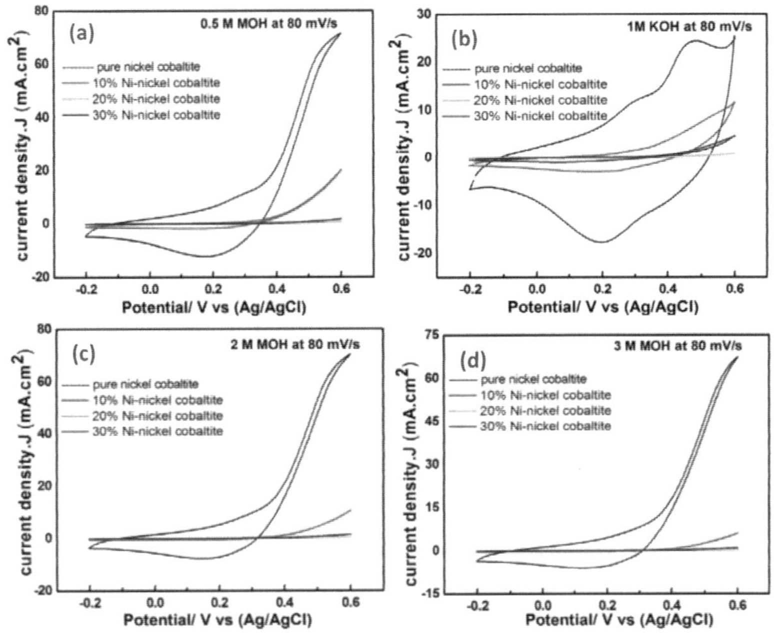

Figure 10. Cyclic voltammograms of pure $NiCo_2O_4$, (10, 20 and 30) % Ni-$NiCo_2O_4$ (a) in absence of methanol (b) 0.5 M, (c) 1 M, (d) 2 M, (e) 3 M methanol all from(a-d) in 1 M KOH and scan rate of 80 mV/s.

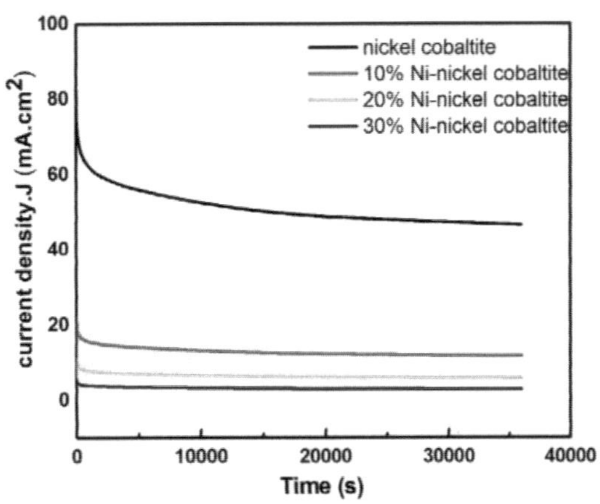

Figure 11. Chronoamperograms of pure $NiCo_2O_4$, 10, 20, and 30% Ni-$NiCo_2O_4$ indicating the catalyst stability in 2 M of methanol, 1 M of KOH and a fixed potential of 0.6 V for a period of a cycle of 36000 seconds.

4. Conclusion

In conclusion, nanosized pure nickel cobaltite, 10, 20 and 30 % Ni- nickel cobaltite was successfully fabricated by using the hydrothermal method. The pure nickel cobaltite nanoparticles were successfully doped with Ni. The prepared nano-particles are used as electrodes for the degradation of methanol depending up on the electro-chemical activity of pure $NiCo_2O_4$, 10, 20 and 30 % nickel-nickel cobaltite (Ni-$NiCo_2O_4$). The obtained results reflected the higher effect of using nickel. The stability of the synthesized electrodes was studied, and the result showed the high stability of the prepared catalysts.

Acknowledgments
Authors are thankful to a co-financed project by the Egyptian and South Africa governments, granted by the Ministry of Scientific Research (MSR) in Egypt and National Research Foundation (NRF) in South Africa, ID no. 18-3-14.

References
[1] El Rouby W M A, El-Dek S I, Goher M E and Noaemy S G 2020 *Environ. Sci. Pollut. Res.* **27** 18985
[2] Gamil S, El Rouby W M A, Antuch M and Zedan I T 2019 *RSC Adv.* **9** 13503
[3] Wennersten R, Sun Q and Li H 2015 *J. Clean. Prod.* **103** 724
[4] Luthra S, Kumar S, Garg D and Haleem A 2015 *Renew. Sustain. Energy Rev.* **41** 762
[5] Wang Q, Yan J and Fan Z 2016 *Energy Environ. Sci.* **9** 729
[6] Khalil M M, El Rouby W M A and Korany M A 2019 *Mater. Sci. Eng. C* **100** 186
[7] Jeena M T, Bok T, Kim S H, Perk S, Kim J Y, Park S and Ryu J H 2016 *Nanoscale* **8** 9245
[8] Zhong Y, Xia X H, Shi F, Zhan J Y, Tu J P and Fan H J 2015 *Adv. Sci.* **3**
[9] Dao V D, Larina L L, Lee J K, Jung K D, Huy B T and Choi H S 2015 *Carbon N. Y.* **81** 710
[10] Al-Hoshan M S, Singh J P, Al-Mayouf A M, Al-Suhybani A A and Shaddad M N 2012 *Int. J. Electrochem. Sci.* **7** 4959
[11] Dong Y and Zhang L 2017 *J. Hazard. Mater.* **333** 23
[12] Liu S, Hu L, Xu X, Al-Ghamdi A A and Fang X 2015 *Small* **11** 4267
[13] Ding R, Qi L, Jia M and Wang H 2014 *J. Power Sources* **251** 287
[14] Yuan C, Li J, Hou L, Yang L, Shen L and Zhang X 2012 *J. Mater. Chem.* **22** 16084
[15] Klissurski D G and Uzunova E L 1991 *Chem. Mater.* **3** 1060
[16] Zhang L, Zhang H, Chu X and Han X 2019 *Ionics (Kiel).* **25** 5445
[17] Karunakaran G, Maduraiveeran G, Kolesnikov E, Balasingam S K, Viktorovich L D, Ilinyh I, Gorshenkov M V., Sasidharan M, Kuznetsov D and Kundu M 2018 *Jom* **70** 1416
[18] Khalid S, Cao C, Wang L and Zhu Y 2016 *Sci. Rep.* **6** 1
[19] Zhao Y, Hu L, Zhao S and Wu L 2016 *Adv. Funct. Mater.* **26** 4085
[20] Pendashteh A, Moosavifard S E, Rahmanifar M S, Wang Y, El-Kady M F, Kaner R B and Mousavi M F 2015 *Chem. Mater.* **27** 3919
[21] Wu Z, Zhu Y and Ji X 2014 *J. Mater. Chem. A* **2** 14759
[22] Feng L, Zhu Y, Ding H and Ni C 2014 *J. Power Sources* **267** 430
[23] Centi G and Perathoner S 2010 *Catal. Today* **150** 151
[24] Esmat M, Farghali A A, Khedr M H and El-Sherbiny I M 2017 *Int. J. Biol. Macromol.* **102** 272
[25] Farghali A A, El Rouby W M A and Hamdedein A 2017 *Res. Chem. Intermed.* **43** 7171
[26] El Rouby W M A, Farghali A A and Hamdedein A 2016 *Water Sci. Technol.* **74** 2325
[27] Holzwarth U and Gibson N 2011 *Nat. Publ. Gr.* **6** 534
[28] Watts J F 1994 *Vacuum* **45** 653
[29] Wagner C D, Davis L E, Zeller M V., Taylor J A, Raymond R H and Gale L H 1981 *Surf. Interface Anal.* **3** 211
[30] Yan K L, Shang X, Li Z, Dong B, Li X, Gao W K, Chi J Q, Chai Y M and Liu C G 2017 *Appl. Surf. Sci.* **416** 371
[31] Khorsand Zak A, Majid W H A, Ebrahimizadeh Abrishami M, Yousefi R and Parvizi R 2012 *Solid State Sci.* **14** 488
[32] Khorsand Zak A, Abd. Majid W H, Abrishami M E and Yousefi R 2011 *Solid State Sci.* **13** 251
[33] Liu C, Bai R and San Ly Q 2008 *Water Res.* **42** 1511
[34] Jia Y Q 1991 *J. Solid State Chem.* **95** 184
[35] Mote V, Purushotham Y and Dole B 2012 *J. Theor. Appl. Phys.* **6** 2
[36] Goodwin A L, Calleja M, Conterio M J, Dove M T, Evans J S O, Keen D A, Peters L and Tucker M G 2008 *Science* **319** 794
[37] González-Elipe A R, Holgado J P, Alvarez R and Munuera G 1992 *J. Phys. Chem.* **96** 3080

[38] Mcintyre N S and Cook M G 1975 *Anal. Chem.* **47** 2208
[39] Kim K S and Winograd N 1974 *Surf. Sci.* **43** 625
[40] Lenglet M, D'huysser A, Bonelle J P, Dürr J and Jørgensen C K 1987 *Chem. Phys. Lett.* **136** 478
[41] Jørgensent C K, Arséne J and Bonnellet J P 1986 *J. Phys. C Solid State Phys.* **19** L363
[42] Grosvenor A P, Biesinger M C, Smart R S C and McIntyre N S 2006 *Surf. Sci.* **600** 1771
[43] Gautier J L, Rios E, Gracia M, Marco J F and Gancedo J R 1997 *Thin Solid Films* **311** 51
[44] Chen Y, Zhang Y and Fu S 2007 *Mater. Lett.* **61** 701
[45] Jiménez V M, Fernández A, Espinós J P and González-Elipe A R 1995 *J. Electron Spectros. Relat. Phenomena* **71** 61
[46] Choudhury T, Saied S O, Sullivan J L and Abbot A M 1989 *J. Phys. D. Appl. Phys.* **22** 1185
[47] Zhang Y, Ma M, Yang J, Sun C, Su H, Huang W and Dong X 2014 *Nanoscale* **6** 9824
[48] Xiao J and Yang S 2011 *RSC Adv.* **1** 588
[49] Wei T Y, Chen C H, Chien H C, Lu S Y and Hu C C 2010 *Adv. Mater.* **22** 347
[50] Xiong G, He P, Liu L, Chen T and Fisher T S 2015 *J. Mater. Chem. A* **3** 22940
[51] Zhang D, Yan H, Lu Y, Qiu K, Wang C, Tang C, Zhang Y, Cheng C and Luo Y 2014 *Nanoscale Res. Lett.* **9** 1
[52] Vijayakumar S, Nagamuthu S and Muralidharan G 2013 *ACS Appl. Mater. Interfaces* **5** 2188
[53] EL-Deeb M M, El Rouby W M A, Abdelwahab A and Farghali A A 2018 n *Electrochim. Acta* **259** 77
[54] Salunkhe R R, Jang K, Yu H, Yu S, Ganesh T, Han S H and Ahn H 2011 *J. Alloys Compd.* **509** 6677
[55] Wang Y, Zhang D, Peng W, Liu L and Li M 2011 *Electrochim. Acta* **56** 5754
[56] Hasanzadeh M and Shadjou N 2013 *J. Nanosci. Nanotechnol.* **13** 4909
[57] Rives V and Ulibarri M A 1999 *Coord. Chem. Rev.* **181** 61
[58] Russell A E 2008 *Faraday Discuss.* **140** 9
[59] Vértes G and Horányi G 1974 *J. Electroanal. Chem.* **52** 47
[60] Golabi S M and Nozad A 2004 *Electroanalysis* **16** 199

Health risk assessment and micro determination of trace elements content in Egyptian olive oil using ICP-OES

A A Yassin [a], M A Ghandour[b] and M M Khalil [a,*]

[a] Chemistry Department, Faculty of Science, Beni-Suef University, Egypt
[b] Chemistry Department, Faculty of Science, Assuit University, Egypt

E-mail: magdy_mmagdy@yahoo.com, mohamed.mahmoud@science.bsu.edu.eg

Abstract. Inductively coupled plasma optical emission spectroscopy (ICP-OES) was applied to investigate multi elements (Zn, Cu, Pb and Cd) of Egyptian olive oil collected from different regions in Egypt. This process involves digestion using 10% nitric acid. The data showed that the concentration ranges were 0.6486-2.4166 μg/g, 0.0092-0.6919 μg/g, 0.0185-0.1026 μg/g and 0.0009-0.0186 μg/g for Zn, Cu, Pb and Cd, respectively. The results indicated that the highly toxic metals content did not exceed that recommended by FAO/WHO. The proposed method exhibited an excellent accuracy, adequate selectivity and high sensitivity. For the first time, this method described the direct determination of the previously mentioned elements and investigated the effect of both temperature and time for different samples in Egyptian olive oil.

1. Introduction

Olive oil is considered a major part of daily food regimes. Olive oils are commonly used in cosmetics, cooking processes, chemical industries and pharmaceutical [1]. Virgin olive oil can be extracted principally by pressure, other mechanical processes and without treatments [2].

Olive oil enhances the human health; it contains sufficient amounts of essential fatty acids and anti-oxidative compounds. The mechanism of olive oil in human body involves penetration rate of fatty acids in to artery wall and therefore, decreasing the risk of cardiovascular diseases and cancers [3, 4]. Spain, Italy and Greece are the major producer for olive oil [5]. The trace metal can be determined for evaluation of its quality [6]. Characterization and adulteration detection of olive oil is affected through trace element analysis [7]. Furthermore, the presence of metals in olive oils can be affected by many factors such as soil, plant genotype, environment, fertilizers and/or metal containing pesticides or by contamination from the metal processing equipment [7, 8]. Also, existing of these elements is essential for the traceability of oil. Therefore, the traceability of the oil is unaffected by transportation, storage, milling and retailing. Consequently, evaluation of trace-element represents a great interest which can be attributed to the potential toxicity of the elements [9].

The human body intakes heavy metals such as copper (3mg), zinc (60mg) and lead (214μg/g) [10]. Although copper metal is necessary for good health, high intake causes adverse health problem such as liver and kidney damage [11]. Simultaneously, the deficiency of copper leads to leucopenia, hypochromic anemia and osteoporosis especially in children [12]. Concerning zinc, it is well known that most metabolic pathways in humans required its presence and skin changes, loss of appetite, growth retardation as well as immunological abnormalities take place in case of its absence [13, 14].

Also, the highly toxic cadmium metal is present in soil and spread in the environment which may be attributed to human activities. High concentration of cadmium may cause rise to renal, skeletal, pulmonary, hepatic, cancer and reproductive effects [14 -17]. In addition, lead metal behaves similar to cadmium metal i.e. it has no useful role in human metabolism and consequently it produces high toxicity [18]. Also, it creates health disorders such as tiredness, sleeplessness, weight and hear loss [19].

It is worthy to mention that classical method ability for heavy metal determination in olive oil can be summarized as follows: wet digestion, dry ashing, acid extraction, closed vessel and focused open-vessel microwave dissolution and dilution [20, 21]. Recently, modern analytical techniques such as ICP-OES and electro analytical method [7, 13, 22, 23, 24] are applied.

The aim of this work is focused on the analysis of trace elements (Zn, Cu, Pb and Cd) in Egyptian olive oils using (ICP-OES).

2. Experimental

2.1. Instrumentation
Thermo Scientific iCAP 6200 inductively coupled optical emission spectrometer (ICP-OES) was used for simultaneous multi-element detection of Zn, Cu, Pb and Cd. The iCAP 6200 ICP-OES (Waltham, Massachusetts, USA) now incorporates a wider wavelength range of 175-847nm. The operating conditions of ICP-OES were reported in table 1.

Table 1. The operating conditions for ICP –OES analysis.

Instrument	ICP6200
RF power (W)	1150W
Gas flow rate (ml/min)	
Auxiliary gas	0.5L/min
Coolant gas	12 L/min
Nebulizer Argon flow	0.6 L/min
Center tube	1.5 mm
Torch	Axial
Elements monitored (Wave length, nm)	Cu (324.754), Zn (213.856), Cd (214.438), Pb (220.353)

2.2. Reagents
All reagents were of analytical reagent grade unless otherwise stated. Deionized water (Milli-Q Millipore 18.2 MX-cm resistivity) was used for all dilutions. Nitric acid (16 M) and HCl (11.65 M) were purchased from Merck, Germany. All containers, including test tubes with stoppers were cleaned with hydrochloric acid 5 % followed by deionized water.

2.3. Olive oil samples
In this study, eleven Egyptian olive oils samples were purchased from Egypt supermarkets. Siwa, North Sinai, El Giza, Marsa Matruh and Alwahat are the most widely accepted and frequently consumed regions in Egypt. The collected oil samples were packed in polyethylene bags and below 4 ^{o}C until analysis.

2.4. Procedure
Aliquots (2.0-3.0 g) of olive oils were weighed directly into the test tube and 1ml nitric acid 10% was added. The oil-acid mixture was shaken at 50 Hz for 60 s with a test tube mixer until the layers were completely mixed. The capped test tube was placed in a shaking water bath at 50 °C for 2 h. After centrifugation at 2800 rpm for 10 min, the lower acid aqueous layer was withdrawn with a pipette and was filled to 25 ml by adding deionized water and then loaded directly into the auto sampler of the (ICP-OES) [25].

IOP Conf. Series: Materials Science and Engineering 1046 (2021) 012028 doi:10.1088/1757-899X/1046/1/012028

3. Results and discussion

3.1. Trace element determination

ICP-OES technique is considered a useful method for micro determination of metals in Egyptian olive oil samples, as shown in Table 2. The concentration of metal found in food could not be modified by traditional techniques and in some cases, it was found that washing may lower their content [26].

In this study eleven brands were collected from five Egyptian regions [Siwa (Altahhan, Al captain, Olive oil), North Sinai (Alqasimuh), Al Giza (Makkah, khoshala), Marsa Matruh (Ashbal, Alkhayrat) and Alwahat (Virgin olive oil,Waha, alwada)] were used for analysis.

The concentration range of copper in the present work is 0.0092-0.6919 µg/g. The concentration of Cu was partially higher than the recommended legal limits. These results are in agreement with that of other authors, whom quantified this metal in different olive oil samples at levels higher than the legislated limits [7, 27]. According to FAO/WHO organization and international requirements, the approved contents of these metals in oils are: 0.1 µg/g (Cu, Pb), 0.05 µg/g (Cd) and 0.04–0.70 µg/g (Zn) [10, 18, 21, 28, 29]. The copper level becomes safe up to 10 mg for a 60 kg adult [30].

Lead and cadmium, very toxic elements, may be responsible for environmental contamination. Specially, the content of lead would be warranted by the presence of highways, metallurgic industries near plantations or streets [17]. The highest content of cadmium and lead in our samples was 0.0186 and 0.1026 µg/g for Cd and Pb, respectively. As shown in table 2 all Cd and Pb concentrations were lower than the legal limits.

Zinc is essential in human nutrition, enhance physical growth and reduce child illness [31]. Zn content in tested samples is ranged from 0.6486-2.4166 µg/g. The recommended daily intake values of Zn are 15 mg for adult males and 12 mg for adult females [14].

Table 2. The results obtained in the analysis of Egyptian olive oil.

Sample	Sample weight	Cd(µg/g)	Cu(µg/g)	Pb(µg/g)	Zn(µg/g)
1	2.72	0.0092	0.6919	0.0738	1.8362
2	2.68	n.d.	0.0092	0.0279	1.8216
3	2.69	0.0186	0.3169	0.0372	1.1245
4	2.70	0.0046	0.2870	0.0185	2.4166
5	2.71	n.d.	0.0184	0.0277	2.0479
6	2.50	0.0020	0.4300	0.0900	2.3900
7	2.61	n.d.	0.2107	0.0670	2.3754
8	2.71	0.0046	0.0331	0.0947	0.6486
9	2.69	n.d.	0.0279	0.0465	1.5242
10	2.68	0.0009	0.1119	0.1026	1.8377
11	2.68	n.d.	0.1306	0.1019	2.0336

n.d.=not determined

3.2. Effect of temperature

Digestion of Egyptian olive oil samples using HNO_3, under safe condition, is necessary and the temperature digestion was examined (40-80 °C). This test was applied for sample no. 8 as an example. It is obvious to note that by increasing the temperature digestion, the digestion efficiency is also improved within the range 40-60 °C. At 70 and 80 °C, the temperature digestion is presented as "0g" [32, 33], except in the case of zinc metal where an improvement in digestion efficiency was observed as shown in table 3. This behavior can be ascribed to the difference in density between zinc and other metals studied. In addition, it was observed that digestion efficiency for all metals was performed at 50 °C as an optimum temperature.

Table 3. The temperature effect on digestion efficiency for sample no. 8

Sample	Sample weight	Cd(μg/g)	Cu(μg/g)	Pb(μg/g)	Zn(μg/g)
T_{40}	2.71	0.0046	0.0191	0.0845	0.5060
T_{50}	2.75	0.0046	0.0331	0.0947	0.6486
T_{60}	2.76	0.0054	0.2355	0.1232	0.8739
T_{70}	2.73	n.d.	n.d.	n.d.	0.8816
T_{80}	2.70	n.d.	n.d.	n.d.	0.9111

nd.=not determined

3.3. Effect of time

It is of great interest to note that, digestion efficiency increases as the time increases (1-4h), except in case of lead and zinc where the digestion decreases when the time reached 4h. This behavior is similar to that published recently according to Manjusha et al., (2019), applying ultrasound-assisted extraction of Pb, Cd, Cr, Mn, Fe, Cu, Zn from edible oils with tetramethylammonium hydroxide and EDTA followed by determination using graphite furnace atomic absorption spectrometer[34]. It is obvious to observe that the digestion efficiency of lead and zinc decreased above 4h as shown in table 4.

Table 4. Effect of time on digestion efficiency for sample no. 8

Sample	Sample weight	Cd(μg/g)	Cu(μg/g)	Pb(μg/g)	Zn(μg/g)
1h	2.71	0.0034	0.0326	0.0838	0.5346
2h	2.73	0.0046	0.0331	0.0947	0.6486
3h	2.69	0.0064	0.0856	0.1445	0.7677
4h	2.75	0.0101	0.1068	0.0727	0.5226

3.4. Comparison study

It is of great interest to note that the Italian olive oil is considered one of the best olive oils in the world, concerning its quality [35]. This behavior can be attributed to the low concentration of toxic heavy metals (Cd and Pb) according to national and international requirements. Whereas, high concentration of lead and cadmium was observed in Greece olive oil; the limits exceeded national and international requirements [36]. In addition, the Egyptian olive oil followed FAO/WHO for the toxic heavy metals (Cd and Pb) and therefore, it is favorable to use the Egyptian olive oil with sufficient safety, as shown in table 5.

Table 5. Comparison between heavy metals (μg/g) content in Egyptian olive oil and those published in European countries besides Turkey

Country	Method	Pb	Zn	Cd	Cu	Ref.
Spain	ICP-MS	-	0.06-0.43	-	0.03-0.33	[21]
China	ICP-AES	0.009-0.016	1.24-1.58	0.0023- 0.0027	0.24-0.27	[18]
Turkey	ICP-AES	0.04-0.10	1.25-1.58	0.043-0.058	0.06-0.11	[36]
Italy	ICP-MS	0.03	-	0.00015	-	[1]
Greece	ICP-MS	1.09 -7.13	0.32-2.54	0.01-0.21	0.01-0.23	[35]
Egypt	ICP-OES	0.018-0.102	0.64-2.41	0.0009-0.0186	0.009-0.691	This work

4. Conclusion

This work describes determination of four heavy metals (Cu, Zn, Cd and Pb) in eleven varieties of Egyptian olive oils and determined using ICP-OES technique. The result indicated that a clear difference existed in some metals concentration among different varieties of Egyptian olive oil samples. The copper levels were found to be higher than the recommended legal limits in some olive oil sample. In addition, the estimated intakes of Cu, Zn, Cd and Pb from daily consumption of 25 g or weekly consumption of 175 g of the investigated olive oils should pose no risk to human health.

References

[1] Dugo G, La Pera L, Torre G L L and Giuffrida D 2004 *Food Chem.* **87** 639
[2] Najafil V, barzegar M and Sahari M A 2015 *J. Agr. Sci. Tech.* **17** 847
[3] Lanteri S, Armanino C, Perri E and Palopoli A 2002 *Food Chem.* **76** 501
[4] Silva, Pinto L J, Carrola J and Paiva-Martins F 2010 *Food Chem.* **121** 1177
[5] Sabbatini V 2014 *Procedia Econ. Financ.* **14** 553
[6] Souza R M, Mathias B M, Silveira C L P and Aucelio R Q 2005 *Acta B* **60** 711
[7] Zeiner M, Stefan I and Cindric I J 2005 *J. Microchem.* **81** 171
[8] Karadjova I, Zachariadis G, Boskou G and Stratis J 1998 *J. Anal. At. Spectrom.***13** 201
[9] Garcia-Gonzalez D, Luna G, Morales M and Aparicio R 2009 *J. Lipid Sci. Tech.* **111** 1003
[10] Food and Agriculture Organization/World Health Organization (FAO/WHO) 1999 Summary and Conclusions of the Fifty-third Meeting of JECFA/63/SC
[11] Ikem A and Egiebor N O 2005 *J. Food Compost. Anal.* **18** 771
[12] Kanumakala S, Boneh A and Zacharin M 2002 *J. Inherit. Metab. Dis.* **25** 391
[13] Cindric I J, Michaela Z and Steffan I 2007 *J.Microchem.* **85** 136
[14] Mendil D, Uluözlü O D, Tuzen M and Soylak M 2009 *J. Hazard. Mater.* **165** 724
[15] Pehlivan B E, Arslan G, Gode F, Altun T and Özcan M M 2008 *Grasas Aceites* **59** 239
[16] Llorent-Martínez E J, Ortega-Barrales P, Fernández-de Córdova M L and Ruiz-Medina A 2011a *Food Control* **22** 221
[17] Llorent-Martínez E J, Ortega-Barrales P, Fernández-de Córdova M L, Domínguez Vidal A and Ruiz-Medina A 2011 *Food Chem.* **127** 1257
[18] Fangkun Z, Wenxiu F, Xuejing W, Li Q and Shuwen Y 2011 *J. Food Chem. Toxicol* **49** 3081
[19] World Health Organization 1993 Guideline for Drinking Water Quality (Geneva:WHO).
[20] Allen L B, Siitonen P H and Thompson H C 1998 *J. Am Oil Chem. SOC.* **75** 477
[21] Garrido D M, Frias I, Diaz C and Hardisson A 1994 *Food Chem.* **50** 237
[22] Hendriskse P W, Slikkerveer F J, Folkersma A and Dieffenbacher A 1991 *Pure Appl. Chem.* **63** 1183
[23] Lo Coco F, Ceccon L, Ciraolo L and Novelli V 2003 *Food Control* **14** 55
[24] Buldini P L, Ferri D and Sharma J L 1997 *J. Chromatogr. A* **789** 549
[25] Pehlivan E, Arslan G, Gode F, Altun T and ozcan M 2008 *Grasas Aceites* **59** 239
[26] Morgan J N 1999 *Adv. Exp. Med. Biol.* **459** 195
[27] Castillo J R, Jiménez M S and Ebdon L 1999 *J. Anal. At. Spectrom.***14** 1515
[28] Kowalewska Z, Izgi B, Saracoglu S and Gucer S 2005 *Anal. Chem.***50** 1007
[29] World Health Organization 1994 Quality Directive of Potable water (Geneva:WHO) P 197.
[30] Ev M 2003 *FSA* **ISBN 1** 904026
[31] Sampath Kuma K P, Debjit Bhowmik and Chiranjib 2010 *Int. J. Pharm. Biomed. Sci.* **1** 5
[32] Edson I M, Juliana P S, Cristiano C M, Aline L H M, Paola A M and Cezar A 2016 *Talanta* **156** 232
[33] Rizwan A, Niyaz A and Adeeb S 2019 *Ind Crop* Prod. **136** 37
[34] Manjusha R, Shekhar R and Sunil Jai Kumar 2019 *Food Chem.* **294** 384
[35] Cacchiarelli L, Carbone A and Sabbatini V 2018 *Agric. Food Econ.* **6** 6
[36] Karabagias I, Michos CH, Badeka A, Kontakos S, StratisI M G and Kontominas 2013 *Food Res. Int.* **54** 1950
[37] Orhan A 2012 *Grasas Aceites* **63** 383

Effect of nanoparticles of feldspar and hematite using different levels of chemical potassium fertilizer on yield and protein production of wheat

Maha M Abd El- Wahab[1], Magdy Y Gabriel[2], Mai Raslan[1,*] and Ahmed A Farghali[3]

[1] Biotechnology and Life Sciences Department, Faculty of Postgraduate Studies for Advanced Sciences, Beni-Suef University, 62511, Egypt.
[2] Soil, Water and Environment Res.Inst., ARC, Egypt.
[3] Materials Science and Nanotechnology Department, Faculty of Postgraduate Studies for Advanced Sciences, Beni-Suef University, 62511, Egypt.

[*]E-mail: m.raslan@psas.bsu.edu.eg

Abstract. The study was conducted to investigate the wheat yield affected by different concentrations of chemical Potassium (K) (0, 25%= 28.6, 75%= 85.7 and 100%= 118.0 kg Kha^{-1} recommended rate, KRR), with and without the application of foliar spraying of feldspar nanoparticles solution and different iron treatments (using foliar spraying of 0.3% twice from ferrous sulphate, or natural hematite nanoparticles). Two field experiments were conducted at two different locations; the first was at the Agricultural farm of Sids Agricultural Research Station (ARC), and the second was at a private farm at Sids village in 2018/2019 season. Grain, straw and biological yields, nitrogen content, protein content, and protein yield were determined. Results showed that increasing K levels up to 118.0 kg Kha^{-1} increased grain, straw, biological yields nitrogen, protein content, and protein yield. Foliar spraying of feldspar nanoparticles or ferrous sulphate or hematite nanoparticles enhanced all the above-mentioned parameters. Combined 75% KRR with feldspar nanoparticles showed yields and quality of wheat equal to those under 100% KRR. Foliar spraying of hematite nanoparticles has equal effect on wheat productivity as ferrous sulphate. The highest wheat yields and protein characteristics were achieved for the treatment of 100% KRR or 75% KRR with foliar spraying of feldspar nanoparticles and foliar spraying of ferrous sulphate or hematite nanoparticles. To conclude; the study revealed the possibility of saving about 25% KRR by foliar spraying of natural feldspar nanoparticles, and using hematite (iron oxide) nanoparticles instead of ferrous sulphate.

1. Introduction

Wheat (*Triticum aestivum* L.) is the most important crops worldwide in respect to the cultivated area and total product. It provides energy and proteins for humans, and animals and provides an almost 20% food calories for people in the world including Egypt. Increasing wheat production is the ultimate goal to reduce the wide gap between production and consumption. Although starch is the major component of the grains, accounting for 60-70% of the total dry weight, the protein content of wheat grains may vary between 10-13% of the total dry matter. Wheat proteins are classified according to their exorability and solubility in various solvents such as: albumins, which are soluble in water, globulins, which are insoluble in pure water but soluble in diluted NaCl and insoluble at high NaCl concentrations, gliadins, which are soluble in 70% ethyl alcohol, and glutenins, which are soluble in dilute acid or NaOH solutions.

The foliar application method is usually preferred for fertilization, because very small amounts of fertilizers are applied per unit area and decrease ground water pollution. Many investigators concluded that foliar application of nutrients sources during growth stage increased grain and straw yields, grain weight, biological yield, grain nutrients concentration and uptake, as well as, grain protein content [1, 2]. Moreover, foliar application in combination with soil application of elements produced the highest values of plant height, tillers numbers/ m^2, spikes number/ m^2, spike length, number of spikelets/spike, number of grains/spikes, grain weight, grain yield, straw yield, biological yield and harvest index % [3, 4].

Nanotechnology is an interdisciplinary research field. In recent past, efforts have been made to improve agricultural yield through exhaustive research in nanotechnology. Nanotechnology has originated from a Greek Word "Nanos" which means "Dwarf". Nanotechnology is providing different Nano devices and nanomaterial which have a unique role in agriculture such as: Nano biosensors to detect moisture content and nutrient status in the soil and also for site specific water and nutrient management. Also, nanotechnology have greater role in crop production with environmental safety, ecological sustainability and economic stability. The nanoparticles produced with the help of nanotechnology can be exploited in the value chain of entire agriculture as Nano fertilizers.

It is well known that potassium (K) has a specific role in developing cell acidification, cell extension and osmotic regulation. Potassium plays critical roles in metabolism and water regulation through controlling the salt and osmotic cell balance. It activates enzymes, maintains tissue turgor pressure, and regulates the opening and closing of stomata and balance the charge of anion. It is an essential mineral element for photosynthesis, the basic function in plant growth and development and encourages the development of the root system of plants [5]. Thus, potassium is regarded as one of the major nutrient elements which affect the yield and quality of grain and fruits. This element is needed for plants in high quantities, so it must be added as a fertilizer to the soil. Feldspars are a group of rock-forming tectosilicate minerals that make up about 41% of the Earth's continental crust by weight. Feldspars are sparingly soluble, usually bio-geochemically inert, and are well known to be used as potash biofertilizers [6].

Iron is essential for the proper functioning of chlorophyll and related to photosynthetic activity. Iron is essential nutrient; it is required for a healthy growth and life cycle completions. Iron plays a role in many plant functions in plant growth and development. This function includes chlorophyll synthesis, and chloroplast development. It also plays a role in energy transfer within the plant, it is a constituent of certain enzymes and proteins, it enters root cells, involved in nitrogen fixation and several enzymatic activities of photosynthesis as well as respiration [7, 8]. Iron application, resulting in better crop production and improved human health [9, 10].

This study aimed to evaluate the effect of potassium and iron fertilizers compared with mineral iron and feldspar fertilizers on wheat production and its protein quality.

2. Materials and methods

2.1. Plant material and soil analysis

The present investigation was conducted in two locations in the Agricultural Farm of Sids Agricultural Research Station, ARC and private farm at Sids village both sites located in Beni-Suef Governorate in 2018/2019 season using wheat (cultivar Sids 12) to study the possibility to reduce the using of chemical fertilizers by using natural fertilizer and nanotechnology. The soil of the two studied locations are clay loam in texture, having pH 7.9 and 8.2, EC 1.20 and 1.24 (dS/m), soil organic matter 1.72 and 1.65%, respectively. The soil N, P and K available in the first location were found to be 24, 25, and 12 μgg^{-1}, respectively; while in the second location were 10, 185 and 176 μgg^{-1} respectively [11].

2.2. Field experimental design

The experimental design was spilt-spilt plot in completely randomized black in four replications. The treatments were chemical potassium fertilizer with different concentrations (0.0, 25% potassium recommended rate (KRR) =28.6 kg K*ha* $^{-1}$, 75% KRR=85.6 kg K*ha* $^{-1}$ and 100% KRR=114 kg K*ha* $^{-1}$ potassium sulphate fertilizer); feldspar nanoparticles (natural rock), i.e. with and without; and iron treatments (0, chemical ferrous sulphate and hematite nanoparticles (iron oxide in natural rocks).

Chemical potassium treatments were applied to the soil at two equal doses; the first dose before the first irrigation and the second one before second irrigation. Whereas, feldspar nanoparticles and different iron treatments were added as foliar spraying twice, the first after first irrigation and the other after one month later at rate of 2% and 0.3% for the two treatments, respectively. The volumes of the spraying solutions were about 500 and 1000 L/*ha* for the two time of spraying, respectively. Wheat grains were planted (at rate of 145 kgK*ha* $^{-1}$) at 15th and 20th of November in the two locations, respectively in plots (3×3.5 m^2=1/950 *ha*). The preceding crop in the two locations was maize. All other cultural practices for wheat were done in district. At maturity, the plots were harvested, and grain and straw yields were determined and converted to *ha* $^{-1}$.

2.3. Protein determination
Nitrogen concentration in grains was determined [11] and converted to protein content by multiplying by 5.75. The protein yield calculated by multiplying protein percentage by grain yield. Results were statistically analyzed [12]. Means of treatment were compared using Duncan Multiple Range Test (DMRT) at 5% level of probability.

2.4. Preparation and Characterization of fertilizers nanoparticles (feldspar and hematite)
The nanoparticles of feldspar and hematite were prepared in Faculty of Postgraduate Studies for Advanced Sciences, Beni-Suef University, Egypt. The required nanoparticles were prepared separately using ball milling for 10 hours under the conditions mentioned in Table 1 [13 14, 15].

The size and shape of nanoparticles was observed using high resolution transmission electron microscope (HRTEM JEOL-JEM 2100, Japan) using an accelerating voltage of 200 kV. X-ray diffraction (XRD) patterns were recorded on a PANalytical (Empyrean) X-ray diffract meter using Cu Kα1 radiation (λ=1.5406 Å) at an acceleration voltage of 40 kV, with current of 30 mA, in the angle range from 5–80°, with step scan of 0.02°.

Table1. Ball milling conditions for preparing nanoparticles

Condition	Description
vessel size	7.5cm diameter
Balls diameters	ranged from 1.11 to 1.75 cm diameter
Materials of vessels	Stainl steel
Materials of balls	porclien
Ball/precipitate mass ratio	8.1 mass ratio
Speed	300 rpm
Time	10h

3. Results and discussion

3.1. Characterization of feldspar nanoparticles
XRD pattern of feldspar nanoparticles prepared by ball milling (figure 1) revealed excellent crystallinity as strong reflections and large number of peaks are observed. By comparing the obtained pattern and matching with ICDD number 01-071-6218, the pattern was indexed and the nanoparticle adopted anorthic structure with space group C-1 (2). Here the crystal symmetry seems to be of triclinic structure.

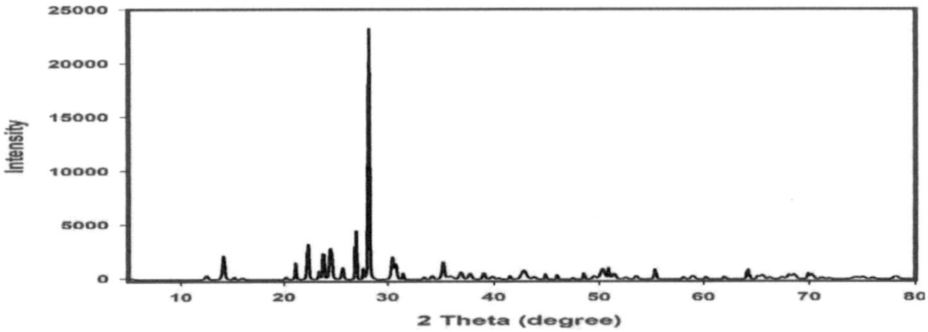

Figure 1. XRD pattern of $NaAlSi_3O_8$ Nanoparticles (Feldspar)

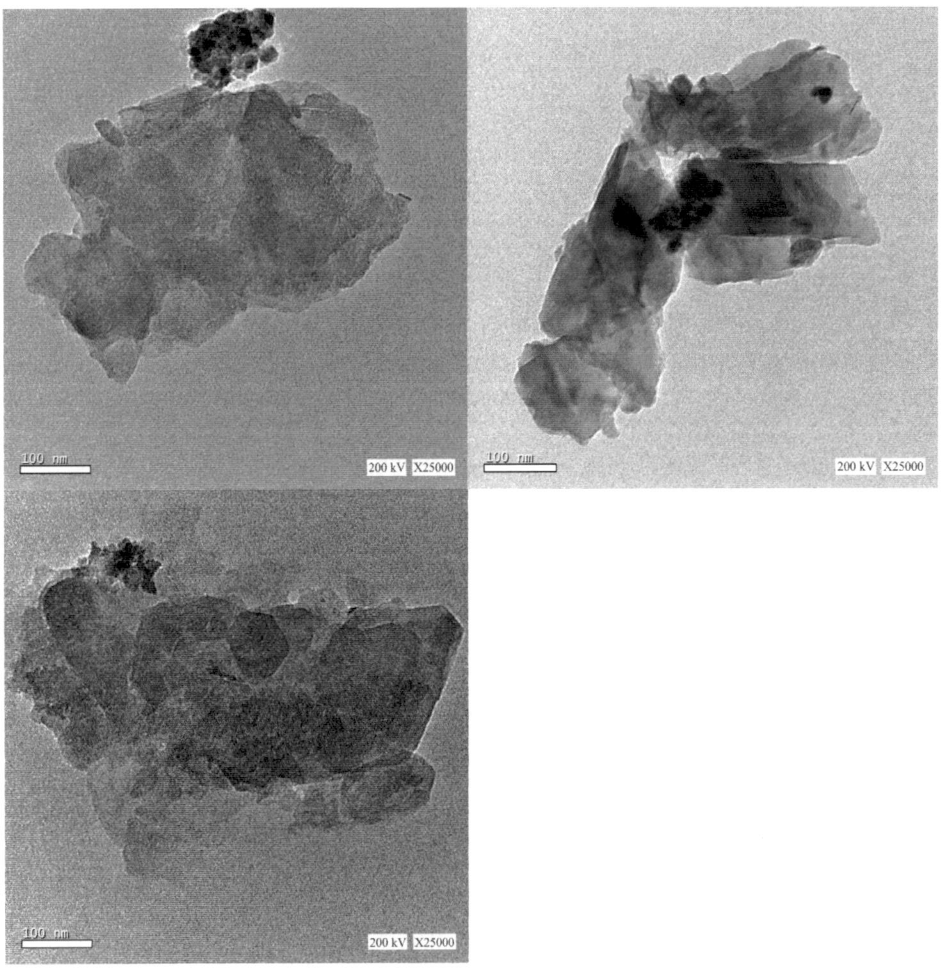

Figure 2. TEM images of $NaAlSi_3O_8$ Nanoparticles (Feldspar)

The feldspar is one of the well-known aluminum sodium silicates of the general formula $NaAlSi_3O_8$. The crystallite size was calculated from the Debge Scherer's equation:
$L=0.9\ \lambda\ /\ \beta\ cos\theta$.

When λ is the target wave length of the x-ray, β is the corrected full width at half maximum and θ is the diffraction angle. The feldspar prepared nanoparticles are in the 264.87 nm average range.

High resolution transmission electron microscope micrographs of feldspar nanoparticles are represented in figure 2. It is clearly seen that the feldspar is formed in a network of sheets stacked together. These sheets are of triclinic symmetry with sharp geometric boundaries. The sheets are of dimensions of few hundreds of nanometers but with very small thickness. No preferred orientation of the nanosheets was remarked. The observed microstructure agreed well with the XRD analysis of the powder.

3.2. Characterization of hematite nanoparticles

XRD pattern of hematite nanoparticles prepared by ball milling (figure 3) revealed excellent crystallinity as strong reflections and large number of peaks are observed. By comparing the obtained pattern and matching with ICDD number 01-071-6218, the pattern was indexed and the nanoparticle adopted rhombohedra structure with space group R-3c (167). Here the crystal symmetry seems to be of triclinic structure. The hematite is one of the well-known iron oxide of the general formula Fe_2O_3. The hematite prepared nanoparticles are in the 287 nm average range.

High resolution transmission electron microscope micrographs of hematite nanoparticles are represented in figure 4. It is clearly seen that the hematite is formed in a network of sheets stacked together. The sheets are of dimensions of few hundreds of nanometers but with very small thickness. No preferred orientation of the nanosheets was remarked. The observed microstructure agrees well with the XRD analysis of the powder.

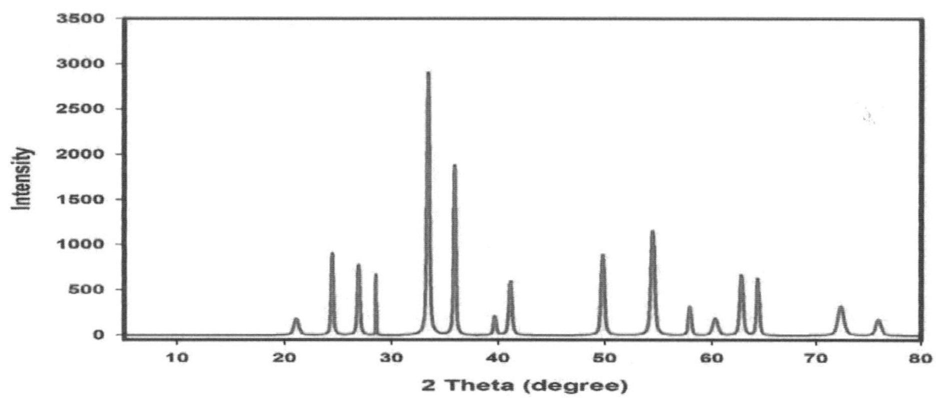

Figure 3. XRD pattern of Fe_2O_3 Nanoparticles (Hematite)

3.3. Effect of different fertilization practices on wheat yield

The data in Tables 2-5 represents the effect of each of chemical potassium fertilizer, feldspar nanoparticles, chemical ferrous sulphate and hematite nanoparticles and their interactions on wheat yields, namely; grain, straw and total biological yields.

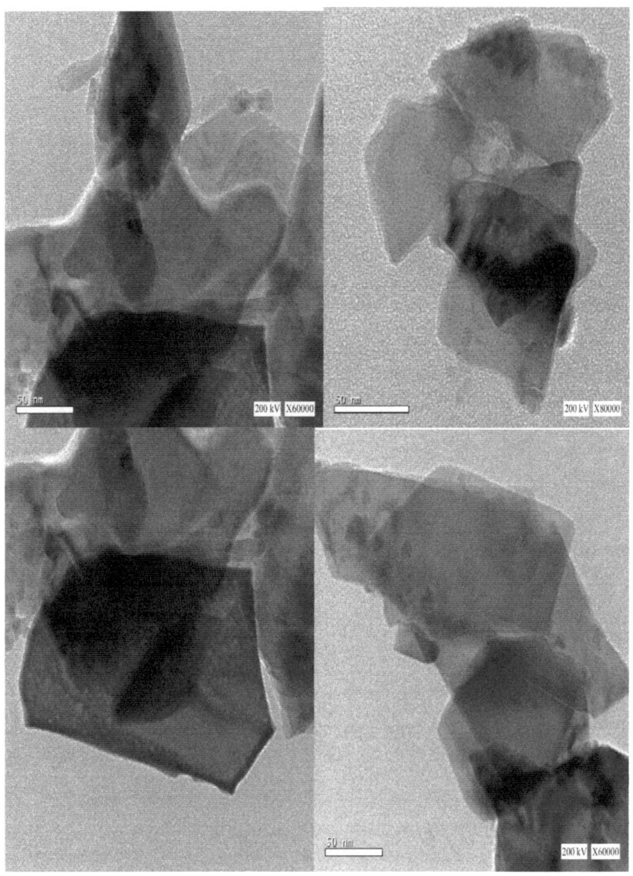

Figure 4. TEM images of Fe_2O_3 Nanoparticles (Hematite)

Table 2. The main effect of chemical potassium fertilizers.

	Grain yield (Mgha^{-1})		Straw yield (Mgha^{-1})		Biological yield (Mgha^{-1})	
	1st location	2nd location	1st location	2nd location	1st location	2nd location
0.0 chemical kgK*ha^{-1}*	7.34^d	7.25^d	15.88^d	15.84^d	23.22^d	23.09^d
28.6 chemical kgK*ha^{-1}*	7.64^c	7.55^c	16.73^c	16.69^c	24.37^c	24.37^c
85.7 chemical kgK*ha^{-1}*	8.28^b	7.79^b	17.89^b	17.85^b	26.17^b	25.92^b
114.0 chemical kgK*ha^{-1}*	8.50^a	8.02^a	18.03^a	17.99^a	26.53^a	26.01^a
L.S.D at 0.05	0.13	0.12	0.15	0.14	0.26	0.24

Means in column followed by the same letter are not significantly differed ($P<0.05$) using L.S.D test

Table 3. The main effect of feldspar nanoparticles.

	Grain yield ($Mgha^{-1}$)		Straw yield ($Mgha^{-1}$)		Biological yield ($Mgha^{-1}$)	
	1st location	2nd location	1st location	2nd location	1st location	2nd location
Without feldspar nanoparticles	7.81^b	7.70^b	16.91^b	16.86^b	24.86^b	24.56^b
With feldspar nanoparticles	8.06^a	7.95^a	17.36^a	17.33^a	25.42^a	25.28^a
L.S.D at 0.05	0.16	0.14	0.21	0.20	0.37	0.36

Means in column followed by the same letter are not significantly differed ($P<0.05$) using L.S.D test

Table 4. The main effect of iron treatments (chemical ferrous sulphate and hematite nanoparticles)

	Grain yield ($Mgha^{-1}$)		Straw yield ($Mgha^{-1}$)		Biological yield ($Mgha^{-1}$)	
	1st location	2nd location	1st location	2nd location	1st location	2nd location
Without iron	7.58^b	7.49^b	16.39^b	16.35^b	23.97^b	23.85^b
With chemical ferrous sulphate	8.12^a	7.99^a	17.51^a	17.47^a	25.63^a	25.46^a
With hematite nanoparticles	8.12^a	8.00^a	17.52^a	17.47^a	25.64^a	25.45^a
L.S.D at 0.05	**0.10**	**0.10**	**0.13**	**0.13**	**0.23**	**0.21**

Means in column followed by the same letter are not significantly differed ($P<0.05$) using L.S.D test.

Table 5. The effect of the interaction between chemical potassium and feldspar nanoparticles

		Grain yield ($Mgha^{-1}$)		Straw yield ($Mgha^{-1}$)		Biological yield ($Mgha^{-1}$)	
Chemical K	Feldspar	1st location	2nd location	1st location	2nd location	1st location	2nd location
0.0	without	7.21^e	7.21^e	15.51^e	15.46^e	23.21^e	22.58^e
kgK*ha*$^{-1}$	*with*	*7.46*^{d}	*7.37*^{d}	*16.25*^{d}	*16.22*^{d}	*24.20*^{d}	*23.59*^{d}
28.6	without	*7.51*^{d}	*7.43*^{d}	*16.32*^{d}	*16.27*^{d}	*24.33*^{d}	*23.70*^{d}
kgK*ha*$^{-1}$	*with*	*7.77*^{c}	*7.67*^{c}	*17.14*^{c}	17.11^c	25.54^c	24.78^c
85.7	without	8.06^b	7.88^b	17.76^b	17.72^b	26.47^b	25.59^b
kgK*ha*$^{-1}$	*with*	8.49^a	8.38^a	18.03^a	17.98^a	27.08a	26.36^a
114.0	without	8.49^a	8.38^a	18.03^a	17.99^a	27.09^a	26.37^a
kgK*ha*$^{-1}$	*with*	8.50^a	8.38^a	18.03^a	17.99^a	27.09^a	26.37^a
L.S.D at 0.05		0.16	0.16	0.20	0.21	0.34	0.33

Means in column followed by the same letter are not significantly differed ($P<0.05$) using L.S.D test.

As for the main effect of chemical potassium fertilization, the data in table 2 clearly show that wheat grain, straw and biological yields were significantly affected by increasing K levels up to 114.0 kgK*ha*$^{-1}$. The highest values of wheat yields obtained using the highest K level in both locations. Suppling wheat plants with 114.0 kgK*ha*$^{-1}$ of K increased grain yield by about 2.7, 11.3 and 15.8 over 85.7, 28.6 and zero (control) kgK*ha*$^{-1}$ respectively in the first location. Same trends were obtained in the second location and for straw and biological yields. The primitive effect of potassium on wheat yields may be due to the vital role of K on many biological functions of plant such as activating various enzymes, improvement of protein, carbohydrates and fat concentration developing as well as tolerance against drought, lodging, bests and disease attack [16]. These results are in line with those obtained in other researches [1,4].

Concerning the main effect of feldspar nanoparticles, the data in table 3 indicate that feldspar nanoparticles application has significantly increased wheat yields in both locations. The increment in grain, straw and biological yields caused by feldspar nanoparticles reached to 3.2%, 2.7% and 2.3% when compared with the absence of feldspar in the first location. Same trends were obtained in the second location. In this concern, it is needed to increasing the availability of K from natural feldspar by using nanotechnology as documented before [17]. In general, applying nanotechnology to fertilizers improves the efficiency of fertilizers and reduces soil pollution caused by using chemical fertilizers, as the prepared nanoparticles release their nutrients slowly and steady [18]. These results are similar to those obtained by previous studies [19].

With regard to the main effect of iron treatments (table 4), the results show that foliar spraying of chemical ferrous sulphate and hematite nanoparticles increased grain when compared with the absence of iron application. The relative increasing of grain yields caused by chemical ferrous sulphate and hematite nanoparticles reached to about 7.1 and 7.1% over without iron addition, respectively in the first location. Same trends were obtained for the second location. It is obvious to notice that the difference between the effect of chemical ferrous sulphate and hematite nanoparticles on wheat yields did not reach to the significant value which means the possibility to use hematite nanoparticles instead of chemical iron sulphate. The beneficial effect of iron, whether chemical or nanoparticles is mainly due to that iron is an important component in many plant enzymes like cytochrome oxidase. It required for nitrate and sulphate reduction, N_2 assimilation and energy reduction. It plays as a catalyst of an enzyme system associated with chlorophyll formation [20]. The beneficial effect of insoluble natural hematite when treated by nanotechnology is mainly because nanoparticles have high surface area in comparison to its volume; consequently they exhibited quantum mechanical behavior [21]. These results agree with those obtained by [10].

Respecting to the interaction among the treatments (table 5), the results show that wheat yields did not respond to the interaction between treatments, except the interaction between chemical potassium fertilizer and feldspar. The obtained data indicate that combining 75% KRR (85.7 kg Kha^{-1}) with feldspar nanoparticles showed grain, straw and biological yields of wheat, statistically equal to those under 100% KRR (114.0 kg Kha^{-1}) which means the possibility of saving about 25% KRR (28.5 kgKha^{-1}) by spraying wheat plants with 2% feldspar nanoparticles solution twice.

3.4. Effect of different fertilization practices on protein yield in wheat grains

The effects of chemical K, feldspar nanoparticles and both iron treatments, as well as, their interaction on nitrogen, protein content and protein yield in grains of wheat are shown in tables 6-9.

Table 6. The main effect of chemical potassium fertilizers.

Chemical potassium (kgKha^{-1})	N content (g kg^{-1}) 1st location	2nd location	Protein content (g kg^{-1}) 1st location	2nd location	Protein yield (g kg^{-1}) 1st location	2nd location
0.0 (kgKha^{-1})	14.86^d	14.52^d	82.34^d	83.78^d	627.23^d	605.21^d
28.6(kgKha^{-1})	15.70^c	15.45^c	90.29^c	88.79^c	690.73^c	671.04^c
85.7(kgKha^{-1})	16.36^b	16.54^b	96.05^b	95.09^b	802.65^b	774.14^b
114.0(kgKha^{-1})	16.97^a	17.30^a	100.46^a	99.47^a	854.34^a	834.24^a
L.S.D at 0.05	0.19	0.18	1.13	1.12	9.65	9.56

.Means in column followed by the same letter are not significantly differed (P<0.05) using L.S.D test.

Table 7. The main effect of feldspar nanoparticles.

	N content (g kg^{-1})		Protein content (g kg^{-1})		Protein yield (g kg^{-1})	
	1st location	2nd location	1st location	1st location	2nd location	1st location
Without feldspar nanoparticles	15.66^{b}	15.65^{b}	91.46^{b}	89.97^{b}	718.21^{b}	696.19^{b}
With feldspar nanoparticles	16.53^{a}	16.25^{a}	95.06^{a}	93.44^{a}	768.97^{a}	746.13^{a}
L.S.D at 0.05	0.17	0.16	1.10	0.10	9.13	9.05

Means in column followed by the same letter are not significantly differed (P<0.05) using L.S.D test.

Table 8. The main effect of iron treatments (chemical ferrous sulphate and hematite nanoparticles.

	N content (g kg^{-1})		Protein content (g kg^{-1})		Protein yield (g kg^{-1})	
	1st location	2nd location	1st location	1st location	2nd location	1st location
Without iron	15.96^{b}	15.75^{b}	91.74^{b}	90.53^{b}	697.91^{b}	681.19^{b}
With chemical ferrous sulphate	16.36^{a}	16.08^{a}	94.03^{a}	92.43^{a}	766.65^{a}	747.16^{a}
With hematite nanoparticles	16.36^{a}	16.03^{a}	94.03^{a}	92.16^{a}	766.22^{a}	740.13^{a}
L.S.D at 0.05	0.08	0.07	0.52	0.50	8.61	8.57

Means in column followed by the same letter are not significantly differed (P<0.05) using L.S.D test

Table 9. The effect of the interaction between chemical potassium and feldspar nanoparticles.

		N content (g kg^{-1})		Protein content (g kg^{-1})		Protein yield (g kg^{-1})	
Chemical K	Feldspar	1st location	2nd location	1st location	2nd location	1st location	2nd location
0.0 (kgK*ha*$^{-1}$)	Without	14.59^{d}	14.30^{e}	83.91^{d}	82.24^{d}	604.98^{e}	585.73^{e}
	With	15.13^{c}	14.73^{d}	86.98^{c}	84.69^{c}	649.47^{d}	624.69^{d}
28.6 (kgK*ha*$^{-1}$)	Without	15.36^{c}	15.18^{c}	88.32^{c}	87.27^{c}	663.22^{d}	648.92^{d}
	With	16.04^{b}	15.71^{b}	92.25^{b}	90.31^{b}	717.03^{c}	693.14^{c}
85.7 (kgK*ha*$^{-1}$)	Without	16.23^{b}	15.81^{b}	93.30^{b}	90.91^{b}	750.01^{b}	716.20^{b}
	With	17.47^{a}	17.26^{a}	100.43^{a}	99.27^{a}	853.29^{a}	832.08^{a}
114.0 (kgK*ha*$^{-1}$)	Without	16.45^{a}	17.29^{a}	100.32^{a}	99.44^{a}	852.62^{a}	833.87^{a}
	With	17.49^{a}	17.30^{a}	100.59^{a}	99.4^{9a}	856.05^{a}	834.60^{a}
L.S.D at 0.05		0.21	0.20	1.24	1.22	11.65	11.39

Means in column followed by the same letter are not significantly differed (P<0.05) using L.S.D test.

As the main effect of chemical potassium treatments, the data in table 6. Indicate that nitrogen content, protein content and protein yield were gradually increased as potassium levels increased. The highest values of these characteristics were achieved using potassium concentration of 114.0 Kg Kha^{-1}, while the treatment without potassium application exhibited the lowest ones. The plants treated with 114.0 kg Kha^{-1} yielded nitrogen, protein content, and protein yield exceeded that without potassium application by about 14.2, 21.4 and 36.2%, respectively in the first location. The corresponding increases in the second location were 19.1, 18.7 and 36.2% in the above-mentioned order. The primitive effect of potassium on protein characteristics may be because potassium has many functions in plant, like N uptake, protein synthesis and photosynthesis and translocation of assimilates [22]. In addition, potassium encourages the development of the root system of plant, hence influencing the uptake of nutrients from the soil [5]. Our results agree with other findings [1, 4].

Concerning the main effect of feldspar nanoparticles, the data in table 7. Reveal that using feldspar nanoparticles as a potassium source had a positive effect on nitrogen and protein content, as well as, protein yield. The increments of the above parameters due to spraying wheat plants with feldspar nanoparticles reached to 5.6, 3.9 and 7.1% compared to the treatments in the absence of feldspar in the first location. K-bearing rocks have low effectiveness in agriculture due to their slow availability, so increasing K availability in feldspar by using nanotechnology resulted in enhancing its effect on growth, nitrogen uptake, protein content and yield in wheat grains as aforementioned [23]. These results are in harmony with those obtained earlier [24, 25].

With reference to the main effect of iron treatments, the obtained results in table 8. Declared that nitrogen, protein content, and protein yield were positively affected by iron application, whether chemical ferrous sulphate or hematite nanoparticles. The increment of these characteristics due to foliar spraying of ferrous sulphate or hematite nanoparticles reached to 2.5 and 2.5, 2.5 and 2.5, and 9.8 and 9.8% over the treatment without iron application, respectively in the first location. Same trends were obtained in the second locations. It can be noticed that the effect of natural insoluble hematite when converted to nanoparticles has equal effect on nitrogen and protein characteristics as soluble chemical ferrous sulphate. The primitive effect of iron from soluble ferrous sulphate or hematite nanoparticles is mainly because applications of iron increase total carbohydrate, starch, chlorophyll and protein content in grains [5]. These results are similar to previously obtained results [1, 26, 27].

Respecting the interaction between treatments (table 9.), the data indicate that nitrogen, protein content, and protein yield were significantly affected only by the interaction between chemical potassium levels and feldspar nanoparticles, where combined 75% KRR of chemical K with foliar spraying of feldspar nanoparticles solution gave the abovementioned characteristics equal to those under 100% KRR. These finding means that the possibility of replacement of 25 KRR with spraying feldspar nanoparticles solution. In general, the highest nitrogen, protein content, and protein yield were recorded for wheat plants fertilized with 100% KRR or 75% KRR + foliar spraying of feldspar nanoparticles + foliar spraying of ferrous sulphate or hematite nanoparticles. On the other hand, the plants without chemical K, feldspar nanoparticles and chemical ferrous iron or iron nanoparticles exhibited the lowest ones.

4. Conclusion

From the results of this research, it could be concluded that nanotechnology application to fertilizers such as feldspar nanoparticles and natural iron oxide nanoparticles (hematite) have greater role in improving wheat production and protein yield, which will reduce the cost of chemical fertilizers and minimize the environmental pollution. It could be substituted by about 25% from the recommended rate of potassium by spraying 2% feldspar nanoparticles solution. Also, natural iron oxide nanoparticles can be used instead of the more expensive ferrous sulphate. Finally, it could be recommended to treat wheat plants with 85.7 kg Kha^{-1} of potassium sulphate along with spraying the plants twice with 2% feldspar nanoparticles and 0.3% hematite nanoparticles to attain maximum quality and quantity of wheat.

References

[1] Hamouda H A, El-Dahshouri M F, Manal F M and Thalooth A T 2015 *Growth* **8** 1473
[2] Rahman I U, Afzal A, Iqbal Z and Manan S 2014 *J. Agric. Allied Sci.* **3** 19
[3] Ismail S A, Galal A M and Sarham G R 2016 *Monufia J. Soil. Sci.* **1** 93
[4] Zafar S, Ashraf M Y, Anwar S, Ali Q and Noman A 2016 *Soil. Env.* **35** 1
[5] Mengel K and Kirkby E A 1987 *Principles of Plant Nutrition.* International Potash Institute, Worblaufen-Bern, Switzerland.
[6] Rashad R T and Hussein R A 2020 *Commun. Soil. Sci. Plant. Anal.* **51** 1078
[7] Brittenham G M 1994 *Cur. Opino. Hemato.* **1** 101
[8] Taiz L, Zeiger E 2010. Plant physiology, 5th edn. Inc., Massaschusetts, 781pp.
[9] Ali M E, Ismail S A, El-Hussieny O H M and Abd El-Hafez A M 2009 *Mansoura Univ. Agric. Sci.* **34** 1901.
[10] Mohamadipoor R, Sedaghathoor S and Khomami A M 2013 *Eur. J. Exp. Biol.* **3** 232
[11] Association of Official Analytical Chemists 1995 *Official Methods of Analysis*, 16th Edition, AOAC International, Washington, DC, USA.
[12] Gomez K A and Gomez A A 1984 *Statistical Procedures for Agricultural Research* (2 ed). John wiley and sons, NewYork
[13] Farghali A A, Khedr M H and Khalek A A A *J. Mater. Process. Technol.* **181** 81
[14] Mahmoud R, Moaty S A, Mohamed F and Farghali A 2017 *J. Chem. Eng.* **62** 3703
[15] Shahien M G, Khedr M M, Maurice A E, Farghali A A and Ali R A 2015 *J. Basic Appl. Sci.* **4** 207.
[16] Marschner H 1997. Functions of Mineral Nutrients: Macronutrients. Mineral Nutrition of Higher Plants 2nd ed. Academic Press, N.Y., p299.
[17] Scovino J S and Rowell D L 1988 *Fer. Res.* **17** 71
[18] Naderi M A A, D Shahraki and R Naderi 2011 *Iran. J. Nanotech.* **12** 16
[19] Ghahremani A, Akbari K, Yousefpour M and Ardalani H 2014 *Int. J. Pharm. Res Schol.* **3** 00167
[20] Jones Jr J B and Case V W1990 Soil Science Society of America, Madison, WI, USA p.389
[21] Banotra B M, Kumar A, Sharma B C, Nandan B, Verma A, Kumar, Gupta V and Bhagat S 2017 *Int. J. curr. Microbial Appl. Sci.* **6** 1541
[22] Pettigrew W T 2008 *Physiologia Plantarum.***133** 670
[23] Harley A D and Gilkes R J 2000 *Nutr. Cycling Agroecosyst.* **56** 111
[24] Kandil E E and Marie E 2017 *Alex. Sci. Exch. J.* **38** 53
[25] Singh M D, Jayadeva H M, Chirag G and Mohan M H 2017 *Int. J. Micro. Res.* **9** 897
[26] Abd El-Hafez A M, Awadalla H A and Ismail S A 2013 *J. Soil. SCI. Agric. Eng.* **4** 1313
[27] Bakhtiari M, Moaveni P and Sani B 2015 *Biol. Forum.* **7** 679

AUTHOR INDEX

Institute of Physics
Dirac House, Temple Back
Bristol BS1 6BE UK

ISSN: 1757-8981
ISBN 978-1-7138-3334-5

CURRAN ASSOCIATES INC.
proceedings.com